STRUCTURAL ELECTRON CRYSTALLOGRAPHY

STRUCTURAL ELECTRON CRYSTALLOGRAPHY

DOUGLAS L. DORSET
Hauptman-Woodward Medical Research Institute, Inc.
Buffalo, New York

PLENUM PRESS • NEW YORK AND LONDON

Library of Congress Cataloging-in-Publication Data

On file

ISBN 0-306-45049-6

A Division of Plenum Publishing Corporation
233 Spring Street, New York, N. Y. 10013

10 9 8 7 6 5 4 3 2 1

Printed in the United States of America

Preface

Arbeitshypothesen sind revidierbar, deklarierten Wahrheiten nicht, sie verkalken zum System; Arbeitshypothesen passen sich den Menschen an, den deklarierten Wahrheiten wird der Mensch angepaßt; die ersten kann mann verwerfen, von den anderen wird man verworfen.

FRIEDRICH DÜRRENMATT, *Nachgedanken*

Working hypotheses can be revised; declared truths cannot—they calcify into dogma. Working hypotheses adapt to people—people adapt to declared truths. One can reject the first but be rejected by the latter.

The concept of electron crystallography, i.e., the quantitative use of electron diffraction intensities to solve crystal structures, is by no means new. Based on extensive pioneering efforts on organic and inorganic substances, two major works on electron diffraction structure analysis (or "electronography" as it was then known in Moscow) appeared in English translation during the 1960s. These books are B. K. Vainshtein, *Strukturnaya Elektronografiya* (*Structure Analysis by Electron Diffraction*, translated by E. Feigl and J. A. Spink, Pergamon Press, Oxford, 1964), and B. B. Zvyagin, *Elektronografiya i Strukturnaya Kristallografiya Glinistykh Mineralov* (*Electron-diffraction Analysis of Clay Mineral Structures*, translated by S. Lyse, Plenum Press, New York, 1967). While there is much in these monographs that still merits serious study, they are unfortunately long out of print. This book has been written, therefore, in part, to provide an updated account of a technique which continues to gain importance in modern structural research. Because it emphasizes the quantitative determination of crystal structures from electron diffraction data, it is called *Structural Electron Crystallography*, which is a modern translation of the Russian title of Vainshtein's opus. (The enormous breadth and impact of Vainshtein's pioneering work should also be noted since he is referenced in every chapter of this book!) It should be made clear, furthermore, that this book is written by a structural chemist interested first in investigating molecular packing and geometry in the solid state.

The work described in this text is based on many efforts around the world. From a personal standpoint, collaborators and colleagues from several countries are thanked heartily for their insights and contributions and continued dialogue—and especially for addressing the very important matter of specimen preparation. Particularly I would like to thank John Fryer, Fritz Zemlin, Bill Tivol, and Chris Gilmore for good collaborations in electron crystallographic structural research. Others, e.g., Jean-Claude Wittmann, Bernard Lotz, Bob Snyder, Leo Mandelkern, Jürg Rosenbusch, and Henri Chanzy, are thanked for suggesting interesting structural problems. Elmar Zeitler is thanked for his generosity in providing access to unique equipment, and encouragement. I am grateful to Jim Turner for encouraging use of the high-voltage electron microscope facility in Albany for electron crystallographic research, aided in a large way by the sample stages designed and fabricated by him. Sven Hovmöller and Xiadong Zou have been very helpful in making useful image processing software available for this work and I thank them for their patient assistance and willingness to make alterations where they are needed. Many thanks are also due to those who have worked with me in this laboratory. I am grateful for the support of the Medical Foundation of Buffalo, Inc. (recently renamed the Hauptman-Woodward Medical Research Institute, Inc.) and especially for the encouragement and advice of its president, Herb Hauptman, one of the pioneers of direct phase determination in crystallography and recipient of the 1985 Nobel Prize in Chemistry. Much of the work described in this book would have been impossible without his method for finding crystallographic phases. I would also mention fond memories of valuable discussions with the late David Harker, who had been at the Medical Foundation for many years after his "retirement." I am grateful to Professor John M. Cowley for his kind advice (and patience) when I was beginning work in this field. Finally, none of my own work would have been possible without the significant financial support of several research grants, including those from the National Institute of General Medical Sciences (GM-21047, GM-46733) and the National Science Foundation (PCM78-16401, CHE79-16916, CHE/DMR81-16318, INT82-13903, INT84-01669, DMR86-10783, CHE91-13899, CHE94-17835), for which I am most grateful. Many thanks are due to Ms. Melda Tugac for her herculean effort and high standards in preparing the figures for this book and for designing its cover.

Introduction

Structural electron crystallography is the quantitative use of electron scattering data to determine the average structure of crystalline objects at atomic or molecular resolution. From such studies one can obtain information about bonding geometry and conformation, in addition to spatial requirements for the close packing of atomic clusters. There are two major potential advantages in using electron scattering data, compared to x-ray or neutron diffraction for structure analyses. One is that electrons, unlike x-rays or neutrons, can be focused by an electromagnetic lens; thus both images and diffraction patterns can be utilized for structural analysis. The other is that the scattering cross section of matter for electrons is approximately 10^3 times as great as x-rays and 10^4 times as great as neutrons (Vainshtein, 1964a). This means that much smaller objects can be studied as single crystals with electron beams than with the other two commonly employed radiation sources capable of resolving interatomic distances. A third potential advantage over x-ray diffraction is the enhanced detectability of some light atoms in the presence of heavier ones, e.g., in principle, the presence of hydrogen covalently bonded to carbon, nitrogen, or oxygen (Vainshtein, 1964b).

Despite these possible advantages, and also the fact that electron crystallography has a long history of applications to structure analysis, the results from such determinations have not been so well accepted by structural chemists as have those from x-ray and neutron diffraction analyses. It has been pointed out by some crystallographers that many early structure analyses based on electron diffraction intensity data employed the results of contemporary x-ray determinations for construction of the initial phase model (e.g., Lipson and Cochran, 1966). Hence, it (unfortunately!) is a commonly held view that ab initio structure analyses are, in fact, not possible and that no unknown structures have ever been determined with such data. Similar criticism is heard from diffraction physicists. Since the physical model for the multiple scattering of electrons by thin crystals (important because of the relatively larger scattering cross section mentioned above and also the very small wavelength of fast electrons) is very complicated in the worst case, it is argued that only an indirect match of data, based on a known crystal structure, can be permitted.

Some materials (e.g., linear polymers, lipids and other amphiphiles, polymorphic forms of some pigments, integral biomembrane proteins) are so difficult to crystallize that a thin microcrystal is the only practical means for obtaining an ordered periodic structure for diffraction studies. The need to obtain information about such substances has motivated some researchers to attempt structure analyses based on readily available single-crystal electron diffraction data or electron microscope images from ordered crystalline regions, in spite of the skeptical viewpoints mentioned above. Obviously, experimental conditions must be manipulated to favor the collection of data that can be directly analyzed.

It is the intent of this book, therefore, to demonstrate that, despite the somewhat complicated theoretical framework for electron scattering, ab initio structure analyses are, in fact, permitted. These analyses, furthermore, can lead to new structures, using no additional information from other structural probes. On the other hand, it is not the intent of this book to refute the perfectly valid multiple-beam dynamical scattering theory for electrons. Knowledge of this theory helps one to design the optimal experimental parameters for data collection, i.e., those conditions that best approach the single scattering approximation. Given this optimization, it will be shown with copious examples that standard techniques for crystallographic phase determination are very successful (and, indeed, enhanced by the use of electron micrographs as additional sources of phase information), and that the derived structures can often be refined to find a chemically reasonable geometry for the atomic array making up the unit cell.

By necessity, this book is incomplete. Many techniques for structure analysis and refinement with electron scattering data are still being developed or improved. Also, the discussions in this volume are largely confined to the use of diffraction intensities and images from single thin crystals, although some texture diffraction data from early studies are also analyzed. Certain types of ordered arrays, such as helical tubes, are not considered and very little mention is made to fiber diffraction information, despite the importance of such samples to the study of certain materials. Although a brief chapter has been inserted on the quantitative study of inorganic structures, most of the book is devoted to molecular organic crystals. Furthermore, even though its development has been quite important for recent progress in structure biology, a somewhat narrow viewpoint of protein electron crystallography will be presented in this volume, specifically discussing strategies for determining crystallographic phases, consistent with the discussion of small-molecule applications.

The book has two parts. In the first five chapters an overview of quantitative crystallography, as it is practiced using the electron microscope, is presented. This includes the preparation of specimens and their preservation in the electron microscope vacuum during data collection. Limitations of the technique are also frankly discussed. In the last seven chapters, specific areas of structure research are covered. Generally, an appreciation of what has been accomplished so far is given but, more importantly, concrete examples of actual structure analyses are reviewed in some detail so that they can be repeated by the interested reader with data provided.

Contents

Part I. Background

1. The Electron Microscope as a Crystallographic Instrument

1.1. The Thin Lens 3
1.2. Fourier Transform Pairs 4
1.3. The Electron Microscope 17
1.3.1. Illumination System 18
1.3.2. Functions 19
1.3.2.1. Imaging 19
1.3.2.2. Diffraction 22
1.4. Geometrical Aspects of Electron Diffraction 26

2. Crystal Symmetry

2.1. The Unit Cell 31
2.2. Symmetry Groups 36
2.2.1. Point Groups 37
2.2.2. Plane Groups 43
2.2.3. Space Groups 49
2.2.4. "Two-Sided Plane Groups" 54
2.3. Unit Cell and Space-Group Identification 55
2.4. Preferred Crystal Packing Motifs for Organic Molecules 63

3. Crystallization and Data Collection

3.1. Crystallization of Organic Compounds 67
3.1.1. Growth from Dilute Solution 67
3.1.2. Crystallization by "Self-Seeding" 68
3.1.3. Crystallization by Sublimation 68
3.1.4. Langmuir–Blodgett Layers 69
3.1.5. Epitaxial Orientation 69
3.1.5.1. Growth from the Vapor Phase 70
3.1.5.2. Growth from a Co-Melt 70

3.1.6. Sonication 76
3.2. Crystallization of Globular Macromolecules 76
3.2.1. Crystallization from Solution 76
3.2.2. In Situ Crystals 76
3.2.3. Reconstitution of Transmembrane Proteins 77
3.2.4. Surface Orientation of Proteins 77
3.3. Crystallization of Inorganic Structures 77
3.4. Preservation of Samples in the Electron Microscope Vacuum 78
3.4.1. Environmental Chambers 78
3.4.2. Cryostages 80
3.4.3. Solvent Replacement 80
3.5. Data Collection and Processing 81
3.5.1. Goniometry 81
3.5.2. Data Collection 83
3.5.2.1. Electron Diffraction 83
3.5.2.2. Electron Microscopy 86

4. Crystal Structure Analysis
4.1. Solution of the Phase Problem 95
4.1.1. Crystallographic Phases via Image Analysis 95
4.1.2. Trial-and-Error Methods 99
4.1.3. Patterson Function 103
4.1.4. Direct Phasing Methods 105
4.1.4.1. Sayre Equation 107
4.1.4.2. Phase Invariant Sums (Symbolic Addition) 109
4.1.4.3. Tangent Formula 119
4.1.4.4. Patterson Search Techniques 122
4.1.4.5. The Minimal Principle 123
4.1.4.6. Density Modification 124
4.1.4.7. Maximum Entropy 124
4.2. Structure Refinement 126
4.2.1. Identification of a Structure 126
4.2.2. Fourier Refinement 127
4.2.3. Least-Squares Refinement 130
4.2.4. Continuous Density Maps 131
4.3. Derived Quantities 133

5. Data Perturbations
5.1. Dynamical Scattering 135
5.2. Secondary Scattering 149
5.3. Diffraction Incoherence due to Crystal Bending 153
5.4. Radiation Damage 159
5.5. Conclusions 166

Part II. Applications

6. Molecular Organic Structures

6.1. Background 169
6.2. Early Data Sets from Moscow 169
6.2.1. Diketopiperazine 169
6.2.2. Urea 176
6.2.3. Thiourea, Paraelectric Form 179
6.2.4. Thiourea, Ferroelectric Form 184
6.3. Recent Analyses Based upon Selected Area Diffraction Data 188
6.3.1. Copper Perchlorophthalocyanine 188
6.3.2. Copper Perbromophthalocyanine 198
6.3.3. C_{60} Buckminsterfullerene 202
6.3.4. Graphite 206
6.4. Conclusions 207

7. Inorganic Structures

7.1. Background 209
7.2. Structures Solved from Electron Diffraction Data 211
7.2.1. Boric Acid 211
7.2.2. Celadonite 215
7.2.3. Muscovite 219
7.2.4. Phlogopite-Biotite 220
7.2.5. λ-Alumina 224
7.2.6. Basic Copper Chloride 225
7.2.7. High T_c Superconductor 231
7.2.8. Potassium Niobium Oxide 231
7.2.9. Aluminum–Germanium Alloys 231
7.3. Structures from High-Resolution Electron Micrographs 232
7.3.1. Potassium Niobium Tungsten Oxide 232
7.3.2. Staurolite 234
7.3.3. Sodium Niobium Fluoroxide 234
7.3.4. Zeolites 237
7.4. Conclusions 238

8. The Alkanes

8.1. Background 239
8.2. Contemporary Structure Analyses 240
8.2.1. Even-Chain Paraffins 240
8.2.2. Odd-Chain Paraffins 258
8.2.3. Thermotropic Phase Transitions of Linear Paraffins 261
8.2.4. Binary (and Multicomponent) Phase Behavior in Paraffins . . . 268
8.2.4.1. Solid Solutions 268
8.2.4.2. Binodal Phase Boundary 278

8.2.4.3. Eutectics 281
8.2.5. Cycloalkanes 283
8.2.6. Perfluoroalkanes 287
8.3. Conclusions 292

9. Alkane Derivatives
9.1. Background 293
9.2. Contemporary Structure Analyses 294
9.2.1. Fatty Alcohols 294
9.2.2. Fatty Acids 297
9.2.3. Ketoalkanes 298
9.2.4. Wax Esters 298
9.2.5. Alkyl Halides 307
9.2.6. Detergents 308
9.3. Conclusions 310

10. The Lipids
10.1. The Methylene Subcell: Its Significance for Electron Diffraction from Lipids 311
10.2. Structure Analyses of Glycerolipids 316
10.2.1. 1,3-Diglycerides 316
10.2.2. 1,2-Diglycerides 317
10.2.3. Triglycerides 321
10.2.4. Phospholipids and Glycolipids 323
10.2.4.1. Phosphatidylethanolamines 327
10.2.4.2. *N*-Methylphosphatidylethanolamines 333
10.2.4.3. *N,N*-Dimethylphosphatidylethanolamines 337
10.2.4.4. Phosphatidylcholines 340
10.2.4.5. Phosphatidic Acids 343
10.2.4.6. Monogalactosyl Diglyceride 343
10.3. Cholesteryl Esters 344
10.4. Conclusions 360

11. Linear Polymers
11.1. Background 361
11.2. Crystal Structure Analyses 361
11.2.1. Two-Dimensional Data Sets 361
11.2.1.1. Poly(ethylene Sulfide) 362
11.2.1.2. Poly(3,3-bis(chloromethyl)oxacyclobutane (BCMO) 363
11.2.1.3. γ-Poly(pivalolactone) 365
11.2.1.4. Poly(p-xylylene) 366
11.2.1.5. Mannan I 368
11.2.1.6. Cellulose Triacetate II 370

11.2.1.7. Chitosan 371
11.2.1.8. Anhydrous Nigeran 372
11.2.1.9. Poly(hexamethylene Terephthalate) 373
11.2.1.10. β-Form of Poly-γ-methyl-*L*-glutamate 374
11.2.2. Three-Dimensional Data Sets 376
11.2.2.1. Poly(1,4-trans-cyclohexanediyl Dimethylene Succinate)(Poly-t-CDS) 377
11.2.2.2. Mannan I 379
11.2.2.3. Polyethylene 383
11.2.2.4. Poly(ε-caprolactone) 390
11.2.2.5. Poly(1-butene), Form III 396
11.3. Conclusions 403

12. Globular Macromolecules
12.1. Background 405
12.2. Structure Analyses of Membrane Proteins at High Resolution 409
12.2.1. Bacteriorhodopsin 409
12.2.2. Outer Membrane Porins from Gram-Negative Bacteria . . . 410
12.2.3. Light-Harvesting Chlorophyll *a*/*b*-Protein Complex 420
12.3. Prospects for the Use of Direct Methods in Protein Electron Crystallography 420
12.4. Conclusions 426

References 429

Index 447

Part I

Background

—insolange die Theorie nicht selber ihren Bankrott erklärt, so lange wird sie vom Vertrauen getragen und die Wirklichkeit ordnet sich ihr unter.
—HERMANN BROCH, *Die Schlafwandler*

—as long as theory does not declare its own bankruptcy, it will be trusted and reality will adapt itself to it.

1

The Electron Microscope as a Crystallographic Instrument

1.1. The Thin Lens

Of all the possible scattering methods capable of resolving interatomic distances, only those employing electrons permit the relationship between an object and its diffraction pattern to be directly visualized. This is because the paths of electrons can be changed by a magnetic field without a change in their energy; hence electron lenses exist and, therefore, electron microscopes can be constructed (see, e.g., Reimer, 1984).

In order to conceptualize this relationship between object and diffraction pattern, consider a thin lens in an ideal phase contrast optical microscope. A nearly transparent thin object is illuminated by a source of radiation and this radiation is scattered by local variations in refractive index within the thin sample. In electron microscopy, these variations are, in fact, local changes in electrostatic potential (Cowley, 1981). Suppose a thin convergent lens is placed in the optical path of the device and a magnified, in-focus image of the object is produced some distance from this lens on a given *image plane*. Between the lens and image planes, there exists another *back focal plane* (Figure 1.1), where all radiation, not scattered by the object, is focused onto a point and all scattered radiation falls outside this central point on the plane. This distribution of scattered radiation is called a *diffraction pattern*, whether or not the object is periodic.

Suppose also that there is a one-to-one mapping of information between the object and its image formed by the thin lens. The relationship of scattering distribution between the image and back focal planes can be described by an operation known as the *Fourier transform*. For example the object density f(**r**) is related to its diffraction pattern amplitude $F(\mathbf{s})$ by:

$$F(\mathbf{s}) = \int_{-\infty}^{\infty} f(\mathbf{r}) \exp 2\pi i\, \mathbf{r} \cdot \mathbf{s}\, dr = FT\, f(\mathbf{r}) \qquad (1)$$

Conversely, the object wave function can be expressed by:

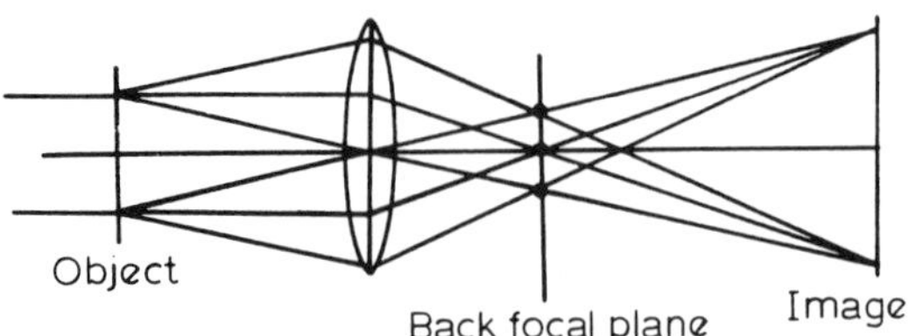

FIGURE 1.1. Geometrical construction of a thin lens, showing principal planes where the diffraction pattern and image of the object are formed. (Reprinted from D. L. Dorset (1989) "Electron diffraction from crystalline polymers," in *Comprehensive Polymer Science, Volume 1*, Sir G. Allen, ed., Pergamon Press, Oxford, 1989, pp. 651–668; with kind permission of Elsevier Science Ltd.)

$$f(\mathbf{r}) = \int_{-\infty}^{\infty} F(\mathbf{s}) \exp(-2\pi i\, \mathbf{r} \cdot \mathbf{s})\, ds = FT^{-1}F(\mathbf{s}) \qquad (2)$$

Note that the form of the integral equations is very similar to a Laplace transform (Rainville, 1963), except that the kernel is imaginary in this case. The two entities related by the Fourier transform are known as *Fourier transform pairs* (Cowley, 1981; Gaskill, 1978; Champeney, 1973), and it is worth considering the characteristics of these in some detail.

1.2. Fourier Transform Pairs

If a semitransparent object (Figure 1.2a) is placed ~~on the~~ at the proper *focal length* from the lens in Figure 1.1, then its diffraction pattern can be visualized at the back focal plane (Figure 1.2b). Note that this pattern has a continuous distribution of density and that its symmetry is higher than that of the object, in that every density point u,v is matched by another at $-u$, $-v$. Such Friedel symmetry (see Chapter 2 for a general discussion of symmetry and its expression in diffraction patterns) is characteristic of all diffraction patterns from thin objects, where multiple scattering events are not important. The relationship between two functions related by a Fourier transform is reciprocal. That is to say, if the distance between two points on an object is given as length $|\mathbf{r}|$, the same measurement on the diffraction pattern must be expressed in terms of reciprocal length $|\mathbf{s}|$, sometimes called the spatial frequency. Hence, on the object, the minimum linear distance between two just discernible points is a direct expression of spatial resolution; in the diffraction pattern, this resolution is expressed by the extent of the scattering function, i.e., the reciprocal distance from its center to a radial limit where a measurable signal can still be detected.

If two identical objects are now placed at some distance $\mathbf{a}$ between their centers (Figure 1.2c), the diffraction pattern (Figure 1.2d) will still retain characteristics of the one from a single object (Figure 1.2b), except it will be modulated by an interference function in a reciprocal direction corresponding to the object repeat. If this repetition is extended to include more identical units spaced by $\mathbf{a}$ (Figure 1.2e), then the diffraction pattern is shown to become discrete in the reciprocal direction corresponding to this repeating motif (Figure 1.2f), while it remains continuous in the orthogonal direction. Stacking another row of objects at a spacing **b** in the orthogonal direction (Figure 1.2g) causes an interference function to appear along the second reciprocal axis (Fig. 1.2h) which also becomes discrete (Figure 1.2j) as enough repeats are added along **b** (Figure 1.2i). Nevertheless, the diffraction pattern always retains the scattering distribution of the single object at points which are sampled in reciprocal space by the reciprocal lattice. (Other pictorial examples can be found in texts by Taylor and Lipson (1964) and Harburn et al. (1975).)

Given the reciprocal net in Figure 1.2i, it is possible to assign indices to the "reflections" or diffraction points in the reciprocal lattice. For the projection, two reciprocal axes are chosen and the spots are sequentially numbered integrally according to a grid (h,k). These integral *Miller indices* actually have a conceptual meaning. For example, a space lattice with spacings **a**, **b** to define an array of *unit cells* can be subdivided into a set of parallel planes in any direction corresponding to simple fractional intercepts of the axial repeats $\mathbf{a}/h$, $\mathbf{b}/k$ (and also $\mathbf{c}/l$ for a three-dimensional lattice). The corresponding indices $(h,k,(l)$ describe a diffraction point originating from a specific set of planes, and, as we shall see in the next chapter, the distance of the reflection (h,k) from the center of the diffraction pattern is inversely related to the spacing of these planes in the crystal. In certain cases, it is also useful to employ a triple index $[UVW]$ to describe a vectorial direction in a space lattice, where U, V, W are multiples of the unit cell axes, **a**, **b**, **c**. For example, such a triple would define a *zone axis* within a crystal, where a projection along $[UVW]$ would be parallel to crystallographic planes obeying the dot product, known as the zone axis equation (McKie and McKie, 1986):

$$\mathrm{U}\cdot h + \mathrm{V}\cdot k + W\cdot l = 0$$

For example, the reflections $(hk0)$ would be excited in the crystal projection along zone axis [001].

The simple illustration above demonstrates the consequence of building up a periodic array of an object on its diffraction pattern. Assuming, for example, an orthogonal space lattice with repeat vectors $n\mathbf{a}$ and $m\mathbf{b}$, this repetitive motif can be thought of as a gridwork, upon which the object is repeated at any intersection point. Given enough repeats, the Fourier transform becomes a reciprocal lattice with spacing $|1/\mathbf{a}|$ and $|1/\mathbf{b}|$, which samples the original scattering function at its intersection points. However, neither the space nor reciprocal lattices need be based on an orthogonal coordinate system. Thus, the reciprocal relationships in distances must also account for the oblique angles between axes in nonorthogonal lattices. The relationship between a crystal and its diffraction pattern has just been described, and can now be stated more formally in terms of Fourier transform pairs.

There are many areas of physics where a basic knowledge of Fourier transforms can aid immeasurably in the intuitive analysis of a problem. Obviously, Fourier transforms are of great importance in optics, of which structural crystallography can

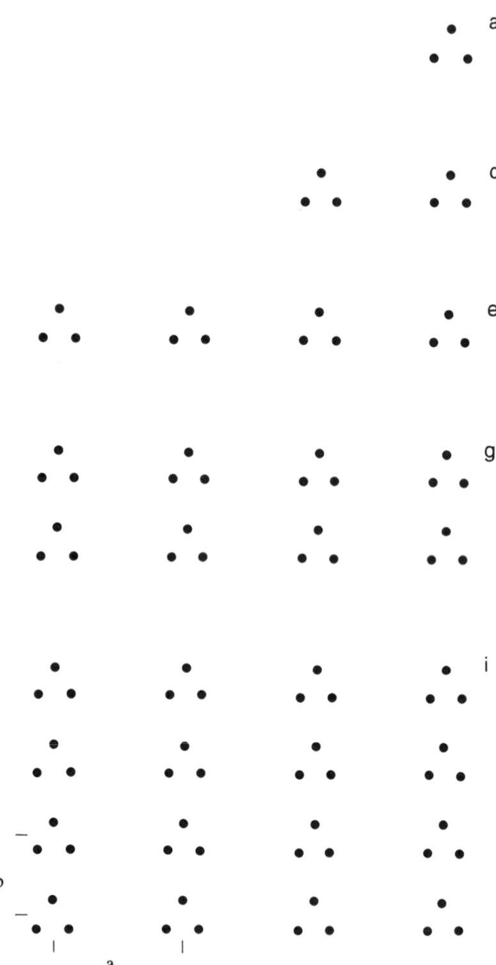

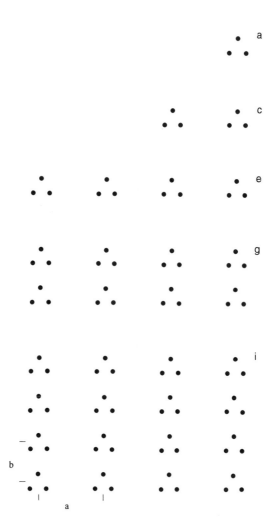

FIGURE 1.2. Diffraction from an object: (a) with three-fold symmetry. Its continuous diffraction pattern (b) has six-fold symmetry because of Friedel's law, i.e., $|F_h| = |F_{-h}|$. When two identical objects are placed in a line as in (c), the diffraction pattern (d) shows the effect of the interference function from interacting waves scattered from the individual motifs. When the regularly spaced row is continued (e), the diffraction pattern becomes more and more discrete (f) in the direction corresponding to the repeat. Stacking another row (g) introduces a second interference function (h). As these repeats are continued in both directions (i), the diffraction pattern (j) becomes a grid of delta functions. However, in all cases, the intensity of the scattering from the fundamental repeating unit is always sampled by the diffraction pattern. (Reprinted from D. L. Dorset (1989) "Electron diffraction from crystalline polymers," in *Comprehensive Polymer Science, Volume 1*, Sir G. Allen, ed., Pergamon Press, Oxford, 1989, pp. 651–668; with kind permission of Elsevier Science Ltd.)

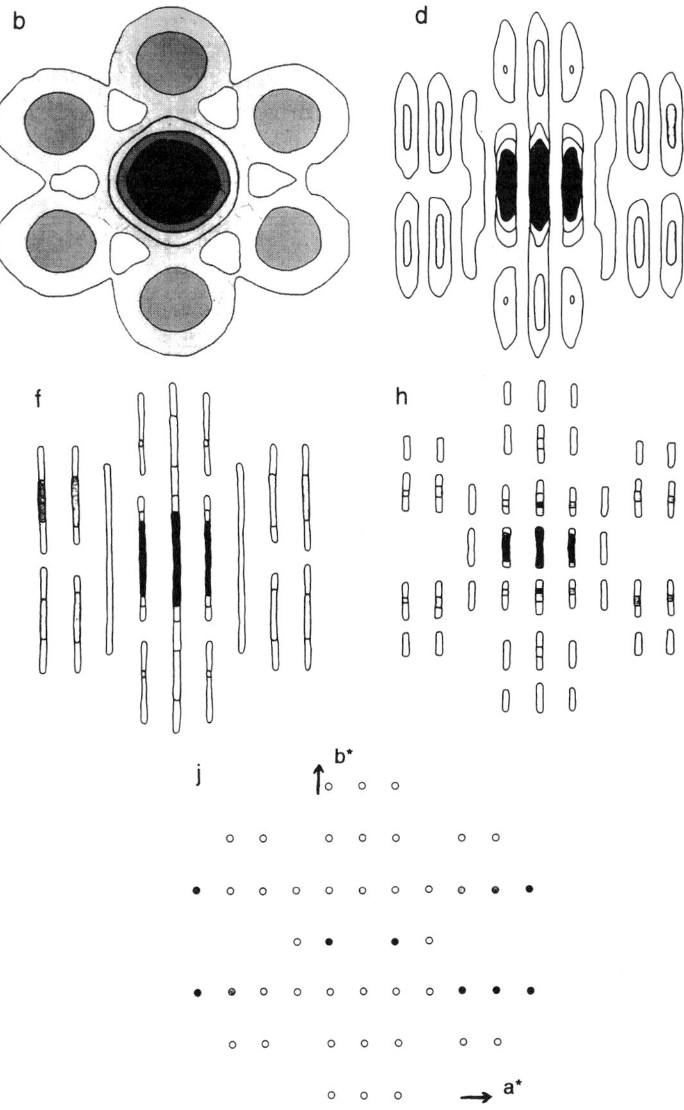

FIGURE 1.2. (*Continued*)

be regarded as a subfield. In electrical engineering, Fourier transform pairs have been of great importance to signal analysis (e.g., Blackman and Tuckey, 1959; Mason and Zimmerman, 1960), where the dimensions in length and inverse length are replaced by time and frequency. In x-ray crystallography, one can cite the book on optical transforms by Taylor and Lipson (1964) as an early text where such a clear approach was used extensively to demonstrate how different diffraction patterns arise from various distributions of scatterers. Since electron crystallography utilizes information in both principal planes of an electromagnetic lens, it is easy to understand directly how Fourier analysis relating a diffraction pattern to an image would be readily adapted to solve structural problems. As will be shown, there are only a few transform pairs that need to be understood in order to analyze a wide range of problems. One can, therefore, decompose a linear system into its components, analyze each component separately, and then combine these solutions to arrive at the complete answer.

In order to illustrate the operation of Fourier transformation, consider a rather simple one-dimensional "image" of a pulse of width *Na* and height *K* (Figure 1.3a) (see Lipson and Lipson, 1969). In equation (1) above $f(\mathbf{r}) = K$, a constant in the interval $(-Na/2, Na/2)$ and zero elsewhere. The following integral is evaluated:

$$F(u) = \int_{-Na/2}^{Na/2} K \exp(2\pi iux)\, dx$$

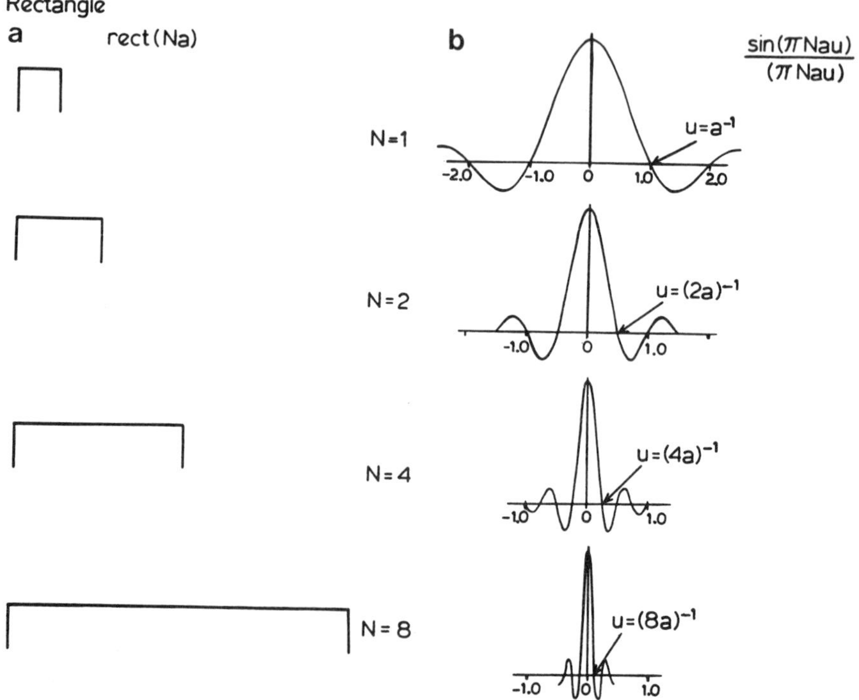

FIGURE 1.3. Fourier transform of rect (*Na*) as *N* becomes large. The width of the sinc(πNau) transform accordingly becomes narrower until it approximates a delta function. (Reprinted from D. L. Dorset (1989) "Electron diffraction from crystalline polymers," in *Comprehensive Polymer Science, Volume 1*, Sir G. Allen, ed., Pergamon Press, Oxford, 1989, pp. 651–668; with kind permission of Elsevier Science Ltd.)

which is simply:

$$F(u) = (K/2\pi iu)\exp(2\pi iux)\Big|_{-Na/2}^{Na/2}$$

$$= (K/2\pi iu)\{\exp(\pi iNau) - \exp(-\pi iNau)\}$$

Now since $\exp(\pi iNau) = \cos(\pi Nau) + i\sin(\pi Nau)$

$$F(u) = K\sin\pi Nau/\pi u = KNa\sin\pi Nau/\pi Nau = KNa\,\mathrm{sinc}(\pi Nau)$$

This function is plotted in Figure 1.3b. Its maximum height at $u = 0$ is KNa and it crosses the abscissa at $1/Na$. In some texts (e.g., Gaskill, 1978), the original pulse function is termed rect(x). Thus:

$$\mathrm{rect}(x;a,K) = 0,\ |x| > a/2$$

$$= K,\ |x| \le a/2$$

For $K = 1.0$, the Fourier transforms of several values of rect(Na), with increasing N, are plotted in Figure 1.3. Note that the normalization to give the function sinc(πNau) will ensure that this function has a maximum value of 1.0.

What is the Fourier transform (FT) of sinc(ξ)? In general, reverse transforms can be found in the following way (where we now use the enclosed variables only to signify respective inverse or direct spaces). Following Gaskill (1978):

$$FT\{F(\eta)\} = G(\xi) = \int_{-\infty}^{\infty} g(\alpha)\exp(-2\pi i\,\xi\alpha)\,d\alpha$$

Let $\mathrm{F}(\eta) = g(\eta)$ and let $g(\eta) = FT\,G(\xi)$. Then

$$G(\xi) = \int_{-\infty}^{\infty} F(\alpha)\exp(-2\pi i\,\xi\,\alpha)\,d\alpha$$

$$= \int_{-\infty}^{\infty} F(\alpha)\exp(2\pi i(-\xi)\alpha)\,d\alpha = f(-\xi)$$

From this result one can write $FT\{\mathrm{sinc}(\xi)\} = \mathrm{rect}(-\eta)$. This is identical to rect(η), given the evenness (one-dimensional mirror symmetry at the origin) of this function. This illustrates the principle of Fourier transform pairs where, in general, the transform of one function to another can be reversed to reproduce the original, a result which is of

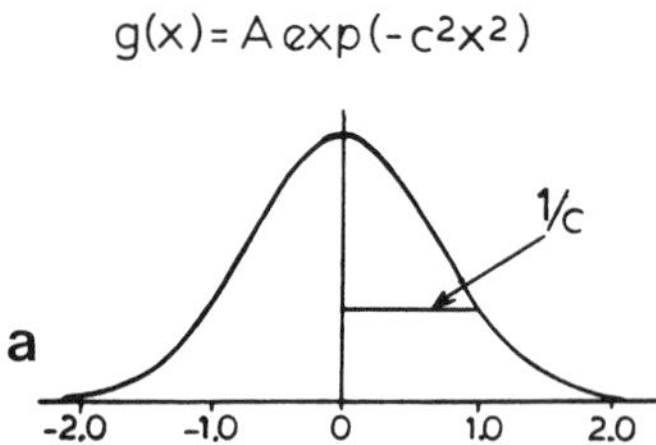

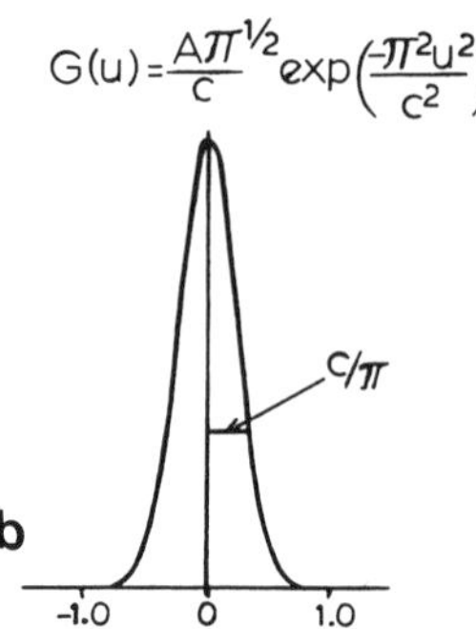

FIGURE 1.4. Fourier transform of the Gaussian function $g(x) = A \exp(-c^2x^2)$. (Reprinted from D. L. Dorset (1989) "Electron diffraction from crystalline polymers," in *Comprehensive Polymer Science, Volume 1*, Sir G. Allen, ed., Pergamon Press, Oxford, 1989, pp. 651–668; with kind permission of Elsevier Science Ltd.)

immeasurable aid to intuitive analysis. The reversibility of this transform can be illustrated directly with the Gaussian function.

A Gaussian function can be defined:

$$g(x) = \exp(-a^2x^2)$$

and is graphed in Figure 1.4a. Its Fourier transform is expressed:

$$G(u) = \int_{-\infty}^{\infty} \exp(-a^2x^2)\exp(2\pi iux)\,dx$$

$$= \int_{-\infty}^{\infty} \exp(-a^2x^2)\cos(2\pi ux)\,dx + i\int_{-\infty}^{\infty} \exp(-a^2x^2)\sin(2\pi ux)\,dx$$

Now, as mentioned by Cowley (1981), since the Gaussian is an even function and the sine function is odd, the last integral disappears. The first integral is found in many tables:

$$\int_{-\infty}^{\infty} \exp(-a^2x^2)\cos(bx)dx = (\sqrt{\pi}/a)\exp(-b^2/4a^2)$$

Since $b = 2\pi u$,

$$G(u) = (\sqrt{\pi}/a)\exp(-\pi^2u^2/a^2),$$

which is another Gaussian function, as graphed in Figure 1.4b. Note the reciprocal relationship between the half-widths of these two Gaussian functions. Now, what happens when one computes the reverse Fourier transform of $G(u)$? This can be written:

$$g(x) = \int_{-\infty}^{\infty} G(u) \exp(-2\pi iux)\, du = \int_{-\infty}^{\infty} (\sqrt{\pi}/a) \exp(-\pi^2u^2/a^2) \exp(-2\pi iux) du$$

which can be expanded as before, omitting the second integral with the sine function. If $c = \pi/a$, using the standard integral above where c is substituted for a and $b = 2\pi x$, it is easy to show that:

$$g(x) = \exp(-a^2x^2)$$

the Gaussian function we started with. Fourier transforms are again shown to be reversible.

The Gaussian function will be useful in diffraction applications for the simulation of disorder. It also serves as a useful approximation for several other functions, (Gaskill, 1978; Cowley, 1981) and, because of its convenient properties under Fourier transformation, will facilitate intuitive analysis of many problems. However, consider what happens to the function $G(u)$ above (Figure 1.4b), as the constant a becomes smaller and smaller. This limit is known as a delta function $\delta(u)$, also defined $\delta(u) = \int_{-\infty}^{\infty} \exp(2\pi iuv)\, dv$, and the same function is reproduced from the Fourier transform of the rectangle function rect(Na) as Na becomes very large. Thus, it can be readily understood that the Fourier transform of a delta function should be a constant amplitude over all space. We shall also make extensive use of this function for analysis of diffraction problems.

Next it is necessary to define two mathematical operations on two functions $f(x)$ and $h(x)$ that are important for relating images to diffraction patterns. These are *convolution*, defined:

$$f(x) * h(x) = \int_{-\infty}^{\infty} f(r_1)\, h(x - r_1)\, dr_1$$

and *correlation*, defined:

$$f(x) \otimes h(x) = \int_{-\infty}^{\infty} f^*(r_1)\, h(r_1 + x)\, dr_1$$

where $f^*(\mathbf{r})$ is the complex conjugate of $f(\mathbf{r})$. Both functions are identical for centrosymmetric objects. To illustrate the convolution function for a rectangle function rect(x), it can be shown in Figure 1.5 how this leads to a triangle function tri(x), i.e., rect(x)*rect(x) = tri(x). We can state:

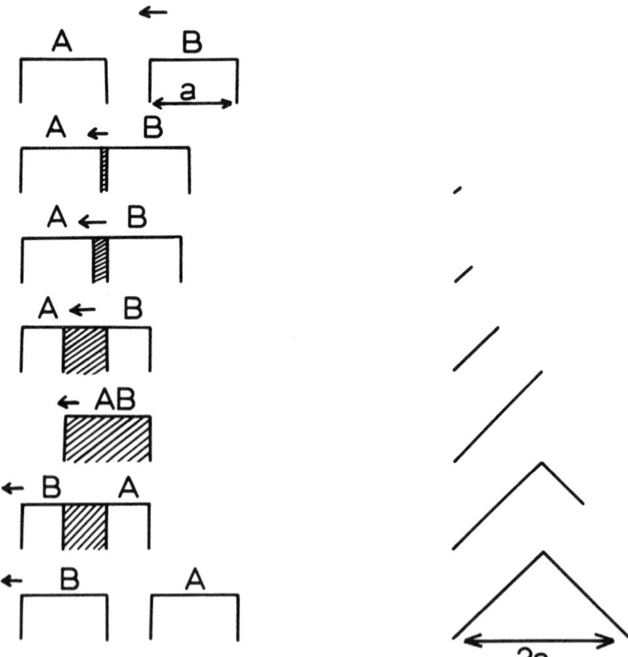

FIGURE 1.5. Graphical representation of the convolution operation with the example: rect (x) * rect (x) = tri (x).

$$\mathrm{rect}(x) * \mathrm{rect}(x) = \int_{-\infty}^{\infty} \mathrm{rect}(r)\,\mathrm{rect}(x-r)\,dr$$

The contribution to this integrand is only not zero when $(x - r) > -a/2$ and $(x - r) < a/2$, or $r < x + a/2$ and $r > x - a/2$.

Case I: Let $x \leq 0$. Then:

$$\mathrm{rect}(x) * \mathrm{rect}(x) = \int_{-a/2}^{x+a/2} K^2\,dr = K^2\,(x+a)$$

Case II: Let $x \geq 0$. Then:

$$\mathrm{rect}(x) * \mathrm{rect}(x) = \int_{x-a/2}^{a/2} K^2\,dr = K^2(a-x)$$

By definition:

$$\text{tri}(x) = \text{rect}(x;\, a,K) * \text{rect}(x;\, a,K)$$

$$= K^2(x + a) \qquad x \leq 0$$

$$= K^2(a - x) \qquad x \geq 0$$

(This derivation is courtesy of H. Hauptman.) Alternatively, using a graphical method (Figure 1.5), the operation involves a shift of one function by another, followed by the plot of the product area at each shift increment.

What is the Fourier transform of tri(x)? We already know that $FT(\text{rect}(x))$ = sinc(u). In order to solve this problem, the Fourier transform of the convolution operation must be found. Suppose

$$G(u) = FT(f(x) * g(x))$$

Then:

$$G(u) = FT\{ \int_{-\infty}^{\infty} f(r_1)\, h(x - r_1)\, dr_1 \}$$

$$= \int_{-\infty}^{\infty} \{ \int_{-\infty}^{\infty} f(r_1)\, h(x - r_1)\, dr_1 \} \exp(-2\pi iux)\, dx$$

$$= \int_{-\infty}^{\infty} f(r_1) \{ \int_{-\infty}^{\infty} h(x - r_1) \exp(-2\pi iux)\, dx \}\, dr_1$$

However:

$$FT\{h(x - r_1)\} = \int_{-\infty}^{\infty} h(r - r_1) \exp(-2\pi iur)\, dr$$

$$= \int_{-\infty}^{\infty} f(\mu) \exp(-2\pi i(\mu + r_1)u\, d\mu$$

$$= \exp(-2\pi ir_1u) \int_{-\infty}^{\infty} f(\mu) \exp(-2\pi i\, u\mu)\, d\mu$$

$$= \exp(-2\pi ir_1u)\, H(u)$$

meaning that the shift constant r_1 becomes transformed to an exponential function with a quantity known as *phase* as the argument, a result of great importance to crystallography. Substituting above, we obtain

$$G(u) = \int_{-\infty}^{\infty} f(r_1)\, H(u) \exp(-2\pi i u r_1)\, dr_1$$

$$= H(u) \int_{-\infty}^{\infty} f(r_1) \exp(-2\pi i u r_1)\, dr_1$$

$$= F(u)\, H(u)$$

In other words, the Fourier transform of a convolution is simply a multiplication. Hence

$$FT(\text{tri}(\eta)) = FT(\text{rect}(\eta) * \text{rect}(\eta)) = FT(\text{rect}(\eta))\, FT(\text{rect}(\eta))$$

$$= \text{sinc}(\xi)\, \text{sinc}(\xi) = \text{sinc}^2(\xi)$$

(Again the enclosed variables are used here to denote shifts from one space to another.) It is easy to find the Fourier transform of a correlation function from the above example. (The reader is referred to the excellent text by Gaskill (1978) for further discussion.) To summarize:

$$f(x) * g(x) \Leftrightarrow F(u)\, G(u)$$

$$f(x)\, g(x) \Leftrightarrow F(u) * G(u)$$

$$f(x) \otimes g(x) \Leftrightarrow F^*(u)\, G(u)$$

$$f(x)\, g(x) \Leftrightarrow F^*(u) \otimes G(u)$$

where "$\Leftrightarrow$" denotes a reversible Fourier transform pair. Now, consider the properties of the delta function.

As mentioned above, the delta function can be imagined to express the limit of a Gaussian function or a sinc function as they become very narrow. For purposes of definition:

$$\delta(x - x_0) = 0 \text{ for } x \neq x_0$$

$$f(x)\, \delta(x - x_0) = f(x_0)\, \delta(x - x_0)$$

$$f(x) * \delta(x) = \int_{-\infty}^{\infty} f(\alpha)\, \delta(x - \alpha)\, d\alpha = f(x)$$

$$f(x) * \delta(x+x_0) = \int_{-\infty}^{\infty} f(\alpha)\delta(x+x_0-\alpha)d\alpha = f(x+x_0)$$

The first relationship states that the delta function exists only at one point, x_0. For $A\,\delta(x-x_0)$, the area of this function is A. The second relation is the "sifting" property of the delta function under multiplication with another function, i.e., it assumes the magnitude of this function at the point x_0. Finally, a convolution of a function with a delta function simply reproduces that function. If can also be shifted by a constant in the argument of the delta function under the convolution operation.

Before the previous discussion of Fourier transforms can be used for the analysis of a crystal, it is necessary to define a comb function:

$$\text{comb}[(x-x_0)/b] = |b| \sum_{n=-\infty}^{\infty} \delta(x-x_0-nb)$$

As shown in Figure 1.6, this is an array of delta functions spaced at a distance $|b|$ from one another. Multiplication of some function f(x) by the normalized comb function (i.e., one that is multiplied by $|b|^{-1}$), according to the sifting property given above, samples this function at intervals $|b|$. As demonstrated, e.g., by Gaskill (1978), the Fourier transform of comb(x) = comb(u):

$$FT\{\text{comb}(x)\} = \sum_{n=-\infty}^{\infty} FT\{\exp(2\pi inx)\}$$

$$= \sum_{n=-\infty}^{\infty} \delta(u-n) = \text{comb}(u)$$

Now the concepts required to interpret Figure 1.2 are assembled. It is obvious, first of all, that the comb function has the property of a crystal space lattice in one

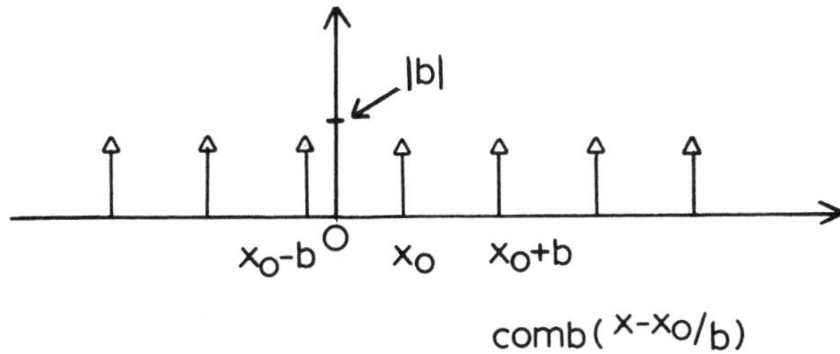

FIGURE 1.6. The comb function.

dimension, i.e., a repeat of a delta function at a regular interval. It is possible also to define two-dimensional and three-dimensional versions of the comb function to express these regularly repeating lattices of delta functions. Again, the spacing along any particular axial row does not have to be the same as that of another axis; neither do the angles between lines of regularly spaced delta functions have to be 90°. It is also clear that, for any of the one-dimensional functions defined above, there are two- and three-dimensional analogs to be used in our analysis.

To analyze Figure 1.2, the motif in (a) can be represented by $f(\mathbf{r})$ and its continuous Fourier transform in (b) is therefore $F(\mathbf{s})$. Repeating the motif along one axis with a spacing $|a|$ can be accomplished by

$$g(\mathbf{r}) = f(\mathbf{r}) * |a^{-1}| \operatorname{comb}(x/a)$$

but since this is an infinite array, it must be sampled (i.e., multiplied) by a finite rectangle function rect(Na) to limit the dimensions of a real crystal, i.e.:

$$h(\mathbf{r}) = f(\mathbf{r}) * |a^{-1}| \operatorname{comb}(x/a) \operatorname{rect}(Na)$$

The Fourier transform $H(\mathbf{s})$, therefore is immediately written

$$H(\mathbf{s}) = F(\mathbf{s})\, |a| \operatorname{comb}(as) * \operatorname{sinc}(\pi Nas)$$

In other words, the continuous transform of $f(\mathbf{r})$ is sampled by the comb function in one direction but these sampled points are each spread by a convolution with the sinc function. As rect(Na) becomes larger, then the sinc(πNas) function becomes narrower and, therefore, more and more like a delta function. These arguments can be repeated in two dimensions. The two-dimensional comb function can be defined:

$$|ab|^{-1} \operatorname{comb}(x/a,y/b) = |a|^{-1} \operatorname{comb}(x/a)\, |b|^{-1} \operatorname{comb}(y/b)$$

This is sampled by rect(Na,Mb), which becomes larger in the two respective directions until we have a lattice of delta functions sifting the two-dimensional continuous transform of $f(\mathbf{r})$.

Suppose now that the precise positions of $f(\mathbf{r})$ at individual lattice sites are uncertain. They can be distributed about space lattice points defined by the delta functions, according to a Gaussian function $g(\mathbf{r}) = \exp(-c^2r^2)$. It was shown above that the Fourier transform of a Gaussian is also a Gaussian, with an inversely related half-width. Hence,

$$FT\{f(\mathbf{r}) * |ab|^{-1} \operatorname{comb}(x/a,y/b) * g(\mathbf{r})\} = F(\mathbf{s})\, |ab| \operatorname{comb}(au,bv)\, G(\mathbf{s})$$

In other words, the crystal is built up from the convolution of the space lattice with the repeating motif $f(\mathbf{r})$. This function is also convoluted with a Gaussian $g(\mathbf{r})$ to express the disorder at each lattice site. The resultant Fourier transform (diffraction pattern), then, is the continuous transform of $f(\mathbf{r})$ sampled (via the "sifting" property mentioned above) by the reciprocal lattice, all of which is attenuated in resolution by another multiplicative Gaussian function (via another sifting operation), the Fourier transform of $g(\mathbf{r})$. An example of a Gaussian resolution attenuation would be the result of isotropic thermal vibrations or static positional disorder of molecules in a crystal lattice.

1.3. The Electron Microscope

Now it will be shown how the principles considered above are relevant to the analysis of data from an electron microscope. The optical path of an electron microscope is illustrated in Figure 1.7 and is shown to consist of an electron gun, a condenser lens arrangement to control the illumination, an objective lens (enclosing some facility for manipulating the specimen), an array of projector lenses below it and finally a

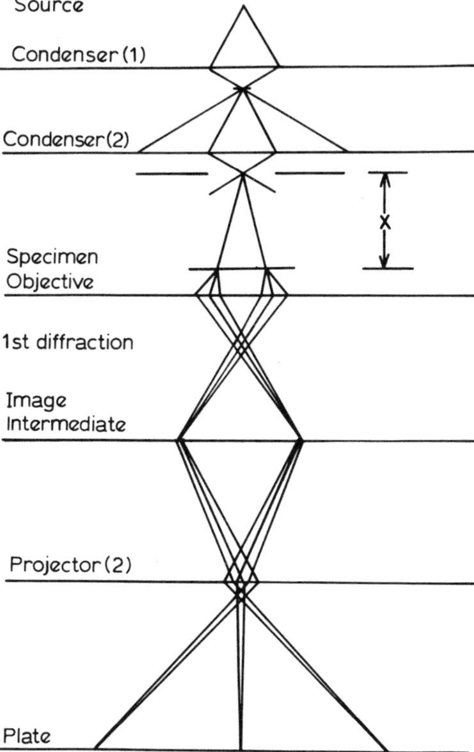

FIGURE 1.7. (a) Ray tracing in an electron microscope illustrating selected area diffraction geometry. (Reprinted from D. L. Dorset (1985) "Crystal structure analysis of small organic molecules," *Journal of Electron Microscopy Technique* **2**, 89–128; with kind permission of Wiley–Liss Division, John Wiley and Sons, Inc.) (b) Appearance of electron diffraction pattern and image at principal objective lens planes. (Reprinted from D. L. Dorset (1994) "Electron crystallography of linear polymers," in *Characterization of Solid Polymers*, S. J. Spells, ed., Chapman and Hall, London, pp. 1–17; with kind permission of Chapman and Hall, Ltd.) *(continued)*

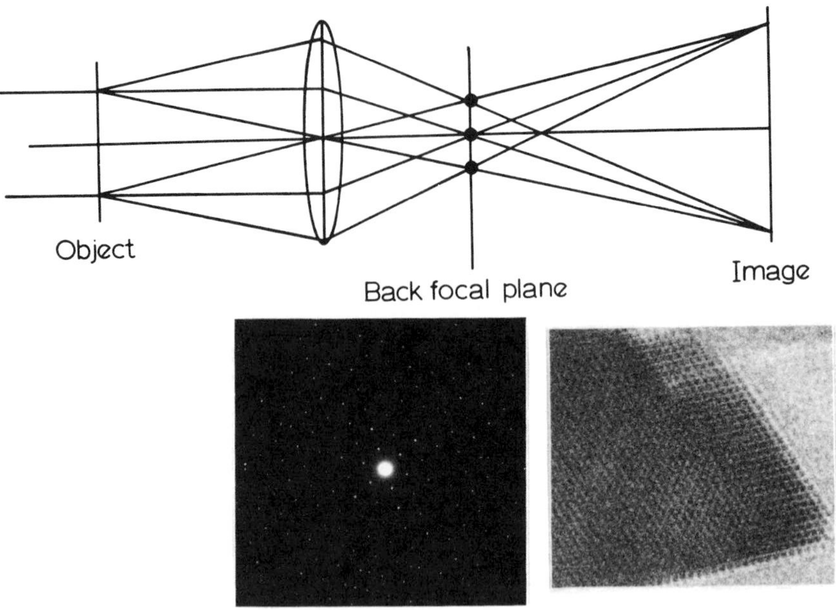

FIGURE 1.7b. *(Continued)*

viewing screen and recording system to produce a permanent copy of an image or electron diffraction pattern. It is not within the scope of this book to consider optimal design features for these various parts nor to consider the merits of various kinds of vacuum systems. Rather, optimal conditions for carrying out crystallographic experiments on (mostly) organic crystals will be established. Also, it is important to demonstrate how the recorded information is interpreted, i.e., to determine crystal structures. Thus, only a brief discussion of the instrument is necessary and one is referred to other books, e.g., that of Reimer (1984), for more detailed discussions of important design considerations.

1.3.1. Illumination System

The microscope acceleration voltage should be high enough to produce electrons with low enough wavelength that a "quasi-kinematic" data set can be obtained. While usual laboratory sources have been traditionally 80 to 100 kV, higher voltages have now become commonplace. The relativistic electron wavelength λ_{rel} is calculated from the accelerating voltage E_a by:

$$\lambda_{\mathrm{rel}} = \frac{h}{mv} = \frac{12.27}{\sqrt{E_a}}\,[1 + 0.978 \times 10^{-6}\,E_a]^{-1/2}$$

The Wehnelt cap of the electron gun, holding, e.g., a tungsten hairpin filament (or a lanthanum hexaboride crystal), can be adjusted to control the illumination of the source. Most control of the illumination is given at the level of the condenser lens system, however. With a double condenser systems, both lenses are nearly maximally excited to spread the illumination into an essentially parallel source. A small aperture (e.g., 20 μm diameter) is inserted at the second condenser system to attenuate the current density to less than 10^{-5} A/cm^2. These illumination conditions, which were described by Glaeser and Thomas (1969), not only provide an acceptable level for examination of unprotected organic crystals and other beam-sensitive samples but also provide a source that has high spatial coherence (Young's fringe experiment in optics).

1.3.2. Functions

1.3.2.1. Imaging. The objective lens is the most important part of the instrument, the part for which most money is spent by manufacturers to improve its characteristics for obtaining images at the highest possible resolution. Electrons are scattered by the potential field of the crystal $\rho(\mathbf{r})$, so that the scattering factors for electrons $f_e(\mathbf{u})$ are related to the x-ray form factors $f_x(\mathbf{u})$ by the Mott formula (e.g., Vainshtein, 1964a):

$$f_e(\mathbf{u}) = (Z - f_x(\mathbf{u}))(8\pi^2 me^2/h^2u^2)$$

Here Z is the atomic number of the atom being considered. Both the distributions of electrons around an atomic nucleus and the potential field between the electron cloud and nucleus can be regarded as being approximately Gaussian in cross section. The form factors, which are the Fourier transforms of these distributions, are also approximately Gaussian (by the arguments presented above). The nucleus, on the other hand, can be modeled approximately by a delta function weighted by its positive charge Z, so that its Fourier transform is a constant level over all reciprocal space, again, as shown above. Thus one can understand intuitively how the Mott equation relates these entities. Compilations of electron form factors have been given by Doyle and Turner (1968). The difference in relative scattering factors for electrons and x-rays is shown in Figure 1.8 for atoms commonly found in organic crystals.

From Figure 1.8a, it is apparent why, originally, there was considerable interest in using electron diffraction data to determine organic crystal structures. While the ratios of scattering factors of hydrogen to carbon are not greatly different at $\sin\theta/\lambda = 0$ for electron and x-ray diffraction experiments, the relative x-ray values decrease appreciably at higher scattering angles while the electron values remain virtually constant. Since the ratio of f_O/f_C and f_N/f_C are actually less than 1.0 for electrons at small angle (while they are greater than 1.0 for x-rays), the detectability of hydrogen bonded to N and O is therefore theoretically much greater in electron diffraction than it is in x-ray diffraction experiments. Thus, Vainshtein (1964b) and his coworkers attempted to use electron diffraction analyses to obtain more accurate hydrogen-heavy atom bonding parameters than were possible by x-ray crystallography. Also, striking differences are found for the scattering factors from ionized atoms in a crystal, particularly at low angle in electron diffraction. For cations, the scattering factor becomes strongly positive, while for anions, it is strongly negative. This may have

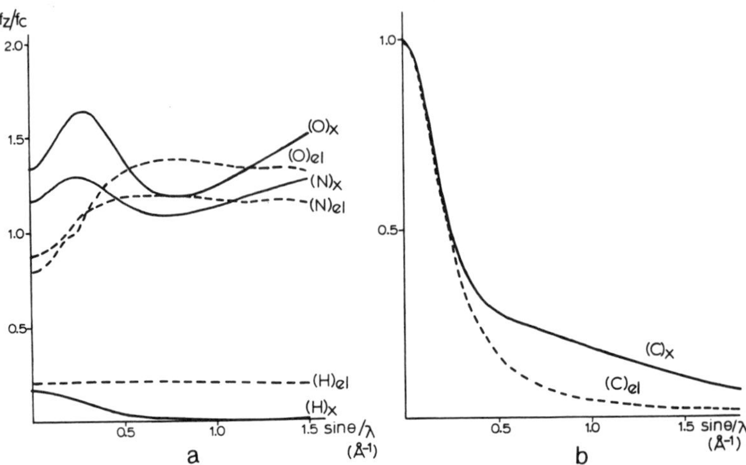

FIGURE 1.8. Comparison of electron and x-ray scattering factors. (a) Relative values for O, N, H compared to C. (b) Normalized values for C. Note the greater relative scattering power for x-rays at high angles (which can be demonstrated for other atomic species such as N, O, and Cl).

particular importance for the study of globular macromolecules. A disadvantage of electron diffraction is illustrated in Figure 1.8b. Although the scattering cross section of matter for electron is much greater than for x-rays, the scattering falloff for x-rays is actually not so abrupt as for electrons (see Vainshtein and Zvyagin, 1993). Given good single crystals for either experiment, then it would be possible to collect relatively higher intensity data at wide angles in x-ray diffraction experiments on carbon-containing compounds.

The emergent electron wave (traveling along z) from the exit face of a crystal is related to the projected potential field $\rho(x,y)$ by the expression (Cowley, 1981):

$$q(x,y) = \exp\{-i\sigma\ \rho(x,y)\}$$

where σ expresses the interaction of the electron beam with the sample, i.e.:

$$\sigma = (2\pi/E\lambda)\ \{1 + (1-\beta^2)^{1/2}\}^{-1}$$

Here E is the accelerating voltage of the electron, λ is its wavelength, and the denominator is a relativistic correction (since electron velocities are near the speed of light). If $\sigma\rho(x,y)$ is small then one can assume:

$$q(x,y) \approx [1 - i\sigma\ \rho(x,y)] * t(x,y)$$

a condition known as the "weak phase object approximation." The latter real space spread function, $t(x,y) = c(x,y) + is(x,y)$ (where c and s stand for Fourier transforms of the reciprocal space cosine and sine components, respectively) accounts for the phase contrast transfer of the lens, as will be discussed in detail later. The power of this signal is recorded by any detector placed in the image plane of the lens. Since $1 * c(x,y) = 1$; $1 * s(x,y) = 0$; $\psi(x,y) = 1 + \sigma\rho(x,y) * s(x,y) - i\sigma\rho(x,y) * c(x,y)$, the image intensity is:

$$\Phi(x,y) = \psi(x,y)\ \psi(x,y)^* = 1 + 2\sigma\ \rho(x,y) * s(x,y)$$

neglecting higher-power terms (see Cowley, 1988). Hence, under favorable conditions, the image intensity is linearly related to the object potential. The Fourier transform of $q(x,y)$ will be found at the back focal plane of the objective lens, i.e.:

$$FT\{q(x,y)\} \approx \delta - i\sigma F(h,k)$$

For a crystal, $F(h,k)$ represents the *structure factors* at delta functions in the reciprocal lattice sampling the continuous transform of $\rho(x,y)$. These will be observed as diffracted intensities: $I(h,k) = |F(h,k)|^2$.

It was stated above that a transfer function plays an important role in determining the distribution of image density for any object observed in the electron microscope. An understanding of spherical and chromatic aberrations of the microscope objective lens, in fact, is very important for interpretation of the image (e.g., see Li (1991)), leading to the transfer functions:

$$\Phi(x,y) = 1 + 2\sigma\rho(x,y) * FT^{-1}\{A(u) \sin \chi_1(u) \exp{-\chi_2(u)}\}$$

In other words, the projected potential is convoluted with the reverse Fourier transform of a term that includes three parts. First is the aperture function $A(u)$, the cross section of which is a rectangle function rect(u). Second is the phase contrast transfer function, including the term for spherical aberration, C_s:

$$\chi_1(u) = \pi\Delta f\lambda|u|^2 + (1/2)\pi C_s\lambda^3|u|^4$$

Finally, there is another transfer function for chromatic aberration:

$$\chi_2 = (1/2)\pi^2\lambda^2|u|^2D^2$$

The first transfer function is the most important since it serves as a "band-pass filter" for diffracted beams contributing to the image. The second function serves as an envelope function to attenuate information at higher resolution. The function $\chi_1(u)$ is plotted in Figure 1.9 for various values of Δf (where a negative value denotes underfocus), given a C_s, which is a constant for the instrument. It is seen that the spherical aberration is compensated for by changes in lens defocus Δf until there is an optimal condition known as the Scherzer focus, defined as:

$$\Delta f = -(C_s\lambda)^{1/2}$$

and the image resolution is found to be 0.43 $(C_s\lambda^3)^{1/4}$. For interpretation of experimental images, an accurate estimate of this function is required. Methods for deter-

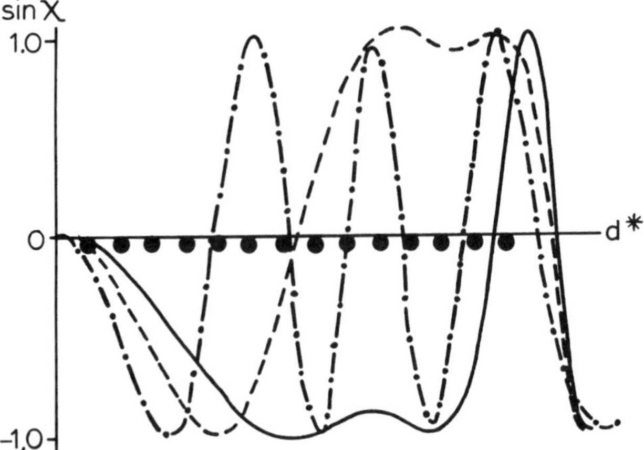

FIGURE 1.9. Plot of the transfer function sin $\chi_1(u)$ versus $d^* = |u|$. A spherical aberration term $C_S = 1.90$ mm is assumed for a 100-kV source. The optimal (Scherzer) defocus $\Delta f = -950$ Å is shown as a solid line. (Dashed lines represent other defocus values.) The dark spots represent hypothetical diffraction maxima spaced (50 Å)$^{-1}$ apart. Alternatively, other authors define $\Delta f = -(\frac{4}{3}C_s\lambda)^{1/2}$ so that the resolution limits is at 0.66 $(C_s\lambda^3)^{1/4}$

mining this function experimentally will be discussed in a later section. Figure 1.9 shows that information within the first envelope is passed with the same contrast. The unscattered beam, however, is phase-shifted by $\pi/2$ from the scattered beam within this envelope. Thus, the aberration itself, when compensated by defocus, establishes the condition of a phase contrast microscope. Without the phase retardation of the unscattered beam from the scattered beam, there would be virtually no contrast in the electron micrograph (Cowley, 1981). While C_s is an important quantity for any objective lens, the electron wavelength is also important for the resolution of the image. Since higher voltage lowers the wavelength of the electron beam, the scattered cone of electrons is closer to the optic axis and, thus, within a more favorable region of the transfer function.

1.3.2.2. Diffraction. The Fourier transform of the transmission function through a crystal was shown, within the phase object approximation, to be directly related to the structure factors in the electron diffraction pattern, which is found at the back focal plane of the objective lens. The structure factor is defined:

$$F(\mathbf{s}) = \sum f_i \exp(2\pi i\, \mathbf{s}\cdot\mathbf{r}) = \sum f_i \exp(2\pi i[hx + ky + lz])$$

Note that this expression resembles the general Fourier transform expression (1) given early in this chapter except that the integral is approximated by a summation. As mentioned already, the terms f_i are the scattering factors for electrons and, actually, these are evaluated in the summation as a Fourier transform of the atomic potential distribution in units of reciprocal length. The fractional coordinates of atoms in the crystals, referred to the origin of a unit cell in the crystal, are x,y,z. The Miller indices

h, k, l, of reflections or points in the diffraction pattern were defined above. The reverse transform to the potential distribution of the unit cell is given:

$$\rho(\mathbf{r}) = V^{-1} \sum F(\mathbf{s}) \exp(-2\pi i \mathbf{s} \cdot \mathbf{r})$$

Note, however, that an important computational difficulty is implied by this pair of Fourier transforms that leads to one of the major analytical problems in crystallography.

If the relative spatial arrangement of mass points in an object is known, it is easy to calculate the Fourier transform. In general, if image intensity in an electron micrograph is measured in terms of pixels $A_i\delta(r_i)$, the Fourier transform of these pixels, weighted by a density value A_i for each position r_i, can be used in the above expression. It was already shown above that, to a first approximation, image intensity is linearly related to the potential distribution. On the other hand, if the intensities of a diffraction pattern, defined as

$$I(\mathbf{s}) = F(\mathbf{s})\, F^*(\mathbf{s})$$

are measured, then only the magnitude $|F(\mathbf{s})|$ can be derived from the square root. However, to calculate e.g., $\rho(\mathbf{r})$, one needs

$$F(\mathbf{s}) = |F(\mathbf{s})| \exp(i\phi)$$

In other words, the term ϕ is lost when one measures the diffracted intensity. This term is the Fourier transform of all relative translational vectors between pairs of mass centers in a unit cell, with respect to one other and, in aggregate, to the origin chosen for the unit cell. This quantity was already mentioned above and was called a *phase term*. Because the phase values are lost in the recorded diffraction pattern, the experimental unidirectional limitation to transforming from one lens plane to another, starting with the intensity distributions of these planes, is known as the phase problem in crystallography. It is immediately seen that an electron microscope has potential advantages for crystallographic investigations because, in principle, it is possible to obtain an image of a crystalline object, however limited the resolution may be by the objective lens aberrations. X-ray crystallographers cannot obtain such high-resolution images of their crystals and so are forced to find other ways of estimating the values of the phase terms ϕ. In addition, the resolution limits imposed by the lens aberrations disappear when recording the diffraction pattern. This is because the phase contrast transfer function generates just another set of phases that vanish when the diffracted intensities are measured. The advantages of using electron diffraction intensities for resolution enhancement will be a major theme of this book.

Each lens in the electron microscope column has its own image and back focal planes and optical manipulations can be employed very useful purposes. For example, if an aperture is inserted at the image plane of the objective lens (or any other lens, for that matter) so that a limited area of the magnified image is sampled, the diffraction pattern observed will also originate from this limited area. This is the basis for *selected*

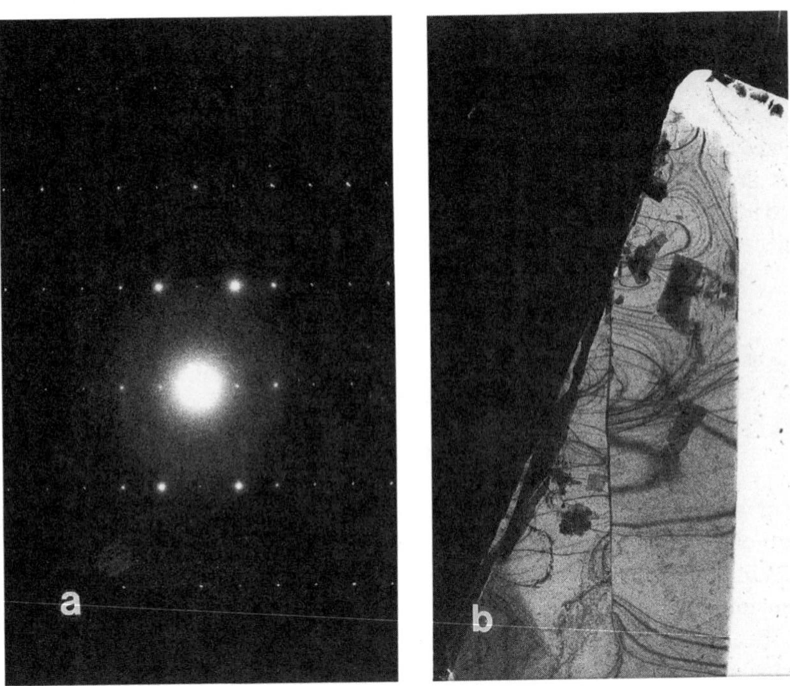

FIGURE 1.10. (a) Selected area electron diffraction pattern from a thin crystal of cytosine. (b) Bright-field electron micrograph obtained by isolating the unscatttered beam. (c) Dark-field micrograph obtained by isolating the (020) beam. (d) Dark field micrograph obtained by isolating the ($\bar{1}$,1,0) beam. (Reprinted from D. L. Dorset (1985) "Crystal structure analysis of small organic molecules," *Journal of Electron Microscopy Technique* **2**, 89–128; with kind permission of Wiley–Liss Division, John Wiley and Sons, Inc.)

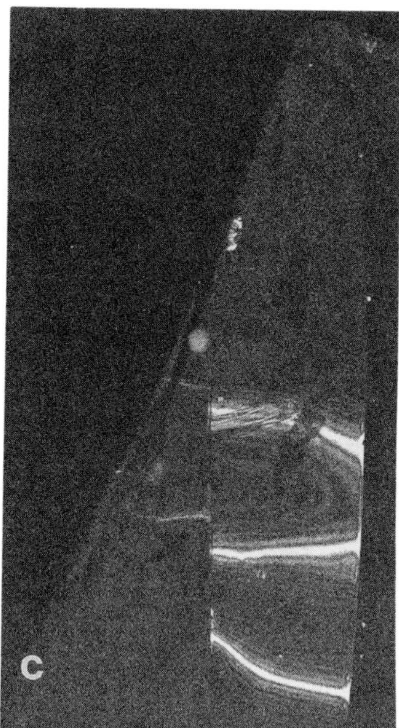

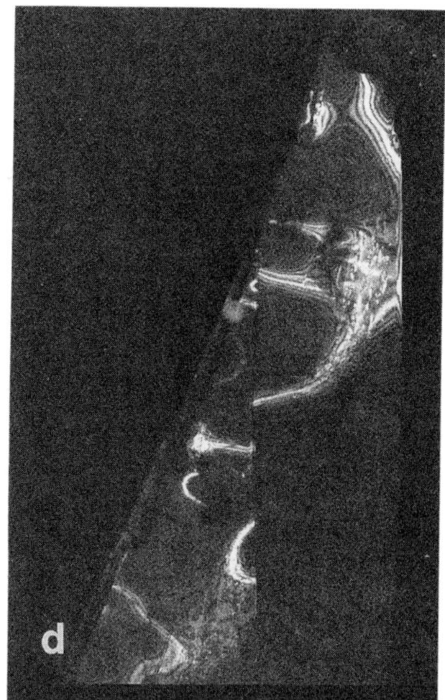

Figure 1.10. (*Continued*)

area electron diffraction (Figure 1.10a). (It is also possible to use the condenser system to limit the illumination area on the specimen and, in fact, it is sometimes more convenient to use this alternative method to isolate smaller crystalline regions.) It is also possible to place an aperture in the back focal plane of any appropriate lens. This can be used to limit the resolution of the image to the useful uniform transfer region of the contrast function (i.e., the $A(u)$ term in the above expression), for example. A smaller aperture can be used to isolate just the central beam, so that all the strongly diffracting beams will be scattered into the metal foil and provide contrast in a *bright field image*, showing details such as crystal bends (Figure 1.10b). Similar isolation of a single diffraction spot to produce a *dark field image*, can show, for example, in which direction of the crystal the various bend contours are running (Figure 1.10c,d). As discussed by Hirsch et al. (1971), the accuracy of area selection is also dependent upon the spherical aberration of the objective lens and the focal position of the specimen, with apparent shifts of detail that is most serious at higher scattering angles.

Magnification of a selected area diffraction pattern is given by the expression (Ferrier, 1969):

$$M_{dp} = f \prod_i M_i$$

where, as usual, f refers to the focal length of the objective lens to the crystalline object and M_i represents the magnifications of successive projector lenses. Now, if the objective lens is shut off so that a lower projector serves as the effective objective to produce a much longer f value, then very large values of M_{dp} can be obtained, e.g., useful for diffraction experiments on proteins or superlattices.

1.4. Geometrical Aspects of Electron Diffraction

A conventional representation of diffraction is given in Figure 1.11, where the reciprocal lattice net (here shown on edge) is sampled by an Ewald sphere of radius $k_o = 1/\lambda$. Since the wavelength of a 100-kV electron beam is some 40 times as small as that of a Cu $K\alpha$ x-ray, it is often a sufficient approximation to say that the Ewald sampling surface is a plane. For a thin perfect crystal, the appropriateness of this approximation is enhanced by the shape function broadening of the diffraction spots in the beam direction, i.e., if the crystal lattice in the beam direction is limited by

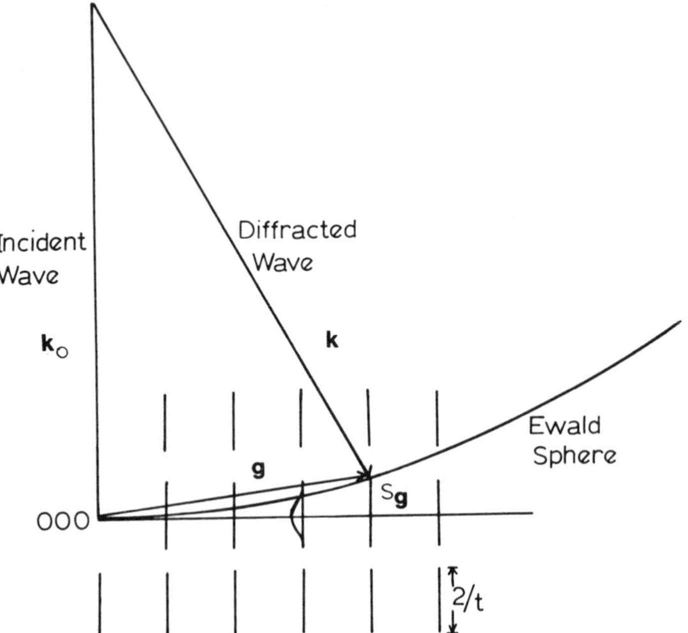

FIGURE 1.11. Ewald sphere representation of diffraction. The vectors $\mathbf{k}_0$ and $\mathbf{k}$, respectively, are incident and diffracted waves. The reciprocal lattice vector $\mathbf{g}$ is related to these by a simple vector sum. The deviation parameter s_g represents the distance of the Ewald sphere from the reflection center. A perfectly flat crystal plate is assumed so that the diffraction maxima will be spread along the reciprocal lattice rods by the sinc(πNau) shape transform described in the text. (Reprinted from D. L. Dorset (1989) "Electron diffraction from crystalline polymers," in *Comprehensive Polymer Science, Volume 1*, Sir G. Allen, ed., Pergamon Press, Oxford, 1989, pp. 651–668; with kind permission of Elsevier Science Ltd.)

rect(η), then the diffraction maxima will be spread by sinc(ξ), as shown above. The undistorted reciprocal net given in Figure 1.2i is actually often observed in selected area electron diffraction experiments on single crystals. It corresponds to the representation of the reciprocal lattice obtained, with greater difficulty, by a precession x-ray camera.

By analogy to x-ray diffraction, electron diffraction patterns need not only be obtained in an electron microscope. Given a distant point source of radiation, it is also possible to obtain patterns without the benefit of an objective lens. Earliest electron diffraction studies were made with an electron diffraction camera and the only lens used was a condenser to control illumination of the specimen (e.g., see a description

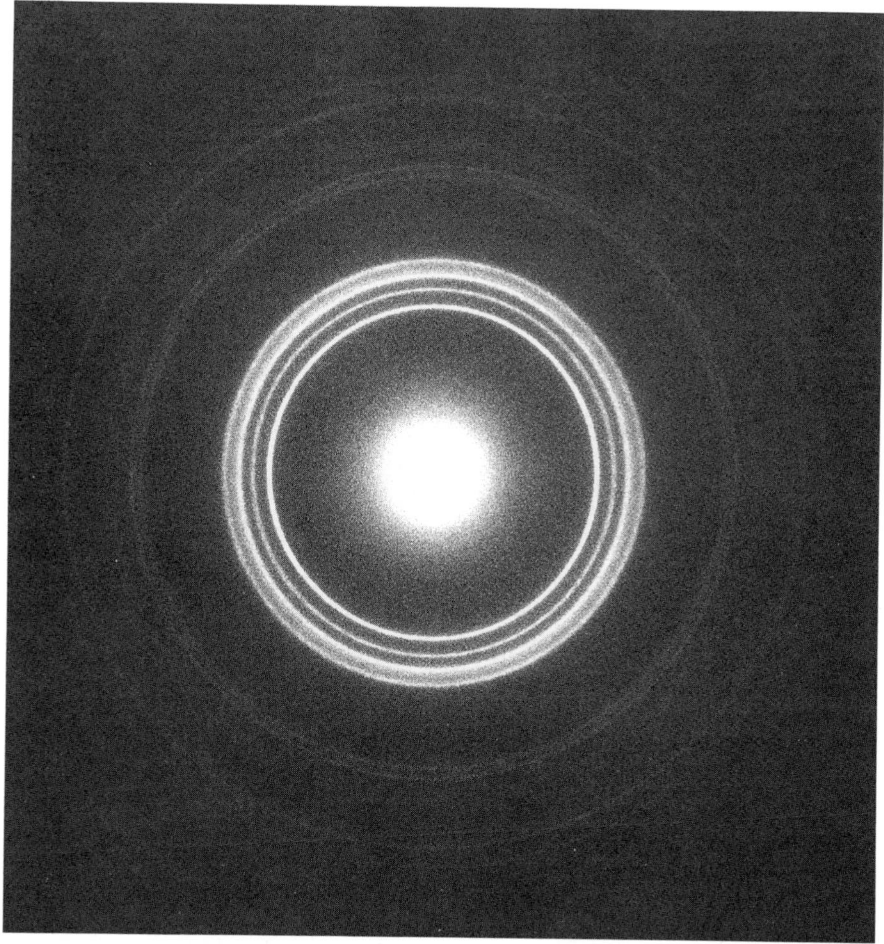

FIGURE 1.12. Oriented powder diffraction pattern from cyclohexanonacontane. (Reprinted from D. L. Dorset and S. L. Hsu (1989) "Polymethylene chain packing in epitaxially crystallized cycloalkanes: an electron diffraction study." *Polymer* **30**, 1596–1602; with kind permission of Butterworth–Heinemann, Ltd.)

by Zvyagin, 1993). Until recently, reflection diffraction holders, available for many electron microscopes, were mounted below all of the electromagnetic lenses in the column so that there was a fixed distance to the photographic plate (see Pinsker, 1953). This fixed-camera-length geometry is the only simple way to obtain accurate electron diffraction measurements of lattice constants, since small variations in the focal distance to the specimen for the lens arrangement discussed above will result in a distribution of camera length values for a sample mounted on a support that is not perfectly flat. Thus it is mandatory that any diffraction experiment be accompanied by an internal calibration with a known material. An easy one to use is an evaporated layer of gold, preferably placed on a segment of the support grid for a representative specimen.

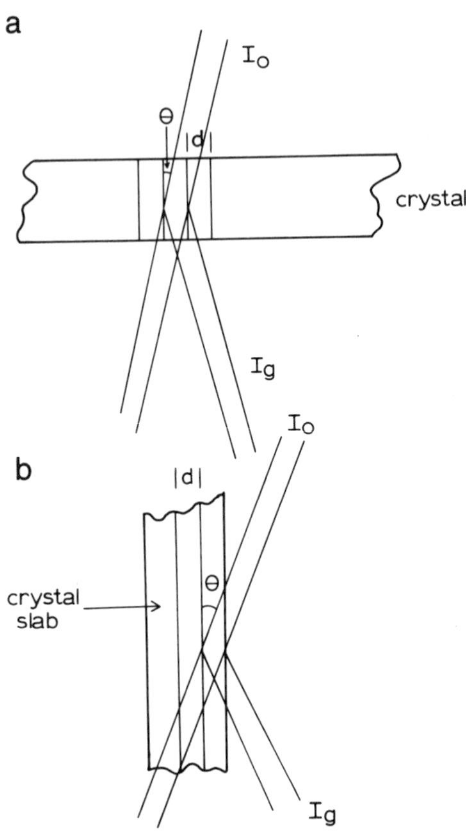

FIGURE 1.13. Geometrical representation of (a) Laue (transmission) and (b) Bragg (reflection) electron diffraction. (Reprinted from D. L. Dorset (1989) "Electron diffraction from crystalline polymers," in *Comprehensive Polymer Science, Volume 1*, Sir G. Allen, ed., Pergamon Press, Oxford, 1989, pp. 651–668; with kind permission of Elsevier Science Ltd.)

FIGURE 1.14. Reflection electron diffraction pattern from a lipid monolayer. Indices are given for individual diffraction lines.

If plate crystals being examined are very small and oriented randomly with respect to a layer normal, a powder diffraction pattern is observed, in which the traces of the diffracted spots describe rings in the pattern (Figure 1.12). These rings also exist for upper layers of the reciprocal lattice from a three-dimensional crystal. If it is possible to tilt a sample at a large angle so that the nearly flat Ewald sampling plane intersects arcs of the rings, then an *oblique texture pattern* is obtained (see Vainshtein, 1964a). Patterns of this kind were very important for the collection of three-dimensional intensity data in the early years of electron crystallography and were often obtained with electron diffraction cameras in which millimeter diameter areas of the specimen were illuminated by the electron beam.

Many of the quantitative diffraction analyses described in this book will be based on transmission selected area experiments (Figure 1.13a) through a crystal slab (Figure 1.10a) as described above—a phenomenon called Laue diffraction. It is also possible to carry out *reflection electron diffraction* (RHEED) from a crystal surface in the geometry shown in Figure 1.13b, also known as Bragg diffraction (Figure 1.14). The advantage of this technique is that it is an effective probe of the crystal surface, with a very low penetration of the crystal bulk (Pinsker, 1953). For certain linear molecules, it will sample a space lattice in the outermost part of the structure. For example, it was a very important technique for the study of lubricant films on metal surfaces and is again gaining popularity for the study of Langmuir–Blodgett layers. It is easier to treat Laue diffraction intensity data, resulting from an excitation of a volume element, quantitatively than it is to use reflection diffraction intensities, arising from the excitation of a surface, for structure analysis. Some progress has been made with the quantitation of RHEED by Yagi (1993). As will be shown below, RHEED can be used effectively to measure the tilt of linear molecules to a surface.

Other significant kinds of diffraction geometry which have proved to be of great worth to the study of inorganic materials will not be treated in this book. This is because they require rather large current densities and are therefore applied to the study of radiation-sensitive organic objects only with great difficulty. For techniques such as *convergent beam electron diffraction*, which is treated in several significant recent reviews (Eades, 1992; Spence and Zuo, 1993), this is a pity, because one can obtain precise information about the point group symmetry not ordinarily accessible to usual crystallographic study.

2

Crystal Symmetry

2.1. The Unit Cell

As was shown in Chapter 1, a crystal can be thought to be made up of a space lattice which is convoluted by the average motif of the repetitive structure. The parallelepiped that describes the vector repeat of this space lattice is the unit cell. It need not be right-angled and, indeed, in two-dimensional projection, it is readily seen that five types of planar space lattice are permitted, as illustrated in Figure 2.1. For organic electron crystallography, oblique and rectangular arrays are most important.

In three dimensions, these two-dimensional arrays can be extended to form 14 unique Bravais lattices which are illustrated in Figure 2.2. The respective axial and angular relationships are listed below. It must also be noted in this figure that, for symmetries higher than triclinic, it is possible to have so-called nonprimitive lattices, where a central insertion of mass can occur within a face of a unit cell (hence prefixes *A*, *B*, *C* for specific faces, *F* for all faces) or within the body of the unit cell (prefix *I*). Requirements for the different unit cells are:

triclinic	$a \neq b \neq c$ $\alpha \neq \beta \neq \gamma$
monoclinic	$a \neq b \neq c$ $\alpha = \gamma = 90° \neq \beta$
orthorhombic	$a \neq b \neq c$ $\alpha = \beta = \gamma = 90°$
tetragonal	$a = b \neq c$ $\alpha = \beta = \gamma = 90°$
cubic	$a = b = c$ $\alpha = \beta = \gamma = 90°$

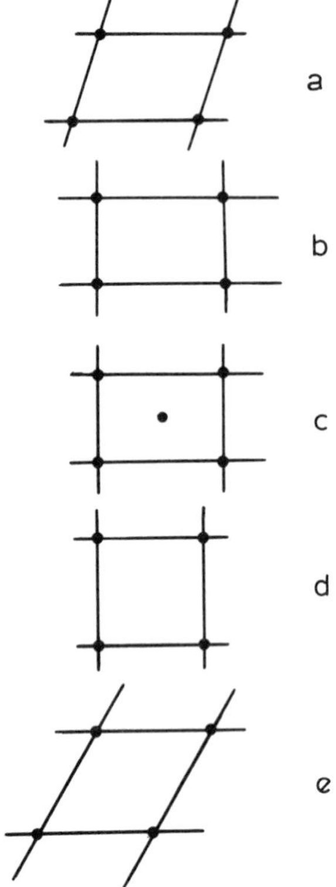

FIGURE 2.1. Two-dimensional nets: (a) primitive oblique; (b) primitive rectangular; (c) centered rectangular; (d) primitive square; (e) primitive hexagonal.

trigonal	$a = b = c$ $\alpha = \beta = \gamma \neq 90°$
hexagonal	$a = b \neq c$ $\alpha = \beta = 90°, \gamma = 120°$

In organic crystals, triclinic, monoclinic, and orthorhombic unit cells are the most common (Stout and Jensen, 1968). Higher-symmetry unit cells can be seen in inorganic or protein crystals.

There is an inverse relationship between a space lattice and a reciprocal lattice, as was discussed in Chapter 1. For example, the superimposed spatial and reciprocal geometry of a monoclinic unit cell repeat is represented in Figure 2.3. Note that the **a*** and **c*** axes are, respectively, perpendicular to the (100), i.e., **bc**, and (001), **ab**, planes of the crystal in the projection down [010]. Conversely, the **a** and **c** axes of the crystal are perpendicular to the **b*c*** and **a*b*** reciprocal nets, respectively. The choice

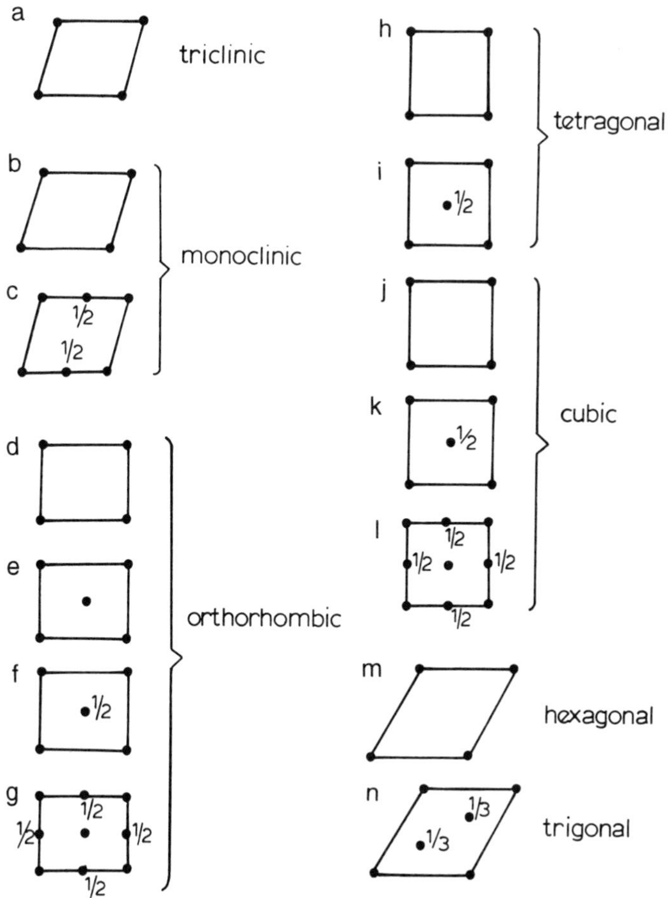

FIGURE 2.2. Projections of the 14 Bravais lattices: *triclinic* (a) primitive ($P\bar{1}$); *monoclinic* (b) primitive ($P2/m$); (c) face-centered ($C2/m$); *orthorhombic* (d) primitive ($Pmmm$); (e) face-centered ($Cmmm$); (f) body-centered ($Immm$); (g) all face-centered ($Fmmm$); *tetragonal* (h) primitive ($P4/mmm$); (i) body-centered ($I4/mmm$); *cubic* (j) primitive ($Pm3m$); (k) body-centered ($Im3m$); (l) all face-centered ($Fm3m$); *hexagonal* (m) primitive ($P6/mmm$); *trigonal* (n) primitive (i.e., the rhombohedral lattice) ($R\bar{3}m$).

of unit cell is not necessarily unique, since several kinds of space lattice might be drawn to account for the periodic repeat of the crystal. Obviously, advantage is taken of any symmetry displayed by the repeating motif to define the unit cell. Any space lattice can be described as a repeat of a primitive triclinic cell but higher symmetry cells, if justified, will be useful for simplifying certain calculations (such as Fourier transforms), as will be described below. For the monoclinic example considered here, there are often many possible choices of β-angle that will satisfy the lattice repeat. Sometimes, certain geometrical features of the molecular packing will be used to specify the **c**-axis direction. (In lipid and synthetic polymer crystallography, for example, this

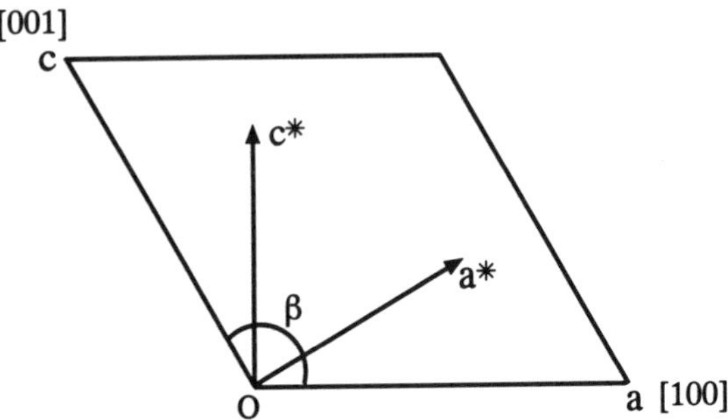

FIGURE 2.3. Direct and reciprocal axial relationships for a monoclinic unit cell projected down **b** ([010]). The reciprocal axis **b*** lies along **b**. However, because of the nonorthogonal β-angle, **a** and **a*** or **c** and **c*** cannot share common directions.

is often defined as the molecular chain axis direction in the unit cell.) Other compendia of crystal structures may chose a convention where β is closest to 90°. Thus, for a thin crystal, where the electron beam is perpendicular to an **ab** face, the angle of tilt around **b** needed to find an "($hk0$)" diffraction pattern will depend on the unit cell definition. Of course, this will also affect the zone axis definition for the untilted crystal projection. These points are of considerable importance, e.g., for comparing electron diffraction data to a known structure model derived from an x-ray determination, especially since, nowadays, x-ray crystallographers rarely seem to be concerned about the orientation of the crystal habit with respect to the unit cell defined for a structure analysis. (However, the work by Winchell (1987) harks back to a time when they were interested in such things. Revived interest in crystal habit has been the result of recent work in organic solid state chemistry, e.g., the book edited by Desiraju (1987).)

In general, for any unit cell, the direction $\mathbf{d}^*_{hkl} \perp (hkl)$ where

$$\mathbf{d}^*_{hkl} = (h/V)\mathbf{b} \times \mathbf{c} + (k/V)\mathbf{c} \times \mathbf{a} + (l/V)\mathbf{a} \times \mathbf{b} = \mathrm{h}\mathbf{a}^* + \mathbf{k}\mathbf{b}^* + l\mathbf{c}^*$$

The unit cell volume is given as $V = \mathbf{a} \cdot \mathbf{b} \times \mathbf{c} = \mathbf{b} \cdot \mathbf{c} \times \mathbf{a} = \mathbf{c} \cdot \mathbf{a} \times \mathbf{b}$ and thus,

$$\mathbf{c}^* = \mathbf{a} \times \mathbf{b}/\mathbf{c} \cdot \mathbf{a} \times \mathbf{b}$$

or, using any suitable permutation of the axes, one can find the other reciprocal axes. A list of geometric relationships for all of the unit cells is given in Table 1 in terms of scalar quantities. In electron diffraction studies, it is often useful to know the angle between two zone axes $[U,V,W]$ to orient the goniometer stage for the rotational search for a particular diffraction pattern (hkl) satisfying the zone axis equation $h \cdot U + k \cdot V + l \cdot W = 0$. For any two zone axes $[U(1),V(1),W(1)]$ and $[U(2),V(2),W(2)]$, the angle between them is found from:

$$\cos\omega = \mathbf{t}_{U(1),V(1),W(1)} \cdot \mathbf{t}_{U(2),V(2),W(2)} / |\mathbf{t}_{U(1),V(1),W(1)}|\,|\mathbf{t}_{U(2),V(2),W(2)}|$$

Table 2.1. Geometric Relationships for Unit Cells

orthorhombic:

$a^* = 1/a,\ b^* = 1/b,\ c^* = 1/c$

$\alpha = \beta = \gamma = \alpha^* = \beta^* = \gamma^* = 90°$

$V^* = 1/V = a^*b^*c^*;\ V = 1/V^* = abc$

monoclinic:

$a^* = 1/a \sin\beta;\ b^* = 1/b;\ c^* = 1/c \sin\beta$

$a = 1/a^* \sin\beta^*;\ b = 1/b^*;\ c = 1/c^* \sin\beta^*$

$\alpha = \gamma = \alpha^* = \gamma^* = 90°;\ \beta^* = 180° - \beta$

$V^* = 1/V = a^*b^*c^* \sin\beta^*;\ V = 1/V^* = abc \sin\beta$

triclinic:

$a^* = bc \sin\alpha/V;\ b^* = ac \sin\beta/V;\ c^* = ab \sin\gamma/V$

$a = b^*c^* \sin\alpha^*/V^*;\ b = a^*c^* \sin\beta^*/V^*;\ c = a^*b^* \sin\gamma^*/V^*$

$\cos\alpha^* = (\cos\beta \cos\gamma - \cos\alpha)/\sin\beta \sin\gamma$

$\cos\beta^* = (\cos\alpha \cos\gamma - \cos\beta)/\sin\alpha \sin\gamma$

$\cos\gamma^* = (\cos\alpha \cos\beta - \cos\gamma)/\sin\alpha \sin\beta$

$V = 1/V^* = abc(1 - \cos^2\alpha - \cos^2\beta - \cos^2\gamma + 2\cos\alpha\cos\beta\cos\gamma)^{1/2}$

When all axes **a**, **b**, and **c** are mutually orthogonal (orthorhombic, tetragonal, or cubic unit cells), the result is particularly easy to find, hence:

$$\cos\omega = U(1)U(2)a^2 + V(1)V(2)b^2 + W(1)W(2)c^2 \times$$

$$(U(1)^2a^2 + V(1)^2b^2 + W(1)^2c^2)^{-1/2}\ (U(2)^2a^2 + V(2)^2b^2 + W(2)^2c^2)^{-1/2}$$

For unit cells with oblique angles, it is more convenient to use the metric tensor (see McKie and McKie, 1986). This matrix of dot products between cell edges is defined:

$$\mathbf{T} = \begin{vmatrix} \mathbf{a}\cdot\mathbf{a} & \mathbf{a}\cdot\mathbf{b} & \mathbf{a}\cdot\mathbf{c} \\ \mathbf{b}\cdot\mathbf{a} & \mathbf{b}\cdot\mathbf{b} & \mathbf{b}\cdot\mathbf{c} \\ \mathbf{c}\cdot\mathbf{a} & \mathbf{c}\cdot\mathbf{b} & \mathbf{c}\cdot\mathbf{c} \end{vmatrix}$$

The quantities defined above are then determined from:

$$\mathbf{t}_{U(1),V(1),W(1)} \cdot \mathbf{t}_{U(2),V(2),W(2)} = [U(1)V(1)W(1)]\ \mathbf{T} \begin{vmatrix} U(2) \\ V(2) \\ W(2) \end{vmatrix}$$

When the unit cell is triclinic:

$$\mathbf{T} = \begin{vmatrix} a^2 & ab\cos\gamma & ac\cos\beta \\ ab\cos\gamma & b^2 & bc\cos\alpha \\ ac\cos\beta & bc\cos\alpha & c^2 \end{vmatrix}$$

and all other examples can be generated from this general case.

When dealing with a crystal structure packing in an oblique unit cell, it is often convenient to transform the atomic or group coordinates to an orthonormal axial system. This can be done with the upper triangular matrix (see Prince, 1982):

$$M = \begin{vmatrix} a & b\cos\gamma & c\cos\beta \\ 0 & b\sin\gamma & c(\cos\alpha - \cos\beta\cos\gamma)/\sin\gamma \\ 0 & 0 & c[1-(\cos^2\alpha + \cos^2\beta - 2\cos\alpha\cos\beta\cos\gamma]^{1/2}/\sin\gamma \end{vmatrix}$$

with appropriate adjustments if the unit cell is monoclinic and not triclinic.

2.2. Symmetry Groups

One of the most striking properties of all crystals is their symmetry. It has already been shown that part of the crystal symmetry is inherent in the type of unit cell, repeated by translation in all directions to form a crystal. However, the packing of so-called asymmetric units of atomic arrays within the unit cell often includes other symmetry operations, partly because these fundamental units themselves have their own internal symmetry. The complete description of unit cell or molecular symmetry, therefore, must account for interactions among all individual symmetry elements used by the object. It should also be possible to show that this description is complete. Such a description of symmetry operators utilizes a concept from modern algebra, i.e., the mathematical set known as a *group* (McCoy, 1960; Sands, 1969; Cotton, 1971).

Formally speaking, a group is defined to be a nonempty set G on which a binary operation "◊" is defined so that the following properties are satisfied:

(i) If $a, b, c \in G$, then $(a \lozenge b) \lozenge c = a \lozenge (b \lozenge c)$

That is to say, the binary operation is associative.

(ii) G contains an element e such that $e \lozenge a = a \lozenge e = a$ for every element a of G.

This merely states that an identity element exists.

(iii) If $a \in G$, then another element $x \in G$ can be found such that $a \lozenge x = x \lozenge a = e$

In other words, each element has an inverse. (Above, "$\in$" denotes "an element of.")

There are many examples of groups in mathematics, e.g., the set of integers under the operation of addition, itself an example of an infinite group. Generally, for descriptions of symmetry, one is constrained to finite groups. One adds a fourth condition of *closure* to the above list:

(iv) The binary operation between any two elements of a group must itself produce another element of the group.

The operation "◊" in symmetry groups is taken to be the combination of symmetry elements, which can be represented by the multiplication of $d \times d$ matrices of dimension d (plus column matrices in some cases, which are added, under the operation "◊"). For n elements of a group, a multiplication table of n^2 products can be constructed in which no column or row is ever repeated but where each contains a rearrangement of all elements of the group (Sands, 1969; Cotton, 1971). Note also that we have not assumed that the identical multiplication of elements $a \times b = b \times a$ necessarily exists. In general, multiplication of symmetry operators is not commutative. When, however, this property is obeyed, the symmetry group is called *Abelian*.

2.2.1. Point Groups

The symmetry of molecules is described by a point group in which four types of symmetry elements (operators) are possible:

(1) the identity element 1 (or E):

$$1 = \begin{vmatrix} 1 & 0 & 0 \\ 0 & 1 & 0 \\ 0 & 0 & 1 \end{vmatrix} \quad \begin{vmatrix} x \\ y \\ z \end{vmatrix} = \begin{vmatrix} x \\ y \\ z \end{vmatrix}$$

(2) inversion $\bar{1}$ (or i):

$$\bar{1} = \begin{vmatrix} -1 & 0 & 0 \\ 0 & -1 & 0 \\ 0 & 0 & -1 \end{vmatrix} \quad \begin{vmatrix} x \\ y \\ z \end{vmatrix} = \begin{vmatrix} -x \\ -y \\ -z \end{vmatrix}$$

(3) mirrors m (or σ), e.g., perpendicular to y:

$$m_{\perp y} = \begin{vmatrix} 1 & 0 & 0 \\ 0 & -1 & 0 \\ 0 & 0 & 1 \end{vmatrix} \quad \begin{vmatrix} x \\ y \\ z \end{vmatrix} = \begin{vmatrix} x \\ -y \\ z \end{vmatrix}$$

(In the parenthetic notation for the mirror, σ_h denotes a mirror perpendicular to a rotation axis while σ_v is one parallel to the rotation axis.)

There are also rotation axes n (or C_n) with rotation $\theta = 2\pi/n$. If this is parallel e.g., to the z-axis then one seeks, for a clockwise rotation:

$$n_{\parallel z} = \begin{vmatrix} \cos\theta & \sin\theta & 0 \\ -\sin\theta & \cos\theta & 0 \\ 0 & 0 & 1 \end{vmatrix} \quad \begin{vmatrix} x_1 \\ y_1 \\ z_1 \end{vmatrix} = \begin{vmatrix} x_2 \\ y_2 \\ z_2 \end{vmatrix}$$

For the above operations, the first symbol given is the Hermann–Mauguin international notation familiar to crystallographers. The parenthetic symbol is in the Schoenflies notation familiar to spectroscopists. As described, the operations are exactly equivalent. These notations deviate, however, in the description of the improper rotation operation. For the Hermann-Mauguin system, this is symbolized as $\bar{n}$ and describes a rotation 360°/n followed by inversion. For example, $\bar{2}$ can be expressed by the sequence:

$$2 \cdot (\bar{1}) = \begin{vmatrix} -1 & 0 & 0 \\ 0 & -1 & 0 \\ 0 & 0 & 1 \end{vmatrix} \begin{vmatrix} -1 & 0 & 0 \\ 0 & -1 & 0 \\ 0 & 0 & -1 \end{vmatrix} = \begin{vmatrix} 1 & 0 & 0 \\ 0 & 1 & 0 \\ 0 & 0 & -1 \end{vmatrix} = \mathrm{m}_{\perp z}$$

In other words, this is equivalent to a mirror plane. The Schoenflies operator $S_n \neq \bar{n}$, in general, is the n-fold rotation followed by a mirror operation. Thus for S_2, we have:

$$C_2 \cdot \sigma_h = \begin{vmatrix} -1 & 0 & 0 \\ 0 & -1 & 0 \\ 0 & 0 & 1 \end{vmatrix} \begin{vmatrix} 1 & 0 & 0 \\ 0 & 1 & 0 \\ 0 & 0 & -1 \end{vmatrix} = \begin{vmatrix} -1 & 0 & 0 \\ 0 & -1 & 0 \\ 0 & 0 & -1 \end{vmatrix}$$

which, in this case, is equivalent to a center of inversion. The mirror operation σ_h is reproduced by S_1 and the equivalence $S_n = \bar{n}$ holds only when n is some multiple of 4.

Now consider some examples. (In this book, the Hermann–Mauguin international notation is generally employed.) The simplest point group is represented by the symbol 1. The group multiplication table is given:

	1
1	1

An example of a molecule with this (absence of) symmetry is any one with handedness or chirality. One example is in Figure 2.4. These molecules are optically active. A representation of this plane group is shown in Figure 2.5 . The next highest symmetry is $\bar{1}$, containing a center of inversion. The group multiplication table is given below:

	1	$\bar{1}$
1	1	$\bar{1}$
$\bar{1}$	$\bar{1}$	1

and its planar representation is shown in Figure 2.5. The final matrix multiplication in this table is easily verified:

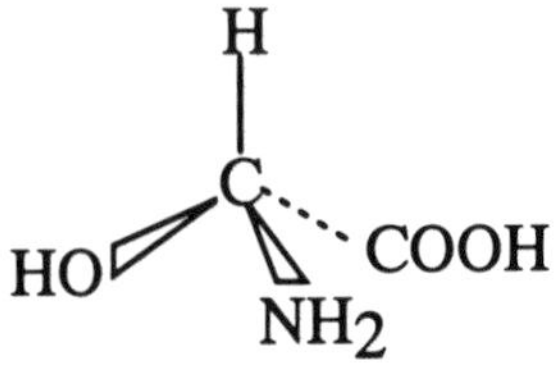

FIGURE 2.4. Structure of a chiral molecule.

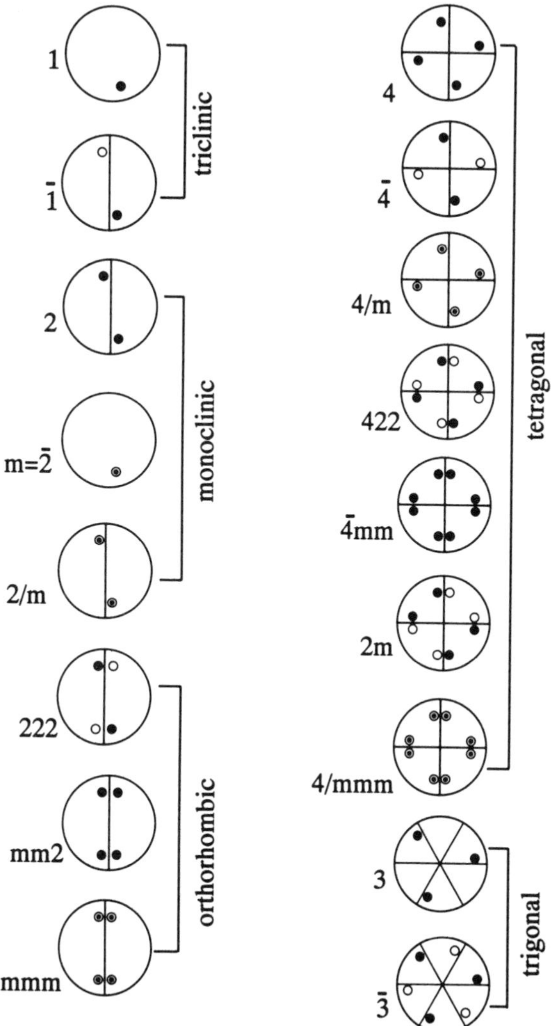

FIGURE 2.5. The 32 point groups. Above the plane through the equator of a sphere (parallel to the paper), an arbitrary point is rendered "•." Below the plane, it becomes "o." Both are equidistant from this plane when related by a symmetry operator. *(continued)*

$$\begin{vmatrix} -1 & 0 & 0 \\ 0 & -1 & 0 \\ 0 & 0 & -1 \end{vmatrix} \begin{vmatrix} -1 & 0 & 0 \\ 0 & -1 & 0 \\ 0 & 0 & -1 \end{vmatrix} = \begin{vmatrix} 1 & 0 & 0 \\ 0 & 1 & 0 \\ 0 & 0 & 1 \end{vmatrix}$$

i.e.: $\bar{1} \cdot \bar{1} = 1$. An example of a molecule with just a center of inversion is represented in Figure 2.6. Often this center occurs with other symmetry elements.

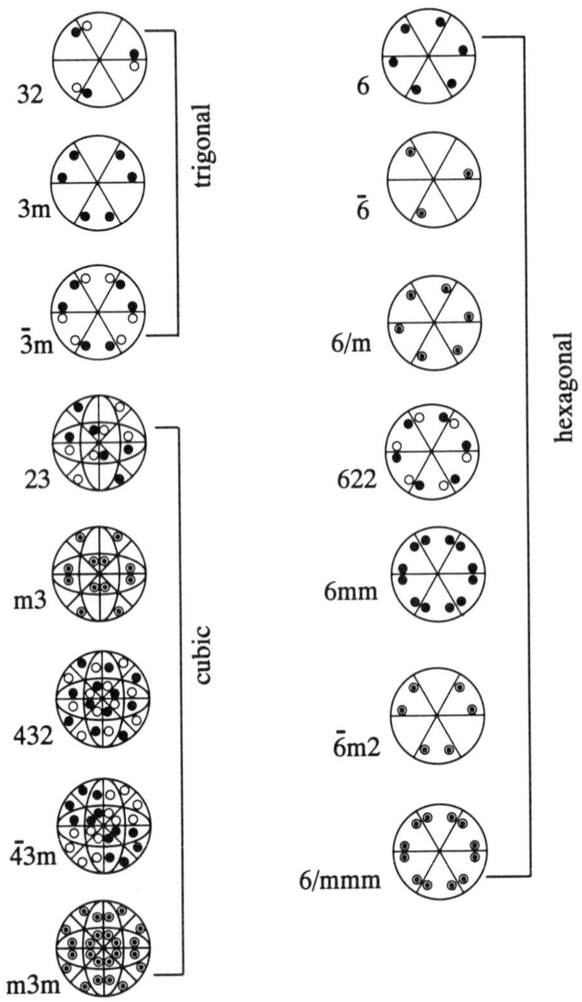

FIGURE 2.5. *(Continued)*

The next point group is a twofold rotation, represented as 2 (Figure 2.5). If we assign the axis of rotation to lie along z, then we can write:

$$2 = \begin{vmatrix} -1 & 0 & 0 \\ 0 & -1 & 0 \\ 0 & 0 & 1 \end{vmatrix}$$

It is easy to show that $2 \cdot 2 = 1$. From this, one constructs the multiplication table:

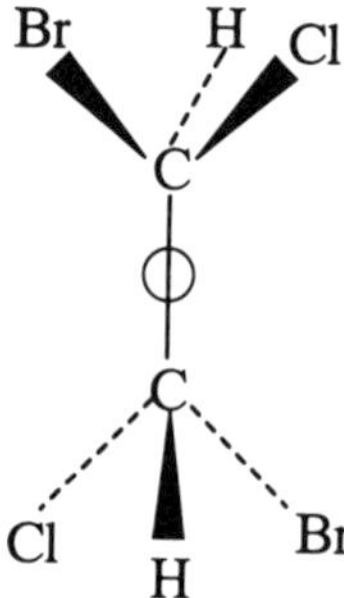

FIGURE 2.6. A centrosymmetric molecule.

	1	2
1	1	2
2	2	1

This point group is also chiral. A molecule with a twofold axis is shown in Figure 2.7. The next point group is $\bar{2} = m$. Its multiplication table is:

	1	m
1	1	m
m	m	1

The planar representation of m is shown in Figure 2.5 and an example of a molecule with this symmetry given in Figure 2.7. The next point group in Figure 2.5 is $2/m$, where the slash "/" means that the twofold is perpendicular to the mirror plane. Its multiplication table is:

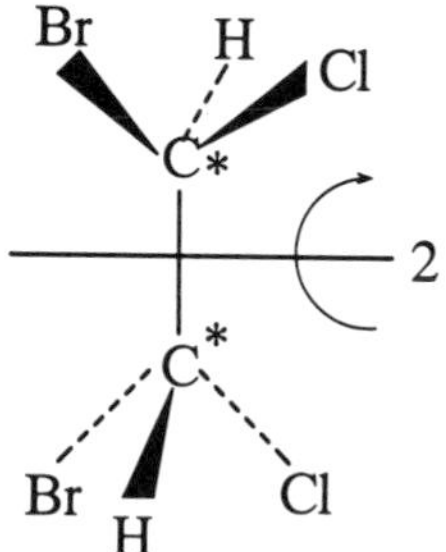

FIGURE 2.7. A molecule with a twofold symmetry axis.

	1	2	m	$\bar{1}$
1	1	2	m	$\bar{1}$
2	2	1	$\bar{1}$	m
m	m	$\bar{1}$	1	2
$\bar{1}$	$\bar{1}$	m	2	1

As shown above, the multiplication $2 \cdot m_{\perp z}$ generates the center of inversion $\bar{1}$ when the twofold axis is perpendicular to the mirror plane. Thus the element $\bar{1}$ must be added to the multiplication table to form a closed group. A molecule with the $2/m$ symmetry is shown in Figure 2.8 and is seen to be centrosymmetric.

This process can be continued until 32 unique crystallographic point groups are constructed. To find the Schoenflies notation for the point groups in Figure 2.5, the following rules should be followed:

- C_n groups: These contain only the identity element E and a right-handed rotation C_nn, as well as the left-handed rotation C_n^2.
- C_{nh} groups: These contain a mirror plane σ_h perpendicular to the rotation axis C_n. For $n = 1$, the resultant plane group m (Hermann–Mauguin notation) would be written C_{1h}, or, more often C_s.
- C_{nv} groups: These contain n mirror planes σ_v parallel to the rotation axis C_n. Again: $C_{1v} = C_{1h} = C_s$.
- S_n groups: These contain the improper rotation axis S_n. When n is odd, $S_n = C_{nh}$. The group S_6 is frequently written C_{3i}.

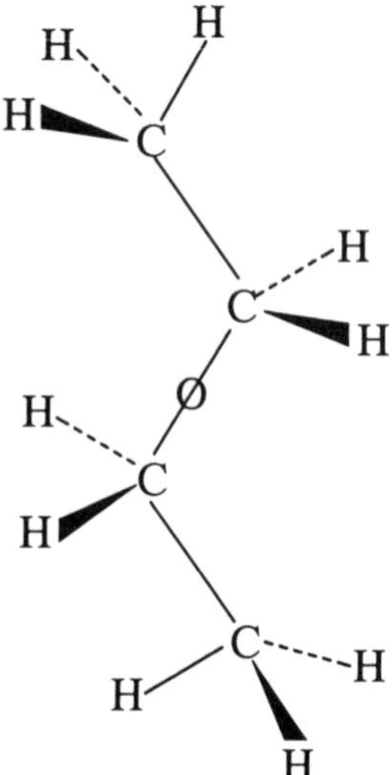

FIGURE 2.8. A molecule with $2/m$ symmetry.

- D_n groups: These have n C_2 axes perpendicular to one C_n rotation.
- D_{nh} groups: These contain the rotation axis C_n as well as n C_2 axes perpendicular to C_n, as before, and an additional mirror plane σ_h perpendicular to C_n.
- D_{nd} groups: These have n C_2 axes perpendicular to C_n and, in addition, n σ_v mirrors bisecting the angles between the C_2 axes.

There are also special Schoenflies notations for higher cubic and icosahedral point groups, e.g., as described by Sands (1969).

Using the component symmetry operations with the multiplication tables will readily demonstrate whether or not two alternative representations are equivalent or if perhaps one point group is a subgroup of another. For the multiplication tables shown above, it will be observed that the point groups are all Abelian. That the commutative property does not necessarily hold for all symmetry groups can be shown for the point group C_{3v} ($\equiv 3m$). The multiplication table in the Schoenflies notation follows:

	E	C_3	C_3^2	σ_v	σ_v'	σ_v''
E	E	C_3	C_3^2	σ_v	σ_v'	σ_v''
C_3	C_3	C_3^2	E	σ_v'	σ_v''	σ_v
C_3^2	C_3^2	E	C_3	σ_v''	σ_v	σ_v'
σ_v	σ_v	σ_v''	σ_v'	E	C_3^2	C_3
σ_v'	σ_v'	σ_v	σ_v''	C_3	E	C_3^2
σ_v''	σ_v''	σ_v'	σ_v	C_3^2	C_3	E

In the above, the elements σ_v, σ_v', and σ_v'' are mirrors parallel to C_3 at the three positions allowed. Note that the symmetry elements are not mirrored across the diagonal of the multiplication table. Thus the multiplication operation is not commutative for this example.

2.2.2. Plane Groups

Now that the symmetry of single molecules is understood, consider the symmetry of the unit cells in which they are packed. It is instructive first to consider the projections of the unit cells, since these are often encountered in electron crystallography (given the diffraction geometry described above). In fact, the *central section theorem* (DeRosier and Klug, 1968; Crowther et al., 1970) states that the diffraction from any crystal orientation reduces to the projection of all mass along the projection axis onto a plane. The symmetry of that plane will, logically, be described by a *plane group.* (These concepts are also important for tomographic analysis.)

As illustrated in Figure 2.9, there are 17 unique crystallographic plane groups. Plane group symmetries describe the interrelation of asymmetric mass points that are distributed through the unit cell projection. Some of the symmetry elements used in

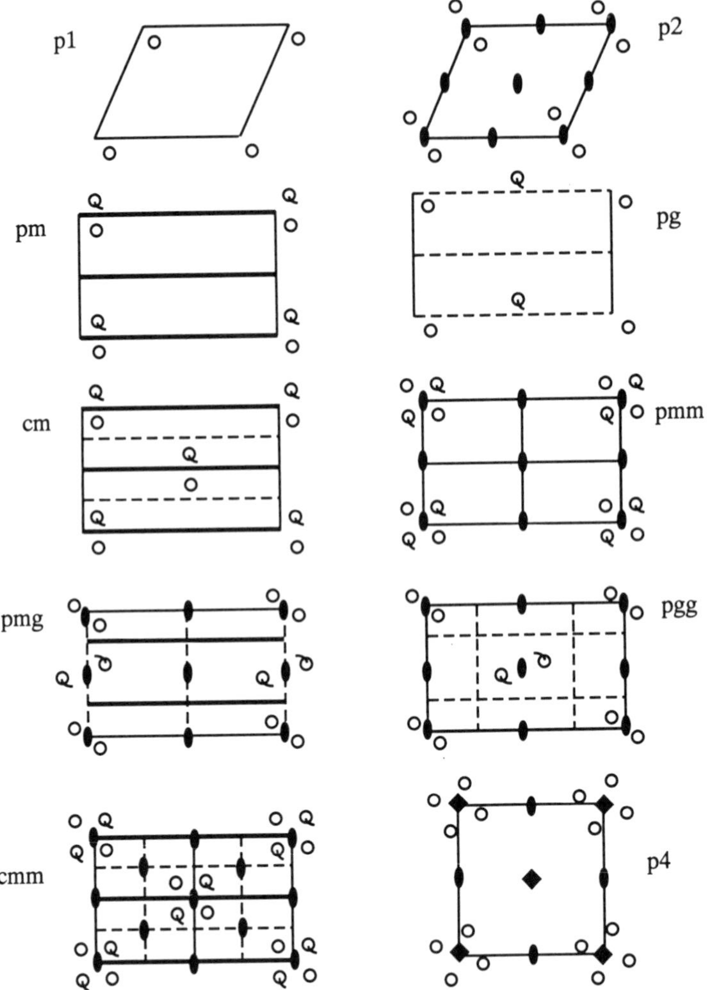

FIGURE 2.9. The 17 plane groups.

these representations are similar to the ones used in the discussion of point groups above. Multiplication tables for the first three plane groups, accordingly, will resemble those for the point groups (however, reduced to two dimensions). When we get to the fourth plane group, however, a new type of symmetry operation, known as a glide element g, is observed. This glide element not only has an axial component:

$$R = \begin{vmatrix} -1 & 0 \\ 0 & 1 \end{vmatrix}$$

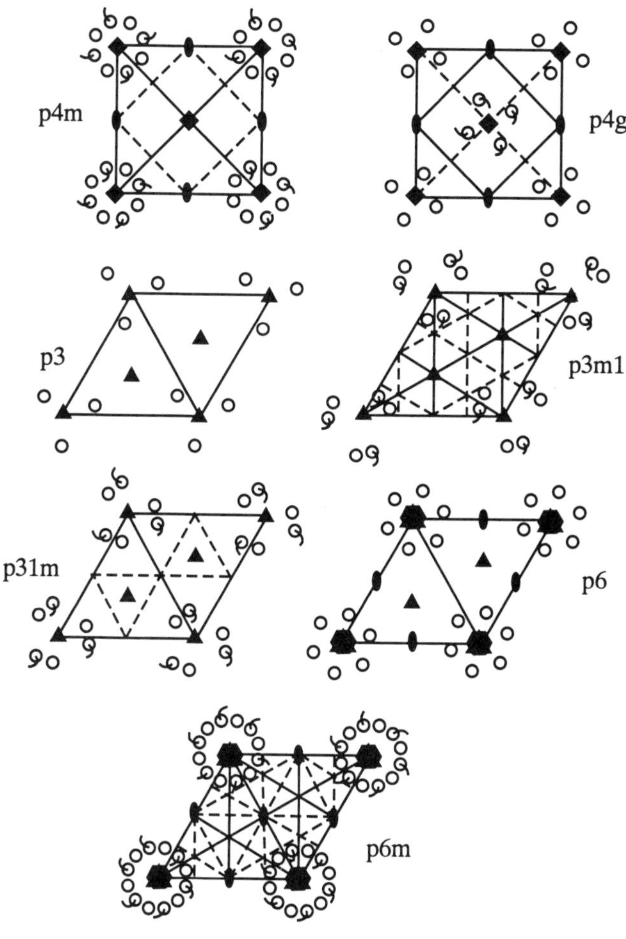

FIGURE 2.9. *(Continued)*

but also a translational component represented by the column vector:

$$t = \begin{vmatrix} 0 \\ 1/2 \end{vmatrix}$$

As already indicated, when multiplication tables of symmetry elements are constructed for any plane group, the products R_mR_n of the axial transformations are carried out, in addition to the sum $t_m + t_n$ of the translational components (e.g., see Hovmöller, 1981). This can be illustrated by the multiplication table for plane group *pgg*, often encountered in crystallographic work.

For the plane group *pgg*, the following symmetry operations are found. Identity 1:

$$R(1) = \begin{vmatrix} 1 & 0 \\ 0 & 1 \end{vmatrix} \qquad t(1) = \begin{vmatrix} 0 \\ 0 \end{vmatrix}$$

inversion $\bar{1}$:

$$R(\bar{1}) = \begin{vmatrix} -1 & 0 \\ 0 & -1 \end{vmatrix} \qquad t(\bar{1}) = \begin{vmatrix} 0 \\ 0 \end{vmatrix}$$

two glide elements:

$$g_+\text{: } R(g_+) = \begin{vmatrix} 1 & 0 \\ 0 & -1 \end{vmatrix} \qquad t(g_+) = \begin{vmatrix} 1/2 \\ 1/2 \end{vmatrix}$$

$$g_-\text{: } R(g_-) = \begin{vmatrix} -1 & 0 \\ 0 & 1 \end{vmatrix} \qquad t(g_-) = \begin{vmatrix} 1/2 \\ 1/2 \end{vmatrix}$$

Note that, for unit cells or their projections, any sum of fractions in any axial direction that adds up to unity is equivalent to zero translation (since this combination arrives at an equivalent point on the next, identical, unit cell). From the above symmetry elements and the defined multiplication operations, it is easy to demonstrate:

	1	$\bar{1}$	g_+	g_-
1	1	$\bar{1}$	g_+	g_-
$\bar{1}$	$\bar{1}$	1	g_-	g_+
g_+	g_+	g_-	1	$\bar{1}$
g_-	g_-	g_+	$\bar{1}$	1

Similar multiplication tables can be generated for the other plane groups. Some will show the Abelian property of this example; others will not.

Considerations of plane group symmetry are not merely exercises in aesthetics; they have a profound effect on the Fourier transform of a repeating mass distribution to its diffraction pattern. It was mentioned above, for example, that relative translational arrangements of mass in a crystal are expressed as a phase term in reciprocal space. For plane group *pgg*, the consequence of the symmetry elements $\bar{1}$ and g_+ on the structure factor expression given above can be demonstrated, as follows.

Starting with $\bar{1}$, which produces equivalent mass at $-x$, $-y$:

$$F(hk) = \sum f_i \left[\exp 2\pi i(hx + ky) + \exp 2\pi i(-hx - ky)\right]$$

$$= \sum f_i [\cos 2\pi(hx + ky) + i \sin 2\pi(hx + ky)$$

$$+ \cos 2\pi(-hx - ky) + i \sin 2\pi(-hx - ky)]$$

Given the properties of the cosine and sine functions, then:

$$F(hk) = 2 \sum f_i \cos 2\pi(hx + ky)$$

In other words a transform with any possible phase value:

$$\phi = \tan^{-1} B/A$$

where $A = \Sigma f_i \cos 2\pi(hx + ky)$, $B = \Sigma f_i \sin 2\pi(hx + ky)$, is now restricted to the phase choices $0,\pi$ by a center of symmetry. This is a very significant result. If the glide element that produces an equivalent mass point at $1/2 + x$, $1/2 - y$ is included:

$$F(hk) = 2 \sum f_i [\cos 2\pi(hx + ky) + \cos 2\pi(hx + h/2 - ky + k/2)]$$

The first term can be expanded:

$$\cos 2\pi(hx + ky) = \cos 2\pi hx \cos 2\pi ky - \sin 2\pi hx \sin 2\pi ky$$

and also the second:

$$\cos 2\pi[(hx - ky) + (h + k)/2] = \cos 2\pi(hx - ky) \cos \pi(h + k)$$

$$- \sin 2\pi(hx - ky) \sin \pi(h + k)$$

Since $(h + k)$ is integral, the last term of the expansion disappears, leaving:

$$\cos 2\pi[(hx - ky) + (h + k)/2] = \cos 2\pi(hx - ky) (-1)^{h+k}$$

If $(h + k)$ is even:

$$\cos 2\pi(hx - ky) = \cos 2\pi hx \cos 2\pi(-ky) - \sin 2\pi hx \sin 2\pi(-ky)$$

Substitution in the structure factor expression leaves:

$$F(hk) = 4 \sum f_i \cos 2\pi hx \cos 2\pi ky$$

For various combinations of Miller index, including negative signs:

$$F(hk) = F(\overline{h}\overline{k}) = F(\overline{h}k)$$

meaning that these structure factors all have the same magnitude and phase.

If $(h + k)$ is odd, it is seen from the above that:

$$F(hk) = -4 \sum f_i \sin 2\pi hx \sin 2\pi ky$$

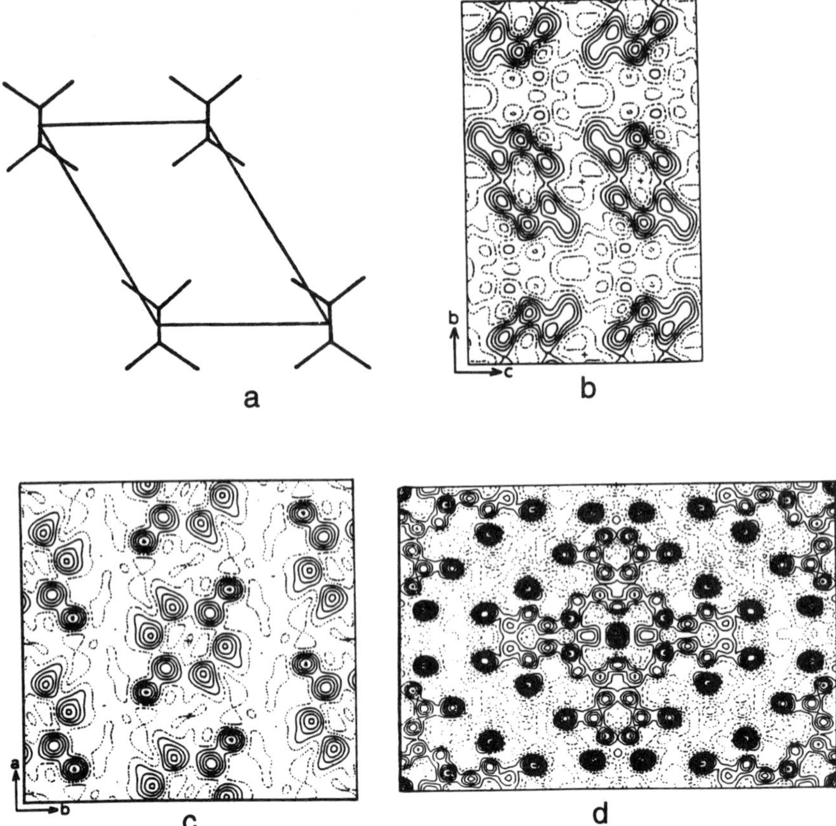

FIGURE 2.10. Molecular packing in common plane groups: (a) methylene triclinic $T_{\parallel}$ subcell (p2) (reprinted from D. L. Dorset (1983) "Electron crystallography of alkyl chain lipids: identification of long chain packing," *Ultramicroscopy* **12**, 19–28; with kind permission of Elsevier Science B.V.); (b) diketopiperazine [100] projection (*pmg*); (c) diketopiperazine [010] projection (*pgg*) (reprinted from D. L. Dorset and M. P. McCourt (1994) "Automated structure analysis in electron crystallography: phase determination with the tangent formula and least squares refinement," *Acta Crystallographica* **A50**, 287–292; with kind permission of the International Union of Crystallography); (d) copper perbromophthalocyanine (*cmm*) (reprinted from D. L. Dorset, W. F. Tivol, and J. N. Turner (1992) "Dynamical scattering and electron crystallography—ab initio structure analysis of copper perbromophthalocyanine," *Acta Crystallographica* **A48**, 562–568; with kind permission of the International Union of Crystallography); *(continued)*

This means that, for a given (h,k), $F(hk) = F(\overline{h}\overline{k}) = -F(\overline{h}k)$. In other words, the insertion of one negative Miller index for this kind of reflection in *pgg* shifts the phase by π, even though the magnitudes are not changed. Note also that, if h is odd or k is odd for $h00$ or $0k0$, then $F(hk) = 0$. This is a consequence of the glide elements in the plane group, causing odd-order reflections along the reciprocal axes **a*** and **b*** to be systematically absent. Other translational operations can also lead to systematic

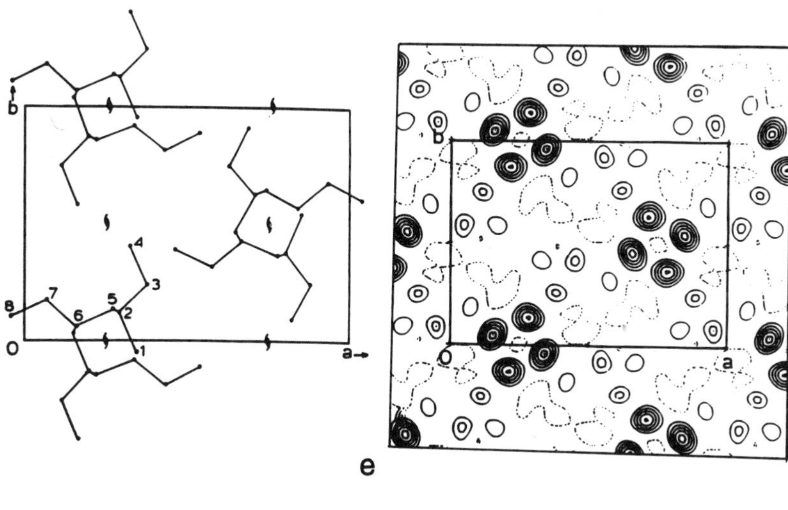

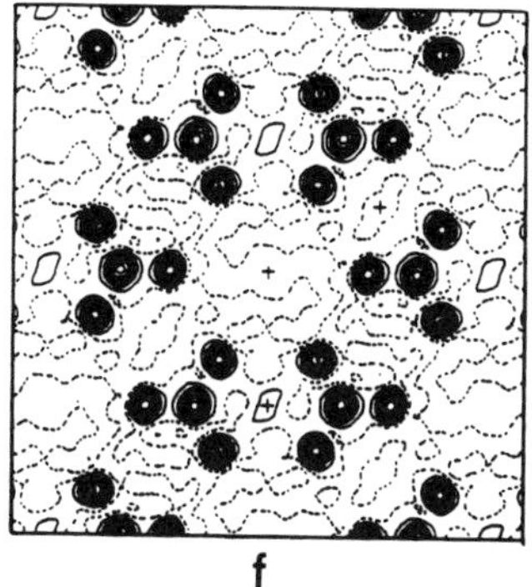

FIGURE 2.10. (*Continued*) (e) poly(1-butene), form III (*pgg* - molecular fourfold symmetry not used by the plane group) (reprinted from D. L. Dorset, M. P. McCourt, S. Kopp, J. C. Wittmann, and B. Lotz (1994) "Direct determination of polymer structures by electron crystallography—isotactic poly(1-butene) form III," *Acta Crystallographica* **B50**, 201–208; with kind permission of the International Union of Crystallography); (f) boric acid (approximate *p6m*)(reprinted from D. L. Dorset (1992) "Direct methods in electron crystallography—structure analysis of boric acid," *Acta Crystallographica* **A48**, 568–574; with kind permission of the International Union of Crystallography).

absences. For example, the centering operation in plane group *cmm* (Figure 2.9) causes all reflections with index $h + k = 2n + 1$, where n is an integer, to vanish. This result can be derived in a manner similar to the one shown for *pgg*. Examples of molecular packing in commonly observed plane groups are depicted in Figure 2.10.

2.2.3. Space Groups

It will come as no surprise that the symmetry group for a three-dimensional unit cell is called a *space group*. Since there are 230 of these, they will not be depicted in a separate figure in this book but they are found as the first volumes of two different

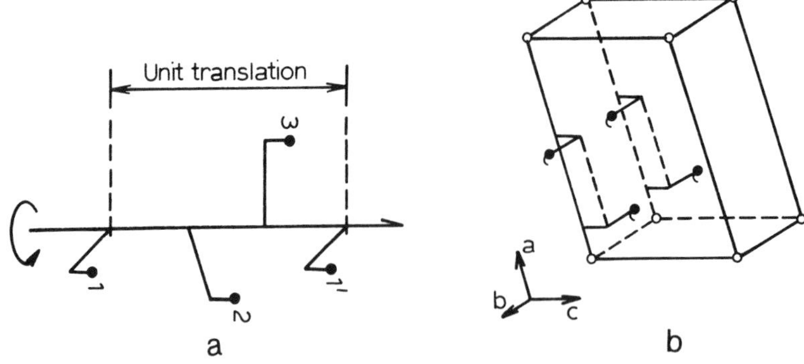

FIGURE 2.11. (a) 3_1- Screw operation; (b) glide plane.

sets of crystallographic tables (Henry and Lonsdale, 1969; Hahn, 1992). They were first derived by Federov (1949). For small organic crystals, the most often represented space groups occur in triclinic, monoclinic, and orthorhombic unit cells. Rules for the most likely molecular packing of any organic compound will be discussed below.

Like the plane groups, space groups are made up of symmetry elements involving an axial component and a translational vector. The glide elements found in two-dimensional representations become either screw axes or glide planes, as depicted in Figure 2.11. The screw axis n_m, where n and m are integers, will rotate a mass point by $2\pi/n$ and translate it along an axial distance m/n through the unit cell. The allowable values for $n = 1, 2, 3, 4$, and 6. A glide plane is designated by a, b, or c if the translation is $a/2$, $b/2$, $c/2$, and by n if it is $(a + b)/2$, $(a + c)/2$, or $(b + c)/2$ (the latter being face diagonals). A mirror plane parallels the translation direction and after the translation, the object is reflected to one with the opposite handedness.

Consider a very common space group $P\,2_1/c$ (Figure 2.12) containing both a screw axis (along b, the unique axis commonly chosen for monoclinic space groups) and a glide plane (perpendicular to b). The symmetry elements are:

Identity:

$$R(1) = \begin{vmatrix} 1 & 0 & 0 \\ 0 & 1 & 0 \\ 0 & 0 & 1 \end{vmatrix} \qquad t(1) = \begin{vmatrix} 0 \\ 0 \\ 0 \end{vmatrix}$$

Inversion:

$$R(\bar{1}) = \begin{vmatrix} -1 & 0 & 0 \\ 0 & -1 & 0 \\ 0 & 0 & -1 \end{vmatrix} \qquad t(\bar{1}) = \begin{vmatrix} 0 \\ 0 \\ 0 \end{vmatrix}$$

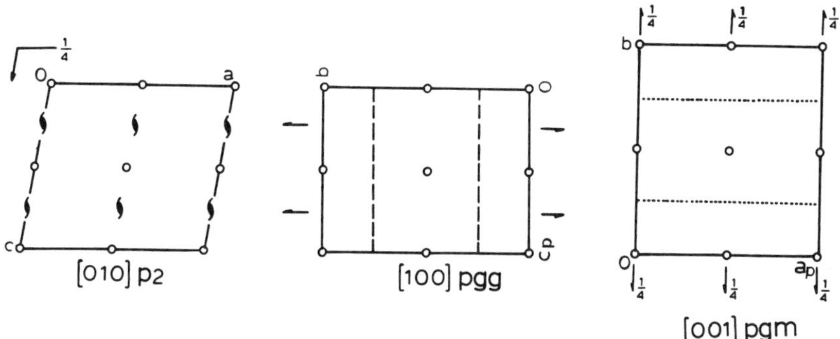

FIGURE 2.12. Space group $P2_1/c$ with projected plane group symmetries marked: Equivalent positions: (x,y,z); $(-x,-y,-z)$; $(-x,1/2+y,1/2-z)$; $(x,1/2-y,1/2+z)$.

2_1 parallel to b:

$$R(2_1) = \begin{vmatrix} -1 & 0 & 0 \\ 0 & 1 & 0 \\ 0 & 0 & -1 \end{vmatrix} \qquad t(2_1) = \begin{vmatrix} 0 \\ 1/2 \\ 1/2 \end{vmatrix}$$

Glide perpendicular to b:

$$R(g) = \begin{vmatrix} 1 & 0 & 0 \\ 0 & -1 & 0 \\ 0 & 0 & 1 \end{vmatrix} \qquad t(g) = \begin{vmatrix} 0 \\ 1/2 \\ 1/2 \end{vmatrix}$$

The group multiplication table is:

	1	$\bar{1}$	2_1	g
1	1	$\bar{1}$	2_1	g
$\bar{1}$	$\bar{1}$	1	g	2_1
2_1	2_1	g	1	$\bar{1}$
g	g	2_1	$\bar{1}$	1

A striking resemblance will be seen between this multiplication table and the one for plane group *pgg*. Indeed, the projection down the [100] axis is *pgg* with the same origin as that of the plane group. Thus, one glide element of the plane group becomes a screw axis and the other a glide plane. (The other two projections of this space group have *pgm* and *p*2 plane group symmetries, respectively.) Note also that alternative axial definitions are permitted for many space groups. For $P2_1/c$, one could just as well state

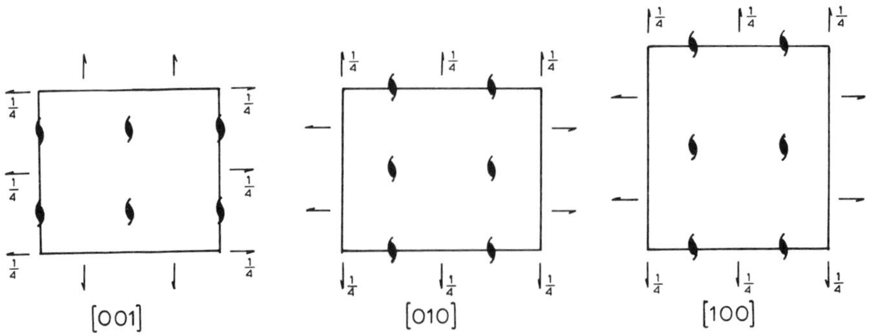

FIGURE 2.13. Space group $P2_12_12_1$. All the axial projections have plane group symmetry *pgg*.

that the glide translation may also be found along **a** to give $P2_1/a$ or along the diagonal (**a** + **c**), to give $P2_1/n$. Orthorhombic space groups are similarly affected and thus one must consult the International Tables for X-Ray Crystallography (Henry and Lonsdale, 1969) to determine how the respective axial orientations are related to one another.

Another commonly observed space group for organics packing without an inversion center is $P\,2_12_12_1$ (Figure 2.13). It consists of an identity symmetry element and three mutually perpendicular screw axes $2_1(x)$, $2_1(y)$, $2_1(z)$, where the parentheses indicate the parallel axis, such that:

$$R(1) = \begin{vmatrix} 1 & 0 & 0 \\ 0 & 1 & 0 \\ 0 & 0 & 1 \end{vmatrix} \qquad t(1) = \begin{vmatrix} 0 \\ 0 \\ 0 \end{vmatrix}$$

$$R(2_1(x)) = \begin{vmatrix} 1 & 0 & 0 \\ 0 & -1 & 0 \\ 0 & 0 & -1 \end{vmatrix} \qquad t(2_1(x)) = \begin{vmatrix} 1/2 \\ 1/2 \\ 0 \end{vmatrix}$$

$$R(2_1(y)) = \begin{vmatrix} -1 & 0 & 0 \\ 0 & 1 & 0 \\ 0 & 0 & -1 \end{vmatrix} \qquad t(2_1(y)) = \begin{vmatrix} 0 \\ 1/2 \\ 1/2 \end{vmatrix}$$

$$R(2_1(z)) = \begin{vmatrix} -1 & 0 & 0 \\ 0 & -1 & 0 \\ 0 & 0 & 1 \end{vmatrix} \qquad t(2_1(z)) = \begin{vmatrix} 1/2 \\ 0 \\ 1/2 \end{vmatrix}$$

Thus the multiplication table can be written:

	1	$2_1(x)$	$2_1(y)$	$2_1(z)$
1	1	$2_1(x)$	$2_1(y)$	$2_1(z)$
$2_1(x)$	$2_1(x)$	1	$2_1(z)$	$2_1(y)$
$2_1(y)$	$2_1(y)$	$2_1(z)$	1	$2_1(x)$
$2_1(z)$	$2_1(z)$	$2_1(y)$	$2_1(x)$	1

Each zone contains two glide elements so that the plane group symmetry is *pgg* for all axial projections. However, there is an important difference from the *pgg* projection given above for $P2_1/c$. The mutually orthogonal screw axes in this space group are nonintersecting and the unit cell origin must, therefore, be chosen halfway between the pairs of screw axes. For example, in the [001] projection shown in Figure 2.13, while the screw axis is located at $y/b = 0$, it is shifted to $x/a = 1/4$ in the other direction. The consequence of this origin shift on the structure factor phases will be to cause, e.g., $hk0$ reflections with odd h-index to have phase values $\pm\pi/2$, instead of $0,\pi$ (which are still the phases of the even h-index reflections in this centrosymmetric zone). Similar adjustments are made to the other two *pgg* projections according to the positions of the plane group origins. Obviously, for any *pgg* projection, it is permissible to redefine the origin so that the phases agree with the setting shown for the plane group, but these will not correspond to the origin of the unit cell itself, as specified by the space group. Mathematically, the consequence of the origin shift for *pgg* can be demonstrated for the [001] projection. The equivalent positions are (x,y), $(1/2 - x, -y)$, $(1/2 + x, 1/2 - y)$, $(-x, 1/2 + y)$. The first two are related by an inversion center translated to (1/4, 0). We can write:

$$F_{hk'} = \sum f_j[\cos 2\pi(hx + ky + h/4) + i \sin 2\pi(hx + ky + h/4)]$$

The first term becomes

$$\cos 2\pi[(hx + ky) + h/4] = \cos 2\pi(hx + ky) \cos 2\pi h/4$$
$$- \sin 2\pi(hx + ky)\, \sin 2\pi h/4$$

and the second

$$\sin 2\pi[(hx + ky) + h/4] = \sin 2\pi(hx + ky) \cos 2\pi h/4$$
$$+ \cos 2\pi(hx + ky)\, \sin 2\pi h/4$$

For different values of h, it can be shown that:

h	$\cos 2\pi h/4$	$\sin 2\pi h/4$
0	1.0	0.0
1	0.0	1.0
2	−1.0	0.0
3	0.0	−1.0
4	1.0	0.0

When $h = 2n$, $F_{hk'} = F_{hk} \exp i(n\pi)$ and when $h = 2n + 1$, $F_{hk'} = F_{hk} \exp i(n\pi/2)$.

Thus, all even h reflections have the same phase as the standard pgg setting when n is also even and a phase shifted by π when n is odd. All odd h reflections have $\pm\pi/2$ values, so that, when n is even, the phases are shifted $\pi/2$ from the standard pgg setting and when n is odd, the shift is $-\pi/2$. In general, the $hk0$ phases for $P2_12_12_1$ can be found from the standard pgg setting by adding $\pi h/2$.

An important point, which will be repeated in a different way when we address the subject of image processing in electron microscopy, is established here: For a given mass distribution, the relative crystallographic phases of the projection are preserved no matter where the unit cell origin is defined. However, in the example considered here, the absolute values of these phases will have identifiably centrosymmetric character (0, π, or $\pm\pi/2$) only for specific origin points in the unit cell. The first part is true because the relative translational elements between the equivalent mass centers are never changed when they are moved the same amount. The latter part is true because this distribution is shifted relative to an arbitrary point.

2.2.4. "Two-Sided Plane Groups"

It is quite possible to carry out electron crystallographic studies on two-dimensional crystals, i.e., those that are only one unit cell thick. In Chapter 1 it was seen that the Fourier transform of such a two-dimensional array should be continuous in the direction corresponding to the layer normal. Nevertheless, the two-dimensional array is still made up of unit cells with space group symmetry, as described above.

In the study of two-dimensional protein crystals, some researchers have been dissatisfied that the plane group description of the projection down the layer normal gives no information about the *sidedness* of the packing array—i.e., the fact that one surface may be different from the other. While a space group notation, itself, would be quite satisfactory for such a description, an alternative "two-sided plane group" symmetry designation (Amos et al., 1982) also can be used, as outlined in Table 2. Here, only the 17 layers permissible for chiral molecules are listed, although, in general, there are 80 such layer groups (see Vainshtein, 1981). In defense of this notation, it will be shown below that the concept of layer packing symmetry is quite important for the prediction of the plane and space groups most likely to be found for organic crystals.

Table 2.2. Two-sided Plane Groups Permitted for Chiral Molecules*

"Two-sided" plane groups	Three-dimensional space group	Conventional plane group
$p1$	$P1$	$p1$
$p21$	$P2$ (c-axis unique)	$p2$
$p12$	$P2$ (b-axis unique)	$p2$
$p12_1$	$P2_1$	pg
$c12$	$C2$	cm
$p222$	$P222$	pmm
$p222_1$	$P222_1$	pmg
$p22_12_1$	$P22_12_1$	pgg
$c222$	$C222$	cmm
$p4$	$P4$	$p4$
$p422$	$P422$	$p4m$
$p42_12$	$P42_12$	$p4g$
$p3$	$P3$	$p3$
$p312$	$P312$	$p3m1$
$p321$	$P321$	$p31m$
$p6$	$P6$	$p6$
$p622$	$P622$	$p6m$

*A complete set of 80 layer groups is given by Vainshtein (1981).

2.3. Unit Cell and Space-Group Identification

During the initial crystallographic investigation of a new material, selected area electron diffraction patterns, representing an undistorted reciprocal net, are generally taken from an untilted thin crystal. The first task is to decide what kind of unit cell is responsible for the diffraction pattern. If the diffraction pattern indicates an oblique layer (without equal axial lengths or a 60° angle between reciprocal axes), the cell can be either triclinic or monoclinic. If the pattern is a rectangular net, the cell can be monoclinic or orthorhombic. Obviously, some information about the third dimension must be obtained to settle these questions about the type of unit cell. While electron diffraction is more convenient than x-ray diffraction for providing a clear view of undistorted reciprocal nets, collection of three-dimensional information by tilting a crystal in the electron microscope is a procedure that is geometrically different from the examination of upper reciprocal layers with screened Weissenberg or precession x-ray cameras. The sampling of three-dimensional diffraction patterns in the electron microscope is a tomographic technique, i.e., the information from a single crystal is referred to axial tilts of a plane parallel to the best-developed crystal face (Figure 2.14). If there is a geometric limitation to this tilt (and the tilt range allowed is generally within ±60° around any axis on the crystal plane), it is not possible to visualize directly the details along a third unit cell axis because of the "missing cone" of information (originally termed the "dead zone" by Vainshtein (1964a)). (An exception to this is found for perfectly flat crystals, from which higher-order Laue zones (HOLZ) can be

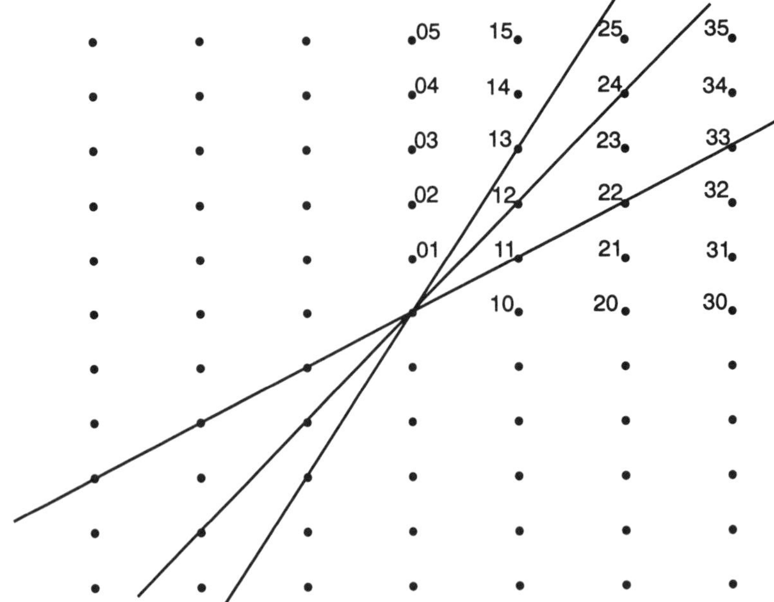

FIGURE 2.14. Tomographic sampling of reciprocal space by a tilt stage in the electron microscope for a hypothetical *hkl* net, with tilt axis parallel to [001]. The (*hk*0) reflections are found at 0° tilt; the *hhl* reflections at 26° tilt; the *h*2*hl* reflections at 45° tilt; the *h*3*hl* reflections at 56° tilt. Beyond ±60°, no reflections (e.g., 0*kl*) can be measured because of the geometrical tilt limitation imposed by the objective lens pole piece gap. This leaves an unsampled "dead zone" if there is only one specimen orientation.

observed in electron diffraction patterns. These are often obtained from inorganic materials, especially when convergent beam diffraction is used. To this author's knowledge, HOLZ diffraction patterns have not been observed for organic specimens due to the plasticity of the crystals, the large area sampled to minimize radiation damage, and possibly also the insufficient intensity at sufficiently high resolution if a flat specimen region can be sampled.)

One of the ways to measure three-dimensional data from microcrystals, starting with the undistorted reciprocal lattice nets found in zonal selected area diffraction patterns, is to use this information to index an powder pattern. After the diffraction rings due to the basal net are identified, other spacings due to the three-dimensional lattice can be found and a unit cell geometry is proposed which best corresponds to these data, by analogy to procedures discussed by Klug and Alexander (1974) for powder x-ray diffraction. However, this presumes that there are no preferred orientations for the microcrystals on the support film. While this may be approximately true for some inorganic structures, organics often crystallize as thin plates which, on average, are only randomly rotated around the normal to the plate face. If a large enough specimen area is sampled for electron diffraction, it is still possible to obtain an oblique texture pattern by tilting the crystal holder at a large angle and sampling diffraction rings from successive layers along the third reciprocal axis. Indexing of such patterns is described in detail by Vainshtein (1964a).

In this book, data collection from oriented single crystals will be considered most often. It will be assumed further that a goniometer stage is available on the electron microscope, capable of ±60° tilt around any axis, which is nearly eucentric, furthermore, and capable of ±180° rotation in-plane. After the initial pattern is obtained from the untilted crystal, reciprocal axes are defined and, with the rotational orientation, any reciprocal axis can be positioned so that it is coincident with the tilt axis of the goniometer stage (using procedures to be defined in Chapter 3). Small tilts are then made in opposite directions, to test for the mirror symmetry of diffracted intensities above and below the untilted net and tilts are gradually increased in both directions to map out as much of the reciprocal lattice as can be accessed within the limits of the goniometer stage, using as many tilt axes as are necessary to solve the problem of determining the reciprocal lattice geometry. For example, if a reciprocal point $hk0$ is defined in a direction normal to the tilt axis with reciprocal spacing d^*_{hk0} and another point hkl is accessed along the same row with spacing d^*_{hkl}, it is very easy to determine the c^* reciprocal spacing if the unit cell is orthogonal (Figure 2.15a) by elementary geometry:

$$(lc^*)^2 + (d^*_{hk0})^2 = (d^*_{hkl})^2$$

where l is the Miller index along this row. If the reciprocal lattice is nonorthogonal, this procedure is not so straightforward. However, using positive and negative tilt

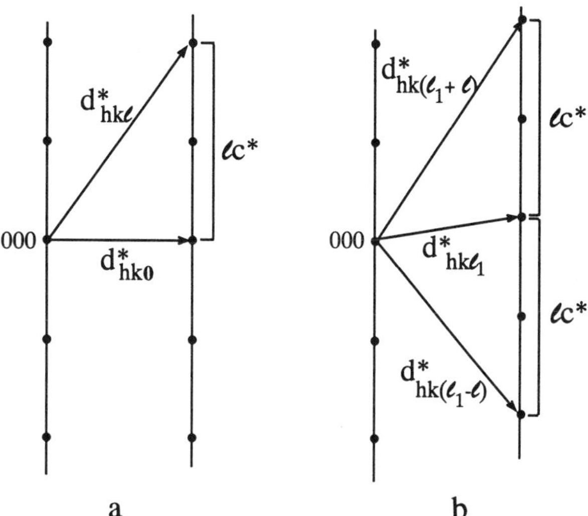

FIGURE 2.15. Determination of reciprocal lattice geometry by a tilt series for a single crystal: (a) orthorhombic structure; (b) oblique structure. The tilt axis is normal to the paper and through (000); usually $l_1 = 0$.

c

b

a

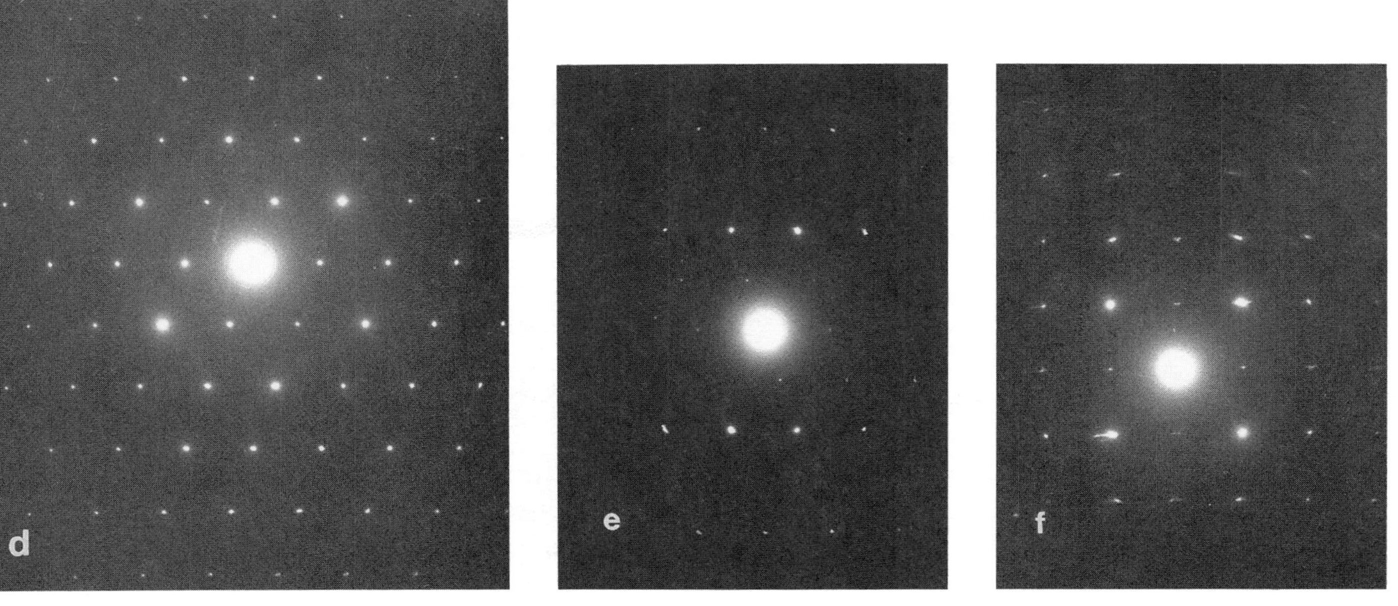

FIGURE 2.16. Electron diffraction patterns obtained from C_{60}-buckminsterfullerene. Crystals grown by evaporation of a dilute solution in benzene were tilted to an indicated value and the zonal pattern was located by rotating the grid: (a) [111], 0° tilt (note weak space group forbidden reflections due to layer stacking disorder); (b) [211], 20° tilt; (c) [123], 22° tilt; (d) [110], 35° tilt; (e) [310], 43° tilt; (f) [100], 55° tilt. Because of the high symmetry of this structure (*Fm3*), all of the reciprocal space could be sampled for one crystal form.

values to access two points *hkl* and *hk,–l*-referred to some *hk*0 on this row (see Figure 2.15b), one can, following Vainshtein (1964a), calculate:

$$c^{*2} = [(d^*_{hkl})^2 + (d^*_{hk,-l})^2 - 2(d^*_{hk0})^2]/2l^2$$

Distinction between a triclinic and a monoclinic unit cell, given an oblique reciprocal net from the untilted crystal, can be made by mapping out the reciprocal net and also noting the merohedral symmetry of intensity across this *hk*0 plane (which would denote a monoclinic cell when the **b**-axis is parallel to the layer normal). Similar arguments can be used to distinguish monoclinic and orthorhombic unit cells, starting with a rectangular reciprocal net from the untilted specimen (i.e., with the monoclinic **b**-axis now parallel to the layer surface plane). It is easy to identify a unit cell as being orthorhombic by observing the merohedry of intensity for equivalent tilt values around reciprocal axes. Experimental illustrations of this procedure have been published in papers describing the structure of linear polymer crystals (Guizard et al., 1985). Figure 2.16 shows a series of projections for the room temperature form of C_{60} buckminster-fullerene, an unusual organic crystal for its high symmetry, but informative for the kind of information that can be obtained by tilting such specimens. The angular values between projections indicated in the figure were calculated from the zone axis relationships given above. However, there are also perturbations to this pattern that will merit consideration in Chapter 6. If the crystal layer is two-dimensional, i.e., only one unit cell thick, then the tilt series will produce a plot of continuous intensity along any reciprocal lattice rod. The symmetry of this continuous intensity plot across the *hk*0 net can still be used to test if the packing is rectangular or oblique (Amos et al., 1982).

So far a restriction to only one orientation, imposed by the specimen, has been assumed. Sometimes, it is possible, by combining two types of crystallization procedures, e.g., solution crystallization and epitaxial growth, to obtain different, even orthogonal, projections of the same crystal packing (Wittmann and Lotz, 1990). This simplifies unit cell identification considerably, especially if the details of the third reciprocal axial spacing can be determined directly by inspection. An example shown in Figure 2.17 is polyethylene where the lattice constants were determined directly from samples crystallized by self-seeding (i.e., from dilute solution) and then by epitaxial orientation on benzoic acid (Hu and Dorset, 1989). If the experiments are constrained to just solution-crystallized samples with preferred projection, the RHEED experiment mentioned above can be used to determine if the **c*** reciprocal axis is oblique or normal to the hk0 net.

Space group symmetry is often not uniquely determined using traditional crystallographic techniques. (However, using convergent beam electron diffraction methods (Buxton et al., 1976; Eades, 1992), this is now possible for samples that can tolerate the high beam doses required; unfortunately, most organic materials are excluded from the CBED experiment as it is normally carried out. Attempts to characterize the symmetry of uniformly bent aromatic films have been made (Vincent, 1985) using larger illumination areas and lower current densities to observe the zero-order Laue

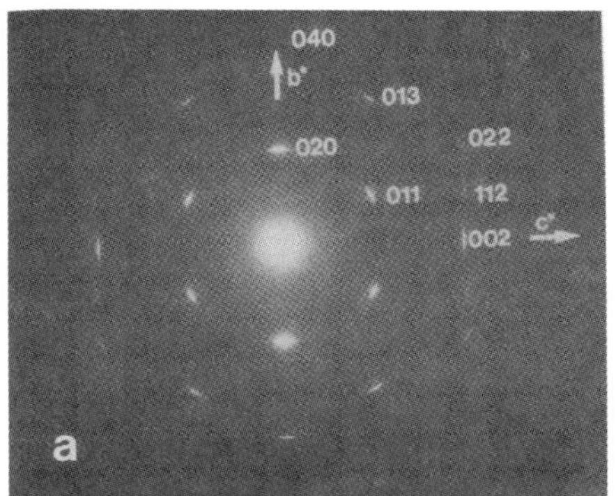

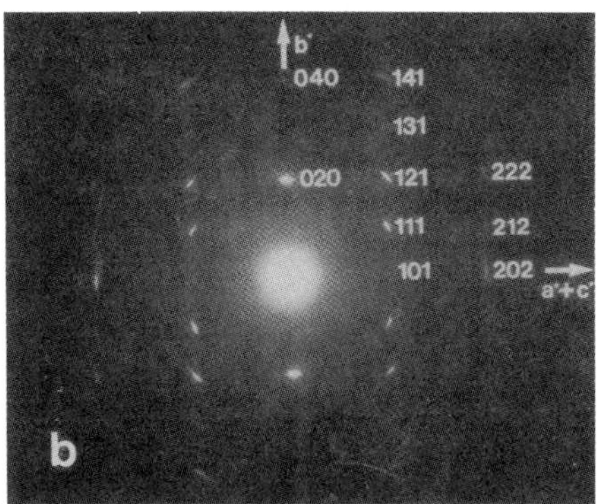

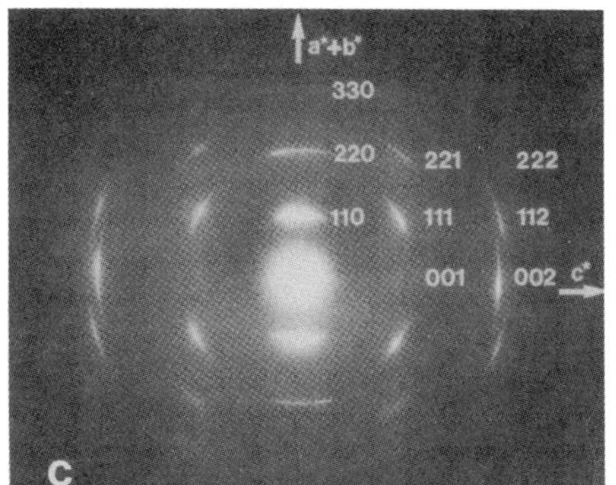

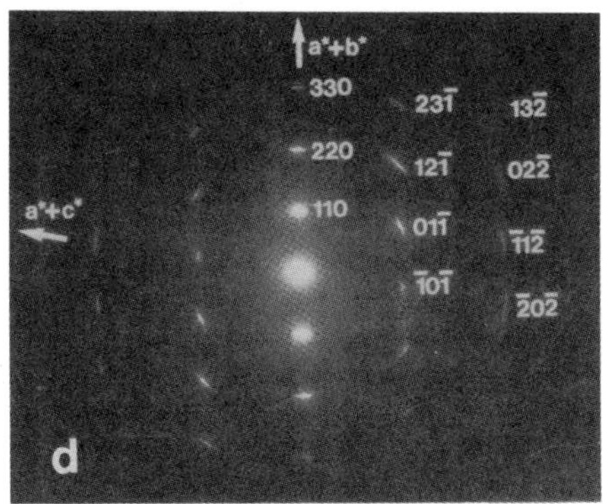

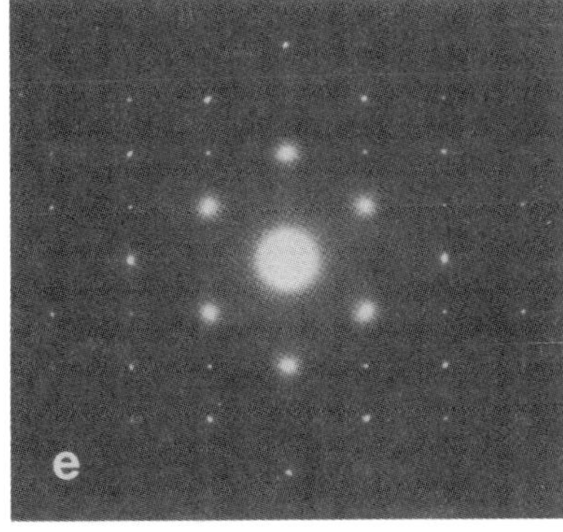

FIGURE 2.17. Electron diffraction patterns from tilted polyethylene crystals: (a) $0kl$; (b) lkl; (c) hhl; (d) $(k + l, k, l)$. All of these patterns were obtained from epitaxially oriented samples. (e) A solution-crystallized sample will provide the $hk0$ pattern. (Reprinted from H. Hu and D. L. Dorset (1989) "Three-dimensional electron diffraction structure analysis of polyethylene," *Acta Crystallographica* **B45**, 283–290; with kind permission of the International Union of Crystallography.)

zone.) In usual practice, however, after the unit cell is identified, the only way to obtain information which can lead to space group identification is to look for systematic absences due to translational symmetry in the unit cell. For example, centering causes certain *hkl* reflections with odd combinations of Miller indices to be absent. Glide planes cause certain zonal layers of reflections to be extinct. Screw axes lead to the extinction of certain reflections along a reciprocal axis. Appropriate rules are outlined in Table 2.3. Since upper reciprocal net layers generally cannot be sampled in electron diffraction as they can in x-ray experiments with a Weissenberg camera, the periodicity of nets from the various tilt projections used to determine the unit cell dimensions must be critically evaluated in order to verify the occurrence of absent reflections. For example, screw axes are most easily identified if projections containing the reciprocal axes are examined. Glide planes require a mapping of the three-dimensional lattice to see if one of the zonal spacings, in fact, is not a fraction (e.g., 1/2) of what it should be. Centering operations similarly must be evaluated for the whole reciprocal lattice, although projections with a face-centering are often obvious, as illustrated above for plane group *cmm*, which also requires mm symmetry of the intensity across reciprocal axes.

Once these translational symmetry elements are identified, several space groups may remain that can equally correspond to these elements, as outlined in Table 2.4. (Fortunately, the two examples discussed above, $P2_1/c$ and $P2_12_12_1$, which are often encountered in organic crystallography, are uniquely determined by these systematic absences.) At times, the type of molecule packing in the unit cell may allow a choice to be made. For example, space groups *Cc* and *C2/c* give the same extinctions. However, the former is noncentrosymmetric and the latter centrosymmetric so that a chiral molecule could pack only in the first. Based on these criteria, only 49 space

Table 2.3. Forbidden Reflections due to Translational Symmetry Elements

Screws	Parallel to axis*	Systematic absences*
2_1, 4_2, 6_3	*c*	$00l$, $l = 2n + 1$
3_1, 3_2, 6_2, 6_4	*c*	$00l$, $l = 3n + 1, 3n + 2$
4_1, 4_3	*c*	$00l$, $l = 4n + 1$, 2, or 3
6_1, 6_5	*c*	$00l$, $l = 6n + 1$, 2, 3, 4, or 5
Glides	**Perpendicular to axis***	**Systematic absences***
a-glide (trans. $a/2$)	*c*	$hk0$, $h = 2n + 1$
b-glide (trans. $b/2$)	*c*	$hk0$, $k = 2n + 1$
n-glide (trans. $[a + b]/2$)	*c*	$hk0$, $h + k = 2n + 1$
d-glide (trans. $[a + b]/4$)	*c*	$hk0$, $h + k = 4n + 1$, 2, or 3
Centering		**Systematic absences**
C-centering		hkl, $h + k = 2n + 1$
F-centering		hkl, $h + k, h + l, k + l, = 2n + 1$
I-centering		hkl, $h + k + l = 2n + 1$

*Any appropriate axis or row or plane may be substituted.

Table 2.4. Some Common Space Groups for Organics with Identical Systematic Absences*

	Absences	Method of selection
Triclinic:		
$P1$, $P\bar{1}$	none	Wilson statistics, chirality
monoclinic:		
$P2_1$, $P2_1/m$	$00l$: $l = 2n + 1$	Wilson statistics, chirality
$P2_1/c$	$h0l$: $l = 2n + 1$	
	$0k0$: $k = 2n + 1$	uniquely determined
$C2/c$, Cc	hkl: $h + k = 2n + 1$	
	$h0l$: $l = 2n + 1$	Wilson statistics. chirality
	$0k0$: $(k = 2n + 1)$	
orthorhombic:		
$P2_12_12_1$	$h00$: $h = 2n + 1$	
	$0k0$: $k = 2n + 1$	uniquely determined
	$00l$: $l = 2n + 1$	
$Pnma$, $Pna2_1$	$0kl$: $k + l = 2n + 1$	
	$h0l$: $h = 2n + 1$	Wilson statistics, chirality

*See also Table 2.6.

groups can be determined unambiguously. Alternatively, there are at least two tests for the centrosymmetry of a measured diffraction intensity distribution. One of these is known as Wilson (1949) statistics, where the number of reflections $N(Z)$, where $I < Z \cdot \langle I \rangle$, is plotted against the fraction Z. Here $\langle I \rangle$ is the average intensity value for the data set. The second test is based on the distribution of normalized structure factors ($|E_h|$) used for direct phase determination (Karle et al., 1965), which will be evaluated in later chapters. In addition, there is another statistical test of specific zonal or central reciprocal lattice rows that, in principle, permit 215 space groups to be uniquely identified (Rogers, 1950). As will be discussed later, observed data must correspond closely to the single scattering approximation for these tests to be meaningful. Additionally, there must be enough measured data, again so that the evaluations are meaningful.

There are many examples of x-ray crystal structures that have been determined in the wrong space group because of a misidentification. Difficulties with refinement to a reasonable geometry suddenly disappear when the correct symmetry is used.

2.4. Preferred Crystal Packing Motifs for Organic Molecules

There are 230 space groups, or, in the triclinic, monoclinic, or orthorhombic range found for most organics, there are, theoretically, still an imposing 74 space groups to choose from for any given structure. Fortunately, nature is not as cruel as that. The number of space groups most often encountered in crystallographic studies of organic molecules is surprisingly very small, for reasons that can be easily understood with a basic understanding of molecular packing.

The great Russian crystallographer A. I. Kitaigorodskii (1961) approached the problem of molecular packing in terms of a hard sphere model for the atomic constituents, i.e., the resultant model was regarded to have a limiting repulsive van der Waals surface. It was just the close packing of these three-dimensional shapes that was most important for the prediction of most likely space group symmetry. (A justification of this assumption has been published very recently by Zefirov (1994).) The first criterion observed to be relevant to these predictions was that the ratio of total molecular volume occupied in a crystal to the total unit cell volume should be very high, similar to what would be expected for the packing of general ellipsoids if the protrusions of such a structure were to fit into the hollows left by a nascent row of these forms. The most efficient packing of a molecular layer, irrespective of any symmetry consideration, was found to be a sixfold coordination, and, in fact, this requirement was found to be more important than any other symmetry constraint, since preservation of many molecular symmetry elements actually compromises attainment of the highest packing density.

Starting with single two-dimensional layers, then, if the molecule has an arbitrary shape to it, all plane groups with a mirror element or a rotational symmetry element greater than 2 are excluded. This leaves only plane groups $p1$, $p2$, pg, and pgg. If the molecular projection has centrosymmetry (i.e., a twofold axis perpendicular to the plane), then $p2$ and pgg are the only possible choices. Incorporation of molecular symmetry into a layer packing is allowed only when it still permits a closely packed array. For example, if m is to be preserved, plane groups pmg and cm are specified; and if mm symmetry is expressed, then the packing occurs in cmm. Examples of molecular packing in several of these plane groups have been shown in Figure 2.10.

Three-dimensional crystals are built up from these close-packed layers, again preserving the packing density. The generation of a crystal packing can occur by simple translation, inversion centers, glide planes, or screw axes but never by the use of mirror planes alone. These arrays will, in most favorable cases, produce a molecular coordination number of 12, with a coordination of 10 or 14 being slightly less efficient. Again, for three dimensions, molecular symmetry in the crystal will be higher than 1 only if

Table 2.5. Use of Molecular Symmetry by a Space Group

Molecular symmetry	Positional symmetry of molecule in crystal
1	1
$\bar{1}$	$\bar{1}$
2	1
m	1
mm	1, 2, or m
$2/m$	$\bar{1}$
mmm	$\bar{1}$
222	1 or 2

Table 2.6. Most Probable Space Groups for Organic Molecules

Inherent molecular symmetry	Molecular symmetry in crystal	Likely space groups*
1, 2, *m*	1	$\mathbf{P\bar{1}}$, $\mathbf{P2_1}$, $\mathbf{P2_1/c}$ *Pca*, *Pna*, $\mathbf{P2_12_12_1}$
$\bar{1}$, 2, *m*, *mmm*	$\bar{1}$	$\mathbf{P\bar{1}}$, $\mathbf{P2_1/c}$, $\mathbf{C2/c}$ $\mathbf{Pbca}$
mm	2	$\mathbf{C2/c}$, $\mathbf{P2_12_12}$, *Pbcn*
	m	*Pmc*, *Cmc*, $\mathbf{Pnma}$
222	2	$\mathbf{C2/c}$, $\mathbf{P2_12_12}$, *Pbcn*

*Boldface space groups are those most commonly found in statistical analysis of organic crystal structures by Nowacki (1967).

it does not compromise the close packing of the molecules. Common expressions of molecular symmetry in the crystal are listed in Table 2.5. Hence, if the molecule has two or more symmetry elements, it will, most commonly, preserve only one of them and only if it is permitted. In general, it is seen that the center of inversion $\bar{1}$ is preserved preferentially, since it does not seriously affect the packing density. The most probable space groups for organic structures are listed in Table 2.6, based on Kitaigorodskii's predictions. Those in boldface are the ones most commonly found in a statistical analysis of organic structures by Nowacki in 1948 (revised in 1967). In other words, Kitaigorodskii's straightforward analysis agrees very well with experimental results. Of course, it is possible to find highly symmetric organic molecules where most of the molecular symmetry is preserved by the space group (which can be of higher symmetry than orthorhombic)—buckminsterfullerene is an example. When these "exceptions" are examined closely, it will be found that close-packing requirements still are not violated and that the symmetry follows, rather than dictates, the packing. Finally, it will be obvious that a molecule with no symmetry cannot pack in a centrosymmetric unit cell. The favored space groups are $P2_1$ and $P2_12_12_1$. Racemates, on the other hand, will pack in space groups with centers of inversion or with glide planes.

More recent surveys of space group frequency (e.g., Brock and Dunitz, 1994) generally concur with the conclusions made above. However, if the crystal space group has a mirror symmetry plane, a molecule with mirror symmetry is nearly always found to lie on the special position. Similarly, threefold axes in the space group and molecular point group are generally coincident.

3

Crystallization and Data Collection

In electron crystallography, preparation of suitably thin, oriented crystalline samples is vital to the success of an ab initio structure analysis. In certain cases, special measures must be taken to protect the crystals from the ultrahigh vacuum of the electron microscope column, else they will not remain ordered. It might also be desirable to carry out dynamic experiments (e.g., heating/cooling) on these samples while preserving their chemical integrity. These requirements dictate the type of preparative procedures that must be followed and, sometimes, call for special attachments to the electron microscope.

3.1. Crystallization of Organic Compounds

3.1.1. Growth from Dilute Solution

The easiest way to crystallize many organic compounds is to place a drop of a dilute solution of the material onto a carbon-film-covered electron microscope grid and then allow the solvent to evaporate. Generally, the grid is gripped between the locked tines of a tweezers and the evaporation can be aided by wicking the edge of the drop with a piece of filter paper. A qualitative measure of the average crystal thickness can be obtained by observing the transmission of the electron beam through these crystals directly in the electron microscope at, e.g., 10,000× magnification at low illumination levels. A more quantitative measure is achieved by obliquely shadowing the grid with evaporated metal at a known angle in a vacuum coater apparatus (Hall, 1966). (Gold leaves a rather coarse coating but is sufficient at low magnifications; carbon-platinum provides a finer coating and is better for studies at high magnifications.) The "shadow length" is related to the crystal height by elementary trigonometry (Figure 3.1).

For typical organic molecules with an oblong shape, growth from solution produces crystalline layers where the longest molecular axis is, more or less, parallel to the layer normal (Jensen, 1970). Often the crystals are very thin plates and the preferred molecular orientation is dictated by the maximal interaction of van der Waals surfaces of nearest neighbors. As will be shown in Chapter 5, this orientation itself can

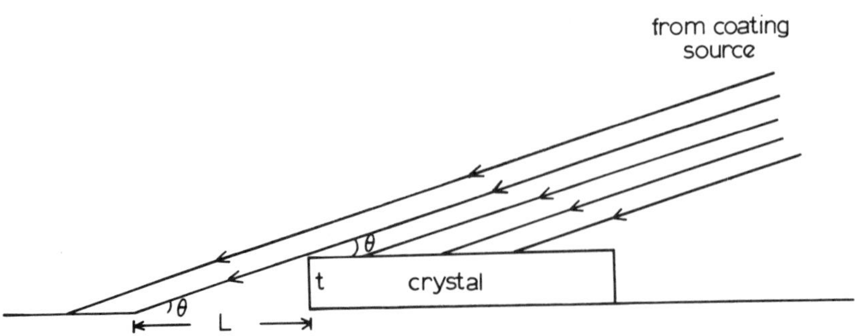

FIGURE 3.1. "Shadow length" L due to the coating of a crystal with a metal at an oblique angle θ. The thickness of the crystal is easily found from $t = L \tan \theta$.

place significant constraints on the collection of useful diffraction intensities, particularly if the projected unit cell axis is longer than, e.g., 20 Å.

3.1.2. Crystallization by "Self-Seeding"

A variant of the above procedure must be employed for crystallization of high-molecular-weight polymers such as thin chain-folded lamellae (Blundell et al., 1966). A poor solvent is selected for the polymer. After raising a dilute suspension of the polymer to a temperature high enough to fully solvate it, the solution is cooled to room temperature to induce growth of ill-formed microcrystals, which are mostly dendrites. This mixture is then reheated until the cloudiness of the suspension just disappears. Small seeds of the polymer crystals still remain in suspension, however, and serve as nuclei when the solution temperature is dropped and held at some point, still well above ambience. Then isothermal crystallization is allowed to occur. When the optimal growth temperature is found, thin lamellae are formed in the suspension and, after crystallization is complete, the solvent temperature can be lowered to ambience to allow the thin microcrystals to be harvested.

To harvest the microcrystals, a drop of the suspension is placed on the carbon-film-covered electron microscope grid, as before, and the solvent evaporation is again aided by use of filter paper held at the grid edge. After drying, the chain-folded lamellae can be studied in the same way as the thin plates of lower-molecular-weight organics.

3.1.3. Crystallization by Sublimation

Certain van der Waals molecules can be sublimed at a low ambient pressure (see Fryer, 1979), e.g., in the vacuum chamber of a coater apparatus. A metal boat containing the material to be sublimed is perforated to allow the organic vapor to escape as it is heated resistively in vacuo. The material is collected nearby on a suitable cold(er) surface on which carbon-film-covered grids are placed. As the layer is formed, the thickness of the sample can be monitored as a change in optical density.

3.1.4. Langmuir–Blodgett Layers

Amphiphilic or detergentlike substances can be spread on a water surface. As the surface pressure is increased by compression with a movable float, a thin monomolecular film is formed with two-dimensional crystallinity (e.g., Lösche, et al., 1984). If carbon-film-covered microscope grids are placed below the water surface before the film is formed, and one of these is pulled through the film afterward, a monolayer is transferred onto the carbon surface. Depending on how the carbon film was treated before this transfer (making it either hydrophobic or hydrophilic), either the nonpolar or polar end of the molecules will attach to its surface (Hui, 1989). The monolayer can then be dried for study in the electron microscope. If the grid is inserted from the top of the monolayer film and then withdrawn, a bilayer is formed or multiple layers with repeated dipping of the grid through this Langmuir film. The crystal properties of monolayers can then be compared to those of multilayers.

It is also possible to study these amphiphile layers in the hydrated state with a suitable environmental stage for the electron microscope, as will be discussed below. The samples are picked up from the Langmuir trough surface by dipping, e.g., a 1000-mesh grid without a carbon support film, so that the bilayer or multilayer is suspended across the grid openings (Hui, 1989). This very fragile sample will then require special humidified transfer to the environmental chamber to prevent drying (Hui et al., 1976).

3.1.5. Epitaxial Orientation

Orientation of a suitably short unit cell dimension along the electron beam path is sometimes required for collection of useful diffraction intensity data from an organic crystal. This is often conveniently realized by epitaxial crystallization of the organic layer onto another crystalline sublayer. Epitaxial growth is thus achieved by lattice matching of the substrate to the crystal lattice (or sublattice) spacings of the desired projection (Wittmann and Lotz, 1990). The substrate can be inorganic or organic. (Representative spacings of some useful substrates are given in Table 3.1.) In the

Table 3.1. Spacings of Common Substrates for Epitaxial Nucleation

Substrate	Plane (direction)		Spacing (Å)
NaCl	(100)	(<110>)	3.98
KCl	(100)	(<110>)	4.44
KBr	(100)	(<110>)	4.65
anthracene (naphthalene, etc.)	(001)	(<110>)	4.93
p-terphenyl (biphenyl, etc.)	(001)	(<110>)	4.60
benzoic acid	(001)	([010])	5.14
		([100])	5.52
phenanthrene	(001)	(<110>)	4.98
p-bromobenzoic acid	(100)	([001])	5.60
p-phenylbenzoic acid	(010)	([001])	7.44
potassium hydrogen phthalate	(001)	(<110>)	5.36

former case, the presence of adsorbed gases and water vapor requires cleaning of the nucleating surface first (e.g., by heating in vacuo). Organic substrates, on the other hand, are not generally susceptible to such contamination and can even be stored in air for long periods before being used. Deposition of the organic sample onto the substrate can also be achieved in several ways.

3.1.5.1. Growth from the Vapor Phase. If the material can be sublimed in vacuo, it is often convenient to have a clean crystalline substrate in the vacuum chamber near the sublimation source to collect the organic vapor deposition. Inorganic crystal substrates (e.g., KBr and KCl) are usually cleaved with a razor blade just before they are inserted into the vacuum system (Uyeda et al., 1972; Ueda and Ashida, 1980) to provide the cleanest possible nucleation surface. The surface is heated to outgas any adsorbed material and cooled for the crystallization experiment. (As discussed by Fryer (1979) and Fryer and Ewins (1992), there may be a need for some warming of the nucleation surface during the deposition to permit sufficiently large crystals to grow by surface diffusion.) After formation of the organic layer is complete, a carbon film can be laid down over the surface to give it additional mechanical strength. When it is removed from the vacuum chamber, the organic film is separated from the salt crystal face by flotation on a water surface. Bare grids are used to pick up the film and, after drying, the sample is suitable for study in the electron microscope. Salt substrates are very useful for many applications but have the disadvantage that they often nucleate rather small crystals oriented in all possible symmetry-equivalent directions of the substrate (e.g., two orthogonal directions for a cubic face—see Figure 3.2).

Organic layers overcome many problems found for inorganic substrates for epitaxial orientation. (Typical molecular interactions between layers have been described by Hoshino et al. (1991).) Potassium hydrogen phthalate can be crystallized as thin plates from aqueous solution onto a glass slide (Zhang and Dorset, 1989). The surface of this organic salt crystal is hydrophobic, so the crystal can be stored in air for months without fear of ambient gases being adsorbed. When the layers are needed for epitaxial nucleation experiments, they are placed in the vacuum chamber to receive the deposited organic vapor. A carbon film backing can again be added and, after the layer plus substrate are removed from the vacuum chamber, the organic sample is again separated from the nucleating layer by flotation onto a water surface (thus dissolving the carboxylic acid salt). Because the organic salt crystallizes in a lower symmetry space group than the cubic inorganic substrates, the nucleating layer is anisotropic and the epitaxially grown crystals are thus larger, with a single orientation extending, sometimes, for millimeters (Figure 3.3).

3.1.5.2. Growth from a Co-Melt. Lattice matching can also be achieved between two components of a eutectic solid. This was originally noted by metallurgists in their description of lamellar alloy eutectics, where it was found that the crystal–crystal interfaces of the two pure components often had a close epitaxial match (Kerr and Lewis, 1971). The following procedures, worked out by Jean-Claude Wittmann and his collaborators (Wittmann and Manley, 1977, 1978; Wittmann and Lotz, 1981a,b; 1989; Wittmann et al., 1983), is based on the use of hypoeutectic solids where the co-melt is dilute in the material which is to be oriented. As illustrated by a phase

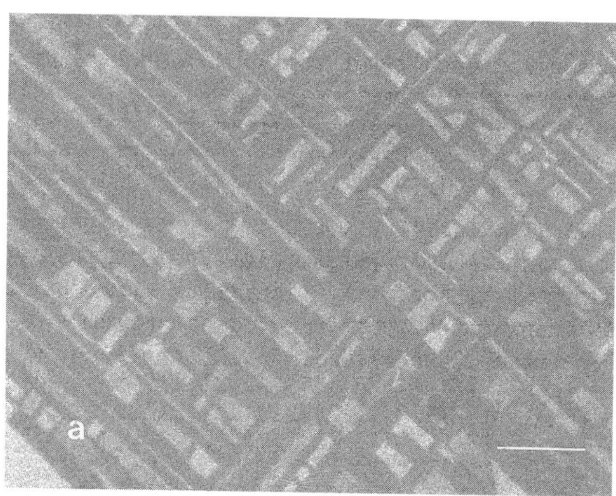

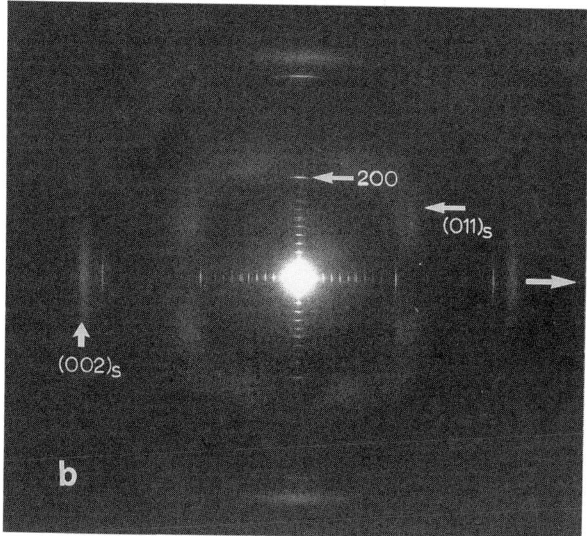

FIGURE 3.2. Epitaxial orientation of n-$C_{24}F_{50}$ from the vapor phase onto KCl: (a) electron micrograph of the crystals showing nucleation in two equivalent directions due to the cubic symmetry of the salt substrate. (b) electron diffraction pattern from a selected area wherein the two equivalent diffraction patterns are superimposed. (Reprinted from W. P. Zhang and D. L. Dorset (1990) "Epitaxial growth and crystal structure analysis of perfluorotetracosane," *Macromolecules* **23**, 4322–4326; with kind permission of the American Chemical Society.)

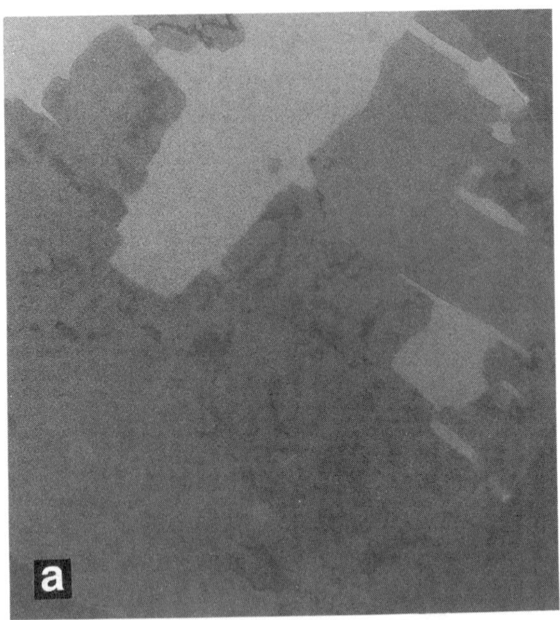
a

b

FIGURE 3.3. Epitaxial orientation of n-$C_{36}H_{74}$ on benzoic acid: (a) electron micrograph showing single orientation of crystals. (b) electron diffraction pattern (0k*l*) from a single crystal orientation. (Reprinted from B. Moss, D. L. Dorset, J. C. Wittmann, and B. Lotz (1984) "Electron crystallography of epitaxially grown paraffin," *Journal of Polymer Science, Polymer Physics Edition* **22**, 1919–1929; with kind permission of John Wiley & Sons., Inc.)

diagram (Dorset et al., 1989; Dorset, 1990) (Figure 3.4), the material to be used as the nucleating substrate crystallizes first from the melt as the liquidus curve is crossed, followed by the dilute sample as the solidus curve is crossed. If lattice matching is possible between the components, the dilute component will be oriented on the surface of the primary crystals (Figure 3.5).

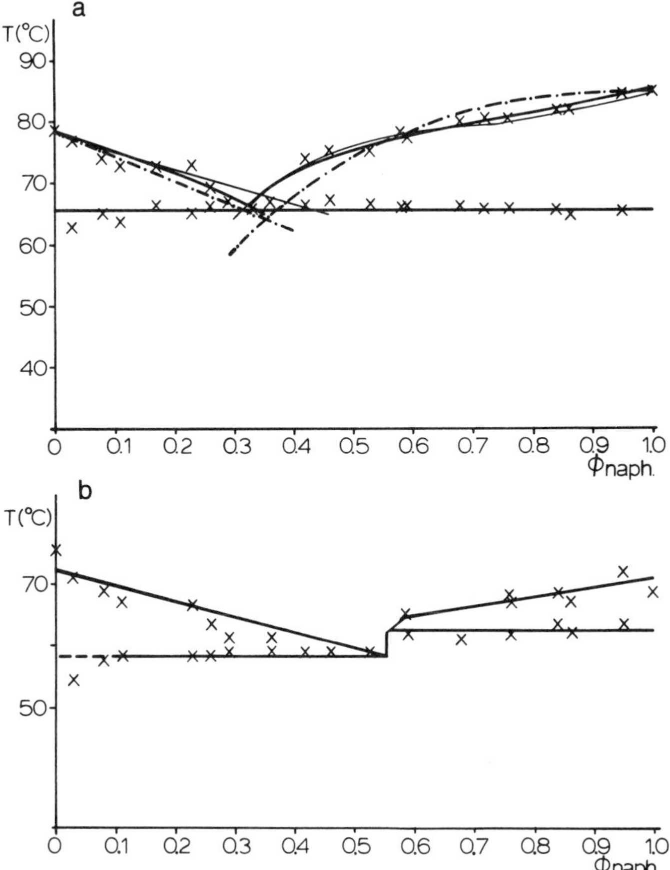

FIGURE 3.4. Phase diagram of n-$C_{36}H_{74}$ in naphthalene: (a) melting of eutectic crystals. (b) growth of crystals from the melt. (Reprinted from D. L. Dorset, J. Hanlon, and G. Karet (1989) "Epitaxy and structure of paraffin-diluent eutectics," *Macromolecules* **22**, 2169–2176; with kind permission of the American Chemical Society.)

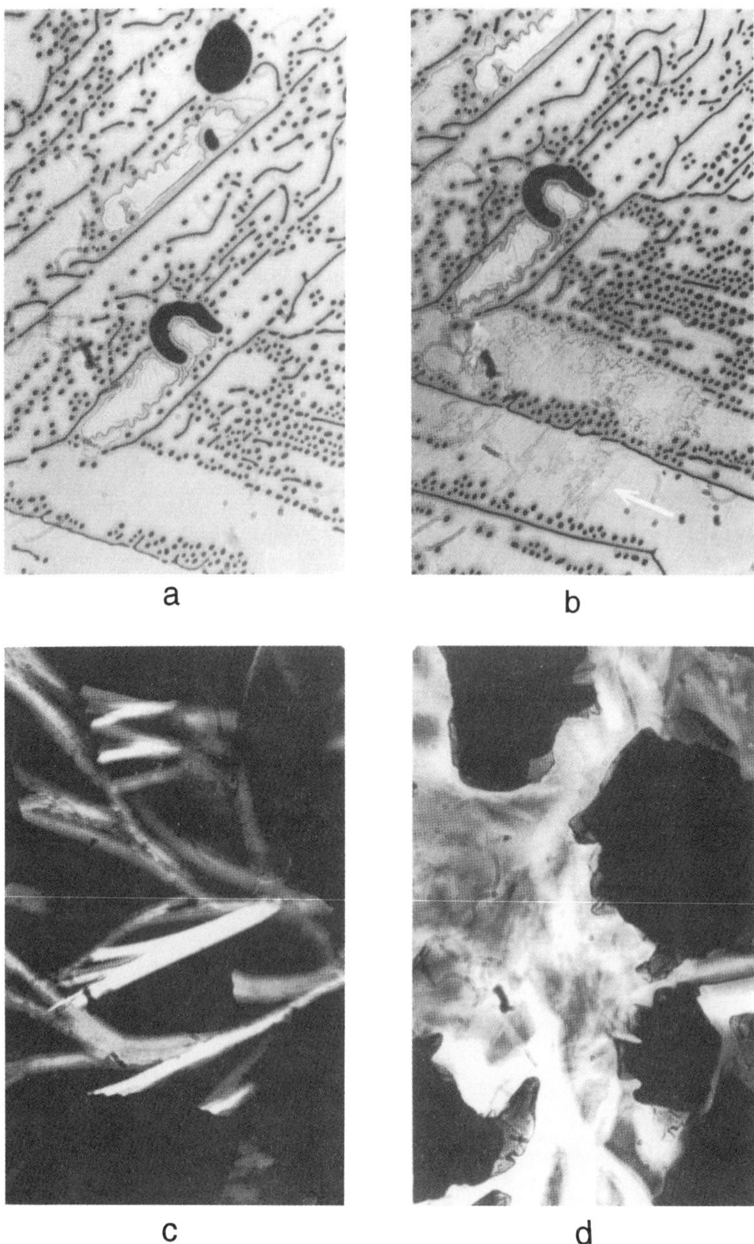

FIGURE 3.5. Observation of *n*-hexatriacontane/naphthalene crystallization from the co-melt by polarizing light microscopy. *High naphthalene concentration*: (a) growth of naphthalene crystals as the liquidus line is crossed. (b) overgrowth of paraffin crystals as the solidus line is crossed (arrow). *High paraffin concentration*: (c) needlelike paraffin crystals grown from the melt as the liquidus curve is crossed. (d) intercrystalline space filled by naphthalene as the solidus line is crossed. (Reprinted from D. L. Dorset, J. Hanlon, and G. Karet (1989) "Epitaxy and structure of paraffin-diluent eutectics," *Macromolecules* **22**, 2169–2176; with kind permission of the American Chemical Society.)

For preparative purposes, the oriented crystal growth can be achieved in many ways. A very simple procedure, used by John R. Fryer (1981) in his preparation of alkane crystals, is to make a dilute solution of the compound in the melt of the substrate (e.g., naphthalene). This can be done in a vial on a hot plate. A carbon-film-covered electron microscope grid is then dipped into the molten solution, withdrawn, and allowed to cool. After the blob of naphthalene is removed by sublimation in vacuo (preferably in a dirty vacuum system—only a floor pump is required), appropriately oriented crystals of the material may be found on the grid surface.

Better control of the crystallization can be achieved if it is carried out on a glass slide or a piece of cleaved mica (Wittmann and Lotz, 1990). A dilute solution of the material to be oriented (in a nonpolar solvent) is evaporated onto the slide surface to leave a thin film residue. Next, a layer of the nucleating agent is spread over this dried film and the two materials are melted together on a hot plate and then cooled by sliding on a thermal gradient (e.g., provided by a Kofler bench or, more crudely, by a metal bar bridging the hot plate and a cold plate). The crystallized eutectic solid can then be examined in a light microscope to look for areas where the sample is well oriented. A carbon film is deposited onto the slide surface and the organic layer floated away from the slide on a water surface, as described above. This procedure is especially effective if the nucleating layer is amphiphilic (e.g., an organic carboxylic acid). The film is then picked up with bare grids, as before.

An alternative procedure is to place carbon-covered electron microscope grids face down over the deposited organic film (formed after evaporation of the solvent) and then to spread the nucleating substrate around the grids. A second glass slide (or the other half of the cleaved mica sheet) is placed over this physical mixture to make a sandwich, after which the organic material is melted and recooled on a temperature gradient. After the slides are separated, the nucleating material can be removed by sublimation in high vacuum (e.g., if naphthalene or benzoic acid is used for this purpose).

Removal of the nucleating substrate by flotation on a water surface is permitted only if the sample is not itself amphiphilic or otherwise polar. Epitaxially crystallized polar lipids require that water not be brought near them (for fear of forming micelles, thus undoing the desired effect of the epitaxial orientation). In this case, sublimation is the only option for substrate removal.

As will be demonstrated in later chapters, epitaxial crystallization has been a very important procedure for realization of successful electron crystallographic structure analyses of many important materials (as well as polydisperse combinations of them, from stable solid solutions to fully fractionated mixtures). In some cases (e.g., linear polymers), it is important to improve the crystal perfection of the sample after initial nucleation on the epitaxial substrate. This can be done by annealing the oriented samples in the presence of the substrate, using a suitable controllable heating stage inside the vacuum chamber (Fryer, 1993) or outside it (e.g., Hu and Dorset, 1990).

3.1.6. Sonication

Sometimes a synthetic source of organic crystals can be broken up into thin plates suitable for electron crystallographic investigation. In general, a suspension of the crystalline clumps is made in a nonwetting solvent and this is sonicated in an ultrasonic cleaning device to break the clusters apart. The stock suspension can be diluted to give an appropriate concentration of microcrystalline particles per unit volume. A drop of the suspension is placed on the carbon-covered grid surface and allowed to dry by evaporation. An example where such preparations have been useful for electron diffraction structure analyses is the triglycerides (Dorset, 1983a).

3.2. Crystallization of Globular Macromolecules

3.2.1. Crystallization from Solution

The study of three-dimensional protein crystals by electron crystallographic methods is often motivated by frustrated attempts to obtain single crystals of suitable size for data collection on an x-ray diffractometer. Often the crystal habit is in the form of thin plates. Crystallization procedures are therefore exactly those used by other protein crystallographers. For example, a protein can be solubilized by a buffer solution at a certain pH and when the pH is changed, thin plates begin to precipitate, as in the case of ox-liver catalase (Sumner and Dounce, 1955). Other proteins require other procedures to remove water of solvation (e.g., with ammonium sulfate or poly(ethylene oxide)). It is beyond the scope of this book to discuss these procedures at length and the interested reader is directed to standard works on protein crystallography (e.g., Blundell and Johnson, 1976).

3.2.2. In Situ Crystals

It is also possible to find reasonably well-ordered two-dimensional arrays of proteins in nature. For example, certain bacteria express what are known as S-layers, for which there have been a number of electron crystallographic structural investigations (Hovmöller et al., 1988). Sometimes the porins in the outer membranes of gram-negative bacteria can also be ordered, as well as those in mitochondria (Mannella and Frank, 1982). Of course, the most famous example is the purple membrane of *Halobacterium halobium* (Henderson and Unwin, 1975), which contains a well-ordered array of bacteriorhodopsin molecules.

The strategy for examining these natural crystals in the electron microscope includes methods for isolating the crystalline patches from other cellular matter that would interfere with data collection. Sometimes the crystallization can be improved if mild digestive agents such as bee venom phospholipase A_2 are used to remove excess lipid from the bilayer matrix surrounding the transmembrane protein (Mannella, 1984). Isolated preparations are usually harvested on the grid by precipitation of a suspension in buffer.

3.2.3. Reconstitution of Transmembrane Proteins

In many cases, however, the transmembrane proteins are not found as ordered two-dimensional crystalline patches in the organism. To study these by electron crystallography, the crystallization must be induced during a reconstitution experiment. There are usually many variables in this procedure, including the use of optimal pH and even the inclusion of certain cations. Thus the general procedure outlined below only sketches the approach to the problem.

The membrane protein is first solubilized by extracting it from the native source with a suitable detergent (see description in Jap et al., 1992). In some cases, ionic detergents, such as sodium dodecyl sulfate, are useful, while, in other cases, these amphiphiles may induce protein denaturation, thus requiring the use of nonionic detergents (Rosenbusch, 1990). The solubilized protein is purified by column chromatography and, after the desired fraction is collected, the detergent suspension is placed in a dialysis bag in the presence of a suitable phospholipid (e.g., 1,2-dimyristoyl-*sn*-glycerophosphocholine, DMPC), perhaps held at a temperature above its mesomorphic transition. As the detergent is removed by dialysis, the protein arrays itself into a lipid bilayer vesicle. The temperature is then lowered to ambience and the vesicles containing (hopefully) two-dimensional crystalline patches are harvested (again by sedimentation from suspension) to be examined in the electron microscope.

If the ordering of the two-dimensional crystals is poor, adjustments can be made to the lipid concentration in the dialysis bag, the type of lipid used, or a variety of other conditions. It is also possible to use the above-mentioned phospholipase A_2 digestion in some cases to remove excess phospholipid from the bilayer, causing the proteins to pack closer together into a more ordered array.

3.2.4. Surface Orientation of Proteins

In a series of papers (Kornberg and Darst, 1991; Ku et al., 1992; Edwards et al., 1994), techniques have been described for forming protein monolayers at a surface, e.g., a lipid monolayer on a glass slide, where the lipid headgroup is attached to some ligand or where the headgroup, itself, contains localized charges. A number of two-dimensional protein crystals have been grown in this way, apparently diffracting to high resolution in the electron microscope. It is also possible to use one of the preformed two-dimensional layers as an epitaxial substrate for three-dimensional growth; as several examples appear in Edwards et al., (1994).

3.3. Crystallization of Inorganic Structures

Although the use of electron crystallography to solve organic and biomolecular structures is the major focus of this book, Chapter 7 will be devoted to quantitative studies of inorganics, in order to demonstrate the general efficacy of the procedures for direct crystallographic phase determination. Some mention of crystallization procedures for inorganic materials is therefore appropriate.

Many of the methods already discussed above also can be used to grow thin inorganic crystals. If the material, e.g., a salt, is soluble in water or another solvent, crystallization can use the evaporation procedure already described. An example would be the preparation of basic copper chloride (Voronova and Vainshtein, 1958). Insoluble materials, if they are layered, can be separated into thin sheets by sonication or other fragmentation procedures, including microtomy. Preparation of clay materials is described by Zvyagin (1967) in a monograph. Other samples can be crushed finely in a mortar and dispersed in a nonwetting liquid (Kihlborg, 1990). For ceramics and metals, the sample is ground and polished to a few hundredths of a millimeter thickness, with a final thinning achieved by electropolishing (Kihlborg, 1990).

3.4. Preservation of Samples in the Electron Microscope Vacuum

A vacuum is a harsh environment for many specimens, especially organics, e.g., if they are stabilized by a volatile solvent of crystallization or if they are prone to sublimation. For this reason, special sample stages have been constructed that allow examination of the crystalline objects in an unaltered state. Alternatively, the solvent can sometimes be replaced by less volatile substances. Various approaches are discussed in the following sections.

3.4.1. Environmental Chambers

Environmental chambers are differentially pumped, aperture-limited devices designed to maintain a microenvironment around the specimen grid, thus isolating it from the vacuum of the electron microscope column but providing a path for the electron beam through the sample (Parsons et al., 1974; Hui et al., 1976). For example, a gas can be introduced to raise the ambient pressure around the sample and, because of the aperture limitation of the outer and inner chambers, few demands are placed on the electron microscope vacuum system. Such a system will permit the study of crystals at a high surrounding vapor pressure. It is also possible to create 100% relative humidity in the atmosphere of water vapor around the sample to permit electron diffraction experiments on fully hydrated crystals (Figure 3.6). An example is the ox-liver catalase microcrystalline preparation, which is very sensitive to solvent loss (Matricardi et al., 1972; Dorset and Parsons, 1975a).

Perhaps the most impressive use of environmental stages has been for the study of suspended lipid bilayers (Hui et al., 1974; Hui and Parsons, 1975). Electron diffraction and low-magnification diffraction contrast imaging can be used, for example, to characterize lateral phase separation of biomembrane components. Inclusion of a heating element in the environmental stage further permits the study of thermotropic phase transitions. The advantage of these studies is that they allow visualization of

→

FIGURE 3.6. Electron diffraction patterns from fully hydrated catalase microcrystals stabilized in a differentially pumped environmental chamber: (a) tilted specimen showing sharp Laue bands; (b) untilted specimen.

a

b

individual components in single microdomains of phase-separated bilayers to produce spot electron diffraction patterns (Hui, 1993).

Although environmental stages require special modifications to be made to the electron microscope, they are not difficult to use. If a hydrated specimen is picked up on a Langmuir trough, a simple humidified transfer device can be used to retain the water content of the specimen as it is being carried from the trough surface to the microscope stage. After the specimen is successfully inserted into the chamber, the procedures for diffraction and imaging are similar to those employed for any other specimen. On the other hand, because of gas scattering, high-resolution imaging of a specimen is very difficult, although energy filtration may be useful here (e.g., see Hui and Parsons, 1978). Designs for environmental chambers nowadays allow for easy insertion into the electron microscope column and also for specimen tilt (Turner et al., 1989, 1991).

3.4.2. Cryostages

Another solution to the study of volatile samples is to freeze the specimen. If the aqueous regions in the samples are extensive, it is possible to plunge the sample-carrying grid into a cryogen, such as liquid ethane, to produce a vitreous form of ice that will not damage the structure (Dubochet et al., 1982; Echlin, 1992). Liquid nitrogen-cooled specimen holders are made for most electron microscope side entry stages and these will maintain the low temperature needed to stabilize the vitreous ice after the sample is successfully transferred to the microscope without accreting ice contamination from the laboratory atmosphere. It is also important that the anticontaminator in the electron microscope column be well designed so that it can be cooled to a temperature lower than that of the specimen, again so that it will serve as a collection surface for freezing out residual water vapor. Cooling stages permit specimen tilting around the main goniometer axis, and, for some designs, an additional orthogonal tilt. A full description of these points has been given by Chiu (1986).

There are many examples where cooling stages have facilitated the study of hydrated specimens by electron diffraction, ranging from chain-folded polysaccharides (Chanzy et al., 1977) to protein crystals (Chiu, 1986). The diffraction patterns from rapidly frozen catalase crystals closely resemble those for the ambient hydrated crystals studied in an environmental chamber (Taylor and Glaeser, 1974). The quality of the diffraction pattern (i.e., removal of solvent scattering) can be improved with the use of energy filtration (Schröder and Burmester, 1993). The major disadvantage of a cryostage is that, obviously, high-temperature phase transitions cannot be studied. The advantage is that high-resolution imaging of the thin crystalline object is possible to permit structure analysis.

3.4.3. Solvent Replacement

Solvent-stabilized structures can be preserved in electron microscope vacuum by replacement of the volatile component by a less volatile one. This is the rationale for the use of "negative stains," which are solutions of heavy atom-containing salts that dry into a glasslike domain within the interstices between globular macromolecules in the crystal. Because of the microcrystallinity of these dried salt solutions, it has often been anticipated that only about 20-Å resolution could be observed for many substances (Vainshtein, 1978). However, electron microscope images, taken at lower beam doses than usually employed, show that higher resolutions are sometimes attainable (Massalski et al., 1987). In other cases, the nature of the protein may allow the visualization of higher-resolution detail (Sherman et al., 1981).

Saccharides such as glucose can be used to replace water. With such specimens, it is possible to obtain high-resolution electron microscope images and electron diffraction patterns if low dose procedures are followed (Unwin and Henderson, 1975). The images have low contrast but detail can be enhanced considerably by lattice averaging techniques (see Chapter 4).

3.5. Data Collection and Processing

3.5.1. Goniometry

A goniometer stage is probably the most important attachment to the electron microscope for crystallographic studies (see Turner et al., 1991). Tilt stages have been available for electron microscopes for many years as self-contained capsules for use in top-entry sample stages. While these ensure the maximum vibrational isolation of the specimens, they are inconvenient for continuous tilting experiments on beam-sensitive specimens because they are not eucentric. Hence, on tilt, the specimen will precess away from the area selected for diffraction and will also be at another focal length to the objective lens, resulting in a different magnification of the diffraction pattern. Often there is also a restricted tilt range, e.g., ±45°.

The most useful goniometer for crystallographic studies is a side-entry rod that can be tilted ±60° around the rod axis. Using a height adjustment for the specimen on this stage, the goniometer can be made nearly eucentric so that, if the specimen moves away from the selected area at high tilt, it will not be far from the starting position. Either a second orthogonal tilt or a ±180° (or full 360°) rotation is available for some specimen rods, allowing complete freedom of orientation of specimens. The tilt-rotation geometry is easy to use, especially if a certain diffraction zone is known to be located at a given specimen tilt. Rotation of the grid at this tilt angle will thus bring the correct zone axis into coincidence with the incident beam direction.

Rotation stages can be used conveniently to locate the position of the goniometer tilt axis with respect to a diffraction film. Electrons travel through magnetic lenses in a helical trajectory (see Reimer, 1984). Hence, unless this is compensated by the microscope design, there will always be a rotational offset of the crystal axes found in an image and in the diffraction pattern. When working only in the electron diffraction mode, the goniometer tilt axis location will change with change in camera length. Thus, the tilt axis position must be defined with a known sample. For this purpose, crystals of linear wax esters or fatty acids in the *B*-polymorph will produce a diffraction pattern resembling those from untilted orthorhombic paraffin crystals if the specimen is tilted

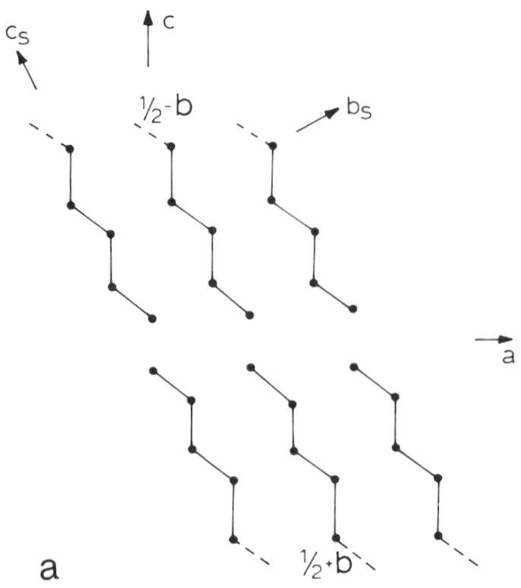

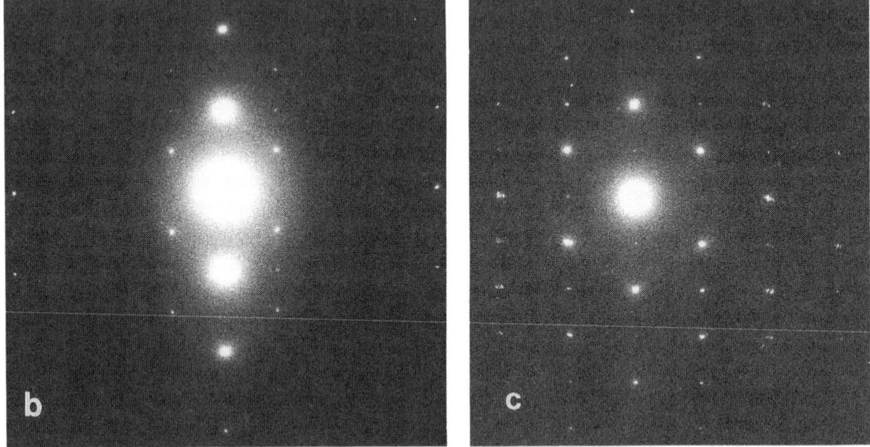

FIGURE 3.7. Calibration of a goniometer tilt axis position in reciprocal space. When crystals such as behenic acid layers are used, the chains (a) are packed at a 27° angle to the layer normal. (b) An $(hk0)$ diffraction pattern contains the subcell $\mathbf{a}_S = 7.5$ Å repeat as strong I_{0k0} reflections. Tilting around this axis by 27° produces the $(hk0)_S$ pattern from the subcell (c) only when $\mathbf{a}_S$ is coincident with the tilt axis of the goniometer. (Reprinted from D. L. Dorset (1976) "A facile location of goniometer tilt-axis position with respect to a single-crystal electron-diffraction pattern orientation," *Journal of Applied Crystallography* **9**, 142–144; with kind permission of the International Union of Crystallography.)

27° and rotated so that the $a \approx 7.5$ Å axis of the crystal (i.e., the inclination axis of chains in the oblique layer packing) coincides with the goniometer rod axis (Figure 3.7) (Dorset, 1976a).

Side-entry goniometer stage geometry is also the most convenient for designing special stages, including the cooling stages discussed above, where the dewar flask containing cryogen is held at the outer end of the rod. Since heating elements are often included near the specimen area in cryostages, it is also possible to follow phase transitions of many organics up to about 150 °C (including subambient transitions), if the specimen does not sublime at higher temperature. Heating stages with a higher temperature range have been designed for the study of inorganic materials.

Compared to x-ray crystallography, the major difficulty in using a goniometer stage in the electron microscope is the limited tilt range, imposed by the need for a support surface for the crystals and also the physical dimensions of the blade holding the grid within the objective lens pole piece gap. Again, the practical limit for tilt is often ±60° in many stages (and is even more restricted when cooling stages are used), but if important structural information is contained in reciprocal lattice regions beyond this limit (i.e., the "missing cone" or "dead zone"), it will have adverse consequences on the structure analysis, as already mentioned. The amount of the reciprocal lattice that can be sampled with only one type of growth-imposed crystal orientation depends on the symmetry of the unit cell. For example, with triclinic structures, to collect a full unique *hkl* data set, one Miller index must be sampled from 0 to ∞, whereas the other two must be sampled from $-\infty$ to $+\infty$. (For any zone where one of the indices is 0, one index still has to be sampled from $-\infty$ to $+\infty$.) Monoclinic cells allow k, as well as either h or l, to be sampled from 0 to ∞. The remaining reciprocal lattice index, however, must be sampled from $-\infty$ to $+\infty$. (In zonal projections, the $h0l$ resembles the triclinic case; for the other zones, the sampling follows the orthorhombic example.) In orthorhombic structures, all indices are sampled only from 0 to ∞. Higher-symmetry unit cells may produce redundant orientations so that the complete data set can be sampled within the tilt limitations of the electron microscope, e.g., the example of buckminsterfullerene shown in Figure 2.16.

3.5.2. Data Collection

3.5.2.1. Electron Diffraction. Selected area electron diffraction experiments on thin organic crystals often utilize a selected area diameter between 1 and 10 μm. Illumination of the object is minimized to $< 10^{-5}$ amp/cm^2 when the first condenser lens is strongly excited and a 20 μm diameter aperture is inserted at the second condenser. Such a diffraction geometry is useful when the unit cell repeat is not extremely large, i.e., a 50-Å phospholipid lamellar spacing can be easily resolved in such experiments. When larger camera lengths are needed, it is convenient to use a high-dispersion geometry where the objective lens is not excited and the first projector is used as the default objective lens to provide a long focal length to the specimen. While it is possible to insert a selected area aperture, smaller diameter areas can be isolated on the specimen (e.g., $< 1.0\,\mu$m) when the condenser lenses are used to focus the source onto the crystal. This alternative high-dispersion geometry is often employed for the study of protein microcrystals (Amos et al., 1982) (see Figure 3.8).

RHEED is a third diffraction mode that is sometimes used to study linear chain molecules (see Pinsker, 1953). Until recently, electron microscopes had a port for insertion of an additional sample holder below all of the magnetic lenses. The complete lens array served as a condenser system. A sample on a flat surface of a polished metal

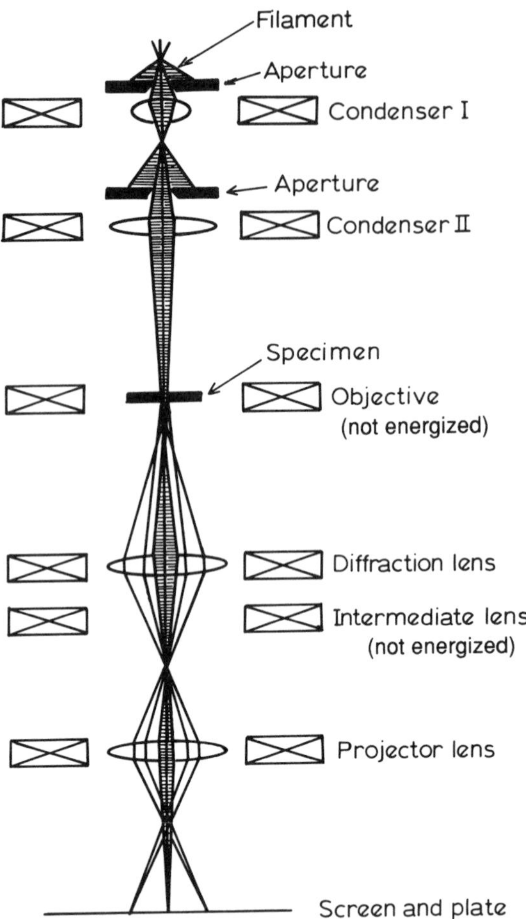

FIGURE 3.8. High-dispersion mode for electron diffraction. Unlike the selected area diffraction geometry (see Figure 1.7), the objective lens is inactivated and the first intermediate lens is used as the actual objective to achieve a very long camera length (see description in Chapter 1). The area is selected by control of the condenser lenses and insertion of a small aperture to restrict the illuminated area. (Reprinted from D. L. Dorset (1985) "Crystal structure analysis of small organic molecules," *Journal of Electron Microscopy Technique* **2**, 89–128; with kind permission of Wiley–Liss Division, John Wiley and Sons, Inc.)

block is translated into the beam and the angle of incidence is adjusted until a shadow edge of the block is moved near to the incident beam spot. Usually another part of the stage contains a grid holder for transmission experiments. With the latter, oblique texture patterns can be obtained when the grid is tilted because millimeter-diameter areas can be illuminated (see Chapter 1).

Electron diffraction experiments require a camera-length calibration, especially when the lenses are placed below the specimen. For smaller unit cells, it is convenient to use microcrystalline samples of gold, made by evaporation of the metal in vacuo onto a part of the grid used for the diffraction experiment (Hall, 1965). The first maximum of the gold rings occurs at $d = 2.355$ Å. (Thallium chloride has also been suggested as a diffraction standard but the reader should be cautioned that this material is extremely toxic!) Larger unit cell spacings are calibrated against standards such as negatively stained catalase (Wrigley, 1968).

At low doses, it is useful to have a photographic film that is sensitive to the electron beam at a useful camera length. While most unscreened films have about the same quantum efficiency per grain (Farnall and Flint, 1969), the larger grained x-ray films (e.g., Kodak DEF-5, DuPont LoDose Mammography) are convenient for visualizing the diffraction pattern at a camera length that facilitates densitometric measurement. Other options for recording electron diffraction intensities exist, such as Vidicon tubes (J. R. Fryer, unpublished data), CCD cameras (J. Minter, unpublished data), and imaging plates (Isoda et al., 1991). However, photographic films are still most convenient for long-term storage of patterns, from which information can always be extracted easily.

When photographing electron diffraction patterns, it is useful to have a film positioned under the screen so that an exposure can be taken quickly. For multiple exposures of the same crystals (as in a tilt series), the beam should be deflected away from the specimen during "dead" time periods when the films are being transported through the camera. This beam offset can be easily implemented by deliberately misaligning the condenser lens deflector coils in the "dark field" setting of most electron microscopes. It is also desirable to have an exactly timed exposure of a diffraction pattern, particularly if intensities from different tilts are to be compared.

To convert the diffraction spot density to an intensity value to be used for crystal structure analysis, some sort of microdensitometer is needed to scan the film (unless this information is directly extracted from pixels in a CCD array mounted in the electron microscope). Flatbed instruments have often been used. Ideally an exposure series is taken for any diffraction orientation so that the film response is not saturated (Vainshtein, 1964a). Difficulties in the extraction of intensity information from films have been discussed by Wooster in 1964. Dark spots are best scanned by a fine raster of small diameter while a weak spot is best measured by a very large window. A compromise has been made to scan patterns with a narrow slit of height equal to the diameter of the darkest spot. Once the maxima are recorded on the graph paper in the double-beam densitometer, the peak area can be approximated by a triangle. Other options for reading films include the use of fast CCD cameras coupled with a frame grabber. Automated software such as the ELD programs in the CRISP package (Zou et al., 1993a) is being developed for this purpose. The main problem in this option is the limited dynamic range of such cameras, again requiring that an exposure series be photographed or that an internal intensity calibration be made (Zou et al., 1993b).

Conversion of the raw intensity I_{meas} to an "observed" intensity may include a correction for crystal texture. Vainshtein (1956a,b) has suggested that a phenomenological Lorentz correction should be used in many cases. For example, for single "mosaic" crystals, he proposed that:

$$|F^{obs}|^2 \propto I_{meas} \cdot d^*_{hkl}$$

and, for oblique texture diffraction patterns, that:

$$|F^{obs}|^2 \propto I_{meas} \cdot d^*_{hkl}\, d^*_{hk0}\, p$$

assuming that the texture axis is [001], where p is a multiplicity factor (accounting for reflection overlap). For polycrystals, this becomes:

$$|F^{obs}|^2 \propto I_{meas} \cdot d^{*2}_{hkl} p$$

Selected area diffraction experiments on single crystals have shown (Dorset, 1976b), however, that:

$$|F^{obs}|^2 \propto I_{meas}$$

since the spreading of the intensity along the reciprocal lattice rods is a rather broad Gaussian function. The dominance of this Gaussian function over the shape transform of a thin plate is due to the bend deformation of the plate over the relatively large selected area. Other situations occur where in-plane paracrystalline distortions lead to arcing of the reflections. Then the densitometer scan cannot detect all of the reflection intensity for high diffraction orders, e.g., from a lamellar lipid layer (Dorset et al., 1986). One can use:

$$|F^{obs}|^2 \propto I_{meas} \cdot l^n$$

where l is the order of the lamellar reflection and n is often near 1.0. Many protein crystals tend to be rather stiff, on the other hand, so that one can see distinct Laue bands in slightly tilted samples (Dorset and Parsons, 1975b). Here one must account for the Ewald sphere passing through the shape transform broadening of all reflections in order to find the true intensity of the reflection. The radius of the sphere is the reciprocal of the electron wavelength $r = 1/\lambda$. For any scattering angle θ, it is convenient to find the deviation parameter s_g for any value of d^* (assuming a known 2θ) by subtracting a y-component from r after converting from polar to rectangular coordinates. The relative intensity of a reflection is then found by comparing values of $\sin^2 \pi t s_g/(\pi s_g)^2$ at higher angles to those found for s_g values near the incident beam. Here t is the crystal thickness. (It is assumed that the center of the Laue circle is at 0,0.) It is apparent that the widths of the Laue zones can be used to determine the crystal thickness to a fair accuracy (Dorset and Parsons, 1975b) and that these values correspond well to measurements based on other criteria (Dorset and Parsons, 1975c).

3.5.2.2. Electron Microscopy.

Low-magnification microscopy. Low-magnification electron microscopy at near 10kX is of great value for characterization of crystal morphology and visualization of some defects. With metal-shadowed preparations, one can operate at "normal" illumination levels used in conventional electron microscopy and, as shown above, such experiments can be useful for measurement of crystal thickness.

Far more interesting results are obtained when specimens are examined at illumination levels used for electron diffraction. Insertion of an aperture into the back focal plane of a projector lens to isolate either the unscattered beam or a diffraction spot will permit observation of bright field or dark field images, respectively. These can be used to characterize bend contours (Figure 3.9, also Figure 1.10), for example (see Cowley, 1967). Some defects such as edge dislocations can be detected by moiré magnification.

At such low magnifications, focusing the objective lenses on the specimen is easily accomplished by use of the "image wobbler" feature found in most modern electron microscopes. It is convenient to intersperse such diffraction contrast images with electron diffraction patterns from the same crystal when working in the selected area mode. Unfortunately, such images are not so conveniently obtained when high-dispersion diffraction experiments are being carried out, particularly when the area is selected by the condenser lenses.

High-resolution, low-dose microscopy. With the pioneering work of Uyeda et al. (1976) and Fryer (1989) on the high-resolution electron microscopy of organic crystals, considerable progress had been made in the direct visualization of crystal lattices at "molecular" resolution. For such studies, the electron microscope is carefully

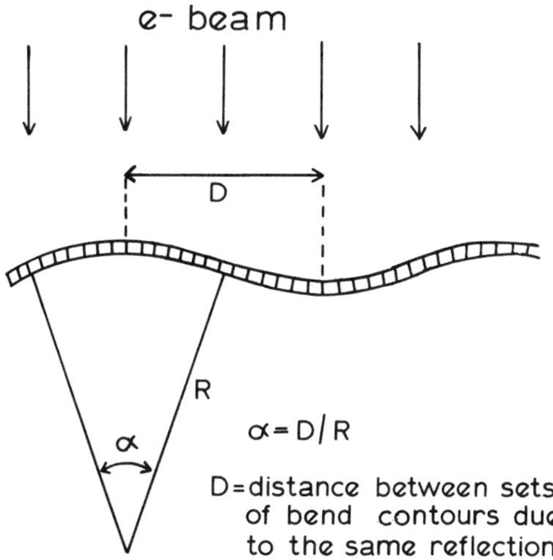

FIGURE 3.9. Geometry of bend contour formation in elastically deformed crystals. Only certain crystal orientations are at a proper incidence to diffract electrons, thus an aperture limitation will allow these suitably oriented regions to be visualized (see Figure 1.10 for examples). (Reprinted from D. L. Dorset (1985) "Crystal structure analysis of small organic molecules," *Journal of Electron Microscopy Technique* **2**, 89–128; with kind permission of Wiley–Liss Division, John Wiley and Sons, Inc.)

aligned at a magnification much higher than that intended for the low-dose experiment, with corrections made for objective-lens astigmatism. After insertion of the sample into the column, the grid is then scanned in the selected area diffraction mode until a crystal diffracting to suitable resolution is found. The grid can then be translated to a nearby area to allow a focal adjustment to be made (Fryer, 1979). After this adjustment is made, the selected area diffraction mode is used again to find the specimen and it is photographed under low-dose conditions in a series of underfocus values. The procedure can also be automated in some microscopes with a deflector system that will cause the illumination to be confined to a smaller area so that the compensations for focus and astigmatism (using the graininess of the carbon support film) are visualized conveniently (Fujiyoshi et al., 1980). After photography with the spread low illumination source at several underfocus values, it is often useful to verify the integrity of the specimen by observing its selected area diffraction pattern again. As in the diffraction work, the illumination is deflected away from the specimen with a misaligned dark field control while film cassettes are being transported through the camera.

Variations of these procedures have been adapted by other laboratories. For example, a "spot illumination" of the specimen has been described that is stated to minimize specimen movement caused by initial beam damage events (Downing, 1991). Very low specimen temperatures (e.g., 4 K) can be achieved on a liquid helium–cooled microscope with a superconducting objective lens (Lefranc et al., 1982). This minimizes translational drifts of the specimen grid during photography of the image, additionally protecting the specimens somewhat against radiation damage.

In general, such low-dose electron microscopy is carried out at direct magnifications between 20,000 and 60,000X. There has been considerable discussion of the optimal type of photographic film to be used in such work. At lower image resolutions (e.g., 10Å), large-grain x-ray films employed for electron diffraction experiments are useful, while at the higher end, a finer-grain film may be desired. Of course, for less beam-sensitive materials, the experimental strictures imposed on high-resolution microscopy are less severe so that direct adjustments for objective lens focus on the specimen can be made. Examples of such materials include the phthalocyanines (Uyeda et al., 1972).

After the low-dose electron micrographs are photographed, it may be possible, in some cases, to identify visually regions of the specimen image that have suitable crystalline order. In other samples, e.g., aliphatic compounds, where one uses the smallest beam dose possible, these regions may not be directly visible. Micrographs then should be placed on an optical bench for further examination. The film is illuminated with a laser source (Figure 3.10) and a screen is inserted at the back focal plane of the glass objective lens to observe the optical diffraction pattern. Translation of the film across the illuminated area will allow crystalline areas to be found and their ordering to be estimated from the resolution of the optical transform.

Densitometry and deconvolution of the phase contrast transfer function. After a high-resolution electron micrographs photographed and a suitable area with sufficient resolution, etc., identified, it is necessary to obtain a digitized representation (x, y,

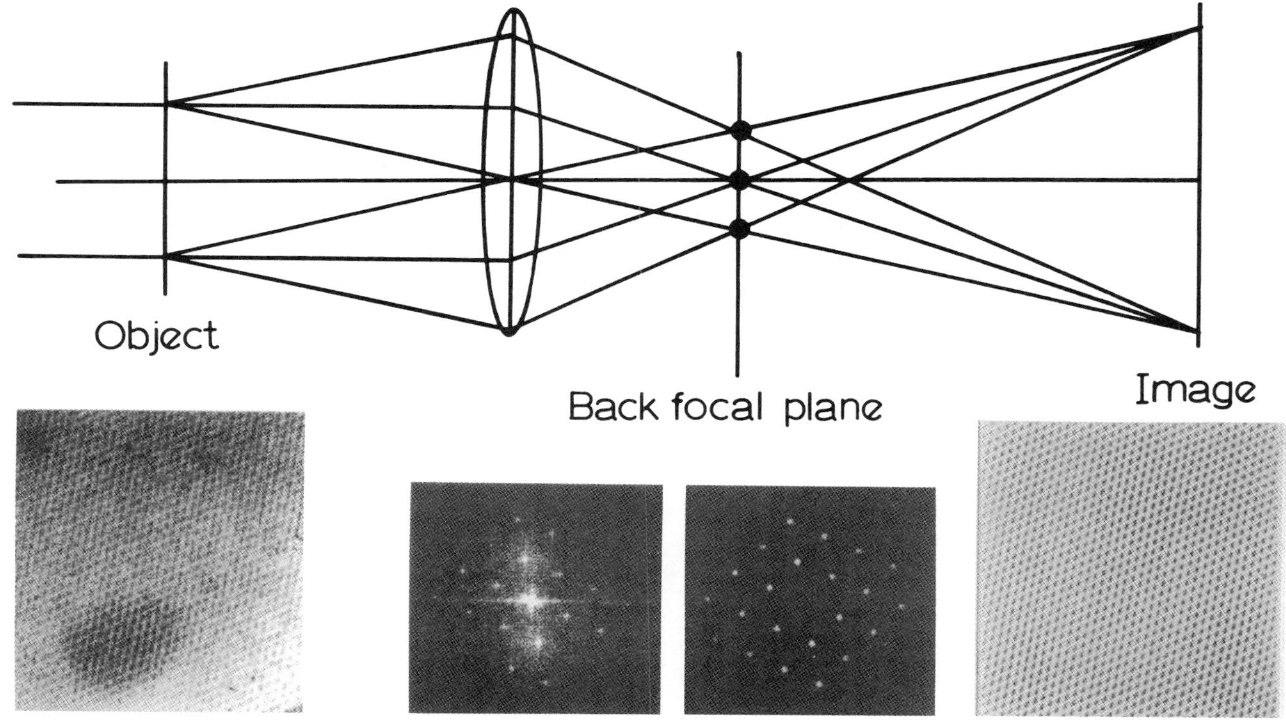

FIGURE 3.10. Geometry of an optical bench used for image reconstruction. An electron micrograph of a crystal is placed at the in-focus object position of the glass lens. Its optical transform is produced at the back focal plane. When a mask with holes spaced at the Bragg peak positions is overlaid on this plane, only amplitudes and phases due to the crystal repeat will form the image. (Reprinted from D. L. Dorset (1994) "Electron crystallography of linear polymers," in *Characterization of Solid Polymers*, S. J. Spells, ed., Chapman and Hall, London, pp. 1–17; with kind permission of Chapman and Hall, Ltd.)

intensity) so that an average representation of the unit cell can be formed by image processing. In early work of this kind, such images were scanned on flatbed or rotating drum microdensitometers. The former is slow but provides the best spatial resolution whereas the latter, while faster, requires special alignment if better than 25-μm detail is to be distinguished on the film. Since, for radiation-sensitive samples, it is preferable to photograph images at the lowest permissible magnification, information can be lost if the latter microdensitometer is used to digitize the film. Recently this resolution problem has been solved rather cheaply by the use of fast CCD cameras. While these do not have the dynamic range of microdensitometers for measuring image intensity, the spatial resolution is very good, so that features, e.g., 5 μm apart on the film, can be easily discerned (Hovmöller, 1982).

Especially when a high-resolution micrograph of a low-contrast object is obtained, it is difficult to say a priori what the objective lens phase contrast transfer function envelope is for a particular exposure. This uncertainty is due to slight local variations of focal length to the specimen as the grid is scanned. As was discussed in Chapter 1, the transfer function, which serves as a band-pass filter for the diffracted waves being combined to form an image, appears as a convolutional term in the measured image intensity:

$$\Phi(\mathbf{r}) = 1 + 2\sigma\rho(\mathbf{r}) * FT^{-1}\{A(s)\sin\chi_1(\mathbf{s})\exp(-\chi_2(\mathbf{s}))\}$$

where,

$$\chi_1(\mathbf{s}) = \pi\Delta f\lambda\ |\mathbf{s}|^2 + 1/2\ \pi C_s\lambda^3|\mathbf{s}|^4$$

and

$$\chi_2(\mathbf{s}) = 1/2\ \pi^2\lambda^2\ |\mathbf{s}|^2\ D^2$$

Deconvolution of the transfer function from the image requires that the experimental defocus value Δf be known, since the spherical (Erickson, 1973) and chromatic (Frank, 1977) aberration terms, C_s and D, respectively, are instrumental constants. Consider the Fourier transform of $\Phi(\mathbf{r}) = T(\mathbf{s})$, where one can write:

$$T(\mathbf{s}) = \delta(\mathbf{s}) + 2\sigma F(\mathbf{s})[\sin\chi_1(\mathbf{s})\exp(-\chi_2(\mathbf{s}))]$$

The total transfer function can be assumed to be included in one term:

$$C(\mathbf{s}) = [\sin\chi_1(\mathbf{s})\exp(-\chi_2(\mathbf{s}))]$$

$C(\mathbf{s})$ can be determined experimentally in several ways. For example, after obtaining an image of a crystal at low dose, it is possible to follow this exposure by one at much higher dose to image the carbon support film. Its optical transform (Unwin and Henderson, 1975) will show the zeroes of the transfer function between the so-called Thon rings (Figure 3.11) and the instrumental constants can be used in the above

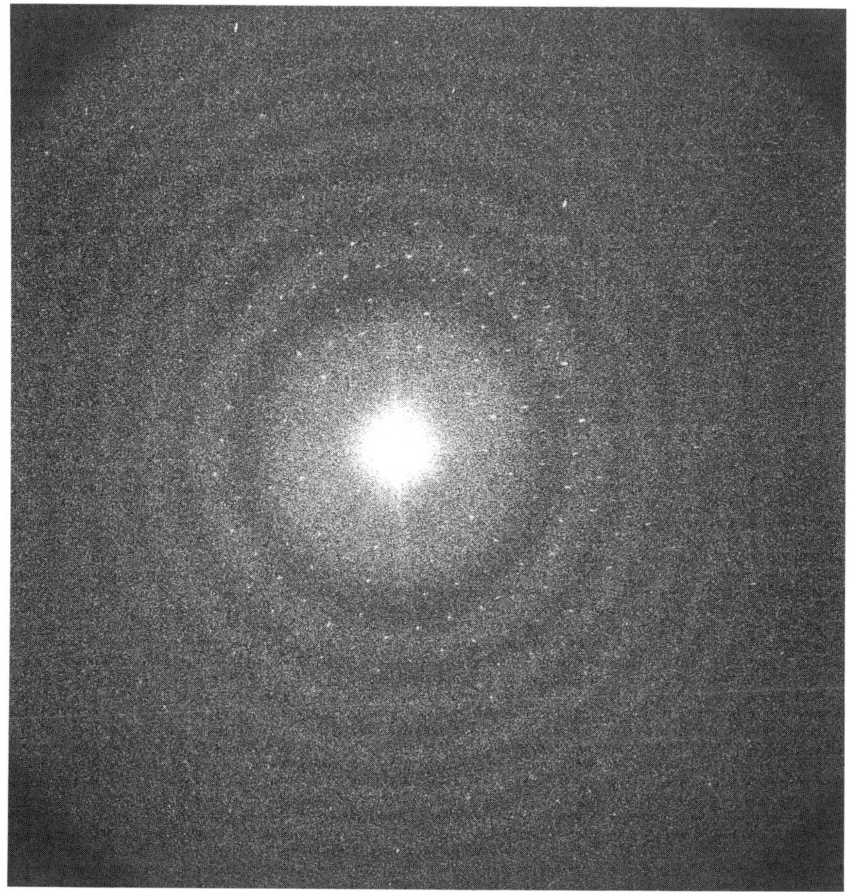

FIGURE 3.11. Thon rings in the optical transform of an electron micrograph of glucose-embedded Omp F porin two-dimensional crystals. (Original micrograph taken by Dr. A. Massalski.)

expressions to determine Δf. Kirkland et al. (1985) have also discussed how a focal series can be used to reconstruct the transfer function. Ideally, however, it would be preferable if Δf could be predicted accurately a priori from a single low-dose image. Recently, such procedures have been worked out by Li and her collaborators.

For example, the above expression for the image transform can be rearranged (Li, 1991), neglecting the delta function:

$$F(\mathbf{s}) = T(\mathbf{s})/2\sigma C(\mathbf{s})$$

Squaring and taking average values of the magnitudes:

$$\langle |F(\mathbf{s})|^2 \rangle = \langle\, |T(\mathbf{s})|^2 \,\rangle /4\sigma^2 C^2(\mathbf{s})$$

From Wilson's (1950a) work in x-ray crystallography:

$$\langle |F(\mathbf{s})|^2 \rangle = \sum f_i^2$$

The squared transfer function becomes:

$$C^2(\mathbf{s}) = \langle |T(\mathbf{s})|^2 \rangle / 4\sigma^2 \sum f_i^2$$

and so

$$|F(\mathbf{s})| = |T(\mathbf{s})| \left[\sum f_i^2 / \langle |T(\mathbf{s})|^2 \rangle \right]^{1/2}$$

At Scherzer focus, the envelope of $C(s)$ will be negative to the practical resolution limit of the electron microscope. From Liu and coworkers (1990):

$$F(\mathbf{s}) = -T(\mathbf{s}) \left[\sum f_i^2 / \langle |T(\mathbf{s})|^2 \rangle \right]^{1/2}$$

Therefore, an estimate of the correctly phased structure factors can be found.

If Δf is not near the Scherzer value, Tang and Li (1988) have devised the following minimization procedure to determine its value. From the arguments given above, the observed function can be written:

$$\sin^2\chi_1(\mathbf{s})_o = |T(\mathbf{s})|^2/4\sigma^2 \, |F(\mathbf{s})|^2 \exp(-2\chi_2(\mathbf{s}))$$

Now, $\chi_1(\mathbf{s})$ changes more rapidly with Δf than does $\chi_2(\mathbf{s})$. Using the transform of the function $|T(\mathbf{s})|$ and electron diffraction intensities $I(\mathbf{s}) = |F(\mathbf{s})|^2$, an estimate of the spherical aberration part of $C(\mathbf{s})$ can be made, since C_s is already known, permitting the calculation of $\sin^2\chi_1(\mathbf{s})_c$ at many possible test values for Δf. Thus one can minimize the quantity:

$$\sum |\sin^2\chi_1(\mathbf{s})_o - \sin^2\chi_1(\mathbf{s})_c|$$

A similar procedure was developed by Unwin and Henderson (1975) and is found to be most successful for densely populated reciprocal lattices. A plot of the ratio $|T(\mathbf{s})|/|F(\mathbf{s})|$ is shown in Figure 3.12 for the experimental image of an epitaxially oriented *n*-paraffin. Other estimates of Δf have been obtained by use of the Sayre (1952) equation (Han et al., 1986) or the maximum entropy figure of merit, in all, yielding at least four procedures for deconvolution of a single image (Li, 1991). All these procedures have been tested with data from copper perchlorophthalocyanine. As will be shown in Chapter 4, the deconvoluted image can be Fourier-transformed to provide an independent source of crystallographic phases for structure determination. However, it is important to realize that these analyses not only depend on the accurate knowledge of Δf, they also assume that the microscope objective lens is well corrected

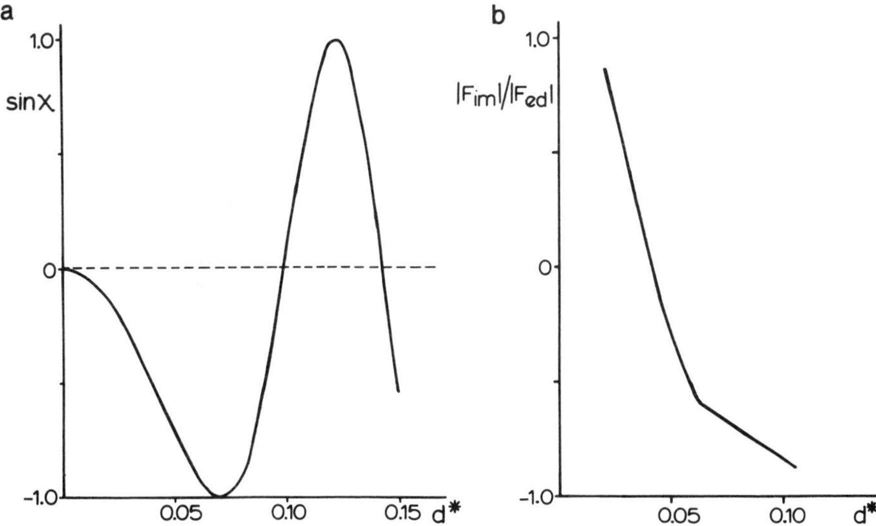

FIGURE 3.12. Comparison of the image transform to the electron diffraction pattern to detect the first zero of the lens transfer function (as in (a) where it crosses the zero contrast line). (b) The ratio of $|F_{im}| = T(\mathbf{s})$ over $|F_{ed}| = |F(\mathbf{s})|$ is plotted, where the former are obtained from the image transform and the latter from the electron diffraction intensities. A minimum of this ratio will be found at the occurrence of the zero contrast point. (Reprinted from D. L. Dorset and F. Zemlin (1990) "Direct phase determination in electron crystallography: the crystal structure of an n-paraffin," *Ultramicroscopy* **33**, 227–236; with kind permission of Elsevier Science B.V.)

for astigmatism, so that the effective transfer function is radially symmetric over the field of view. If some astigmatism is present and remains undetected, it can lead to errors in crystallographic phase determination for reflections near the boundaries of the constant contrast regions of the transfer function, where phase envelopes change sign. In this case, a direct visualization of the Thon ring ellipticity is necessary for a correction to be made.

4

Crystal Structure Analysis

4.1. Solution of the Phase Problem

In crystallography, finding the spatial arrangement of the average unit cell contents requires solution of the phase problem. As described in Chapter 1, the Fourier transform relationship between an image and its diffraction pattern can be experimentally realized in one direction if only intensity data are recorded. Since the image intensity is linearly proportional to the crystal potential, when the weak phase object approximation is valid, the directly computed transform finds the crystallographic phases (the Fourier transform of relative mass shifts). Structure factor magnitudes are all that one can obtain from recorded diffraction intensities, on the other hand, and the phase interactions of the scattered beams are lost. (However, an intensity transform, assuming all-zero phases, can also be useful, as will be shown.) Using high-resolution electron micrographs as an independent source of crystallographic phase information, therefore, demonstrates a potential advantage of electron crystallography over x-ray crystallography. On the other hand, great care must be taken to ensure that all measured data are adequately near the single scattering approximation to guarantee the success of ab initio structure analyses (see Chapter 5).

In Chapter 2 it was shown that crystallographic phases are defined with respect to an allowed unit cell origin, generally (but not always) corresponding to the location of a principal space group symmetry element. It was also shown how space-group symmetry could be used to simplify the trigonometric part of the structure factor expression. Thus, the symmetry operators themselves will lead to relative crystallographic phase relationships between certain classes of reflections. In addition, Friedel symmetry requires that $\phi_h = -\phi_{-h}$.

4.1.1. Crystallographic Phases via Image Analysis

Procedures for the photography of high-resolution images for thin crystals at low-beam doses was discussed in Chapter 3. (Note that the term *resolution* will be used in many ways in this book, e.g., as already introduced in Chapter 1. Diffractionists often use the word to specify the limit of the measurable diffraction information in their recorded patterns, i.e., the $d^*_{\max}$ of the smallest Bragg spacing. The intensity of the diffraction envelope or data sampling to this limit also has a practical impact on the meaning of this resolution definition (Glaeser and Downing, 1992). For the resolving power of the diffraction experiment, in terms of point to point detail observed when phases are supplied to the amplitudes, the number $0.61/d^*_{max}$ has been given by James (1950), based on the Rayleigh criterion. For a phase contrast image taken in an electron microscope at optimum Scherzer defocus, the quantity $0.66(C_s\lambda^3)^{1/4}$ is often used (Cowley, 1988). Problems with these definitions have been discussed by Spence (1980).

It will also be assumed that the actual objective lens phase contrast transfer function can be deconvoluted from the experimental image using one of the methods discussed in Chapter 3. After a pixel array of the experimental image is obtained, image analysis is employed to provide the best average representation of the unit cell repeat. An average representation of the unit cell is generally obtained by either of two ways: Fourier (peak) filtration (Misell, 1978) and correlation analysis (Frank, 1980).

Fourier peak filtration is the easiest image-averaging technique to carry out. The pixel array of the image (Figure 3.10) is Fourier-transformed to a diffraction pattern that contains amplitudes and phases (even though the intensity transform or power spectrum is displayed on the graphics screen). It is assumed that the continuous scattering information outside of the Bragg peaks is mostly due to lattice imperfections. Apertures of appropriately small diameter are placed over each diffraction peak to exclude the continuous signal, and the reverse Fourier transform is computed through this mask to produce an average representation of the structure repeat. (Alternatively, just the phase and amplitude at each Bragg peak center, in the best fit of the average reciprocal lattice, can be used for the reverse transform calculation, in lieu of a mask.) In early work, this averaging procedure was performed with an optical bench, wherein the mask was an opaque plate containing drilled or etched holes to pass the Bragg peaks at the back focal plane of a lens. Thus the average image was observed on a screen before it was photographed.

Extraction of crystallographic phases from the Fourier transform of an electron microscope image must refer to an unit cell origin permitted by the space group. Initially, an image density distribution $\rho(\mathbf{r})$ for a periodic object is commonly referred to an arbitrary origin, i.e., a general position which may be different from the ones specified by the plane group of this projection. If a translation $(\mathbf{r} + \mathbf{r}_o)$ is needed to locate the crystallographic phase origin specified by the space group, then, from the principles outlined in Chapter 1:

$$g(\mathbf{r}) = \rho(\mathbf{r}) * \delta(\mathbf{r} + \mathbf{r}_o) = \rho(\mathbf{r} + \mathbf{r}_o)$$

is sought. In reciprocal space, this transforms to:

$$G(\mathbf{s}) = \mathrm{F}(\mathbf{s}) \exp(2\pi i\, \mathbf{s}\cdot\mathbf{r}_o) = |F(\mathbf{s})| \exp [i(\phi_s + 2\pi i\, \mathbf{s}\cdot\mathbf{r}_o)$$

where ϕ_s represents the crystallographic phases and $2\pi i\, \mathbf{s}\cdot\mathbf{r}_o$ is the phase shift due to the origin translation. As discussed above, it is clear that the value of $|F(\mathbf{s})|$ does not change with image translation, but, for general origin locations, the measured phase values may be quite different from the, e.g., $0,\pi$ values expected for a centrosymmetric projection. Some image-processing computer software packages (e.g., Hovmöller, 1992) seek the best unit cell phase origin based on the presumed symmetry of the projection. The deviation of the average experimentally shifted phases can be compared to one another for various symmetry choices to see which combination is best fitted by the average distribution of image density. After deconvolution of the experimental objective lens transfer function, the resultant low-resolution crystallographic phase set can be combined with other values obtained by any of several techniques for structure solution, as will be outlined below.

Sometimes, the crystalline lattice in electron micrographs is deformed by paracrystalline disorder. In this case, the peak filtration procedure is not particularly useful for finding the highest-resolution phase information, since the distribution of orientations in the image will smear out the higher-angle reflections in the computed

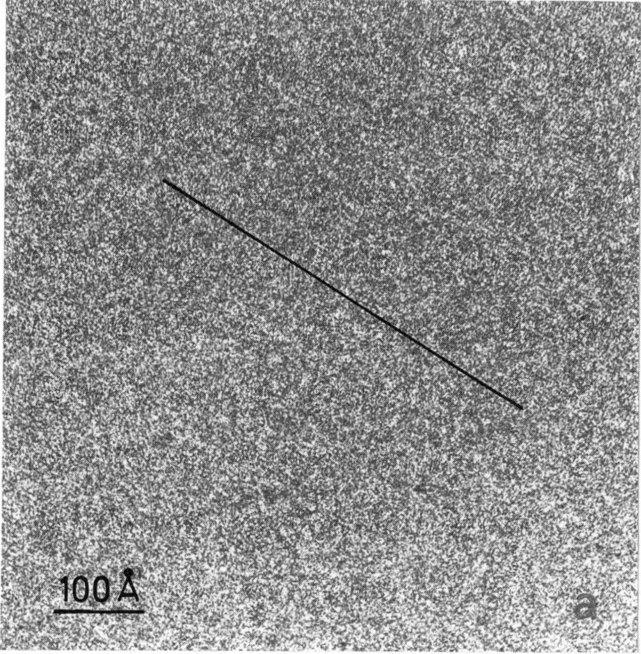

FIGURE 4.1. Image analysis of a low dose electron micrograph from an n-$C_{44}H_{90}$ monolayer crystal. (a) The electron micrograph shows no discernible contrast. (b) Optical transform of subareas reveals the average crystal lattice with different tilt orientations. (c) After correlation of these areas and summation of image subareas with the same tilt orientation, an average representation of the crystal lattice is obtained. (Reprinted from F. Zemlin, E. Reuber, E. Beckmann, and D. Dorset (1986) "High-resolution electron microscopy of beam-sensitive specimens: results with paraffin" *Proceedings of the 44th Annual Meeting of the Electron Microscopy Society of America,*, San Francisco Press, San Francisco, pp. 10–13; with kind permission of the San Francisco Press, Inc.) *(continued)*

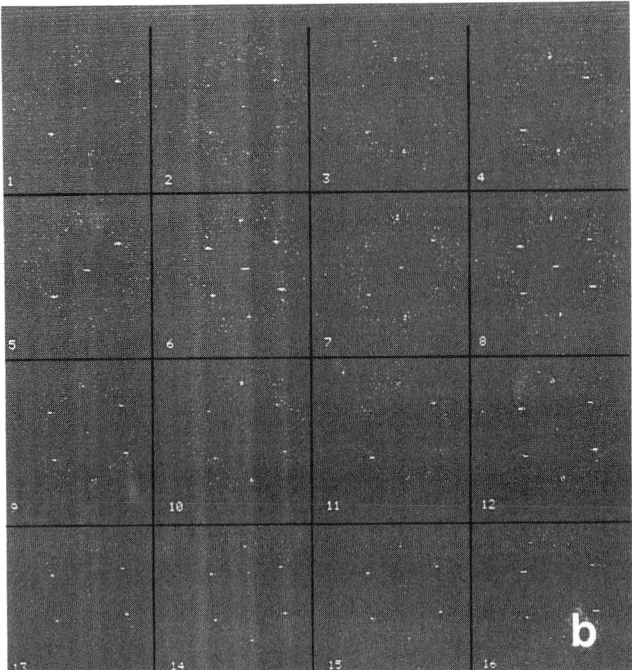

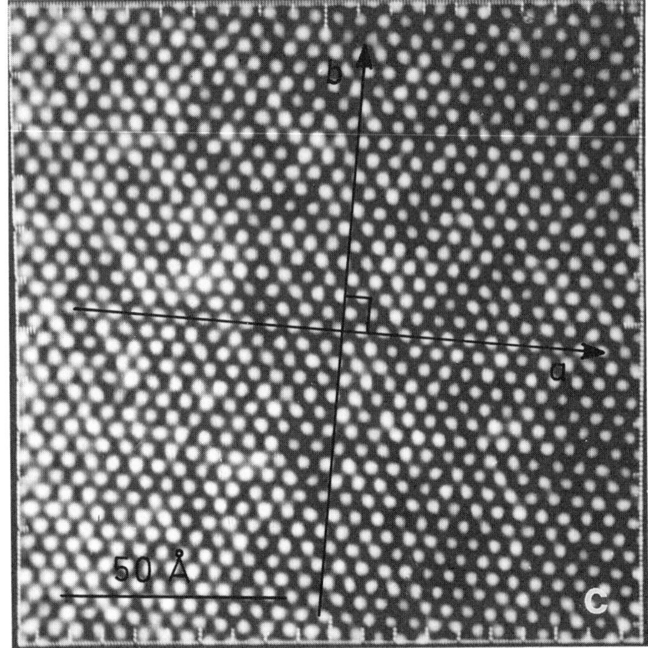

FIGURE 4.1. *(Continued)*

diffraction pattern. This problem is often encountered in micrographs of two-dimensional protein crystals (e.g., see Henderson et al., 1990). For a lattice that is not too badly deformed, an average image of a low-resolution structure can be obtained by peak filtration, as described. This initial low-resolution image, as a test object, can be compared to subareas of the crystalline regions in the micrograph. Rotational reorientation of the subareas is then employed to maximize the cross-correlation function:

$$H(h'k') = N^{-1} \sum_{h} \sum_{k} F(hk)\, G(h + h', k + k')$$

where F and G are Fourier transforms of the test object and image subareas, respectively. After reorientation, all realigned subareas can be combined to generate an average image, i.e., by superposition of the respective average unit cells of each area. This procedure was also important for averaging the first molecular resolution images of an n-paraffin crystal (Zemlin et al., 1985) since the low contrast of the original micrograph did not permit direct visualization of the crystalline lattice (Figure 4.1a). In this case, the computed diffraction pattern of subareas was used first to identify those regions having the same crystal orientation. Areas with the same pattern were then aligned to one another (Figure 4.1b). After the average image was generated (Figure 4.1c), it was then translated to an allowed unit cell origin, again, for extraction of crystallographic phases.

Other techniques for phase determination in electron crystallography are identical to those used in x-ray crystallography. That is to say, only the diffracted intensities are used for the analysis. Although lower-angle phase information from image transforms can be quite beneficial as supplementary data for facilitating such determinations, there are many cases where just the diffraction data are sufficient for an ab initio structure determination.

4.1.2. Trial-and-Error Methods

If only electron diffraction intensity data are available for a structure analysis, it is conceivable that a structure solution might be found by a trial-and-error guess of the atomic or molecular packing in the unit cell. With trial positions, the structure factors

$$F(\mathbf{s}) = \sum f_j \exp(2\pi i\, \mathbf{s} \cdot \mathbf{r}_j)$$

are conveniently computed, since the atomic scattering factors are also known. The success of a trial structure is then evaluated by the crystallographic residual (or R-factor):

$$R = \sum ||F_o| - k|F_c|| / \sum |F_o|$$

where each $|F_o|$ is experimentally observed and each $|F_c|$ is a calculated structure factor for a trial model. As shown by Wilson (1950b), if atoms are placed randomly in the unit cell, this figure of merit assumes values 0.83 and 0.59, respectively, for centrosymmetric or noncentrosymmetric space groups. Stout and Jensen (1968) have stated that, as a rough guide, $R = 0.45$ might indicate a structural model that has merit. When the value 0.35 is reached, a model has been obtained that can refine to the correct structure and when 0.25 is calculated, the atomic positions should be within 0.1 Å of their ideal values. The difficulty in using the trial-and-error approach is that, for any unit cell, there are an infinite number of possible packing arrays for a given molecule or cluster of atoms that will fill the unit cell. However, not all of these are energetically reasonable (Kitaigorodsky, 1973).

In any crystal structure analysis, there is also the question of arriving at a unique solution. Hosemann and Bagchi (1962) have argued that homometric solutions are sometimes possible, i.e., atomic arrays that yield the same autocorrelation function and, hence, the same intensity transform. The most trivial example is the Babinet structure (see discussion in Lipson and Lipson (1969)), which is the negative of the structure responsible for the diffraction pattern (so that all phase values are shifted by π). The constraints of atomicity and positivity in any structure (assumptions often valid in electron crystallography, particularly for van der Waals molecules, and always valid in x-ray crystallography) remove the possibility of a Babinet solution being accepted at high resolution. Other types of homomorphs may be possible, especially for relatively simple structures, but, fortunately, they seem to be rare.

The trial-and-error approach to structure analysis is useful only if adequate constraints can be placed on the molecular or group orientations in the unit cell. For example, there may be a space-group symmetry operator that is shared by the molecule (or cluster) point group, so that its location at a special position in the unit cell is fixed. For most organic structures, this constraint may not be as important as it might be for highly symmetric inorganic structures (see Table 2.5, for example). Therefore, a Patterson function (see below) may also be needed to determine the molecular orientation in the cell. However, centers of symmetry and mirror planes can be useful sometimes for finding fixed reference points for molecular packing (see Chapter 2).

If the molecular orientation is known with respect to the unit cell origin, searches for a lattice energy minimum can be carried out via atom–atom potential functions of the form (Pertsin and Kitaigorodsky, 1987):

$$\varepsilon = Ar^{-n} - Br^{-m}$$

where A and B are constants for repulsive and attractive contributions, $m < n$, and r is the interatomic distance for any atom pair. This approach has been often used in the electron diffraction structure analysis of linear polymers. The polymer chain direction is known and a minimum is sought for the conformational energy of the chain subunits. For such analyses, the x-ray crystal structures of monomeric and oligomeric units are determined to discern what parts of the polymer repeat can be assumed to remain rigid (e.g., Remillard and Brisse, 1982a,b). After the linkage of the molecular chain is defined, flexible junctions are identified for these subunits, around which twists are permitted (Brisse, 1989; Perez and Chanzy, 1989) and these conformational angles are varied while the internal nonbonded potential energy is monitored. Any conformational energy minimum that also corresponds to an R-factor minimum is thus judged to be a likely structure solution.

A generalization of the internal energy minimization for an arbitrary molecular packing is difficult in three dimensions, although some progress has been made (Gavezzotti, 1991). The ab initio analysis of layer packings, based on the principles of close packing (Kitaigorodskii, 1961), has been more successfully realized in recent years (Scaringe, 1991, 1992). Since electron diffraction analyses can be carried out for two-dimensional crystals, this is a significant advance. Scaringe (1992), for example, has been able to determine the structure of a Langmuir–Blodgett film of a phthalocyanine derivative, based on the close-packing principles worked out by him, using additional constraints provided by a small amount of electron diffraction data from the layer.

Translation functions can also be employed for structure searches. For example, in the analysis of a phospholipid lamellar packing, a conformational model of the structure (or maybe two choices) may be suggested (Dorset et al., 1987) from the x-ray crystal structure of a similar molecule (e.g., Elder et al., 1977) (Figure 4.2). (If there are two possible "head-group" conformations (Figure 4.3a), the most likely might be identified by comparison of Patterson functions (see below).) Adjustments are made for the linkage of the polymethylene chain to the glycerol backbone of the lipid and also for the length of the chain. The model is then translated past the unit cell origin and the progress of the structure search monitored by calculation of the crystallographic R-factor (Figure 4.3b).

Stout and Jensen (1968) have also proposed an efficient translational search technique for centrosymmetric structures. Consider a form of the structure factor equation written as

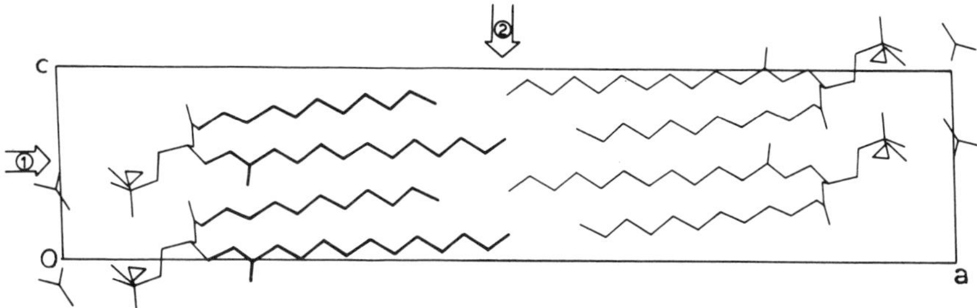

FIGURE 4.2. Crystal structure of 1,2-dilauroyl-*rac*-glycerophosphoethanolamine (Elder et al., 1977). A model for a phospholipid bilayer can be based on the conformational geometry of this compound, with allowances made, e.g., for different chain linkages and lengths. The trial molecule can then be translated in the unit cell (direction 1), assuming centrosymmetry, to seek a best match to the observed lamellar intensities. (Reprinted from D. L. Dorset (1983) "Electron crystallography of alkyl chain lipids: identification of long chain packing," *Ultramicroscopy* **12**, 19–28; with kind permission of Elsevier Science B.V.)

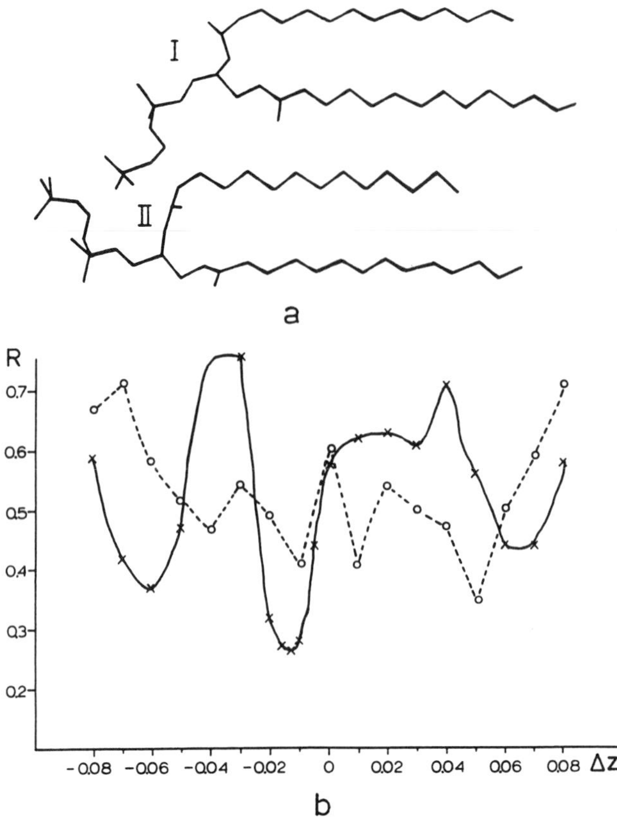

FIGURE 4.3. (a) Two possible head-group conformations for a diacyl phosphatidylcholine based on the crystal structure of Pearson and Pascher (1979). (b) When models of 1,2-dihexadecyl-*sn*-glycerophosphocholine are constructed, based on these two choices, two plots of the crystallographic residual R result as the trial structures are translated along z in the unit cell. The lowest R-factor corresponds to headgroup conformation II. (Reprinted from D. L. Dorset (1988) "Two untilted lamellar packings for an ether-linked phosphatidyl-N-methylethanolamine. An electron crystallographic study," *Biochimica et Biophysica Acta* **938**, 279–292; with kind permission of Elsevier Science B.V.)

$$F_{hkl} = \sum_j f_j \exp 2\pi i[h(x_j + x_a) + k(y_j + y_a) + l(z_j + z_a)]$$

$$+ \sum_j f_j \exp 2\pi i[h(-x_j - x_a) + k(-y_j - y_a) + l(-z_j - z_a)]$$

Here coordinates with subscript a refer to an arbitrary origin and those marked by j subscripts refer to atoms defined with respect to this origin. After factoring a common term:

$$F_{hkl} = \exp 2\pi i(hx_a + ky_a + lz_a) \sum_j f_j \exp 2\pi i[h(x_j) + k(y_j) + l(z_j)]$$

$$+ \exp -2\pi i(hx_a + ky_a + lz_a) \sum_j f_j \exp -2\pi i[h(x_j) + k(y_j) + l(z_j)]$$

Now let $b = \cos 2\pi(hx_a + ky_a + lz_a)$, $c = \sin 2\pi(hx_a + ky_a + lz_a)$. Then:

$$B = \sum_j f_j \cos 2\pi[h(x_j) + k(y_j) + l(z_j)]$$

$$C = \sum_j f_j \sin 2\pi[h(x_j) + k(y_j) + l(z_j)]$$

From the factored expression above we can obtain:

$$F_{hkl} = (b + ic)(B + iC) + (b - ic)(B - iC) = 2(bB - cC)$$

Since only the local origin needs to be changed, shifts only have to be applied to b and c instead of the whole set of atomic coordinates.

There are significant problems with this approach to structure analysis, however. When the crystallographic residual is used as the sole figure of merit to find which possible solution is best, it is statistically significant only when the number n of observed intensity data greatly exceeds the number of variable parameters p. As shown by Hamilton (1964), the quantity

$$\Re_{n,\, n-p,\, \alpha} = [(p/n - p)\, D_{n,\, n-p,\, \alpha} + 1]^{1/2}$$

must be evaluated. Here D is an expected distribution for the data and α is a confidence level for claiming that a particular solution is correct. If there are two possible structure solutions, giving residuals R_1 and R_0, where R_0 is the smaller quantity, then the criterion $R_1 > R_0 \Re$ must be satisfied before the second model can be rejected. For small data sets, such a rejection may require an eased confidence level, e.g., a statement that the model has one chance in ten of being incorrect, rather than one chance in a hundred.

4.1.3. Patterson Function

The Fourier transform of the diffracted intensity $I(s) = F(s)\cdot F(s)^*$ is the autocorrelation function, $\rho(r)\otimes\rho(r)$, of the crystal potential, as we have seen in Chapter 1. This Patterson function can also be expressed by the intensity transform

$$P(uvw) = V^{-1} \sum I(hkl) \cos 2\pi(hu + kv + lw)$$

and is immediately seen to be always centrosymmetric, with all terms assigned a phase value of zero. At atomic resolution, this autocorrelation function is equivalent to

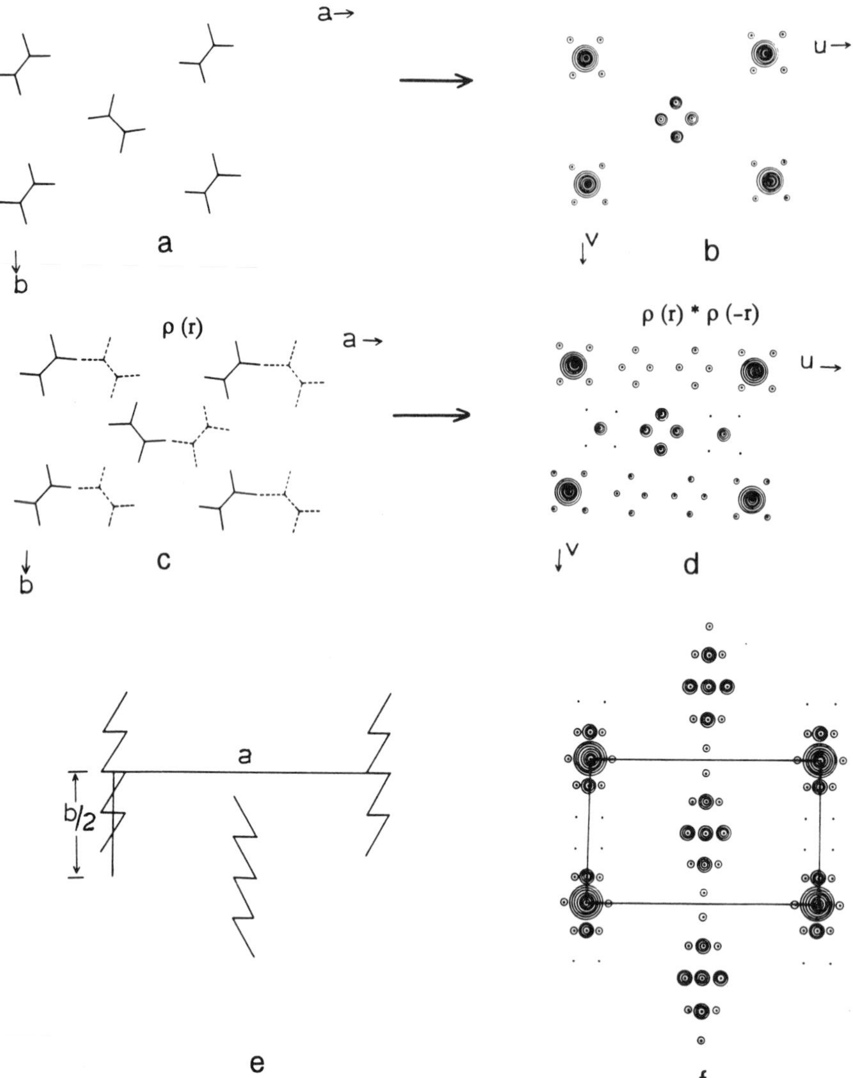

FIGURE 4.4. Representative Patterson functions for paraffin layers in a view onto the chain layer: (a) rectangular monolayer and (b) its Patterson function; (c) rectangular bilayer and (d) its Patterson function; (e) oblique monolayer and (f) its Patterson function. (Reprinted from D. L. Dorset (1985) "Crystal structure analysis of small organic molecules," *Journal of Electron Microscopy Technique* **2**, 89–128; with kind permission of Wiley–Liss Division, John Wiley and Sons, Inc.)

translation of all interatomic vectors of the unit cell to a common origin. For example, the Patterson function of a projected paraffin layer packing in plane group *pgg* is shown in Figure 4.4b. Note that this map itself has symmetry *pmm*. This map can also be generated by translating one unit cell by another while retaining the common orientation. The reference atom of one structure is overlapped on each atom of the second structure and all atomic peaks in the second cell are then mapped on the piece of paper. The largest peak will be the origin, since it has N overlaps when N is the number of atoms in the unit cell. To illustrate how a Patterson function is interpreted in this projection, the pgg plane group has equivalent positions at (x,y), $(-x,-y)$, $(1/2 + x,1/2 - y)$, and $(1/2 - x,1/2 + y)$. By vectorial subtractions of equivalent points, there is the above-mentioned major peak at (0,0), which is of no use for structure determination, as well as peaks at $(2x,2y)$, $(1/2 + 2x,1/2)$, and $(1/2,1/2 + 2y)$, from which atomic coordinates can be obtained. In general, traces of space-group symmetry are expressed in two or three dimensions by so-called Harker lines and planes, where major clusterings of Patterson vectors can be found. For example, in space group $P2_1$, equivalent positions are located at (x,y,z) and $(-x,1/2 + y,-z)$. The Harker plane occurs at $(u,1/2,w)$ because of the vectorial subtractions and the corresponding atomic coordinates are found at $u = 2x$, $w = 2z$.

In addition to its being the major means for determining crystallographic phases in the early days of x-ray crystallography (Buerger, 1959) (it remains so for finding positions of heavy atom derivatives in protein crystallography (Blundell and Johnson, 1976)), the Patterson function has also been often employed for structure analyses in electron crystallography (e.g., for the determination of layer silicate (Zvyagin, 1967) and methylene subcell (Dorset, 1976c) structures). While its use for finding group orientations in light atom structures is still relevant, the major difficulty for its application to electron crystallography is the poor detectability of heavy atom positions. This problem arises because the relative scale of electron diffraction scattering factors is compressed compared to that for x-ray form factors (Doyle and Turner, 1968). It is generally desirable that a heavy atom in a structure can be detected at the level $(\Sigma f^2_{\text{heavy}}/\Sigma f^2_{\text{light}}) \approx 1.0$ (Stout and Jensen, 1968). For the structure analysis of copper DL-α-alaninate from electron diffraction data (D'yakon et al., 1977), for example, this ratio is only 0.47 compared to 2.36 for x-ray diffraction intensities. It is even less favorable for locating heavy atom positions in protein structures (Glaeser, 1985).

As mentioned above, Patterson functions can be very useful for finding a likely molecular orientation in the unit cell, particularly if the molecule is oblong, e.g., a chain, or contains rings of atoms. In the case of the phospholipid layer packing analyses described above, a one-dimensional Patterson map (Dorset, 1987a) can be useful in choosing between two possible polar group conformations represented in the x-ray crystal structure (Dorset, 1987b) (Figure 4.5). Only one compares well with the intensity transform of the experimental sets of data.

4.1.4. Direct Phasing Methods

In a sense, the use of electron microscope images for crystal structure analysis may be considered a direct method for deriving crystallographic phases. On the other hand, the concept of finding enough phase information from just the diffracted intensities to solve a crystal structure was initially regarded with considerable skepticism by the crystallographic community. It was thought at first that the interpretation

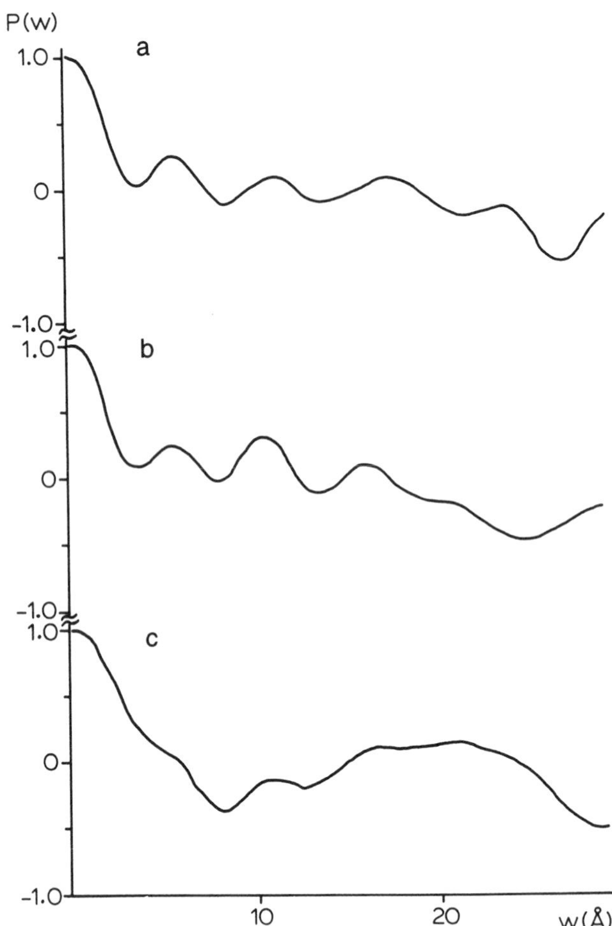

FIGURE 4.5. One-dimensional Patterson functions for 1,2-dihexadecyl-*sn*-glycerophosphocholine: (a) calculated from experimental lamellar electron diffraction intensities. (b) calculated from a model based on headgroup conformation II in Figure 4.3a. (c) calculated from a model based on headgroup conformation I in Figure 4.3a. The close match between (a) and (b) is clear. (Reprinted from D. L. Dorset (1988) "Two untilted lamellar packings for an ether-linked phosphatidyl-N-methylethanolamine. An electron crystallographic study," *Biochimica et Biophysica Acta* **938**, 279–292; with kind permission of Elsevier Science B.V.)

of Patterson maps was the only way that observed intensity data could yield any information about the crystal structure. Fortunately, this criticism is not correct. It will be shown how the structure factors magnitudes themselves can be used in a direct way to determine crystal structures without recourse to any other information. The various direct phasing techniques used in modern x-ray crystallography, therefore, are also appropriate for electron crystallographic analyses.

In x-ray crystallography, the possibility of direct phasing was seen from the Harker–Kasper (1947, 1948) inequality. For a centrosymmetric structure, the structure factor can be defined:

$$F_h = V\int \rho(\mathbf{r}) \cos 2\pi \mathbf{h}\cdot\mathbf{r}\, d\mathbf{r}$$

Use of Schwartz's inequality

$$\left|\int fg\, dt\right|^2 \leq \left(\int |f|^2\, dt\right) \left(\int |g|^2\, dt\right)$$

leads to the expression:

$$F_h^2 \leq V^2 \left[\int \rho\, (\mathbf{r})\, d\mathbf{r}\right] \left[\int (1/2)\, \rho(\mathbf{r})\, \{1 + \cos 2\pi(2\mathbf{h}\cdot\mathbf{r})\}\, d\mathbf{r}\right]$$

via the identity $2 \cos^2\alpha = 1 + \cos 2\alpha$.

Since, for x-rays, atomic number $Z = V\int \rho(\mathbf{r})\, d\mathbf{r}$, then

$$F_h^2 \leq Z((1/2)Z + (1/2)F_{2h})$$

or, defining the unitary structure factor, $U_h = F_h/Z$:

$$U_h^2 \leq (1/2) + (1/2)\, U_{2h}$$

Hence, if U_h^2 is greater than 1/2, then U_{2h} must be positive, etc. Similar relationships can be found for other symmetry operations on zonal reflections. Generally, this result requires some reflections with very high intensity for the inequality to be useful.

4.1.4.1. Sayre Equation. The Sayre (1952) equation:

$$F_h = (\theta/V) \sum_k F_k F_{h-k}$$

where $h = h, k, l$ and $k = h', k', l'$, was one of the first relationships derived for the determination of crystallographic phases from structure factors. This convolution of phased structure factors is weighted by a function of the atomic scattering factors θ. For structures containing just one atomic species the expression is exact, so long as the atomic positions are not overlapped. Since the relative range of electron scattering factors is narrower than for x-rays, it may be of greater potential use for electron diffraction applications, because the additional condition that $\rho(\mathbf{r})$ be everywhere positive is also often satisfied, particularly for covalent molecules.

To use the Sayre equation, it is apparent that some phase information should be known a priori. In electron crystallographic applications, this primary phase information may well be obtained from an electron microscope image, as discussed above, or even from a partial determination with other direct phasing methods, shown below. Thus, the method is best suited for resolution enhancement or phase refinement. For calculation, phases are assigned to structure factor magnitudes for a basis set and zero values are given to amplitudes and phases for indexed reflections within the resolution limit to be considered. If more than one cycle is required to reach the resolution limit of the electron diffraction data, then, within the intermediate shell of newly phased data, structure factor magnitudes are combined with the phase values and the convolution is repeated until all data are accessed. The condition of positivity required in this procedure suggests that the value of the zero order term $F_{000} = \Sigma f_{i(\sin\theta/\lambda=0)}$ be used in the convolutions.

Practical applications of the Sayre equation in electron crystallography were first made mostly by Fan (1991) and his collaborators. For example, given phases from an ideal 2.0-Å resolution image of copper perchlorophthalocyanine (Uyeda, 1978–1979), it was possible to extend to the 1.0 Å resolution of the electron diffraction pattern, using simulated data, to find a structure where practically all light atom positions could be clearly discerned (Liu et al., 1988). More recently, we have shown that this procedure is useful when phases from an experimental 2.3-Å resolution image of this compound, taken at 500 kV, are used to expand to the 0.9-Å resolution of electron diffraction amplitudes obtained at 1200 kV, especially when this is followed by Fourier refinement (see below). This phase extension was successful even with phase errors in the basis set due to uncorrected astigmatism (i.e., leading to a nonradial transfer function). Of course, starting-phase data from ideally corrected images with no errors were more convenient for finding most of the structure in the initial maps. It is also possible to use the partial phase set from the evaluation of phase invariant sums (see below) in the Sayre equation.

The Sayre equation has also been used in four dimensions to solve the structures of incommensurate-phase superconducting materials (Mo et al., 1992). Phases of the most intense reflections from the average structure were found from the electron microscope image (Fan, 1993; Li, 1993) and the superlattice reflections of the incommensurate phase were related to these reflections in a hyperspace unit cell so that the diffraction pattern would be periodic in four dimensions (see Amelinckx and Van Dyck, 1993b). The resultant fourth Miller index m is used to label the superlattice reflections in the reciprocal hyperspace. Given phases of the strong reflections, the superlattice reflections are also assigned values to permit visualization of the structural modulation in the resultant potential map.

Tests of the Sayre equation have also been made for other small organic structures. For example, if a partial phase set is obtained for a paraffin layer by other means (Dorset, 1992a), the complete phase set will be retrieved from the Sayre equation; again, missing reflections can, initially, be either assigned a zero amplitude or a measured amplitude with zero phase angle. The convolution is repeated until the solution converges to a stable phase set. Tests on partly phased data sets from linear polymers indicate that the Sayre equation will be useful if the starting set is large enough. Most important is the accessibility of new phases via the convolution operation from the partial phase set for the success of this endeavor. On the other hand, attempts to use this method to refine phases for one-dimensional lamellar structures, diffracting, e.g., to 3.4 Å, were only partly successful, owing to the limited data resolution (Dorset, 1991a). Successful applications of the Sayre equation to protein structures will be discussed in Chapter 12.

Another interesting application of the Sayre equation has been made for deconvolution of the transfer function from an experimental image. In Chapter 3, it was seen that the Fourier transform of an experimental image can be written:

$$T(\mathbf{s}) = \delta(\mathbf{s}) + 2\sigma F(\mathbf{s})[\sin\chi_1(\mathbf{s})\exp(-\chi_2(\mathbf{s}))]$$

If different values of Δf are used to produce different test values of F_h^{calc} from $T(\mathbf{s})$, they can then be used in the Sayre equation to produce an expanded set of structure factors that can be compared directly to the magnitudes $|F_h^{obs}|$ observed in the electron diffraction pattern (Han et al., 1986).

4.1.4.2. Phase Invariant Sums (Symbolic Addition). For ab initio structure analyses based on electron diffraction intensity data, the evaluation of phase invariant sums has been found to be especially fruitful. For any measured diffraction data set with Miller indices $\mathbf{h}_i = h_ik_il_i$ it is possible to define linear combinations of phases $\phi_{h,(i)}$, i.e.:

$$\psi = \phi_{h(1)} + \phi_{h(2)} + \phi_{h(3)} + \ldots$$

If the constraint $\Sigma\mathbf{h}_i = \Sigma h_ik_il_i = 0,0,0$ is also satisfied, then the sum of phases is known as a *structure invariant*, i.e., it is true no matter where the unit cell origin is defined. These are, in fact, simultaneous equations in phase. As originally proposed by Hauptman and Karle (1953), since there are more equations than unknowns, it may be possible to solve these equations for new phase values (see also Hauptman, 1972).

In order to use such equations to find new phase values, it is first required that a quantity can be predicted for the linear combination of phases $\psi = 0,\pi$. This can be done as follows. Suppose the structure is made up of point scatterers. It would be possible to find normalized structure factors $|E_h|$ from the observed intensities if a correction were made for the scattering factor falloff. Thus:

$$|E_h|^2 = I_h^{obs} / \varepsilon \sum_i f_i'^2$$

The factor ε compensates for special classes of reflections, e.g., those affected by translational symmetry (i.e., those subsets containing space-group forbidden reflections) (see Luzzati, 1972). The scattering factors are also corrected for an overall Debye–Waller factor B_{iso}, which can be determined from a Wilson (1942) plot. Thus:

$$f' = f\exp(-B_{iso}\sin^2\theta/\lambda^2)$$

A plot of $\ln(\langle I_h^{obs} / \Sigma_i f_i^2\rangle)$ versus median values of $\sin^2\theta/\lambda^2$ for successive overlapping shells of reciprocal space will determine the temperature factor, i.e.:

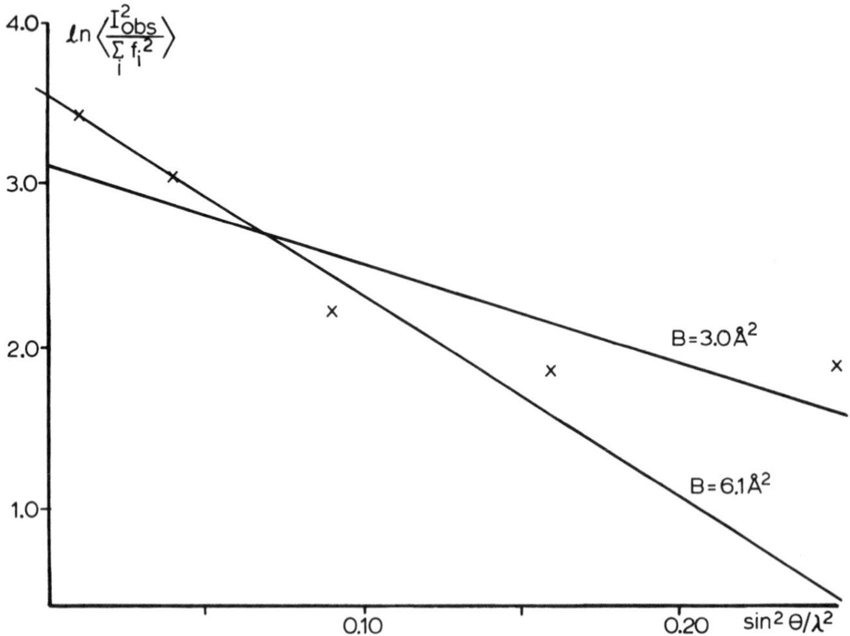

FIGURE 4.6. Experimental Wilson plot for poly(1,4-*trans*-cyclohexanediyl dimethylene succinate) based on electron diffraction data published by Brisse et al. (1984). If the highest resolution point is ignored, the overall temperature factor is $B = 6.1$ Å^2. If this point is included, then $B = 3.0$ Å^2.

$$\ln(<I_h^{\text{obs}} / \sum_i f_i^2 >) = \ln C - 2B_{\text{iso}}\,(\sin^2\theta/\lambda^2)$$

An example of a Wilson plot is shown in Figure 4.6. Resultant $|E_h|$ values are scaled so that $<|E_h|^2> = 1.000$.

There are many kinds of phase invariant sums but, typically, three are most commonly employed in electron crystallography. One that can be used cautiously is the Σ_1-triple (Hauptman, 1972), where

$$\psi = \phi_{h(1)} + \phi_{h(1)} + \phi_{-2h(1)}$$

Using the $|E_h|$ values associated with these reflections, one computes:

$$A_1 = (|E_h|^2 - 1)|E_{2h}|/\sqrt{N}$$

A far more useful three-phase invariant is the Σ_2-triple

$$\psi = \phi_{h(1)} + \phi_{h(2)} + \phi_{h(3)}$$

where $h(1) \neq h(2) \neq \mathrm{h}(3)$. For the equal atom case:

$$A_2 = (2/\sqrt{N})\ |E_{h(1)}E_{h(2)}E_{h(3)}|$$

is contained in the conditional probability distribution (Cochran, 1955):

$$P(\psi \mid |E_{h(1)}|, |E_{h(2)}|, |E_{h(3)}|) = K^{-1} \exp(A_2 \cos \psi)$$

where $K = 2\pi I_o(A_2)$ and I_o is a modified Bessel function. If A_2 is large, then the predicted value of ψ is sharply distributed around $\cos \psi = +1.0$. (For the Σ_1-triple, large positive values of A_1 can also be used to predict values of ϕ_{2h} near 0 and large negative values will find those near π.) It is interesting to note that the *concept* of the Σ_2-phase relationship can be derived from the Sayre equation (e.g., see Fan et al., 1991a). Starting with

$$F_h = \theta(\mathbf{h}) \sum_k F(\mathbf{k})\ F(\mathbf{h} - \mathbf{k})$$

multiplication of both sides by $F(-\mathbf{h})$ gives:

$$|F_h|^2 = \theta(\mathbf{h}) \sum_k\ F(-\mathbf{h})F(\mathbf{k})\ F(\mathbf{h} - \mathbf{k})$$

Since the right-hand side of this equation must be real:

$$|F_h|^2 = \theta(\mathbf{h}) \sum_k |F(-\mathbf{h})F(\mathbf{k})\ F(\mathbf{h} - \mathbf{k})| \cos(\phi_h - \phi_k - \phi_{h-k})$$

If $|F_h|^2$ is large and the triple product of amplitudes is also large, then (remembering the Friedel relationship for phase) the sum of phases must be near zero. (The derivation is not exact, however, since the three-phase invariant estimates are based on probabilistic criteria while the Sayre convolution is an exact expression for equal atom structures.)

Another useful phase invariant is the quartet:

$$\psi = \phi_{h(1)} + \phi_{h(2)} + \phi_{h(3)} + \phi_{h(4)}$$

(Green and Hauptman, 1976). A quantity B is $(2/N)\ |E_{h(1)}E_{h(2)}E_{h(3)}E_{h(4)}| \times [|E_{h(1)+h(2)}|^2 + |E_{h(2)+h(3)}|^2 + |E_{h(1)+h(3)}|^2 - 2]$.

It is used in a fashion similar to the A-values above, i.e.:

$$P(\psi \mid |E_{h(1)}|, |E_{h(2)}|, |E_{h(3)}|, |E_{h(4)}|, |E_{h(1)+h(2)}|, |E_{h(2)+h(3)}|, |E_{h(1)+h(3)}|) = K^{-1} \exp(B \cos \psi)$$

where $K = 2\pi I_o(B)$. If $B > 0$, then the conditional probability predicts phase values to be distributed around $\cos \psi = +1.0$, and when $B < 0$ (i.e., small "cross-terms"), then $\cos \psi = -1.0$ is predicted.

What these A and B values signify, in effect, is that the phase invariant sums can be ranked according to their probability of being correctly determined. Thus, not only do we have simultaneous equations in phase, they can be listed in order of their reliability.

In order to solve these equations for new phases, a few starting values are needed, hopefully those which will interact most often with in the most highly probable phase invariant sums. Such initial phases can be found in a process known as origin definition (see Hovmöller (1981), for a general procedure, or Rogers (1980) for tables appropriate for any space group). Consider plane group *pgg* (Figure 4.7), which is a projection of many important space groups found for organic structures (e.g., $P2_1/c$ and $P2_12_12_1$). We have already shown in Chapter 2 that, because of centrosymmetry, the structure factor expression reduces to:

$$F(h,k) = 2 \sum_j f_j \cos 2\pi(hx + ky))$$

There are four equivalent choices of origin for this space group, corresponding to the positions of the twofold symmetry operators at (0,0), (1/2,0), (0,1/2) and (1/2,1/2). Suppose the origin is moved from position 1 to position 2 at (1/2,0). The new structure factor expression will be written:

$$F'(h,k) = 2 \sum_j f_j \cos 2\pi(hx + h/2 + ky)$$

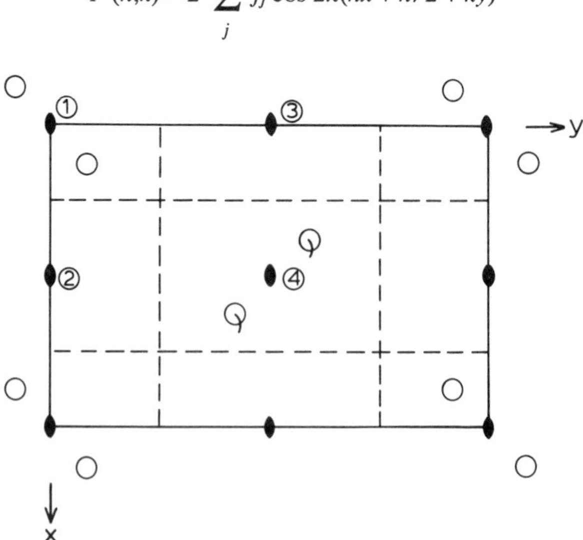

FIGURE 4.7. Plane group *pgg* with the origin defined at the twofold axis. Four equivalent choices of origin are possible for this plane group. (Reprinted from D. L. Dorset (1994) "Electron crystallography of organic molecules," *Advances in Electronics and Electron Physics*, **88**, 111–197; with kind permission of Academic Press, Inc.)

Table 4.1. Effect of Origin Shifts in Plane Group pgg on the Phase of the Structure Factor*

		Phase change for Miller index parity			
Origin	Shift	gg	*gu*	*ug*	*uu*
1	0	0	0	0	0
2	$a/2$	0	0	π	π
3	$b/2$	0	π	0	π
4	$(a+b)/2$	0	π	π	0

*See Figure 4.7.

$$= 2\sum_j f_j \cos[2\pi(hx+ky)+\pi h]$$

$$= 2\sum_j f_j\,[\cos 2\pi(hx+ky)\cos(\pi h)$$

$$-\sin 2\pi(hx+ky)\sin(\pi h)]$$

Now $\cos(h\pi) = (-1)^h$ and $\sin(h\pi) = 0$ for integral h. Thus $F'(hk) = (-1)^h F(hk)$ so that an origin shift of $a/2$ will change the phases of odd h index reflections by π, but not the values of even h index reflections. (The result of shifting this plane group origin to (1/4,0), as in a projection of $P2_12_12_1$, was discussed earlier in Chapter 2; see Figure 2.13.) Similar deductions can be made for the other possible origin positions as reviewed in Table 4.1. Note that, when the Miller indices are both even, nothing happens to the phase values; these reflections are, themselves, structure invariants, commonly termed semi-invariants. However, reflections with at least one odd index ($u \equiv$ ungerade $\equiv$ odd) can undergo a phase shift due to the origin change. If two such reflections with different index parity (which do not add to a combination (g,g), where $g \equiv$ gerade $\equiv$ even) are assigned phase values ab initio, then the origin is defined for this projection. A third phase value with differing index parity, but not leading to a (g,g,g) combination, can be defined for simple primitive space groups when three-dimensional data are available. The origin-defining reflections need not be zonal, phase-restricted terms, but it is convenient to use them if a noncentrosymmetric structure is being determined. These starting phase values are used algebraically with the most probable structure invariant sums to determine new phase terms. If not enough relationships are found, then it is also possible to define additional phase terms algebraically, i.e., $\phi_{h(i)} = a$, $\phi_{h(j)} = b$, etc., with no strictures placed on the Miller index parities. If n such terms are needed, then the solution is sought by calculation, for centrosymmetric structures, of 2^n potential maps, where each $\phi_{h(n)}$ unknown is permuted through $0,\pi$.

It is important to interject here that the origin definition described above illustrates the most convenient case where specific *points* in the unit cell (e.g., centers of symmetry) can be used for this purpose. At times, the space-group symmetry can affect the positions of the unit cell origin so that they are somewhat different from those found for the separate plane group projections. This was already illustrated in Chapter 2 for *pgg* projections in space group $P2_12_12_1$. Also, the origin definition may be much less specific than those defined on centers of symmetry. For example, the space group $P2_1$ (where **b** is the unique axis) allows a specific origin definition to be made only in the **ac** plane, but any point along the **b**-axis (parallel to the 2_1-screw) is permitted as an origin. For *Pm*, the origin lies anywhere on the mirror plane, etc. Nevertheless, the procedures just described are still valid, even for such space groups, but the table listed by Rogers (1980), as well as the International Tables for X-Ray Crystallography, should be consulted to determine which reflection classes are permitted to specify the unit cell origin.

To illustrate how the evaluation of phase invariant sums can be applied to electron crystallography, an *hk*0 intensity data set taken from chain-folded lamellae of the polysaccharide chitosan $[C_6H_{11}O_4N]_n$ by Mazeau et al. (1992) can be considered. The cell constants are $a = 8.07$, $b = 8.44$, $c = 10.54$ Å and the space group is $P2_12_12_1$. It was mentioned already how the origin shift for *pgg* in this projection (Figure 2.13) leads to phases $0,\pi$ when h is even and $\pm\pi/2$ when h is odd. There are still four equivalent unit cell origins as before, so that the origin can be defined in the way just described. (Indeed, it is also correct to solve this structure in the usual *pgg* representation but, if there are three-dimensional data available, it is better to remain with an origin choice that is consistent with the space-group symmetry.) Friedel's law requires that $\phi_{hkl} = -\phi_{-(hkl)}$. Also, for $h\bar{k}l$ reflections, $\phi_{hkl} = \pi - \phi_{h-kl}$ when $h + k = 2n$, $k + l = 2n + 1$. (Other relationships of this kind are listed in the International Tables for X-Ray Crystallography but also can be easily derived using the principles outlined above.)

A list of observed *hk*0 $|E_h|$ values is given in Table 4.2. (For the computation of $|E_h|$ from the original set of 22 $|F_h|$, $\varepsilon = 2$ for *h*00 and 0*k*0 reflections to compensate for the systematic absences due to the 2_1-screw axes. Otherwise, $\varepsilon = 1$.) The phase

Table 4.2. Experimental Electron Diffraction Data for Chitosan*

*hk*0	$\|F\|$	$\|E\|$	*hk*0	$\|F\|$	$\|E\|$
020	3.92	2.17	240	0.46	0.62
040	0.43	0.37	250	0.33	0.55
110	2.63	1.89	310	0.19	0.19
120	2.54	2.08	320	0.34	0.38
130	1.61	1.60	330	0.19	0.24
140	1.01	1.24	340	0.19	0.29
150	0.78	1.20	400	1.34	1.18
200	0.48	0.27	410	0.19	0.24
210	0.35	0.29	420	0.37	0.51
220	0.50	0.45	430	0.19	0.30
230	0.39	0.43	510	0.33	0.54

*Collected by Maseau et al. (1992). $a = 8.07$, $b = 8.44$, $c = 10.54$ Å, space group $P2_12_12_1$.

determination progresses as follows: the origin is defined by assigning values $\phi_{110} = \pi/2$, $\phi_{120} = \pi/2$. This is allowed, since $110 + 120 = gu0$. Looking through the list of high-probability Σ_2-triples, it is required to use an algebraic unknown to access enough reflections. Thus, it is possible to state a priori that $\phi_{140} = a$. We can now start a process known as *symbolic addition*:

A_2	h_1	h_2	h_3	Conclusion
1.69	020	120	–(140)	
		$\pi/2$	$-a$	$\phi_{020} = a - \pi/2$
1.98	020	110	–(130)	
	$a - \pi/2$	$\pi/2$		$\phi_{130} = a$
1.26	020	130	–(150)	
	$a - \pi/2$	a		if $a = \pm\pi/2$, $\phi_{150} = \pi/2$
0.56	110	130	–(240)	
	$\pi/2$	a		$\phi_{240} = a + \pi/2$
0.55	120	130	–(250)	
	$\pi/2$	a		$\phi_{250} = a + \pi/2$
0.51	120	110	–(230)	
	$\pi/2$	$\pi/2$		$\phi_{230} = \pi$
0.35	120	1–10	–(210)	
	$\pi/2$	$\pi/2$		$\phi_{210} = \pi$

A couple of positive quartet relationships are also useful:

B	h_1	h_2	h_3	h_4	Conclusion
0.77	020	–(110)	–(130)	220	
	$a - \pi/2$	$-\pi/2$	$-a$		$\phi_{220} = \pi$
0.63	020	–(110)	–(140)	230	
	$a - \pi/2$	$-\pi/2$	$-a$		$\phi_{230} = \pi$

While carrying out this procedure, phase invariants sums were used above threshold values for A and B, defined by the self-consistency of the derived phases. The analysis leaves us with 11 phases from the original data set of 22 reflections and these are linked through one algebraic unknown, $a = \pm\pi/2$. Two potential maps must be calculated (Figure 4.8). Typical of zonal projections of polymers down the chain axes, there are numerous overlapping atoms so that it is impossible to resolve their individual positions. However, it turns out that one of the two solutions is a good density envelope for the structure determined by a conformational search (Mazeau et al., 1992, 1995). Just how this correct solution can be recognized a priori will be discussed later. However, we have reduced the possible choices of phase combinations from 2^{11} to 2^{1} by this direct phase determination.

Many times, fortunately, the atomic detail can be resolved. In the crystal structure analysis of urea by direct methods in plane group pmg, 15 Σ_2-triples and 8 quartets were evaluated. After origin definition, only four phases were assigned unequivocal

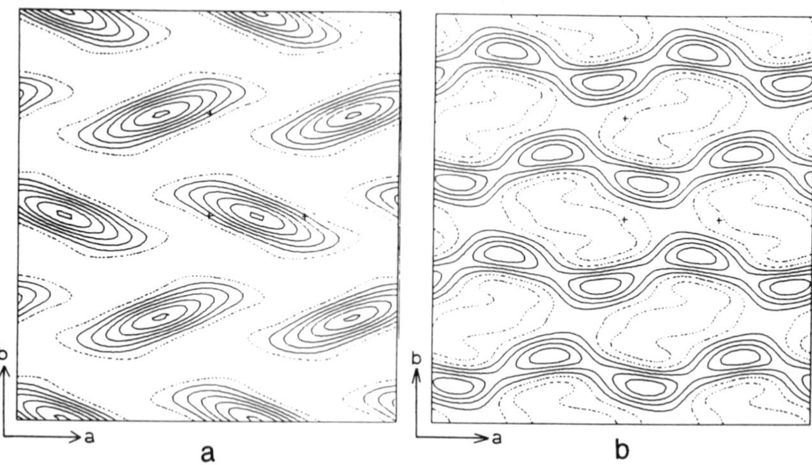

FIGURE 4.8. Two maps generated for alternate choices of phase ambiguity in the direct analysis of the chitosan structure (based on data of Mazeau et al. (1992)): (a) $a = \pi/2$; (b) $a = -\pi/2$. (Reprinted from D. L. Dorset (1994) "Electron crystallography of organic molecules," *Advances in Electronics and Electron Physics*, **88**, 111–197; with kind permission of Academic Press, Inc.)

values—again requiring the use of an algebraic unknown to find phase relationships among 9 other reflections. Permuting $a = 0,\pi$ to calculate two maps (Figure 4.9), the structure of the molecule (Dorset, 1991b) was recognized immediately from its bonding geometry. If atomic positions were taken from the potential map and used to calculate phases for all 60 measured data (Lobachev and Vainshtein, 1961), the definition of the maps was greatly improved. As will be seen in later chapters, the collection of three-dimensional data often ensures that the phase determination can be made without any ambiguity so that the structure is found in the first map.

Yet another example can be given for a one-dimensional data set from a phospholipid, 1,2-dihexadecyl-*sn*-glycerophosphoethanolamine (DHPE), obtained at 1000 kV and corrected for dynamical scattering. The list of structure factor magnitudes and derived $|E_h|$ values is given in Table 4.3. High resolution images taken on a cryoelectron microscope (see Chapter 10) have been used to provide experimental phase values for the first seven reflections (Dorset et al., 1990). Alternatively, it is possible to phase most of the reflections with just the $|E_h|$ values and the Σ_1- and Σ_2-triples assembled from them. First, the origin is defined by setting $\phi_{001} = 0$. There is only one type of reflection that can be used for this definition for a row of reflections and this particular odd-index maximum is chosen because it has a large $|E_h|$ value and a small Miller index; thus it will interact with many other reflections through the phase invariant sums. From the Σ_2-triples we can make the following conclusions:

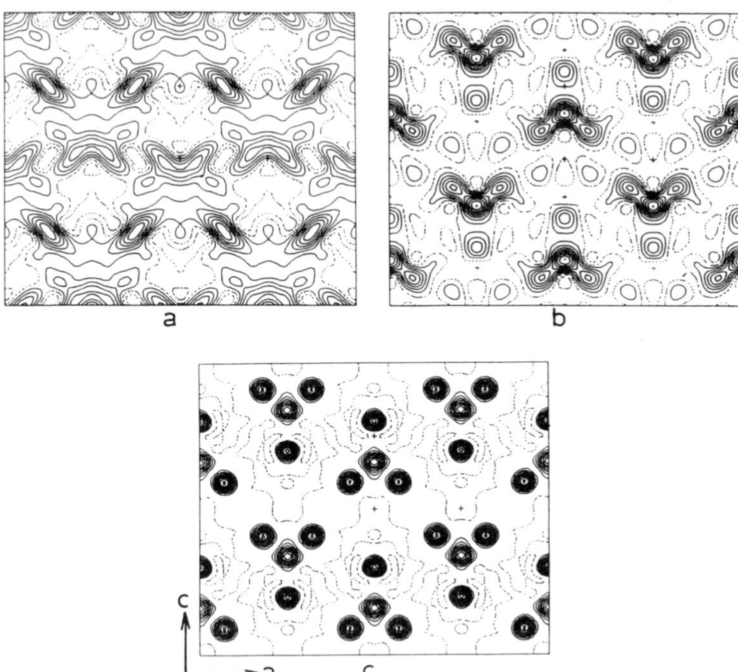

FIGURE 4.9. Phase ambiguity in the structure analysis of urea (based on data of Lobachev and Vainshtein (1961)): (a) incorrect phase choice; (b) correct phase choice in which the structure geometry is readily observed; (c) potential map after all phases are found by a structure factor calculation based on the atomic positions in (b). (Reprinted from D. L. Dorset (1991) "Is electron crystallography possible? The direct determination of organic crystal structures," *Ultramicroscopy* **38**, 23–40; with kind permission of Elsevier Science B.V.)

Table 4.3. Corrected Experimental Data for L-DHPE

l	$\|F\|$	$\|E\|$	l	$\|F\|$	$\|E\|$
1	1.33	0.92	9	0.21	0.52
2	0.34	0.47	10	0.96	0.84
3	1.30	0.96	11	0.95	0.93
4	0.61	0.55	12	1.72	1.52
5	0.93	0.72	13	1.45	1.51
6	0.45	0.43	14	1.75	1.55
7	0.73	0.59	15	1.19	1.49
8	0.34	0.16	16	1.00	1.24

A_2	h_1	h_2	h_3	Deduction
0.60	00,14	00,–1	00,–13	$\phi_{0014} = \phi_{0013}$
0.56	00,13	00,–1	00,–12	$\phi_{0013} = \phi_{0012}$
0.52	00,15	00,–1	00,–14	$\phi_{0015} = \phi_{0014}$
0.34	00,12	00,–1	00,–11	$\phi_{0012} = \phi_{0011}$
0.32	00,16	00,–1	00,–15	$\phi_{0015} = \phi_{0016}$
0.18	00,11	00,–1	00,–10	$\phi_{0011} = \phi_{0010}$
0.12	004	00,–1	00,–3	$\phi_{003} = \phi_{004}$
0.09	005	00,–1	00,–4	$\phi_{004} = \phi_{005}$
0.07	006	00,–1	00,–5	$\phi_{006} = \phi_{005}$
0.07	003	00,–1	00,–2	$\phi_{003} = \phi_{002}$
0.06	007	00,–1	00,–6	$\phi_{006} = \phi_{007}$

Negative Σ_1-triples have been found to be unusually informative for lipid lamellar data sets (Dorset, 1991a,b), for reasons that are not fully understood. Given this situation, we can evaluate the following phases:

A_1	h_1	h_2	h_3	Deduction
–0.16	006	006	00,–12	$\phi_{0012} = \pi$
–0.12	007	007	00,–14	$\phi_{0014} = \pi$
–0.11	008	008	00,–16	$\phi_{0016} = \pi$
–0.04	005	005	00,–10	$\phi_{0010} = \pi$
–0.03	004	004	00,–8	$\phi_{008} = \pi$

A list of phase assignments can be made for 15/16 reflections as reviewed in Table 4.4. Six of these are assigned the symbolic values a. The phase of a can be found from the Fourier transform of the electron microscope image. Alternatively, it is possible to evaluate other Σ_2-triples containing ϕ_{003}, from which we deduce this phase to be 0, thus resolving the ambiguity.

Finally, it should be pointed out that, when convergent beam diffraction techniques can be used (i.e., for sufficiently radiation-resistant samples), Σ_1- and Σ_2-triple

Table 4.4. Phases for L-DHPE*

l	ϕ	l	ϕ
1	0 (origin)	9	–
2	a	10	π
3	a	11	π
4	a	12	π
5	a	13	π
6	a	14	π
7	a	15	π
8	π	16	π

*From image analysis or evaluation of other Σ_2 phase invariants via ϕ_{003}: $a = 0$.

invariant relationships can be evaluated directly for noncentrosymmetric structures. Convergent beam intensity is sensitive to crystallographic phase via dynamical interactions when such phase invariant sets of reflections are excited in the microscope. Thus experimental measurements, where accelerating voltage is a variable, can be used to uncover phase information, as discussed, e.g., by Zuo et al. (1989) and reviewed by Spence and Zuo (1992). This has been particularly useful for finding accurate structure factor values for low-angle reflections.

4.1.4.3. Tangent Formula. Evaluation of individual phase invariant sums is most effectively applied to centrosymmetric structures. Instead of evaluating:

$$\phi_h \approx \phi_k + \phi_{h-k}$$

it is often possible to find a more reliable estimate of ϕ_h when all possible vectorial contributions to it within the data set k_r are considered:

$$\phi_h \approx \langle \phi_k + \phi_{h-k} \rangle_{k(r)}$$

In practice, this phase determination is made via the tangent formula (Karle and Hauptman, 1956):

$$\tan \phi_h = A/B$$

where:

$$A = \sum_{k(r)} w_h \, |E_k| \, |E_{h-k}| \sin(\phi_k + \phi_{h-k})$$

$$B = \sum_{k(r)} w_h \, |E_k| \, |E_{h-k}| \cos(\phi_k + \phi_{h-k})$$

The phase determination, as before, begins with origin definition to build up a basis set. Additionally, for noncentrosymmetric problems, it is also possible to specify an enantiomorph-defining reflection (e.g., see Rogers, 1980). Such a reflection has an appropriate index parity to enable it to form phase-invariant sums with one or more of the origin-defining reflections. The desired sensitivity to a change of enantiomorph, i.e., for the two chiral density distributions, leaves the invariant sum with a value near $\pm\pi/2$. This phase sum is not easily defined a priori for some noncentrosymmetric space groups. However, space group $P2_12_12_1$ is one of the most convenient ones for stipulating such an enantiomorph since, because of the unusual origin for plane group *pgg* (see Chapter 2), phase invariants can easily be found that add to the required values. To complete the basis set, a number of high $|E_h|$ reflections are given algebraic phase signs and allowed to be permuted through values $0,\pi$ (or $\pm\pi/2$, depending on index parity and space group) for centrosymmetric zonal reflections and, e.g., $\pi/4$ +

$n\pi/2$ for unrestricted phases. It is also convenient to include phases from high-resolution electron micrographs in this basis set, if they are available. All phases φ_h are sequenced according to the order of their being derived from precursor phase definitions via available Σ_2-triples and hence, after permutation of the algebraic signs, one ends up with multiple trial solutions. The procedure is cycled several times, improving estimates of the phase variance $V(\phi_h)$ (Ladd and Palmer, 1980), which is directly related to the magnitude:

$$\alpha_h^2 = (\sum_{k(r)} A \cos(\phi_k + \phi_{h-k}))^2 + (\sum_{k(r)} A \sin(\phi_k + \phi_{h-k}))^2$$

Here $A = (2/\sqrt{N})|E_h\, E_k\, E_{h-k}|$, as before.

The initial pass through the tangent formula in its QTAN version (Langs and DeTitta, 1975) starts with the estimate:

$$\alpha_{h,\,\mathrm{est}} = [\sum_k A^2 + 2 \sum_k \sum_{k'} AA' \times I_1(A) I_1(A')/I_0(A) I_0(A')]^{1/2}$$

where I_1 and I_0 are, again, modified Bessel functions. There are several other versions of the tangent formula including MULTAN (Germain et al., 1971) and RANTAN (Yao, 1981), the latter being capable of generating random starting phase sets in case trials with conventional methods become locked into false solutions.

After the multiple solutions are generated, some means must be found to select the most likely phase combinations. With x-ray data, the figure of merit (Karle and Karle, 1966):

$$R_{\mathrm{Karle}} = \sum_h ||E_h| - |E_h|_{\mathrm{calc}}| / \sum_h |E_h|,$$

where $|E_h|_{\mathrm{calc}} = k < |E_k|\, |E_{h-k}| >_{k(r)}$, is often employed. However, this has not been useful for the electron crystallographic applications of the tangent formula. Another term, known as NQEST (DeTitta et al., 1975), where:

$$\mathrm{NQEST} = \sum B \cos(\phi_h + \phi_k + \phi_l + \phi_m) / \sum B$$

has identified correct solutions for two thiourea structures. It is, unfortunately, sensitive to dynamical scattering because of its reliance on negative quartet invariants, as will be shown in future chapters. For example, the structure analysis of diketopiperazine with the tangent formula (Dorset and McCourt, 1994a) found a false solution at the lowest NQEST value (Figure 4.10c,d) but the correct solution (Figure 4.10a,b) was observed within the four possible solutions with the most negative values for this figure of merit.

In a recent work, it was predicted that the tangent formula would not be applicable to electron crystallographic problems (Gilmore et al., 1993). This is a far too pessimis-

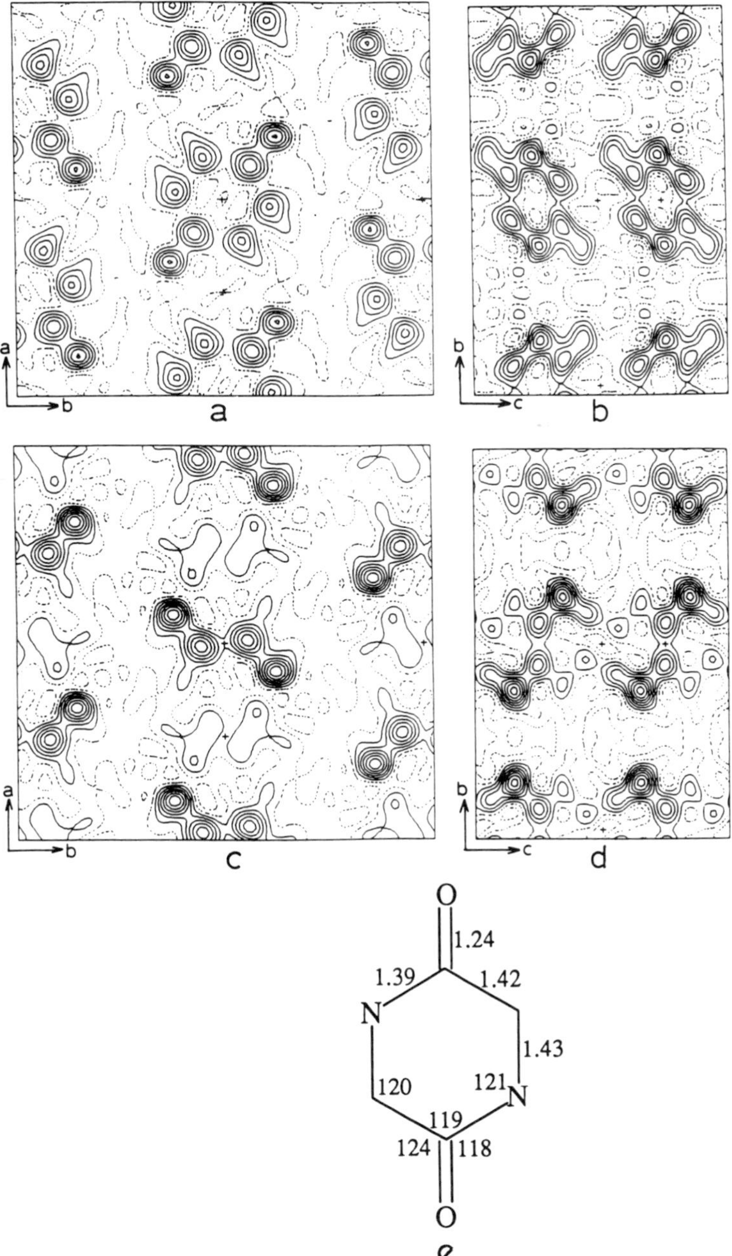

FIGURE 4.10. Determination of the diketopiperazine structure by the tangent formula (based on data published by Vainshtein (1955)). (a, b) The correct structure does not correspond to the lowest value of NQEST whereas (c, d) a chemically unreasonable atomic array does. (e) Derived bond distances and angles are shown from the correct structure. (Reprinted from D. L. Dorset (1994) "Electron crystallography of organic molecules," *Advances in Electronics and Electron Physics*, **88**, 111–197; with kind permission of Academic Press, Inc.)

tic statement to be made without substantiating evidence. As just mentioned, the difficulties lie with the choice of a suitable figure of merit for identification of the best structure solution. Despite these difficulties, the tangent formula has proven to be one of the most effective direct phasing approaches with electron diffraction data, as we shall illustrate with examples below. The phase sets are found to be highly accurate when the best solution can be identified. It is the best method for solving noncentrosymmetric structures and is a convenient way of including phase information from images, e.g., the phase extension of copper perchlorophthalocyanine (Fan et al., 1985, 1991b), although the Sayre equation appears to be better suited for this purpose. More recently, it has been shown that the limitations of individual figures of merit for identifying the correct phase set from a multisolution list may be overcome by a combined figure of merit, as suggested by Cascarano et al. (1987), or by use of the minimal function (see below).

4.1.4.4. Patterson Search Techniques. The premise of heavy-atom methods, which were the mainstay of small molecule x-ray crystallography (and remain so for large globular macromolecules) for many years, is that determination of a heavy-atom location in a Patterson map provides enough initial phase information to retrieve the rest of the structure via Fourier refinement (see below). However, as mentioned above, heavy-atom techniques are not quite as effective in electron crystallography. If the structure of a significant molecular fragment can be constructed a priori from knowledge of chemical architecture, on the other hand, this molecular geometry can be used to provide additional information for phase determination, as if it were a heavy atom.

Given a fragment of a structure p, it is possible to calculate from it a set of trial structure factors $F_p(R,t,h)$ for any rotation R or translation t in the unit cell for all Miller indices h. From this, the partial Patterson function $P_p(R,t,w)$ can be calculated, where w is a vector in Patterson space. With observed data, it is also possible to calculate a complete experimental Patterson map $P_o(w)$. For vectors contained in this experimental map below a certain length $|r|$, mostly intramolecular interactions are expressed. For those beyond this limiting length, mostly intermolecular interactions are seen (see, e.g., Beurskens and Beurskens, 1993). The former can be used to orient a fragment rotationally in the unit cell by comparing the two Patterson functions via the correlation function:

$$C_p(R) = \int P_o P_p(R) dw$$

Equivalent correlation expressions exist for reciprocal space, e.g.:

$$C_p(R)' = \langle |F_o|^2 |F_p(R)|^2 \rangle \text{ or } C_p(R)'' = \langle |E_o|^2 |E_p|^2 \rangle$$

the latter giving a sharpened distribution (which can be very useful). After rotational alignment, translational alignment can be carried out in a similar fashion with the longer vectors. In macromolecular crystallography, these are the concepts behind molecular replacement (Rossmann and Blow, 1962), wherein known structures can be compared to data from similar molecules. Also noncrystallographic symmetry between similar subunits can be exploited for phase refinement. Such techniques have already been utilized in macromolecular electron crystallography, e.g., comparing two protein polymorphs (Rossmann and Henderson, 1982) or from correlations between subunits, for phase refinement (Earnest et al., 1992).

In small-molecule crystallography, orientation of structural fragments provides initial information for direct phase determination (Beurskens and Smykalla, 1991; Beurskens and Beurskens, 1993). Given the total $|F_o|$ from the crystal and the F_p determined from an oriented fragment, the $F_r = |F_o| - F_p$ phases can be sought by the tangent formula or the Sayre equation, i.e.:

$$E_r(\mathbf{h}) \approx c \sum_k E_r(\mathbf{k})\, E_r(\mathbf{h}-\mathbf{k})$$

To begin the process (Beurskens and Smykalla, 1991), the normalized difference structure factors are multiplied with weights $W_p = (2p_1 - 1)^2$, where p_1 is the probability that an input phase ϕ_1 is correct compared to its extreme opposite $\phi_1 + \pi$. This selects an initial set of E_1, based on normalized ΔF values, to be used as an initial E_r set. Resultant phases ϕ_t from the tangent formula are then assigned new weights $W_t = (2p_t - 1)^2$, where p_t is another probability function. When $W_t/W_1 > 1.0$, the phase $\phi_t = \phi_r$ is accepted with weight W_t. If $W_t/W_1 < 1.0$, ϕ_t is only partly accepted when $|\phi_t - \phi_1| < \pi/2$, to give a new value that can be inserted into the tangent formula for the next cycle. In electron crystallography, there is already interest in applying this technique to the analysis of incommensurate phase structures (Beurskens et al., 1993).

4.1.4.5. The Minimal Principle. In order to circumvent the constraints imposed upon direct methods by a large number of atoms N in the unit cell, Hauptman (1993) has developed a minimal function:

$$R(\phi) = \sum_{H,K} A_{H,K}(\cos \phi_{H,K} - t_{H,K})^2 / \sum_{H,K} A_{H,K}$$

where $A_{H,K} = (2/\sqrt{N})|E_h E_k E_{h+k}|$, $\phi_{H,K} = \phi_H + \phi_K + \phi_{-H-K}$, as before, and $t_{H,K} = I_1(A_{H,K})/I_0(A_{H,K}) > 0$ is an expectation value for the phase triple. The residual can be used as a more powerful figure of merit (than, e.g., NQEST) for conventional multisolution direct phasing procedures, e.g., the tangent formula in its many manifestations. Alternatively, trial random structures can be generated to seek an initial minimum of $R(\phi)$. Phase refinements via an annealing process are then carried out (Miller et al., 1993), wherein the constraints of positivity and local "peakiness" of atomic sites are maintained, in order to minimize the residual. This procedure can be recycled until a solution is found. In these tests, likely solutions are found from the minimal principle that $R_t < 1/2 < R_R$ where R_t is a trial phase set and R_R is the known value for a random phase distribution for the assembly of atoms in the space group.

Preliminary tests of this method on problematic electron diffraction data sets have resulted in successful structure determinations for two polymers (including poly(1-butene)) via the so-called "Shake and Bake" algorithm (Miller et al., 1993) (M. McCourt, unpublished data). The correct structure solution could not be identified by the NQEST figure in QTAN (see Chapter 11). Application of $R(\phi)$ as a figure of merit also found a solution with the tangent formula (C. M. Weeks, unpublished data) corresponding closely to the correct structure.

4.1.4.6. Density Modification. With a small number of phased structure factors to begin with, it is also possible to extend the phase set by a simple method known as density modification (Hoppe and Gassmann, 1968; Gassmann and Zechmeister, 1972; Gassmann, 1976). Starting with the small basis set, an initial estimate of the potential map $\rho(\mathbf{r})$ is calculated by the Fourier transform operation. This map is then modified, utilizing criteria such as limiting the heights of peak density and removal of all negative density regions. The modified map $\rho(\mathbf{r})'$ produces a new phased structure factor set for all reflections, i.e., F_h'. Phase values are accepted by a threshold criterion in reciprocal space where $E_{calc}/E_{obs} \geq p$. This set is then transformed to produce a new $\rho(\mathbf{r})'$, and so on, until the structure is found.

The procedure has been tested in the electron crystallographic case with simulated data from copper perchlorophthalocyanine (Ishizuka et al., 1982). Starting with crystallographic phases from an electron micrograph, it was possible to fill in new phase information in regions of the contrast function where $C(s) \leq 0.2$. It was also possible to extend from image resolution to the resolution of the electron diffraction pattern, especially if the former resolution was around 2.0 Å. The method was not very satisfactory, however, when the image resolution was somewhat worse, e.g., 2.5 Å.

As is known in protein crystallography (Schevitz et al., 1981; Wang, 1985), density modification procedures also can be used to improve the initial phase estimate obtained by other means. For example, starting with an initial map calculated from these phases (wherein an estimate of F_{000} is made to minimize the negative density regions), an outline can be defined for the protein structure to establish its boundaries. Within this boundary, the modified density $\rho^{mod} = \rho$ is accepted, where the latter are values found from the map itself, but outside of the boundary $\rho^{mod} = \langle\rho\rangle$, where the average denotes a flattening of the exterior structure. This may be a solvent region, for example. A reverse Fourier transform is then calculated from the modified map to produce a new set of phases ϕ. A new map is then calculated from $(|F_h|, \phi_h)$ where the amplitudes are multiplied by so-called Sim (1959) weights based on the product of normalized structure factor magnitudes (see discussion in Blundell and Johnson, 1976). This procedure is, therefore, similar to the Hoppe–Gassmann method already discussed.

4.1.4.7. Maximum Entropy. The term "maximum entropy" has been found to have several meanings when applied to electron crystallography. In the sense that images are optimized, the entropy term

$$S = -\sum_i P_i \ln P_i$$

where $P_i = p_i/\Sigma_i p_i$ and p_i is a pixel density, has been evaluated for various test images. For crystals, the true projected potential distribution function is often found to have the maximum value S. If the phase contrast transfer function used to obtain a micrograph is unknown, test images (i.e., trial potential maps) can be calculated for different values of Δf_{trial}. The value that corresponds to the maximum entropy would be near the true defocus. In this way, the actual transfer function can be found for a single image (Li, 1991) by a different technique from the ones already discussed.

Another use of the maximum entropy technique is to guide the progress of a direct phase determination (Bricogne and Gilmore, 1990; Gilmore et al., 1990). Suppose that there is a small set H of known phases $\phi_{h \in H}$ (corresponding either to origin definition of the Fourier transform of electron micrograph) with associated unitary structure factor amplitudes $|U_{h \in H}|$. (The unitary structure factor is defined $|U_h| = |E_h|/\sqrt{N}$.) As usual, the task is to expand into the unknown phase set K to solve the crystal structure. From Bayes' theorem, the procedure is based on an operation where: $p(\text{map}|\text{data}) \alpha$ $p(\text{map})\, p(\text{data}|\text{map})$. This means that the probability of successfully deriving a potential map can be estimated, given diffraction data, a so-called posterior probability. This posterior probability is approximately proportional to the product of the probability of generating the map (known as the prior) and the probability of generating the data, given the map (known as the likelihood). The latter probability consults the observed data and can be used as a figure of merit.

Beginning with the basis set H, a trial map is generated from the limited number of phased structure factors. As discussed above, the map can be immediately improved by removing all negative density. The map can be improved further if its entropy is maximized using the equation given above for S. This produces the so-called maximum entropy prior $q^{ME}(X)$.

So far it has been assumed that all $|U_{h \in K}| = 0$. If large reflections from the K set are now added and their phase values are permuted, then a number of new maps can be generated and their entropies can be maximized as before. This creates a phasing "tree" with many possible solutions and individual branch points can have further reflections added via permutations to produce further sub-branches, and so on. Obviously, some figure of merit is needed to "prune" the tree, i.e., to find likely paths to a solution.

The desired figure of merit is the likelihood $L(H)$. First a quantity

$$\Lambda_h = 2NR \exp(-N(r^2 + R^2))\, I_o(2NrR)$$

where $r = |^{ME}U_h|$, the calculated unitary structure factors, and $R = |^{o}U_h|$, the observed unitary structure factors, is defined. From this one can calculate:

$$L(H) = \sum_{h \notin H} \ln \Lambda_h$$

The null hypothesis $L(H_o)$ can also be calculated from the above when $r = 0$, so that the likelihood gain

$$LLg = L(H) - L(H_o)$$

ranks the nodes of the phasing tree in order of the best solutions. A small molecule structure starting with phases from an electron micrograph and extending to electron diffraction resolution has been reported (Dong et al., 1992). An application of this procedure in protein structure analysis will be discussed in Chapter 12.

4.2. Structure Refinement

4.2.1. Identification of a Structure

In any ab initio crystallographic analysis, arriving at a structure "solution" sometimes must rely on the chemical knowledge of the investigator, who can readily determine if an array of density points in any test map can be the basis for a "reasonable" partial atomic model. This is particularly true if the number of phases assigned to structure factor magnitudes is small relative to the size of the structure. A recognizable fragment of a molecule, therefore, becomes more and more complete, with chemically reasonable geometry, as the refinement progresses. Again, as in x-ray crystallography, this fragment will be readily recognized by some investigators and not by others. Sometimes the total structure is seen in the initial map, e.g., as shown in Figure 4.9 for urea. This expectation is least likely to be satisfied for complicated structures.

Proof of the structural model depends on the match of the calculated structure factor magnitudes to their observed values. In modern x-ray determinations, R-factors < 0.05 are often taken to signify a good solution after refinement is carried out. Implicit in this good figure of merit are the criteria of statistical significance stipulated by Hamilton (1964), as discussed above. Because of the several possible perturbations to electron diffraction intensity data (to be discussed in Chapter 5), structure analyses will not be so accurate as those carried out with x-ray data. A very good R-factor may be < 0.20, with typical values somewhat larger than 0.25. Specific bond distances and angles may, accordingly, be slightly distorted, so that high geometrical accuracy may not be a realistic goal of the structure determination, whereas the characterization of molecular conformation and packing are more realistic aims.

A more onerous task awaits when poor data resolution and/or atomic overlap in a projection leads to a continuous density profile as seen in Figure 4.8 for the polymer chitosan. When the very powerful constraint of resolved atomic positions is removed, identification of a density profile as a correct structure solution can be very difficult. Nevertheless, certain criteria for such selection have been suggested. The basis for discrimination depends, moreover, on the resolution of the structure analysis. For example, at low resolution, Luzzati et al. (1988) have suggested a moment calculation where a function of the potential $<\Delta\rho^4>$ is minimized, where $\Delta\rho = \rho - <\rho>$ and $<\rho>$ is the mean density of the map. A criterion of smoothness $<\partial\rho/\partial x>\}$ has also been used successfully for some structures (Dorset, 1991c). With high-resolution data (such as the chitosan example), tests of "peakiness", e.g., $\Sigma\rho^n$, have been used successfully for crystal structures with near atomic resolution (Stanley, 1986). The selection of a correct map for chitosan (Figure 4.8) can be made readily with this latter criterion, e.g., when $n = 5$. This procedure has also been used successfully for analyses of two other polymer structures in similar projections down the chain axis.

4.2.2. Fourier Refinement

Fourier methods for structure refinement have been found to be very useful for electron crystallographic analyses. Identified atomic positions have been used to calculate a potential map, based on phase estimates for all measured reflections; hence:

$$\rho(r) = V^{-1} \sum_h \sum_k \sum_l F_o \exp(i\phi_c) \exp(-2\pi i[hx + ky + lz])$$

Before considering an actual example, it is worth examining this inverse transform in detail to understand its parts. If we define $\kappa = 2\pi(hx + ky + lz)$, then, from the definition of the structure factor given above, $F \exp -i\kappa = (A + iB)(\cos \kappa - i \sin \kappa) = A \cos \kappa + B \sin \kappa - i(A \sin \kappa - B \cos \kappa)$. If a summation is taken over all Friedel pairs (hkl) and ($\bar{h}\,\bar{k}\,\bar{l}$), where the structure factor amplitudes are the same but their phases are negatives of each other, the following expression is derived from the above (see Lipson and Cochran, 1966; Glusker and Trueblood, 1985):

$$\rho(r) = V^{-1} [F_{000} + 2 \sum_h \sum_k \sum_l (A \cos \kappa + B \sin \kappa)]$$

The first term is a scaling parameter to set the level of the ensuing map so that there are no negative regions. (Given a delta function at the index triple (000), its Fourier transform will be an added continuous level over all space, as can be appreciated from the arguments in Chapter 1.) When all structure factors are based an exact scale then, in x-ray crystallography $F_{000} = \Sigma_i Z_i$, or the total electron content of the unit cell. For electron crystallography, $F_{000} = \mu V$, where μ is the mean inner potential of the unit cell (see Vainshtein, 1964b). Now, since $A = |F| \cos \phi$ and $B = |F| \sin \phi$, the above expression becomes:

$$\rho(r) = V^{-1}[F_{000} + 2 \sum_h \sum_k \sum_l |F| \cos (\kappa - \phi)]$$

which is the computational form of this inverse transform.

The procedure of Fourier refinement can be illustrated (Figure 4.11) for the case of copper perbromophthalocyanine (Dorset et al., 1992). In the analysis of this structure from electron diffraction data, an initial map was calculated from the partial phase set, in which the probable positions of the copper and two bromine atoms could be discerned (Figure 4.11a). In this map, the halogen atom densities were reinforced and probable positions for another bromine and a nitrogen are discerned (Figure 4.11b). This procedure was repeated until all heavy-atom positions were located, as

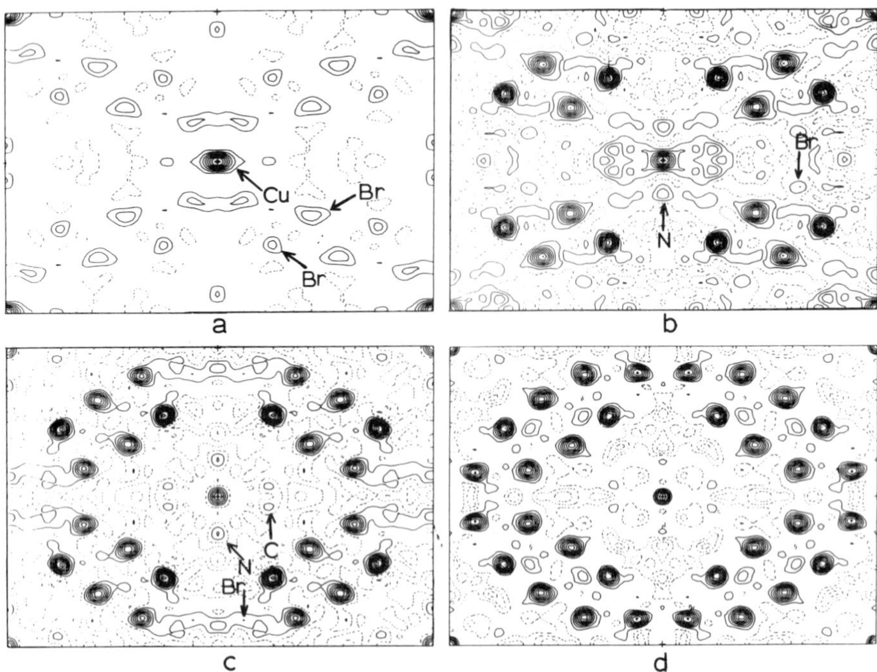

FIGURE 4.11. Initial stages of Fourier refinement for copper perbromophthalocyanine: (a) initial map after direct phase determination in which three heavy-atom positions are found: (b) first map based on total phase set calculated from initially determined atomic positions. Two more atomic positions are identified; (c) subsequent map with final Br position located; (d) potential map for which all heavy-atom positions are identified. (Reprinted from D. L. Dorset, W. F. Tivol, and J. N. Turner (1992) "Dynamical scattering and electron crystallography—ab initio structure analysis of copper perbromophthalocyanine," *Acta Crystallographica* **A48**, 562–568; with kind permission of the International Union of Crystallography.)

well as an additional carbon position. At this stage, difference Fourier maps were used to locate the other lighter atom positions (Figure 4.12), i.e.:

$$\rho(r)' = V^{-1} \sum_h \sum_k \sum_l (2|F_o| - |F_c|) \exp(i\phi_c) \exp(-2\pi i[hx + ky + lz])$$

and

$$\Delta\rho(r) = V^{-1} \sum_h \sum_k \sum_l (|F_o| - |F_c|) \exp(i\phi_c) \exp(-2\pi i[hx + ky + lz])$$

Again, the procedure was repeated until all atoms in the structure were found and optimized in density. A local minimum of the crystallographic R-factor was sought, but, because of data perturbations, this did not correspond to a global minimum which, in fact, corresponded to a structural model with distorted bond geometry. A global

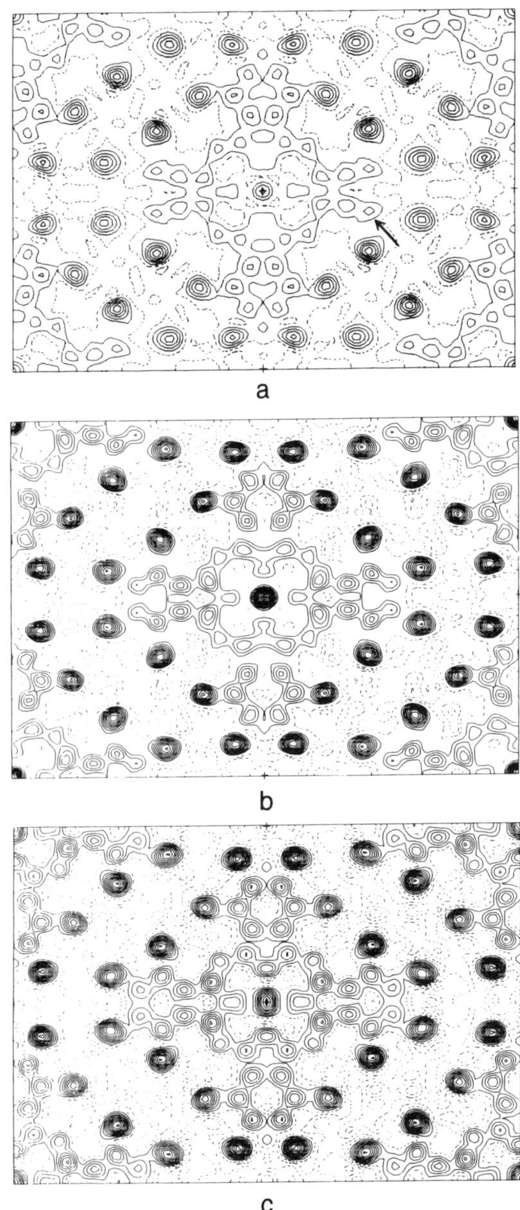

FIGURE 4.12. Final stages of Fourier refinement for copper perbromophthalocyanine: (a) potential map where all but one of the lighter atom positions are located; (b) map corresponding to the lowest kinematical R-factor (0.36) but with distorted bond geometry; (c) map with best bond geometry but where $R = 0.41$. (This, however, is lowered to 0.26 after a multislice dynamical scattering calculation.) (Reprinted from D. L. Dorset, W. F. Tivol, and J. N. Turner (1992) "Dynamical scattering and electron crystallography—ab initio structure analysis of copper perbromophthalocyanine," *Acta Crystallographica* **A48**, 562–568; with kind permission of the International Union of Crystallography.)

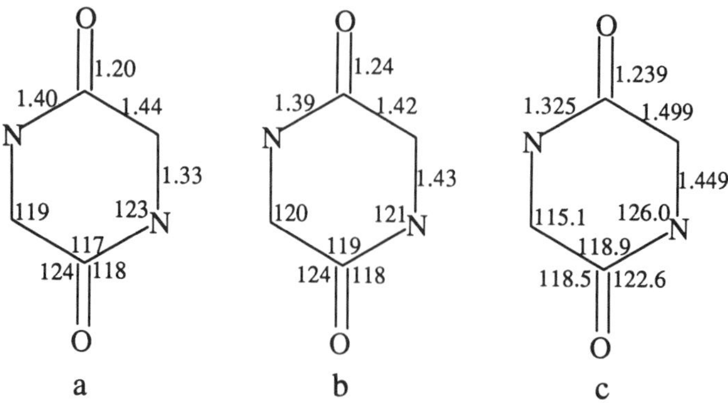

FIGURE 4.13. Bond distances and angles for diketopiperazine: (a) after direct phase determination; (b) after least-squares refinement. (c) x-ray structure. (Reprinted from D. L. Dorset and M.P. McCourt (1994) "Automated structure analysis in electron crystallography: phase determination with the tangent formula and least squares requirement." *Acta Crystallographica* **A50**, 287–292 with kind permission of the International Union of Crystallography.)

minimum can be realized only if the data perturbations are accounted for during the refinement (Sha et al., 1993).

4.2.3. Least-Squares Refinement

Sometimes, if enough data are measured per refinable parameter (e.g., three positional shifts per atom along the respective unit cell axes ($\partial F/\partial r$), as well as isotropic temperature factors ($\partial F/\partial B$), and an overall scale factor), a least-squares refinement can be carried out. It is desirable to have at least three intensity data per variable. Thus, the quantity $M = \Sigma_{hkl}\, w_{hk}(F_o - F_c)^2$ is minimized (Stout and Jensen, 1968) during the refinement.

If x-ray crystallographic software is to be adapted for such refinements, the scattering factor tables must be changed to electron diffraction values (Doyle and Turner, 1968) and the shift of thermal parameters must be uncoupled from the atomic positional shifts. Additionally, the atomic shifts should be dampened by a factor (0.1 to 0.2) to prevent translation away from the desired local R-factor minimum that would lead to a distorted structure. Least-squares refinements have been carried out with three-dimensional data sets from diketopiperazine (Dorset and McCourt, 1994a) and polyethylene. The final bond distances and angles for the cyclic structure after ten cycles are shown in Figure 4.13 to compare favorably to the x-ray determination.

Constrained least-squares refinements have already been mentioned above as a structure seeking technique in a trial-and-error procedure. In this case, the bond distances and angles are held fixed while conformational angles are changed to find a minimum of the crystallographic residual that also corresponds to the lowest atom–atom potential energy.

4.2.4. Continuous Density Maps

If observed electron diffraction data do not permit visualization of the structure at atomic resolution, special measures must be taken to improve the phase set. Sometimes regions of the structure can be identified to have a known feature, e.g., the hydrocarbon chain packing of a phospholipid bilayer or the solvent region of a protein crystal; from this knowledge, an improved phase estimate can be obtained by density flattening (Wang, 1985), as discussed above. After the map (Figure 4.14) is modified, its

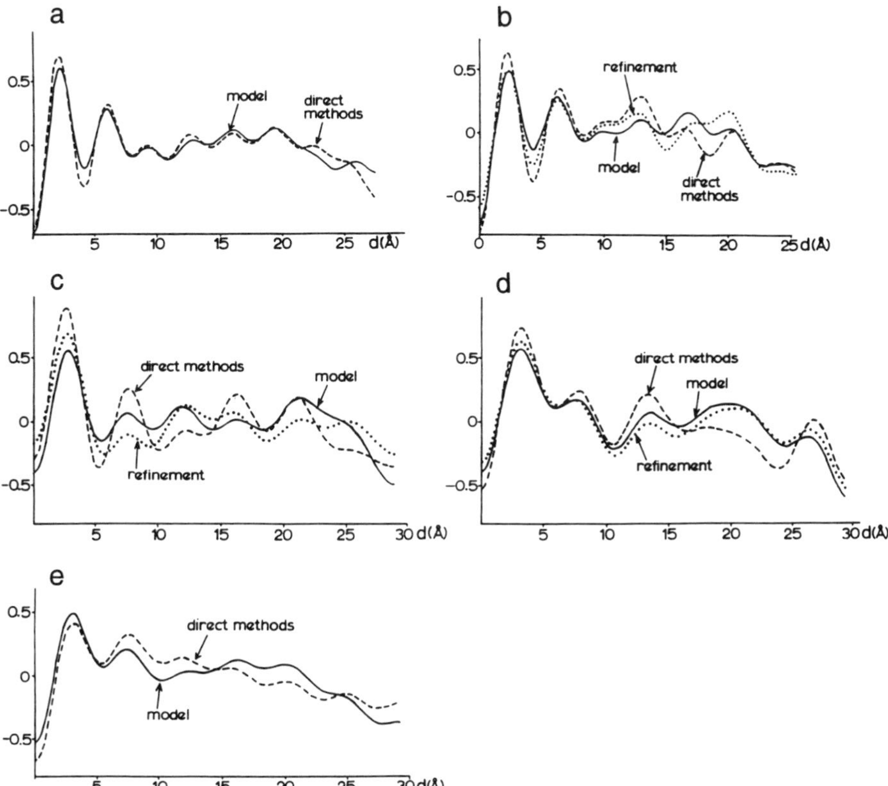

FIGURE 4.14. Process of phase refinement for one-dimensional phospholipid bilayer profiles based on electron diffraction data. An initial partial phase set was obtained by direct methods. Density modification (i.e., flattening the hydrocarbon chain region) is imposed for the reverse Fourier transform of the initial potential map to obtain new phases. This process is repeated to find maps with maximized density smoothness or flatness. When this final solution differs from previous phase determinations (e.g., based on a conformational model), the resultant "refinement" is distinguished from "model": (a) 1,2-Dihexadecyl-*sn*-glycerophosphoethanolamine; (b) 1,2-dimyristoyl-*sn*-glycerphosphoethanolamine; (c) 1,2-dihexadecyl-*sn*-glycerophopho-N-methylethanolamine; (d) 1,2-dihexadecyl-*sn*-glycerophosphocholine; (e) 1,2-dipalmitoyl-*sn*-glycerophospho-N,N-dimethylethanolamine. (Reprinted from D. L. Dorset (1991) "Direct determination of crystallographic phases for diffraction data from lipid bilayers. II. Refinement of phospholipid structures" *Biophysical Journal* **60**, 1366–1373 with kind permission of the *Biophysical Journal*.)

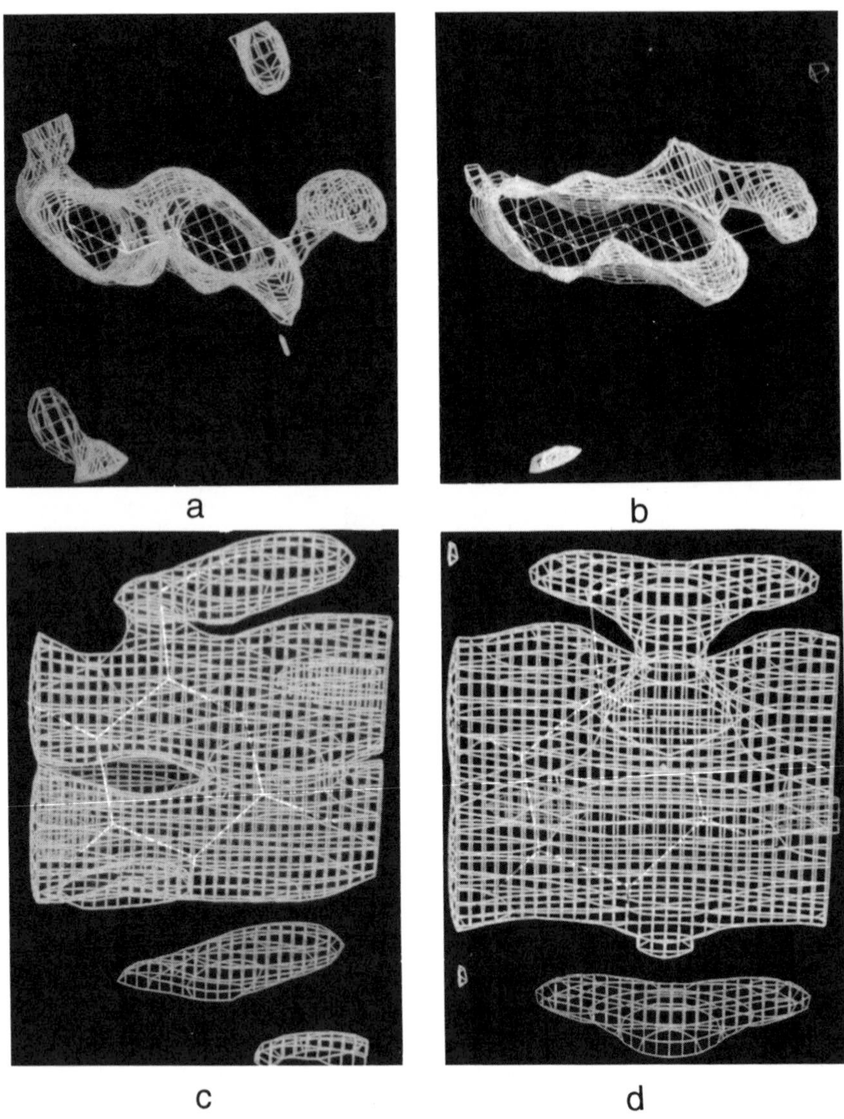

FIGURE 4.15. Solution of the mannan I structure. A three-dimensional data set from a single lamellar crystal was incompletely sampled because of the goniometer tilt limits. After direct methods were used to find crystallographic phases, an algebraic ambiguity remained. The structure was solved by fitting a monomer model to the density. The best fit is for two projections of the structure (a, c), corresponding to one of the choices for the phase ambiguity. (Reprinted from D. L. Dorset and M. P. McCourt (1993) "Electron crystallographic analysis of a polysaccharide structure—direct phase determinations and model refinement for mannan I," *Journal of Structure Biology* **111**, 118–124; with kind permission of Academic Press, Inc.)

reverse Fourier transform will then lead to a new set of phased structure factors (Dorset, 1991c). New phases can be accepted gradually while certain criteria such as density smoothness are monitored (see above). Examples will be given in Chapters 10 and 12.

If a three-dimensional potential map with no atomic detail is the result of a direct phase determination, an atomic model can be fit to it to find their coordinates in the unit cell. Such a procedure is used to fit polypeptide chains to density maps in protein crystallography. For polymer mannan I (Figure 4.15), limited data sampling along the chain axis direction, due to goniometric tilt restrictions, caused the potential map to be elongated (a "series termination effect" in the Fourier transform) so that atomic features could not be discerned. The crystal structure of the mannose repeat is known and, from it, it was possible to find a good fit of the monomer to the map density (Dorset and McCourt, 1993), leading to a structure that closely matched the original model based on a conformational analysis (Chanzy et al., 1987).

4.3. Derived Quantities

After a crystal structure analysis is completed, it is possible to measure chemically relevant geometrical quantities from the final atomic positions in the unit cell. For molecular crystals, bond distances (or nonbonded interatomic distances) are obtained from the expression (Stout and Jensen, 1968):

$$L^2 = (x_2 - x_1)^2a^2 + (y_2 - y_1)^2b^2 + (z_2 - z_1)^2c^2 + 2(x_2 - x_1)(y_2 - y_1)ab \cos \gamma$$

$$+ 2(y_2 - y_1)(z_2 - z_1)bc \cos \alpha + 2 (z_2 - z_1)(x_2 - x_1)ca \cos \beta$$

Here x_n, y_n, z_n, where $n = 1,2$, are fractional coordinates of two associated atoms and a, b, c, α, β, γ are the unit cell parameters.

For any atoms A, B, C forming two consecutive bonds A-B and A-C, the angle θ between them is:

$$\cos \theta = [(AB)^2 + (AC)^2 - (BC)^2]/[2(AB)(AC)]$$

where AB, AC, BC are interatomic distances.

Generally, bond angles and distances obtained from electron crystallographic analyses are not as accurate as those obtained from x-ray structures. On average, however, they should compare well to typical values obtained from x-ray or neutron analyses of similar materials.

5

Data Perturbations

There is a controversial aspect to using electron scattering data for ab initio determination of crystal structures. The observed intensities are often imagined to be so effectively contaminated by various perturbations to obscure the desired, single-scattering, kinematical information. It would be a mistake to pretend that such perturbations do not exist. The approach taken in this book, i.e., the search for experimental conditions to favor collection of quasi-kinematical data, must be based on a realistic understanding of such theoretical constraints. The following is a review of the major causes for deviation of the observed electron microscope images or electron diffraction intensities from a straightforward representation of the unit cell contents.

5.1. Dynamical Scattering

As mentioned in Chapter 1, electrons are strongly scattered by matter, compared to x-rays or neutrons. Although this fact was known from the early days of electron diffraction structure analysis, it was imagined that deviations from the kinematical assumption could be easily corrected by separately considering the effects on individual diffracted beams. To justify this approach, a single crystal was imagined to be an array of mosaic blocks (Figure 5.1), so that each beam in the diffraction pattern would be generated independently from the other ones, i.e., from crystalline blocks with the proper orientation to the incident beam (Vainshtein, 1964a). With this two-beam model for dynamical scattering, which is essentially the same as the primary extinction model used in x-ray crystallography (Blackman, 1939), it can be shown that, on average, the observed structure factor magnitudes will cease to be related to the diffracted intensity as $I^{0.5}$ and will approach something closer to $I^{1.0}$ (Vainshtein and Lobachev, 1956). Using a graphical procedure similar to a Wilson plot, early work attempted to adjust the average measured structure factor magnitudes so that they would agree with the expected fall-off of the atomic scattering factor curves (e.g., see Li, 1963). Further, it was imagined that, with increased accelerating voltage for the electrons, a "kinematical limit" could be reached (Honjo and Kitamura, 1957) eventually at a small enough wavelength. While this procedure can be used to detect the presence of dynamical

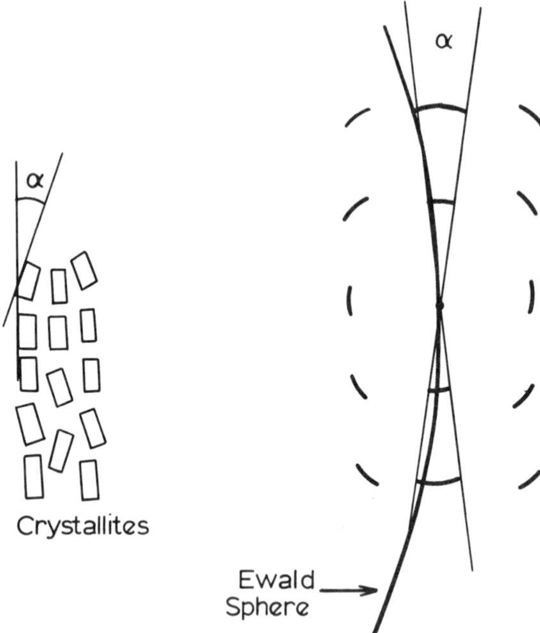

FIGURE 5.1. Concept of a mosaic crystal. The distribution of blocks causes the reflections to be spread along arcs. If each block is separately responsible for a single reflection, then some justification could be given for a two-beam dynamical scattering model.

scattering, the two-beam model also can be shown to be inadequate for explaining how individual intensities are influenced by multiple scattering. Although the two-beam correction may be somewhat useful for polycrystalline samples (Cowley, 1957) (also taking the rather large specimen areas sampled in early electron diffraction experiments into account), it is not sufficient to explain the electron scattering from single crystals, especially data from micron-diameter (or smaller) areas.

The early arguments supporting the two-beam dynamical scattering model fail because the single crystal texture assumed to justify the model does not exist for microareas sampled in selected area diffraction experiments. As mentioned by Cowley (1957), and justified by numerous experimental observations, such selected area diffraction patterns often originate from crystal areas that are much more perfect than the mosaic model, i.e., there are not nearly the number of defects within the selected area needed to justify the construction in Figure 5.1. Since the reciprocal lattice is sampled by an Ewald sphere with a very large radius, the interactions of all simultaneously excited diffracted beams must be considered, not only in their association with the incident beam, but also in their associations with one another. That is to say, any relevant dynamical scattering theory will account for the interactions of *all* beams simultaneously excited for any crystal orientation.

There are several equivalent (Goodman and Moodie, 1974) formulations of the n-beam dynamical theory for electron diffraction/image formation from a thin single crystal (e.g., Cowley and Moodie, 1957; Fujiwara, 1959; Howie and Whelan, 1961; Sturkey, 1962). If crystal thickness is to be monitored as a continuous variable, it is often most convenient to employ the "slice" method of Cowley and Moodie (1957) for calculating dynamical structure factors and phases. In this approach, a crystal is divided up into a series of suitably thin, two-dimensional slices along the beam path with thickness Δz (total crystal thickness $t = n\Delta z$). In Chapter 1, it was shown that the wave function exiting any slice of a crystal is (neglecting absorption):

$$q(r) = \exp[-i\sigma\, \rho(r)]$$

where σ is an interaction term which can be defined:

$$\sigma = 2\pi k\, \Delta z/(\lambda\, E\, [1 + (1 - \beta^2)^{1/2}])$$

with $k = 47.87/V$. Taking the Fourier transform of $q(r)_n$, i.e., $Q(s)_n$, then, after the nth slice, the exit wave is:

$$\Psi_n(s) = \{\Psi_{n-1}(s)\, P_{n-1}(s)\} * Q(s)_n$$

The reciprocal space multiplication of the previous wave function with the propagation function $P_{n-1}(s) = \exp[2\pi ic\zeta(s)]$ describes the Fresnel diffraction of this wave on transmission through distance Δz and includes the Ewald sphere curvature through the so-called "excitation error" $s_g = \zeta(s)$, which is just the distance (in the direction of the incident beam) of a diffracted beam from the Ewald sphere. This treatment assumes that the crystal is flat. An alternative treatment is to expand the projected potential as a Taylor series and then take its Fourier transform:

$$F^{\text{dyn}}(s) = \delta(s) - i\sigma t F(s) - [(\sigma t)^2/2!]F(s)*F(s) + i[(\sigma t)^3/3!]F(s)*F(s)*F(s) + \ldots$$

This is known as the phase-grating expression (Cowley and Moodie, 1959) and assumes that the Ewald sphere has no curvature. Both treatments can be shown to yield equivalent results at high voltage (when the electron wavelength is very small). Because these expressions are complicated, it is readily seen, in general, that it would be difficult a priori to find measured values close to the kinematical structure factor amplitudes $|F(\mathrm{s})|$, if the interaction term σ is large, the crystal is too thick, or the accelerating voltage is too low. Note also that there is a phase term associated with the dynamical scattering that will affect the characteristics of the observed electron micrograph. The complexity of this expression has forced many theoreticians (e.g., Humphreys and Bithell, 1992; Amelinckx and Van Dyck, 1993) to believe that the only possibility for a structure analysis is an indirect one, i.e., to first assume a model for the structure to obtain atomic coordinates. After a dynamical scattering calculation is carried out, based on these coordinates and other experimental variables, a test of convergence with the observed image or diffraction data is made and a match is used

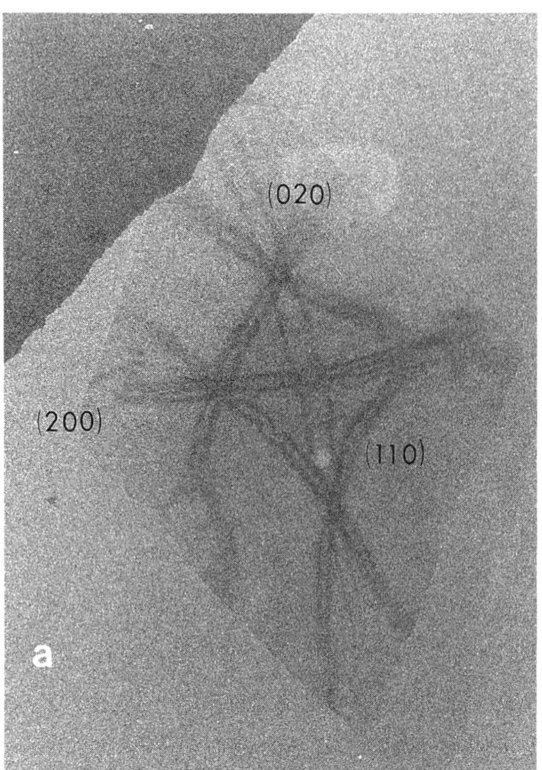

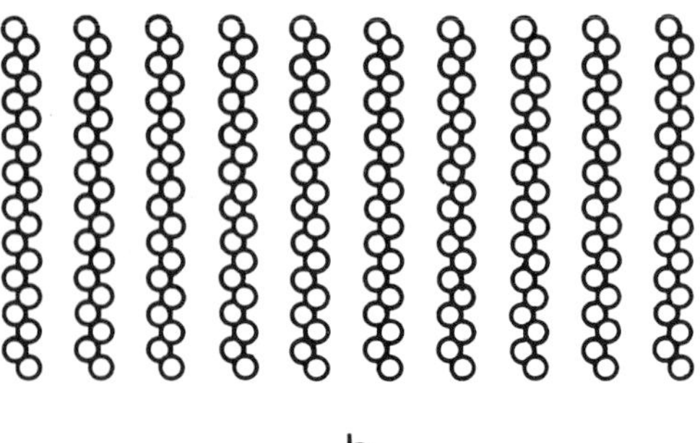

FIGURE 5.2. (a) Bright field image of a monolamellar paraffin crystal with major bend contours labeled. (Reprinted from D. L. Dorset (1985) "Crystal structure analysis of small organic molecules," *Journal of Electron Microscopy Technique* **2**, 89–128; with kind permission of Wiley–Liss Division, John Wiley and Sons, Inc.) (b) Laterally, the structure is a two-dimensional array of chains where the chain length accurately determines the crystal thickness and the 2.55-Å methylene "subcell" repeat along the chain (also the beam direction) can be used in models for *n*-beam dynamical calculations. (Reprinted from D. L. Dorset and F. Zemlin (1984) "Specimen movement in electron-irradiated paraffin crystals—a model for initial beam damage," *Ultramicroscopy* **21**, 263–270; with kind permission of Elsevier Science B.V.)

to justify the structural model. This approach presumes that a crystal structure is known before it is determined, a procedure that can lead to great difficulties, because the prediction of complicated atomic arrangement in unit cells by trial-and-error methods is no easy task (see Chapter 4).

Is it possible to demonstrate experimentally that the *n*-beam model for dynamical scattering is appropriate for organic structures? This demonstration was, in fact, made in the late 1970s involving diffraction experiments on monolamellar *n*-paraffin crystals. These crystals are convenient objects for such studies because (Figure 5.2) they are easy to grow from dilute solution in a suitable organic solvent, their thickness is rigorously controlled by the length of the carbon zigzag chain (and can be varied by the change in chain length for a homologous series packing with the same layer structure), and the methylene sublattice repeat is a convenient unit cell repeat, oriented parallel to the incident beam, with adequately small repeat distance (2.55 Å) for accurate dynamical scattering calculations. How, then, can one demonstrate the presence of *n*-beam scattering with an object that is comprised of light atoms?

It can be shown, first of all, that a primary extinction correction of the type proposed by Vainshtein and Lobachev (1956) or by Li (1963) improves the fit of

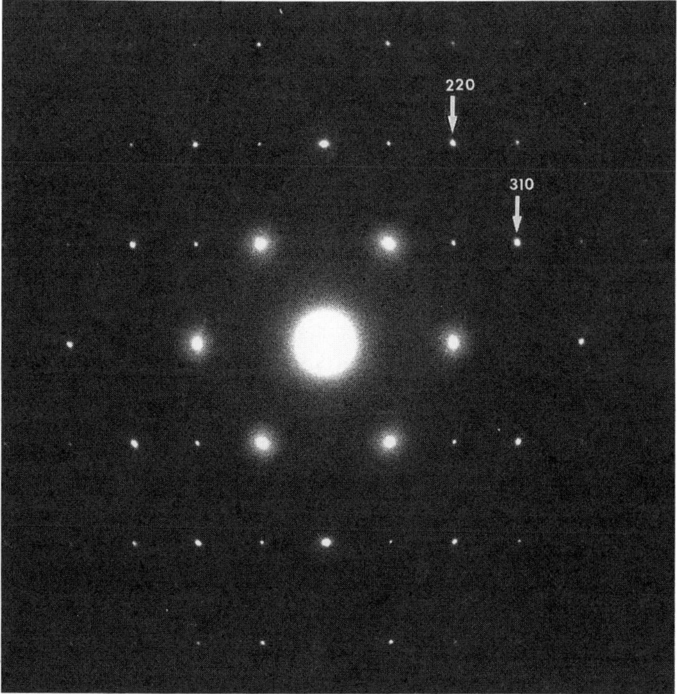

FIGURE 5.3. Selected area electron diffraction pattern from an *n*-paraffin monolayer for which the (220) and (310) reflections, the ones most influenced by *n*-beam dynamical scattering, are identified.

observed and calculated data. (This correction can be particularly useful for polycrystalline specimens, for reasons discussed by Cowley (1967) and Turner and Cowley (1969).) It is assumed that

$$KI = \{F_{obs}^2 \exp(-2B \sin^2\theta/\lambda^2)\}^{\alpha}$$

A graphical solution for the appropriate value of α is found by averaging over reciprocal space shells of measured intensity. However, it is also found that this correction is not very specific. There is no prediction of *which* reflections that will be most affected by dynamical scattering. A specific agreement to observed data is obtained (Dorset, 1976b) with the phase grating expansion above, with the virtue that the (310) and (220) reflections (Figure 5.3) are accurately found to be the ones in the *hk*0 set most affected by dynamical scattering. (More accurate predictions of this kind will be demonstrated in Chapter 6 for other organic compounds.)

A second test for *n*-beam scattering evaluates a series of diffraction patterns from crystals with the same thickness over a range of accelerating voltages. As mentioned above, the two-beam dynamical theory predicts convergence to the kinematical limit at a suitably high voltage. Electron diffraction patterns were obtained from monolayer crystals of the paraffin *n*-$C_{36}H_{74}$ at accelerating voltages from 100 kV to 1000 kV in increments of 100 kV. "Primary extinction plots" (Honjo and Kitamura, 1957) of ln I_{hk0} versus λ^2, where the intensity is normalized to a summed value for a subgroup of reflections, show no sign of convergence to more "kinematical" values. The same is true for plots of dynamical structure factor magnitudes versus accelerating voltage (Figure 5.4). In fact, dynamical scattering is not found to disappear at high voltage, since a phase grating correction always improves the fit of experimental and calculated structure factors (Dorset, 1976c).

The third parameter that can be varied conveniently in these tests is crystal thickness. An isostructural series of homologous *n*-paraffins from *n*-$C_{24}H_{50}$ to *n*-$C_{44}H_{90}$ was grown as monolayers to provide a series of crystals with exactly defined thickness increments. Electron diffraction intensity data were recorded in single-crystal patterns obtained at 100 kV, and the relative values of the (310) and (220) reflections were compared to their predicted value for the same crystal thicknesses. As shown in Figure 5.5, the fit of the *n*-beam calculation to the experimental data is very good (Dorset, 1980).

Finally, as proposed originally by Cowley and Kuwabara (1962), it is possible to demonstrate the presence of *n*-beam dynamical scattering if the relative intensity values on a continuously excited reciprocal lattice row is found to change as the crystal is tilted about this reciprocal axis. Two-beam dynamical theory or kinematical theory, on the other hand, would not predict any change of relative intensities during this experiment. It is possible to use single monolayer crystals of the linear wax, cetyl palmitate, for this purpose. The crystal packing of cetyl palmitate is virtually the same as that found for the monoclinic form of even-chain paraffins (Kohlhaas, 1938). The chains are tilted with respect to the same methylene orthorhombic subcell found for the paraffin monolayers above by some 27° around the subcell $a_s \approx 7.50$ Å axis, thus

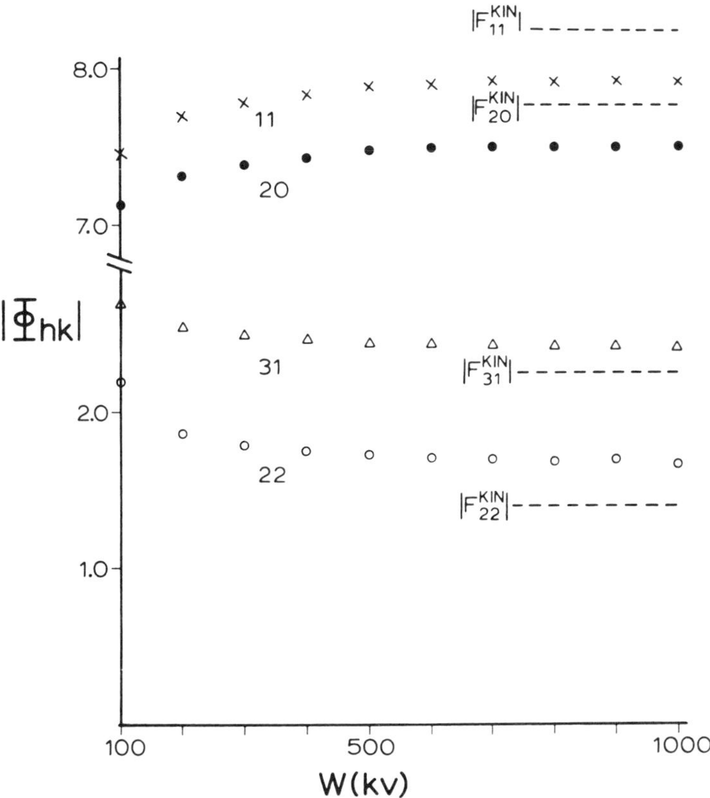

FIGURE 5.4. Plot of n-beam dynamical structure factor magnitude $|\Phi_{hk}|$ for a paraffin versus accelerating voltage W(kV), showing that the kinematical limit is never reached. This is, in fact, observed experimentally. (Reprinted from D. L. Dorset (1976) "Persistence of n-beam dynamical effects in the high voltage electron diffraction from single thin paraffin microcrystals," *Journal of Applied Physics* **47**, 780–782; with kind permission of the American Institute of Physics.)

giving the same electron diffraction pattern as shown in Figure 5.3 in this tilted orientation. Crystals of cetyl palmitate were so oriented so that the $\mathbf{d}^*_{100}$ subcell diffraction row was aligned to the goniometer tilt axis and a plot of I_{200}/I_{400} was monitored for the subcell reflections as the crystal was tilted from 27° to 0°. As shown in Figure 5.6, this ratio does not remain an invariant with tilt, demonstrating that n-beam interactions are important for the subcell diffraction, despite the fact that the subcell itself is only comprised of carbon and hydrogen atoms. N-beam theory anticipates this result because different numbers of reflections interact with the I_{h00} row as the crystal is being tilted through various zonal projections.

From the above, it is apparent that n-beam interactions can be detected experimentally in the electron scattering from organic crystals. Jap and Glaeser (1980) carried out rigorous theoretical multislice dynamical calculations for model organic

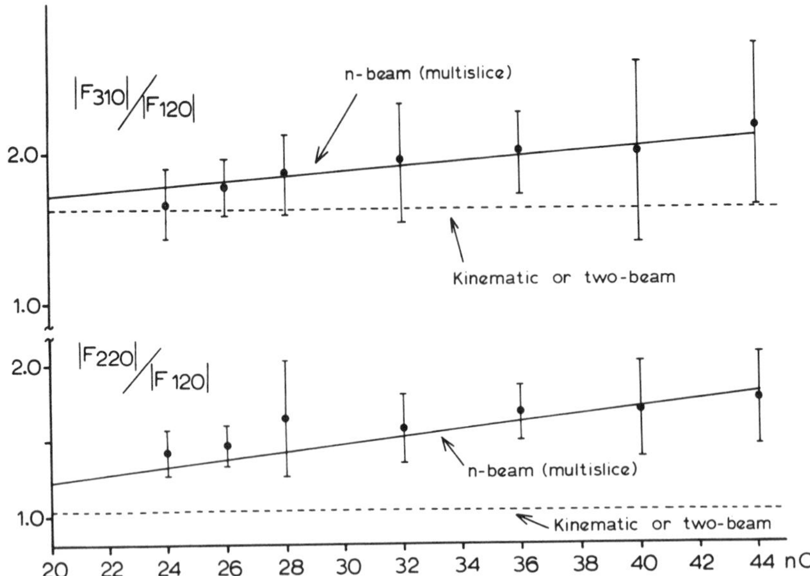

FIGURE 5.5. Plot of relative structure factor magnitudes for a rectangular paraffin layer versus lamellar thickness (expressed as the number of carbon atoms in the chain—e.g., see Figure 5.2). The experimental values are well-matched by an *n*-beam dynamical calculation but they do not agree with the results expected from kinematical or two-beam theory. (Reprinted from D. L. Dorset (1980) "Electron diffraction intensities from bent molecular organic crystals," *Acta Crystallographica* **A36**, 592–600; with kind permission of the International Union of Crystallography.)

crystals (cytosine and a disodium salt of 4-oxypyrimidine-2-sulfinate) to determine how variations of crystal thickness and electron accelerating voltage would affect the type of data available for forming high-resolution images (e.g., comparison of dynamical phases to kinematical crystallographic phases) or for their use in electron diffraction structure analyses (i.e., comparison of structure factor magnitudes). Obviously, higher voltages and thinner crystal layers predispose collection of a useful data set. Similar predictive simulations were carried out for electron diffraction data from native (Ho et al., 1988) and negatively stained (Dorset, 1984) protein microcrystals. More recently, dynamical scattering effects, expressed as a breakdown of Friedel symmetry, have been measured in 20-kV electron diffraction intensities from 45-Å-thick two-dimensional crystals of bacteriorhodopsin (Glaeser and Ceska, 1989). The differences in I_{hkl} and $I_{-(hkl)}$ were much less apparent at 120 kV. Multislice calculations based on a model bacteriorhodopsin structure model (Glaeser and Downing, 1993) indicated that the measured data should be close to the kinematical limit at 100 kV. If stacks of bacteriorhodopsin were used to simulate three-dimensional crystals, then the *R*-factor due to dynamical perturbations for a 200-Å crystal was 0.19 for a 100-kV source.

The utility of intensity data from small organics for ab initio structure analysis was evaluated by using simulated dynamical data (from a multislice calculation) for

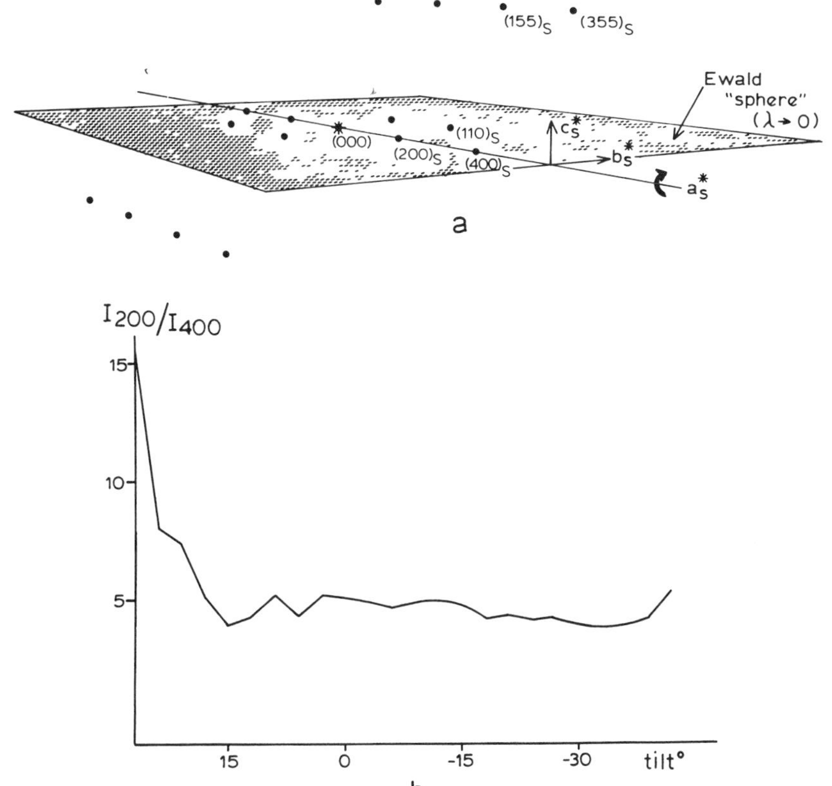

FIGURE 5.6. Continuous excitation of an (h00) row where the methylene subcell $\mathbf{a}_s^*$ axis is used as the tilt axis: (a) geometry of the experiment. (Reprinted from D. L. Dorset (1985) "Crystal structure analysis of small organic molecules," *Journal of Electron Microscopy Technique* **2**, 89–128; with kind permission of Wiley—Liss Division, John Wiley and Sons, Inc.); (b) plot of I_{200}/I_{400} versus tilt angle. The result is only explained by n-beam dynamical theory. (Reprinted from D. L. Dorset (1980) "Electron diffraction intensities from bent molecular organic crystals," *Acta Crystallographica* **A36**, 592–600; with kind permission of the International Union of Crystallography.)

a direct structure analysis with the tangent formula (Dorset et al., 1979). Although only two voltage extremes were considered (100 kV and 1000 kV), as well as a rather sparse sampling of crystal thicknesses, it was clear that the conditions for data collection lay well within the range experimentally attainable in the laboratory. For example, at 100 kV, intensity data from 76-Å thick cytosine crystals would be adequate for a successful structure determination. At 1000 kV, data from 300-Å thick crystals could be used. The statistical distribution of normalized structure factors was found to change from the expected centrosymmetric values to ones more like a noncentrosymmetric structure as dynamical scattering effects became more pronounced.

It is also interesting that Jap and Glaeser (1980) anticipated an optimal accelerating voltage for collection of electron diffraction data as a result of their model calculations. They argued that, as the Ewald sphere curvature is reduced at higher voltage, very high-resolution reflections will be excited more and more strongly as their excitation error is reduced. Thus there is a perceived tradeoff between the reduction of the interaction constant σ and the number of reflections contributing significantly to the convolution of phased structure factors. Certainly this model might be correct for a perfectly flat crystal held at 0 K. However, more recent diffraction experiments on copper perchlorophthalocyanine at different accelerating voltages indicated that such an optimal voltage for data collection in the 500-kV range may not exist (Tivol et al., 1993). Rather, it appeared that higher voltages always produced diffraction intensities more amenable to direct structure analysis. The reason for this discrepancy between simulation and experiment is that thermal motion in organic crystals is the most important factor for determining the actual resolution of the diffraction pattern recorded at, e.g., 20 °C. Although multislice calculations often are carried out to ultrahigh diffraction resolution—e.g., 4.0 $Å^{-1}$—this does not accurately model the actual resolution seen for most crystals at room temperature, e.g., 1.0 $Å^{-1}$. Thus the very high-angle reflections may have no experimental significance.

Also, one must recognize that the crystals used for electron diffraction experiments are not flat but have a finite curvature over the, e.g., 10 μm selected area diameter sampled in many diffraction experiments (Dorset, 1976b, 1978, 1980). In the multislice formulation of dynamical scattering, the Ewald sphere curvature makes an important contribution to the calculation as the crystal thickness increases. The actual excitation errors can be grossly overestimated, therefore, for actual electron diffraction data from nonflat crystals. For example, a comparison of dynamical structure factor magnitudes from multislice and phase grating calculations was made to experimental diffraction amplitudes from an n-$C_{36}H_{74}$ monolayer obtained at 20 kV (Dorset, 1992a), shown in Table 5.1. Although the agreements to experimental data in a suitable low-angle region, where the excitation errors are not very large, are quite similar, the multislice calculation gives a rather poor account of the complete set of observed data at the measured resolution. The phase grating expression does a better job, despite the fact that it quite unrealistically assumes that the Ewald sampling surface is well approximated by a plane. Thus, the observed diffraction resolution from organic crystals is not predicted by the multislice calculation. Because of elastic crystal bending, the experimental resolution is always greater, because, in effect, the higher-angle reflections are not extended along "reciprocal lattice rods" only by the sinc πNau shape transform of a thin crystal, but, more importantly, by some Gaussian function expressing the distribution of lattice sites along z. When the multislice calculation is used to model electron diffraction data from organic crystals—e.g., in the case of copper perbromophthalocyanine (Dorset et al., 1992b)—the comparison of observed and calculated data must be restricted in resolution to the region where the computed excitation error is suitably small. In other words, the dynamical scattering from a bent organic crystal would be accurately simulated if and only if an average could be made over all possible crystal orientations, as suggested originally by Turner and Cowley (1969) for another application to powder samples. However, when numerous crystal orientations and thicknesses are sampled in an electron diffraction experiment, it is

Table 5.1. Use of n-Beam Dynamical Calculations to Model Experimental 20-kV Electron Diffraction Data from n-Hexatriacontane

			$\lvert F_{dyn} \rvert$	
hk	d^*	$\lvert F_{obs} \rvert$	Phase grating	Multislice
20	0.27	4.48	4.79	7.93
40	0.53	2.08	1.64	1.16
11	0.24	5.19	4.98	8.58
21	0.33	1.39	1.33	1.71
31	0.45	2.51	3.15	2.90
41	0.57	1.32	0.85	0.42
51	0.70	0.95	1.02	0.52
02	0.40	3.16	2.70	3.32
12	0.42	1.29	1.06	1.10
22	0.48	2.54	3.21	2.52
32	0.57	1.57	1.42	0.64
42	0.67	1.23	1.91	0.18
52	0.78	1.06	0.37	0.29
13	0.62	1.31	1.26	0.64
23	0.66	1.25	1.02	0.23
33	0.72	1.01	1.37	0.42
43	0.81	1.00	1.21	0.29
		R-factor:	0.18	0.46

also clear that the strong nonsystematic diffraction effects observed in selected area diffraction from single crystals may be suppressed, so that the observed intensities may be adequately modeled by a two-beam or nearly kinematical approximation (Cowley, 1967). Otherwise, the successful use of texture diffraction intensities obtained in the 50- to 60-kV range for ab initio structure analyses by Russian electron crystallographers (see below) cannot be easily explained.

To summarize the above arguments and observations, n-beam dynamical theory is the only accurate description of electron scattering from a crystalline object. For purposes of ab initio structure analysis, however, its effect can be suitably minimized by experimental conditions so that a "quasi-kinematical" data set can be measured. Most of the comments made so far are mainly relevant to collection of useful diffraction intensities. The suitablility of electron microscope images for extraction of accurate crystallographic phases is another matter, particularly since n-beam interactions are first expressed as phase perturbations to scattered waves in the electron microscope column. Multislice calculations have been carried out for some molecular organic crystals by Ishizuka and Uyeda (1977) and O'Keefe et al. (1983). As mentioned above, such simulations, based on a known structural model, are often used to match an experimental image in the study of inorganic materials (Self and O'Keefe, 1988). How quickly these dynamical phase perturbations caused by multiple-beam interactions will degrade the measurement of crystallographic phases from experimental images has not been thoroughly evaluated, however. In a preliminary study of the zeolite structure, M. Pan and P. Crozier (1993) have found that measured phases from 2-Å resolution, 400-kV images from approximately 30-Å-thick crystals were very useful for phase expansion via the Sayre equation or the tangent formula (S. Kopp, unpublished data). When 200-Å crystals were used at this voltage, the phase extension was not possible since 40% of the resultant basis set from the image contained errors. Thus the phase aberrations due to the objective lens transfer function are not the only relevant perturbations to images, despite claims of this kind (Klug, 1978–1979) that multiple scattering is only a secondary complication. However, as will be demonstrated in Chapter 7, it is sometimes surprising how much useful information can be obtained from the image analysis of an experimental micrograph, even for inorganic materials.

Another commonly held misconception is that the appearance of space-group forbidden reflections in electron diffraction patterns is primarily the result of n-beam dynamical scattering. Actually, for many common symmetry operators, such as the 2_1-screw axis, centering operations, and simple glide planes, this is not so, as was demonstrated originally by Gjønnes and Moodie (1965). Since the self-convolution of phased structure factors is the mathematical operation in the n-beam scattering calculations that would account for the appearance of these space-group extinctions, i.e.:

$$F_{hk} * F_{hk} = \sum F_{h(2)k(2)} F_{h(1)-h(2),k(1)-k(2)}$$

the continued observation of forbidden reflections can be easily verified for any model data set with suitable symmetry. The result of the single convolution of $hk0$ structure factors for an n-paraffin monolayer with projected plane group symmetry pgg is listed in Table 5.2. Other screw operators such as a 3_1-axis, on the other hand, *can* lead to

Table 5.2. Single Convolution of Phased Structure Factors for n-$C_{36}H_{74}$

hk	$k\lvert F_h \cdot F_h\rvert$	hk	$k\lvert F_h \cdot F_h\rvert$
10*	0.00	12	0.73
20	4.94	22	2.55
30*	0.00	32	1.32
40	2.35	42	1.40
01*	0.00	52	0.98
11	5.16	03*	0.00
21	0.82	13	1.65
31	2.89	23	0.97
41	0.83	33	1.07
51	1.15	43	0.94
02	3.59		

*Space-group forbidden reflections.

the appearance of forbidden reflections (Cowley, 1981). Such considerations are very important for the interpretation of convergent beam electron diffraction patterns.

If a "primary extinction" correction for electron diffraction data were appropriate, it would have the virtue of permitting the observed intensities to be adjusted a priori, as demonstrated repeatedly in Vainshtein's (1964a) early structure analyses. At the other extreme, significant multiple-beam interactions imply that the crystal structure must be known before a correction can be made a posteriori—hence the commonly held view that a direct structure determination is impossible has led to virtual paralysis in the field, with the result that electron crystallographic structure analyses are rarely attempted for many heavy-atom materials. While the major intent of this book is to show that appropriate experimental conditions can be manipulated to permit collection of directly analyzable data, it can also be shown that, in some special cases, approximate corrections can be made a priori for even n-beam dynamical interactions.

An approximate experimental correction for n-beam scattering is possible when the crystal is made up of light-atom components and is in the thickness range where multiple-beam interactions just begin to become problematic (Dorset, 1992a). The correction is not rigorous and is most useful for obtaining a better estimate of normalized structure factor magnitudes for direct phase determination (Dorset et al., 1993) (see Chapter 4). In the phase grating series above, an imaginary part contains the kinematical structure factor, as well as higher terms:

$$Im(\mathbf{h}) = F^{\text{kin}}(\mathbf{h}) - (\sigma^2/3!)F^{\text{kin}}(\mathbf{h})*F^{\text{kin}}(h)*F^{\text{kin}}(\mathbf{h}) + \ldots$$

At a sufficiently high accelerating voltage, $Im(\mathbf{h}) \approx F^{\text{kin}}(\mathbf{h})$. There is also a real term:

$$Re(\mathbf{h}) = (\sigma/2!)F^{\text{kin}}(\mathbf{h})*F^{\text{kin}}(\mathbf{h}) - (\sigma^3/4!)F^{\text{kin}}(\mathbf{h})*F^{\text{kin}}(\mathbf{h})*F^{\text{kin}}(\mathbf{h})*F^{\text{kin}}(\mathbf{h}) + \ldots$$

and for some lower accelerating voltage, the approximation:

$$Re(\mathbf{h}) \approx (\sigma/2!)F^{\text{kin}}(\mathbf{h})*F^{\text{kin}}(\mathbf{h})$$

will be sufficiently accurate. In terms of measured dynamical intensity $I^{\text{dyn}}(\mathbf{h}) = F^{\text{dyn}}(\mathbf{h})^2$, we can write, therefore:

$$I^{\text{dyn}}(\mathbf{h}) = Im(\mathbf{h})^2 + Re(\mathbf{h})^2 = F^{\text{kin}}(\mathbf{h})^2 + Re(\mathbf{h})^2$$

Supposing that $I^{\text{dyn}}(\mathbf{h}) = F^{HV}(\mathbf{h})^2$, the dynamical intensities to be corrected are observed e.g., at 1000 kV. As has been observed in two cases, another data set measured at a lower voltage, e.g., 100 kV, is well approximated by $Re(\mathbf{h})^2$, i.e., $F^{LV}(\mathbf{h})^2 \approx Re(\mathbf{h})^2$, so that we can seek:

$$F^{\text{kin}}(\mathbf{h})^2 = F^{HV}(\mathbf{h})^2 - m\,F^{LV}(\mathbf{h})^2$$

where m is a trial weight (theoretically formed from the ratios of terms involving the interaction constant σ in the phase grating expressions). Corrections of electron

Table 5.3. Experimental Dynamical Correction for n-$C_{36}H_{74}$

hk	$\lvert F_{hk}^{800\,kV}\rvert$	$\lvert F_{hk}^{20\,kV}\rvert$	$\lvert F_{corr}\rvert$	$\lvert F_{kin}\rvert$
20	5.90	4.48	6.10	6.33
40	2.27	2.08	2.31	2.47
11	6.07	5.19	6.19	6.71
21	1.23	1.39	1.20	1.34
31	1.97	2.51	1.88	1.84
41	1.03	1.32	0.98	0.90
51	0.80	0.95	0.78	0.82
02	3.77	3.16	3.87	3.62
12	1.20	1.29	1.19	1.16
22	1.60	2.54	1.40	1.14
32	1.50	1.57	1.49	1.30
42	1.03	1.23	1.00	0.43
52	1.00	1.06	0.99	0.94
13	1.33	1.31	1.33	1.58
23	1.20	1.25	1.19	1.16
33	0.70	1.01	0.64	0.51
43	0.63	1.00	0.56	0.87

diffraction data from an n-paraffin and an epitaxially oriented phospholipid are shown in Tables 5.3 and 5.4.

For the above correction to be made, intensity data, first of all, must be obtained from a high-voltage electron microscope. Second, the phase grating series must be valid for the voltages used in the diffraction experiments. While this approximation is good for high voltages, when the Ewald sphere is sufficiently "flat," it would not be

Table 5.4. Dynamical Correction for L-DHPE

l	$\lvert F^{HV}\rvert$	$\lvert F^{LV}\rvert$	$F^{kin}\rvert$
1	1.49	1.69	1.33
2	0.62	1.29	0.34
3	1.42	1.39	1.30
4	0.82	1.10	0.61
5	1.00	0.89	0.93
6	0.52	0.64	0.45
7	0.80	0.82	0.73
8	0.40	0.53	0.34
9	0.31	0.58	0.21
10	1.02	0.87	0.96
11	1.04	1.05	0.95
12	1.80	1.29	1.72
13	1.55	1.41	1.45
14	1.84	1.44	1.75
15	1.27	1.12	1.19
16	1.06	0.87	1.00

useful at lower voltages if data were collected from perfectly flat crystals. However, it was already shown above that crystal domains sampled in many selected area diffraction experiments include a measurable bend component, so that the actual resolution of the experimental diffraction pattern is not well predicted by a multislice calculation (see Table 5.1). If the effect of the dynamical scattering is localized to the small-angle region of the diffraction pattern, i.e., for light-atom crystals, then the correction seems to be useful. When the structure contains heavy-atom components, on the other hand, such a correction cannot be made since the atomic scattering factors also contribute significantly to wide-angle regions.

5.2. Secondary Scattering

The major reason for the appearance of space-group forbidden reflections in experimental electron diffraction patterns is secondary scattering, an incoherent multiple scattering phenomenon. Cowley and his coworkers (1951) and also Vainshtein (1964a) had noted its effect on the electron diffraction intensity from multilamellar *n*-paraffin crystals (Figure 5.7). In effect, strongly excited beams from upper crystal-

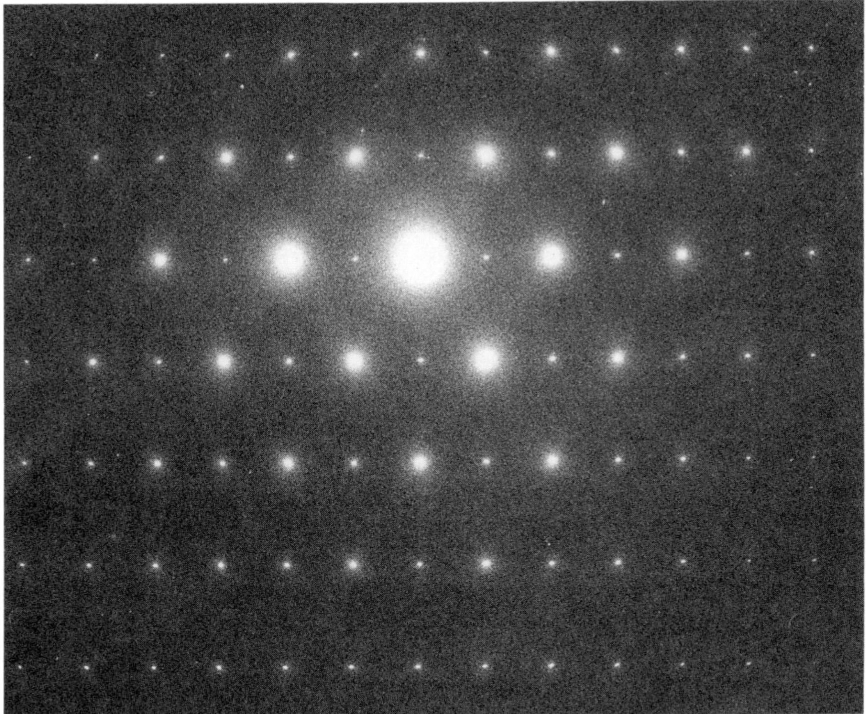

FIGURE 5.7. Electron diffraction pattern (*hk*0) from a paraffin multilamellar crystal with strong secondary scattering contributions. Forbidden reflections along (*h*00) and (0*k*0) are no longer absent and the pattern itself has a greater apparent resolution than the one shown in Figure 5.3.

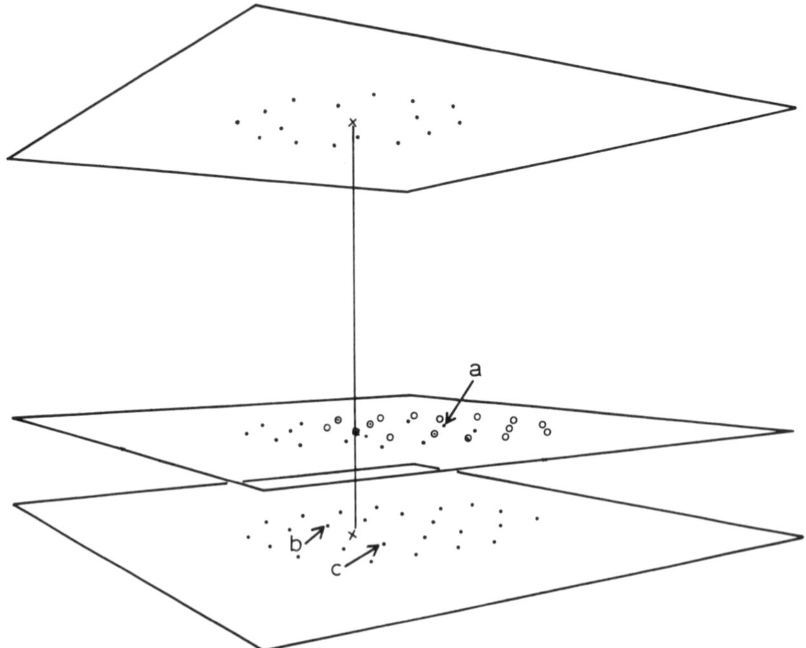

FIGURE 5.8. Geometry of secondary scattering. If successive crystalline layers are incoherently coupled to one another, then strong diffracted beams from upper layers can behave as ancillary primary beams for lower layers. (Reprinted from D. L. Dorset (1985) "Crystal structure analysis of small organic molecules," *Journal of Electron Microscopy Technique* **2**, 89–128; with kind permission of Wiley—Liss Division, John Wiley and Sons, Inc.)

line layers, which are essentially incoherently coupled to lower layers, behave as ancillary primary beams for the lower crystalline layers (Figure 5.8). Hence, instead of the intensity $I(\mathbf{h})$, one measures:

$$I(\mathbf{h})' = I(\mathbf{h}) + m\, I(\mathbf{h}) * I(\mathbf{h}) + n\, I(\mathbf{h}) * I(\mathbf{h}) * I(\mathbf{h}) + \ldots$$

The convoluted terms in the above series all have positive values. Therefore, as shown in Table 5.5, space-group forbidden reflections will not, by necessity, remain extinct under the convolution operations. Extinctions, e.g., due to centering operations can be preserved because only data, e.g., with $h + k = 2n$ occur in a zone and, thus, no vector combination will lead to an odd summation of indices. However, absences due to glide elements (i.e., screw axes or glide planes) are not preserved.

In experiments, secondary scattering can present a serious barrier to space-group identification, especially when it is not expected. To detect this perturbation, it is advisable to obtain many diffraction patterns from a particular crystal zone to determine how consistently reflections, which would lie at suspected absences, are ob-

Table 5.5. Convolution of $hk0$ Intensities for a Paraffin

hk	I	$I*I$	hk	I	$I*I$
10*	0	761	12	1.82	1186
20	46.10	7393	22	1.90	5386
30*	0	1087	32	2.79	1395
40	9.73	3582	42	0.38	1306
50*	0	771	52	2.25	844
60	0.76	1316	03*	0	1249
01*	0	1003	13	4.58	2591
11	50.98	7524	23	2.69	1125
21	2.19	974	33	0.59	963
31	4.75	6345	43	2.10	1030
41	1.42	1051	04	0.71	1323
51	1.46	1677	14	0.90	700
61	1.06	544	24	0.03	753
02	17.22	6659	34	1.64	746

*Space-group forbidden reflection.

served. Sometimes the appearance of such forbidden reflections is obvious when the translational symmetry operators indicated by very intense reflections are violated (Figure 5.7). In more difficult cases, secondary scattering must be detected by careful tilt experiments, in which suspected reflections on, e.g., reciprocal axes are continuously monitored. Plots of intensity along reciprocal lattice rows parallel to the beam direction suddenly will show sharp peaks because of this incoherent multiple scattering.

In many electron diffraction structure analyses, secondary scattering effects do not pose a significant problem. However, there have been cases where this incoherent scattering component has caused the measured data to be misinterpreted. For example, three-dimensional intensities from the room temperature form of C_{60} buckminsterfullerene (Figure 2.16) contain even-order axial ($h00$) reflections with high intensity values. As shown in earlier x-ray analyses (André et al., 1992), these intensities should, in fact, be very weak and the difference in measured intensities is important for determining whether or not the molecules pack in an ordered or random orientation in the crystal. Actual extinction of these reflections was detected only after careful tilt experiments were carried out (Van Tendeloo et al., 1992) and their intense contribution to the three affected zones could be verified after a simplified correction (Dorset and McCourt, 1994b) was made, i.e.:

$$I(\mathbf{h})' \approx I(\mathbf{h}) + m' \, I(\mathbf{h}) * I(\mathbf{h})$$

using the same value of m' for all three zones. The strong intensity of these reflections is due to the type of disorder found in the microcrystals, e.g., a series of stacking faults along [111] and other defects that would create the type of incoherently scattering layers required for this phenomenon to occur.

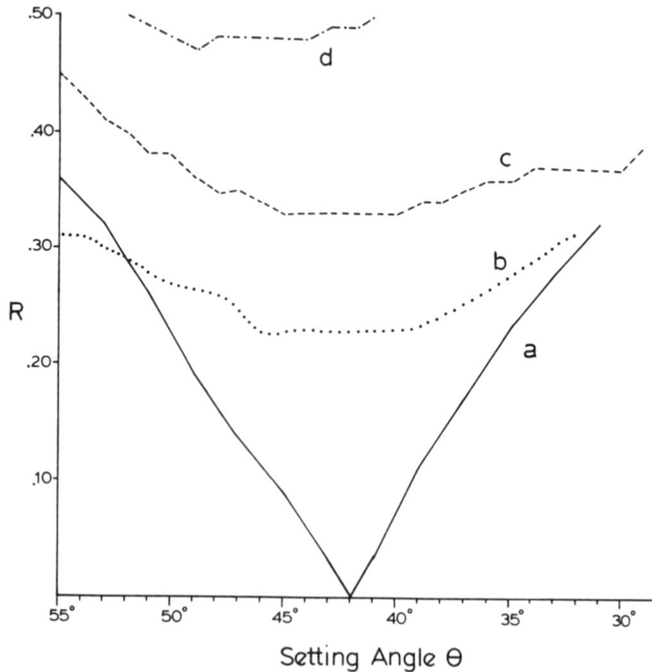

FIGURE 5.9. In-plane rotational search with a paraffin chain model for an R-factor minimum: (a) ideal data: (b) experimental data from a low-molecular-weight polyethylene; (c) simulated data incorporating a secondary scattering component; (d) experimental data from a paraffin where strong forbidden reflections were observed. (Reprinted from D. L. Dorset and B. Moss (1983) "Crystal structure analysis of polyethylene with electron diffraction intensity data. Deconvolution of multiple scattering effects," *Polymer* **24**, 291–294; with kind permission of Butterworths-Heinemann, Ltd.)

The simplified correction given above, involving one weighted convolution term, is appropriate only when the secondary scattering contribution is relatively weak. It would be difficult a priori to include higher-order terms since the additional weighting factors may lead to an overparameterized fit to the data. Even if the multiple incoherent scattering contribution is recognized and the space-group forbidden reflections are removed without any other correction to the data, it may be difficult to find a sharp minimum of the R-factor in a structure refinement (Figure 5.9). Also, secondary scattering can lead to a spurious increase of data resolution (Figure 5.7), and thus, the false conclusion that more structure details can be visualized in an analysis than are actually justified.

Unless there are unusual circumstances that cause a strong secondary scattering perturbation to the electron diffraction pattern, it is generally convenient to minimize its effect by limiting the thickness of the sample used for collection of intensity data.

5.3. Diffraction Incoherence due to Crystal Bending

As mentioned above, areas of molecular crystals selected for electron diffraction experiments are rarely flat, especially when radiation sensitivity requires that the sampled diameter should be large enough to record a useful signal at a suitably low incident beam dose so that there is no discernable damage to the specimen. Curvatures of several degrees can be observed experimentally in low-resolution bright-field micrographs by measurement of bend contours (Figures 1.10 and 5.2).

While the occurrence of crystal bending is widely recognized, the resultant influence on electron diffraction intensities is often overlooked. As discussed by Glaeser and Thomas (1969), electron beam illumination conditions used for low-dose diffraction experiments provide a source that has a high spatial coherence, especially when compared to typical x-ray diffraction experiments (see Chapter 1). The diffraction experiment, therefore, takes a coherent average over various crystal orientations within the several μm^2 selected area.

As noted by Cowley (1961) and Cowley and Goswami (1961), the illumination of an elastically bent crystal by a spatially coherent source results in an effective diffraction incoherence. Although the unit cells of some of the layer silicate samples examined in their early work were actually monoclinic, all the selected area electron diffraction patterns from them were observed to contain hexagonal symmetry, corresponding to the packing of sublayers normal to the incident beam. This result was explained by a kinematical model based on a perturbed Patterson function, that Fourier-transforms to the observed intensity:

$$I(\mathbf{s}) = \Sigma_i W_i(\mathbf{s}) \exp(2\pi i \mathbf{r}_i \cdot \mathbf{s}) \exp(-\pi^2 c^2 s^2 z_i^2)$$

Patterson vectors $w_i(\mathbf{r})$ transform to $W_i(\mathbf{s})$. (In Chapter 1 it was shown that, if the former is a δ-function weighted by A, the latter "scattering factor" is a constant amplitude over all reciprocal space with magnitude A.) The Gaussian function at the end of the expression contains the amount of crystal bending c in radians, the z_i component (along the beam path) of the Patterson vector $\mathbf{r}_i$ and the position of reflection h at $\mathbf{s}$. In effect, the unperturbed autocorrelation function of a crystal will include small and large vector components in the direction parallel to the slowest crystal growth (i.e., the direction normal to a crystal plate which, for untilted samples, is parallel to the incident electron beam). If the crystal is bent, the influence of the larger vector contributions to this Patterson function will be suppressed while the contributions from shorter vector lengths will be emphasized. Hence, the Fourier transform of the distorted Patterson function may no longer correspond to the total unit cell contents. The effect, moreover, is dependent on diffraction resolution. Lower-resolution intensities will be less affected than those measured at high angle.

Unfortunately, solution growth of organic crystals generally predisposes a molecular packing orientation where the effects of bend distortion are most apparent in the resultant zonal electron diffraction patterns. As discussed by Jensen (1970) and also Kitaigorodskii (1961), layers grow so that lateral intermolecular contacts are maximized while those contacts across layers are much less significant. (For example, the interlayer packing accounts for only 3% of the crystal potential energy for the paraffin n-$C_{36}H_{74}$ (Scaringe, 1991).) If the molecule is oblong, the resultant crystal plate habit requires that the longest molecular axes should be oriented more or less perpendicular to the major crystal face.

Experimentally, the effect of crystal bend distortions can be illustrated with data from polymethylene chain compounds, for which chains pack in successive layers but, within each layer, a methylene sublattice is found (Figure 5.10). Consider the orthorhombic polymorph of even-chain n-paraffins (Dorset, 1977, 1980). There are two

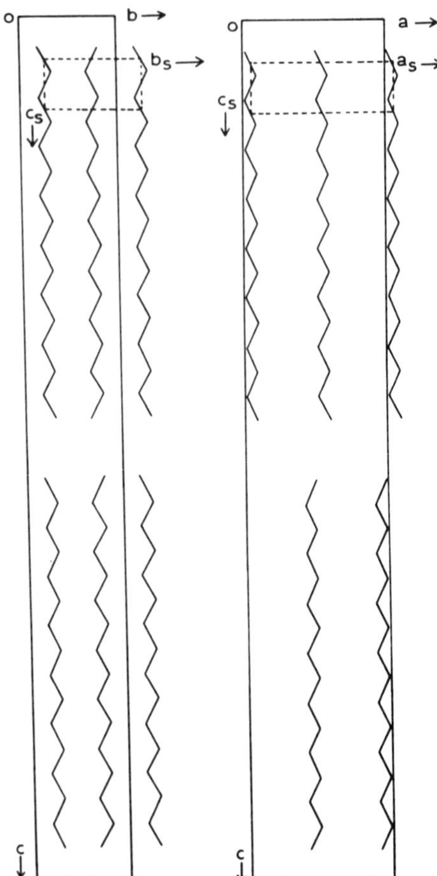

FIGURE 5.10. Schematic diagram of a paraffin multilamellar structure. For solution-crystallized samples, the incident electron beam will be parallel to the chain axes. There are two types of vector interactions in the Patterson that are important for describing the ensuing diffraction intensity: *short intra*molecular interactions due to the methylene "subcell" repeat along the chain and *long inter*molecular interactions due to the layer stacking. (For an orthogonal view, see Figure 4.4c, d.)

FIGURE 5.11. Electron diffraction pattern from multilamellar paraffin crystal in which the (110) reflections have appreciably lower intensity than the (200) spots. This is expected for a true bilayer unit cell. However, the higher angle intensities appear to originate from a chain monolayer and not the true bilayer unit cell. (Reprinted from D. L. Dorset (1977) "Electron diffraction intensities from slightly-bent multilayer paraffin crystals," *Zeitschrift für Naturforschung, Teil A* **32a**, 1166–1172; with kind permission of Verlag der Zeitschrift für Naturforschung.)

mutually shifted chain layers in the unit cell—e.g., for n-$C_{36}H_{74}$, many interlayer vectors at about 47.5 Å will lie parallel to the incident beam (Figure 5.10). The polymethylene sublattice, on the other hand, has a repeat of 2.55 Å, also parallel to the beam direction. At sufficiently large values of crystal bend, only the latter will make an important contribution to the modified Patterson function above; thus the diffraction pattern will appear to originate from single layers, even though a multilamellar crystal is being used for the experiment (Figures 4.4c,d and 5.3). At smaller bend values, the situation is somewhat more complicated. Smaller-angle reflections will have the intensity distribution expected for a flat crystal (e.g., $I_{110} \approx 0.5\, I_{200}$ in Figure 5.11) but the wider-angle reflections will not. These observations are well explained by Cowley's kinematical model.

When the chain axes are inclined to the incident beam, as in the oblique layer packing of symmetric wax esters (Dorset, 1978), different intensity distributions are observed depending on the amount of crystal curvature. A projected chain packing in an oblique layer is represented in Figure 5.12. The methylene carbons are successively aligned in rows so that the diffraction pattern resembles that from a linear grating (Figure 5.13a). For a flat crystal, the Fourier transform of the projected Patterson function, therefore, will correspond to the anticipated intensity distribution (Figure

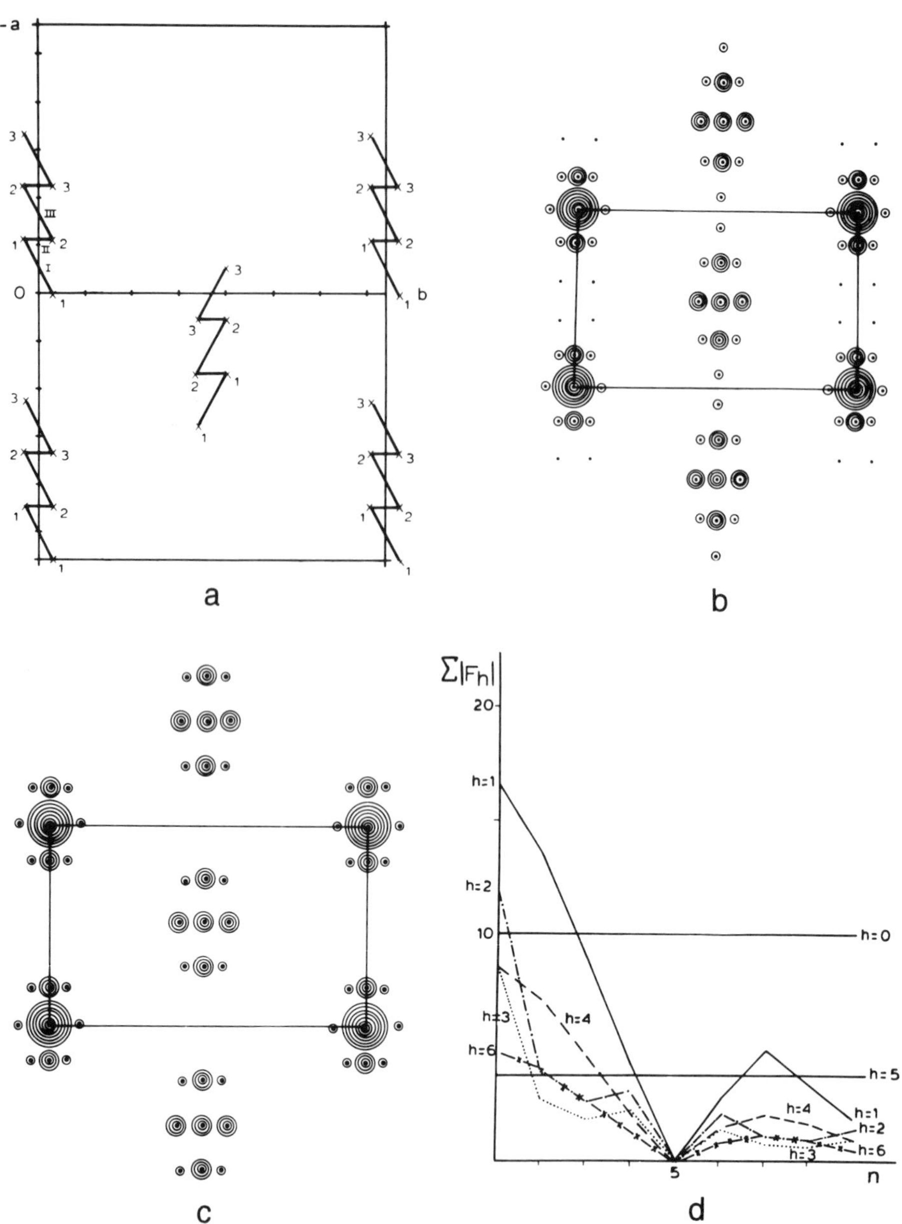

FIGURE 5.12. (a) Projected oblique layer packing of compounds such as symmetric wax esters and fatty acids where the orthorhombic methylene subcell $\mathbf{a}_S$ = 7.50 Å axis is the tilt axis for the chains. (b) Its Patterson function without bend deformation. (c) Its apparent Patterson function when all vectors beyond a z limit are eliminated (the result of bend deformation). this will change the symmetry and intensity distribution of the diffraction pattern. (d) Plot of summed layer line structure factor magnitudes for rows of constant h index, as a function of the number of coherently scattering carbon atoms along z. (Here $\mathbf{a}_{\text{unit cell}}$ = 5.6 Å, $\mathbf{b}_{\text{unit cell}}$ = 7.5 Å = $\mathbf{a}_S$.) All reflections other than h = 0,5 (the major transform of the grating in (a)) will be affected by this consequence of crystal bend. (Reprinted from D. L. Dorset (1978) "Transmission electron diffraction intensities from real organic crystals: thin plate microcrystals of paraffinic compounds," *Zeitschrift für Naturforschung, Teil A* **33a**, 964–982; with kind permission of Verlag der Zeitschrift für Naturforschung.)

5.12b). Bending the layer will alter the Patterson function so that even the projected symmetry is changed (Figure 5.12c). Depending on the amount of crystal bend, intermediate rows of reflections will be predicted to have a much higher intensity than is expected for the diffraction from a flat crystal (Figure 5.13b), a prediction that is matched by experimental observation.

The effect of crystal bending on the electron diffraction from solution-crystallized polymethylene compounds is so universal that the scattering from a particular methylene "subcell" can be isolated from that of the "functional" groups on the molecule if the crystal is tilted to project the electron beam along the chain axes (Dorset, 1983). As will be shown in Chapters 8–10, this result can be used conveniently to identify chain packing for unknown polymethylene chain compounds, since there is only a relatively small number of possible methylene subcells.

Is it also possible to model the effect of crystal curvature on the diffracted intensity more quantitatively? Moss and Dorset (1983a) carried out a multislice dynamical calculation for bent paraffin and polymer crystals by assuming a periodic continuation of the curved component to create a wave-form superlattice on which to superimpose the crystal structure repeat. Essentially, these calculations reproduced the results predicted by Cowley's kinematical model at the limit where the crystal layer is thin and the bend component is sufficiently large. However, as discussed by Turner and Cowley (1969), it is often not easy to include all possible perturbations to electron diffraction intensities in a single computational model.

If no sublattice exists for the structure being studied, the observed diffraction intensities may have no simple relationship to the underlying molecular packing when the crystal is bent, so that it would be pointless to attempt a crystal structure analysis. The amount of crystal curvature required to cause the observation of an uninterpretable data set was investigated by Moss and Dorset (1982) in their use of the tangent formula to determine a structure from simulated bend-distorted intensities for cytosine. The projection used in the model study was the most favorable orientation of the molecular packing down the shortest unit cell axis (3.82 Å). When the crystal was bent over 7.5°, it was no longer possible to find a correct solution.

Linear polymers represent a class of chemical compounds for which the presence of crystal bending can actually be advantageous for structure analysis. These materials typically crystallize as thin chain-folded lamellae, with chain axes often perpendicular to the largest crystal face. Because of crystal bending, electron diffraction intensities correspond to the subunit structure along the chain. Contributions from the chain folding, on the other hand, are not observed since the fold regions can be separated by, e.g., 100 Å. Thus, three-dimensional electron diffraction intensities can be used to determine the chain "stem" packing, as will be shown in Chapter 11. If the axial repeat is somewhat large (say > 20 Å), some correction for crystal bending may be required (Moss and Dorset, 1982) but, most often, the progress of an ab initio structure analysis is not severely compromised.

How is it possible to determine structures for which the projected axial length in solution crystallized samples is simply too large? For several years, this was a matter of great concern, e.g., in our study of lipids and alkane derivatives, since there are other

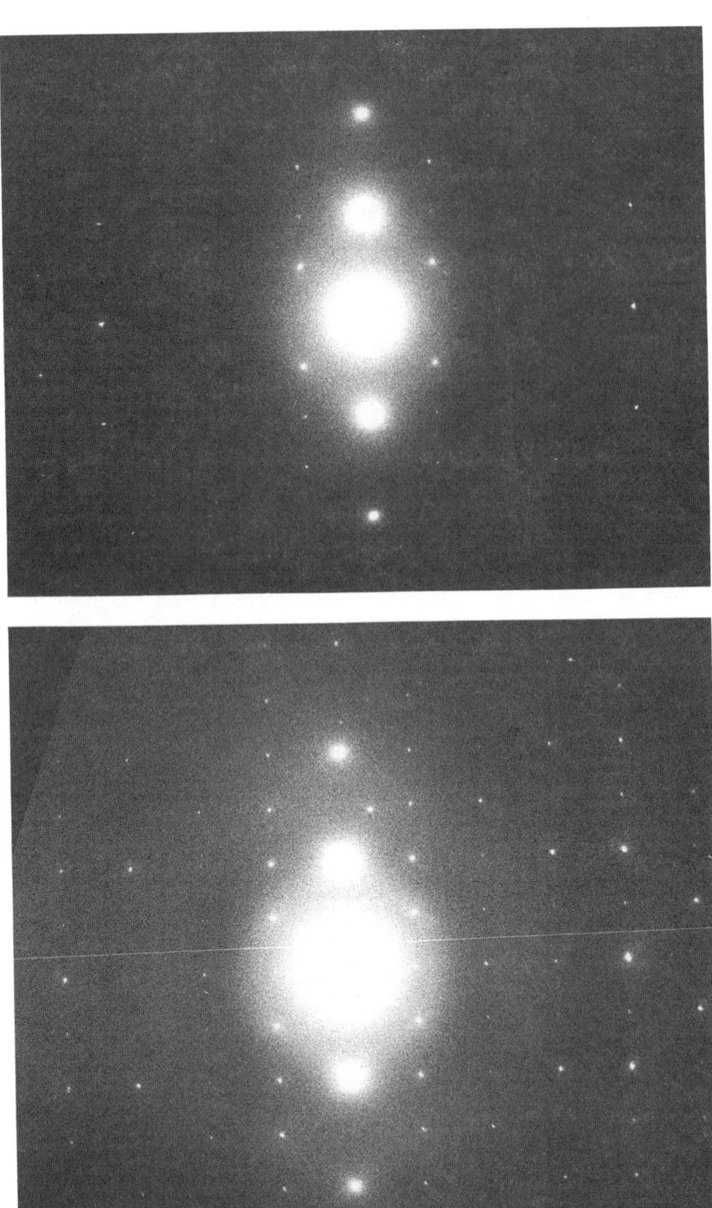

FIGURE 5.13. Electron diffraction from fatty acid crystals in the B-polymorphic form. (a) Crystal representing nearly the ideal condition for major rows of intensity at $h = 0,5$ but with large contributions for $h = 1$. (b) Crystal where large contributions appear for all values of $h \neq 0,5$, even when they should be vanishingly small for a perfectly flat lamellar plate. (Reprinted from D. L. Dorset (1978) "Transmission electron diffraction intensities from real organic crystals: thin plate microcrystals of paraffinic compounds," *Zeitschrift für Naturforschung, Teil A* **33a**, 964–982; with kind permission of Verlag der Zeitschrift für Naturforschung.)

important aspects of the molecular packing than just the methylene subcell packing! As will be shown in later chapters, diffraction data from conformationally locked analogs of glycerolipids were obtained to compare to those from the actual compounds, but this qualitative comparison often did not lead to a satisfactory conclusion. Such a method also would not be suitable for other classes of organic compounds.

Conceptually, the solution of this problem was simple. It required that a suitably short unit cell axis be oriented parallel to the incident beam direction so that the effects of crystal bending would have no serious effect on the recorded intensity data. This result could be achieved in two ways. First, the sample could be tilted to a large enough angle so that the unit cell projection would fall within the allowable z_i limits. Although it was not intentionally used for this purpose, the high tilt orientation of polycrystalline films for collection of texture diffraction data achieved this goal, even though very large sample areas (e.g., 1-mm diameter) were selected. (In Chapter 9, the utility of using such diffraction data from a lead soap for an ab initio structure analysis will be mentioned. Selected area diffraction data from untilted samples are useless for the complete analysis of the crystal structure, on the other hand.)

Second, if selected area diffraction is to be used for data collection, orient the material epitaxially on some suitable substrate (see Chapter 3), i.e., to force the longest unit cell axis to lie on one of the best developed crystal faces so that the projected unit cell length is sufficiently small. Thereby single crystal diffraction patterns are obtained where the intensity data correspond to the total unit cell contents. Numerous examples of successful crystal structure analyses of linear and aromatic molecules, taking advantage of this epitaxial orientation, will be discussed in later chapters.

Bend distortions can also impose a restriction on the structure analysis of globular proteins at near atomic resolution although, as discussed in Chapter 3, such crystals seem to be somewhat stiffer than plastically deformed molecular organic crystals. This possible limitation is particularly relevant because *all* unit cell axial spacings (and thus many important Patterson vector lengths) in protein crystals are large. A simulation of the effect of elastic bend distortions on the diffracted intensity from a protein crystal was made (Dorset, 1986) using atomic coordinates from the x-ray crystal structure of rubredoxin. Modest amounts of bending could have devastating effects on the use of high-angle electron diffraction intensities for resolution enhancement, e.g., by direct phasing methods. Fortunately, during the collection of diffraction data from bacteriorhodopsin, Henderson et al. (1986) demonstrated that strict controls could be imposed (by monitoring the spread of diffraction maxima at high tilt) to prevent the acceptance of data from crystals that are significantly bend-deformed.

5.4. Radiation Damage

One of the most discussed perturbations of an organic specimen by an electron beam is the inelastic scattering interaction which leads to damage. It is difficult to write generally about such interactions, since they can lead to different results for different types of materials. If the inelastic event produces an excited electronic state in a

Table 5.6. Observed Saturation Dose D for Radiation Damage of Organic Compounds*

Compound	D (C/cm^2)†
halogenated polynuclear aromatics and phthalocyanines	10^1
polynuclear aromatics and phthalocyanines	10^0
conjugated molecules	10^{-1}
aliphatics	10^{-3} to 10^{-2}

*From Fryer (1992). 100 kV at 25 °C.
†1.0 C/cm^2 = 624.2 e/Å^2.

molecule, then the consequences are largely *chemical* in nature (Fryer, 1987). Thus, for an alkane, beam damage might be expressed by hydrogen abstraction (Patel, 1975; Downing, 1983) or cross-linking of the carbon chain (Kobayashi and Sakaoku, 1965). For an aliphatic ether, a stable leaving group will lead to mass loss (Heavens et al., 1970). Numerous π-delocalized molecules will distribute the effect of the beam damage throughout the whole structure so that the overall geometry is preserved at higher electron doses than would be tolerated in the aliphatic examples (Fryer, 1984). Early work on the electron beam damage to organic specimens has provided information about doses that can be tolerated by the specimen (Table 5.6) but, until recently, changes in the electron diffraction pattern have been expressed solely by an exponential decay of individual reflection intensities (Reimer, 1965). This, unfortunately, provides little useful information. Following the progress of the damage in terms of a Patterson function would be more realistic since this corresponds to changes in molecular structure, for example. Another type of quantitative structural description has been given for a beam exposure series from a thin paraffin crystal (Dorset and Zemlin, 1985) (Figures 5.14, 5.15), wherein one can monitor the degradation of the molecule as it changes its cross-sectional detail. Circumstantially, this change has been attributed to an actual heating of the specimen by the electron beam, a false interpretation that has deceived many unwary workers in the field, including this author (Dorset and Turner, 1976). The similarity of this structural change to a thermotropic phase transition with the hexagonal chain packing as an endpoint (Thomas and Sass, 1973) can, therefore, be understood in chemical terms. The electron beam induces abstraction of hydrogen from the polymethylene chain and the creation of *trans* double bonds thereby lowers the melting point of the crystalline layer. There is also evidence for overall specimen movement from the initial intensity increase for lowest-angle reflections (Dorset and Zemlin, 1985), e.g., as might be imagined for the flattening of a bend-deformed crystal plate. Again, the specific details vary from specimen to specimen. Fryer (1984) describes how damage is nucleated at defect sites for aromatic molecular crystals, for example. Damage propagation may also proceed according to the ability that neighboring molecules have to interact chemically with one another, and this diffusion-limited process may take much longer than the first electronic change. In fact, it seems possible, in some cases, to capture an image from a specimen rapidly illuminated by a large dose of radiation before significant damage occurs via the slower chemical processes that follow (Fryer, 1987). It is also clear that temperature can have a major effect on stabilizing the crystalline order of the damaged sample. Although the primary damage caused by the beam is not very temperature-sensitive (Siegel, 1972), the "freezing-in" of damage (i.e., the slowing down of diffusion by chemically reactive products) at low temperatures is well documented (Siegel, 1972; Knapek, 1982; Fryer et al., 1992). From a practical standpoint, a five- to sevenfold increase of stability may be realized at liquid nitrogen cooling of a specimen and this can be increased to better than tenfold when liquid helium is used.

Concerns about radiation damage to organic specimens are much less important for electron diffraction experiments than they are for recording high-resolution images. In diffraction work on sensitive specimens, one becomes used to working at very low electron illumination levels and being able to carry out all required adjustments to the microscope in the dark or under subdued lighting. A photographic film is used that is very sensitive to electrons at the desired magnification of the diffraction pattern (i.e., the film has a rather large grain size). With practice, taking care that illumination is deflected from the sample at "dead periods" when it is not necessary to observe or photograph the diffraction pattern, it is entirely possible to collect a tilt series from a single aliphatic molecular crystal without significant change occurring in the sample. If there are concerns about such damage, it is also possible to monitor individual intensities as a function of dose so that a correction to zero-time exposure can be made (Perez and Chanzy, 1989).

High-resolution microscopy of organic specimens is much more difficult because one is trying to record details as point-to-point spatial resolution in an image instead of having this information conveniently confined to discrete δ-functions in a diffraction pattern. Such work often demands use of a statistically noisy image so that the perfection of the crystalline lattice is relied upon for producing an average representation of the unit cell after image averaging (see Chapter 4). However, appropriate low-dose procedures are now widely used by most labs (including the use of cooling stages) so that the earlier concept of "normal" illumination levels, equivalent to exploding a 10-MT nuclear bomb 30 m from the specimen (Grubb, 1974), is no longer thought to be a prerequisite for such work, even though such extreme conditions still may be used in the study of inorganic materials.

In order to predict the ultimate resolution that can be obtained from an organic crystal, there was once much discussion of the Rose equation

$$Cd \geq (S/N)/\sqrt{(fN_{cr})}$$

Here C denotes the inherent image contrast (defined as the difference in image intensity between two points, divided by the local average image intensity), d is resolution, S/N (≈ 5) is the signal-to-noise ratio, f accounts for detection efficiency of the film, and N_{cr} is the critical exposure for damage. For paraffins, it was stated that perhaps 40- to 100-Å detail would be attainable (Glaeser, 1975; Thomas and Ast, 1974). Unfortunately, this kind of prediction had the same demoralizing effect on many researchers as did the n-beam dynamical theory discussed above, even though intentions were good. Crucial experiments were not attempted until John Fryer (1981) published 16-Å

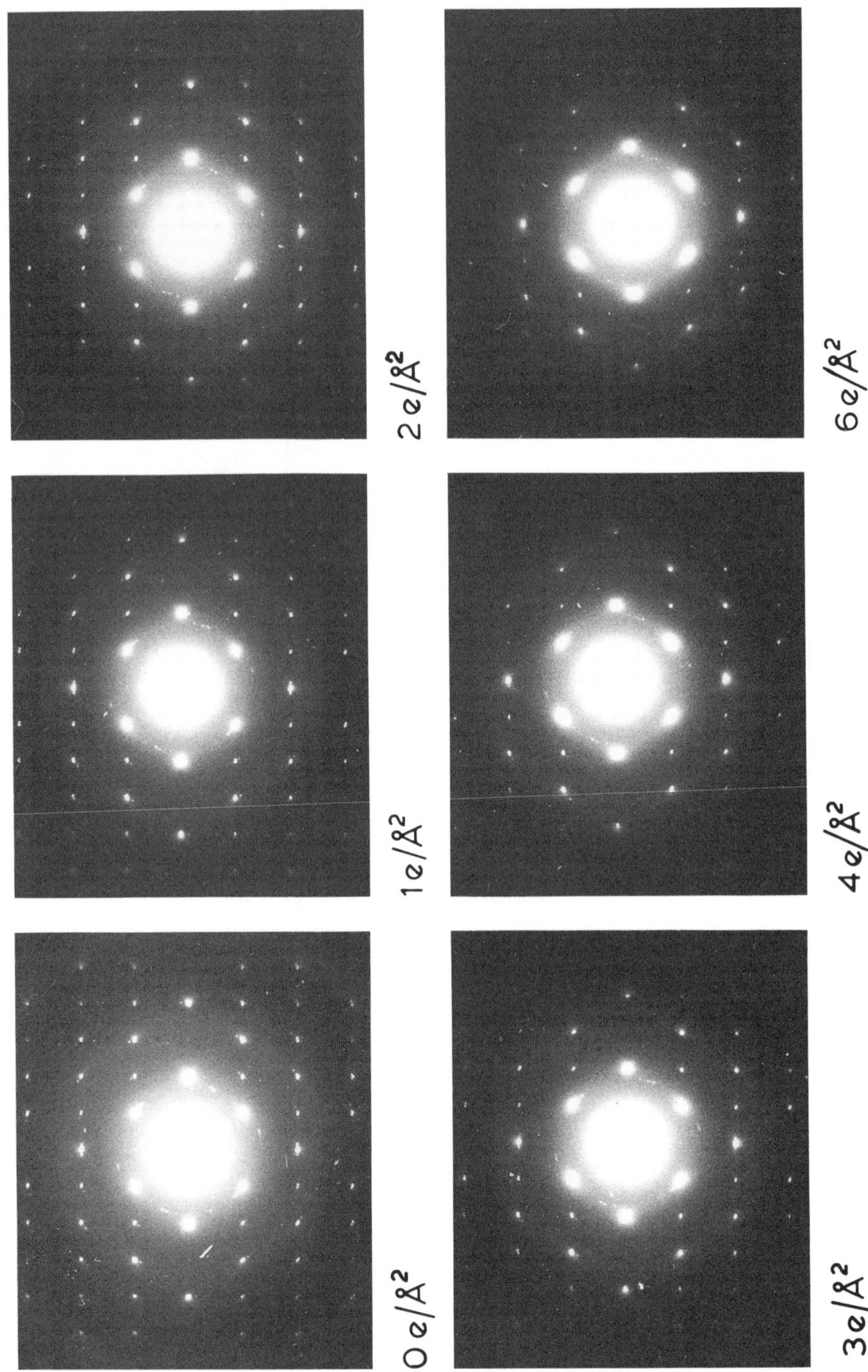
0e/Å²
1e/Å²
2e/Å²
3e/Å²
4e/Å²
6e/Å²

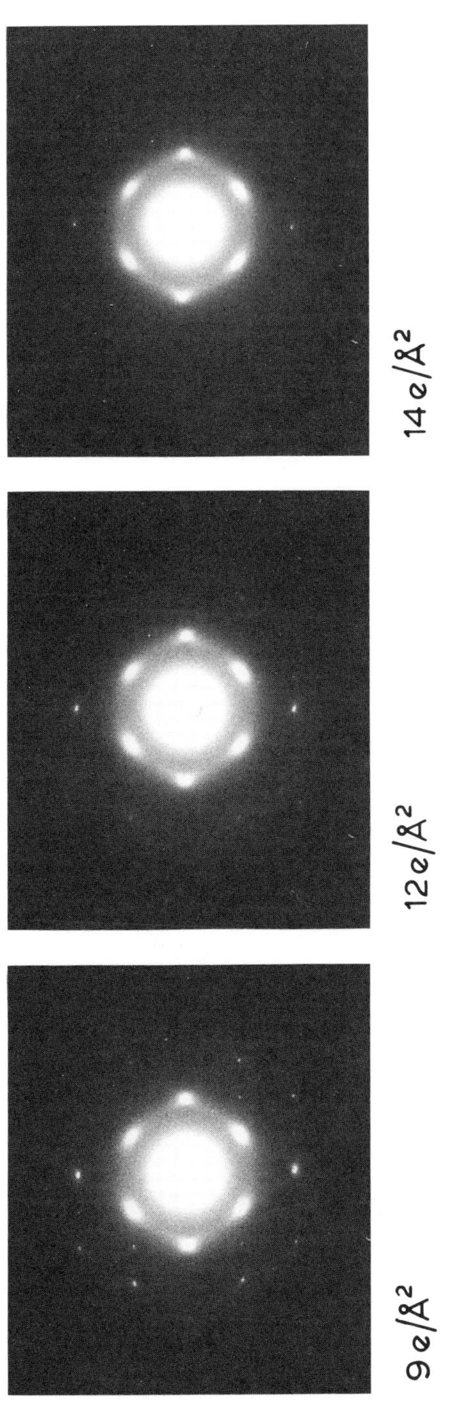

FIGURE 5.14. Change in the electron diffraction pattern from a paraffin layer with increasing exposure to the electron beam. (Reprinted from D. L. Dorset and F. Zemlin (1985) "Structural changes in electron-irradiated paraffin crystals at <15 K and their relevance to lattice imaging experiments" *Ultramicroscopy* **17**, 229–236; with kind permission of Elsevier Science B.V.)

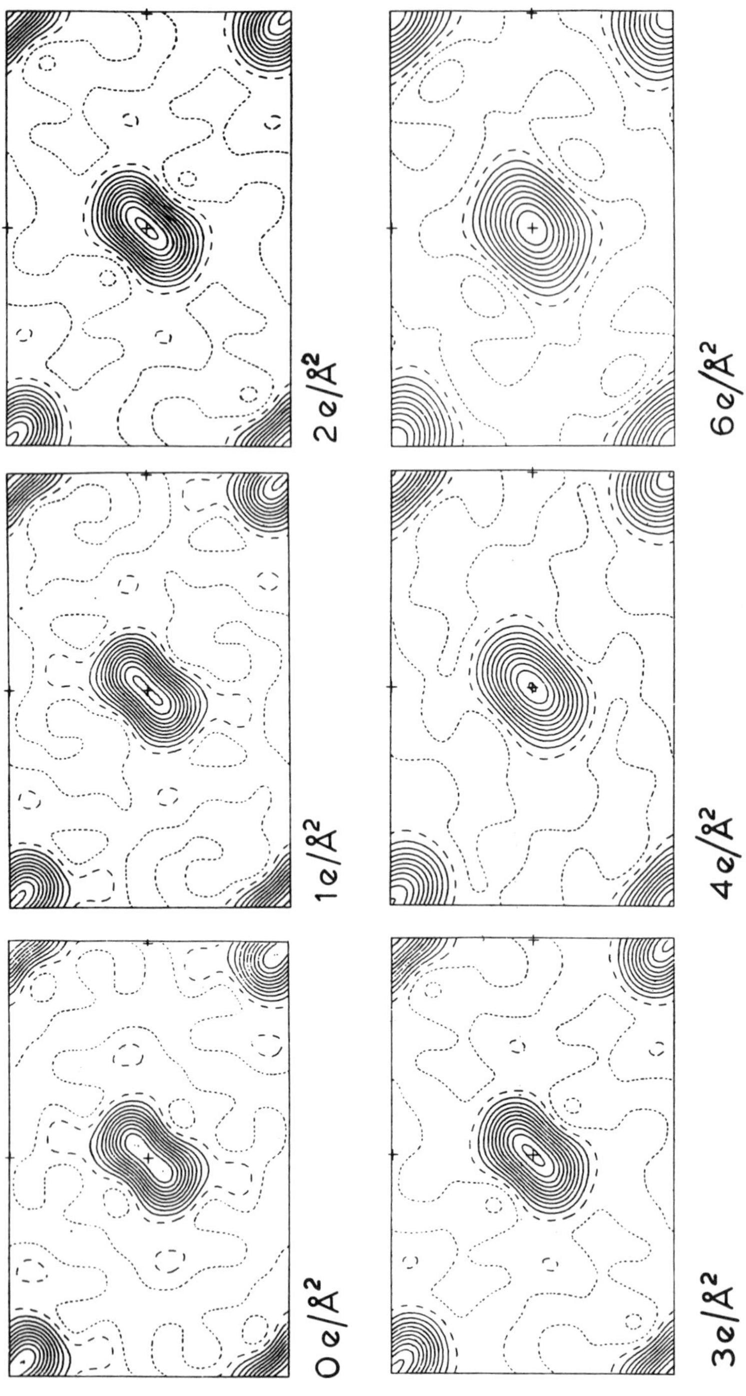
0e/Å²
1e/Å²
2e/Å²
3e/Å²
4e/Å²
6e/Å²

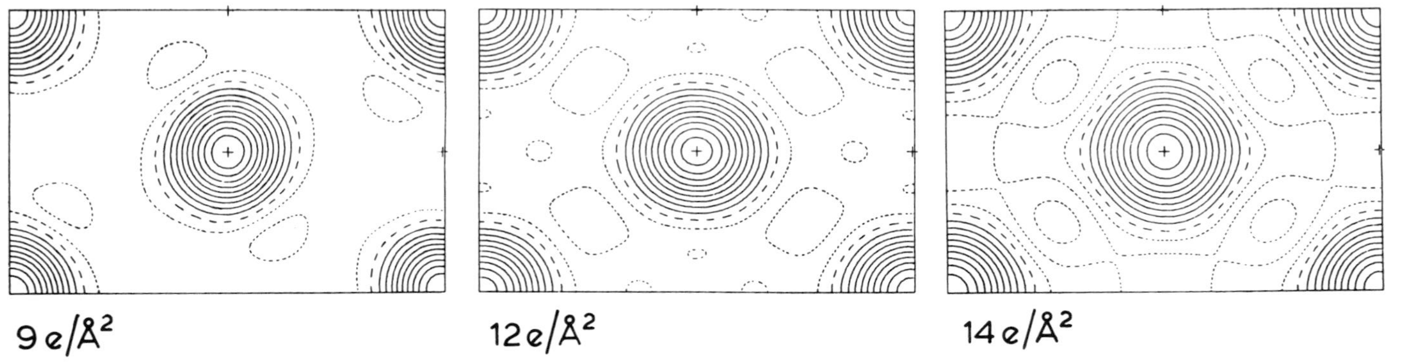

FIGURE 5.15. The radiation-induced changes in the electron diffraction patterns from a paraffin layer (Figure 5.14) correspond to structural alterations, i.e., as if the chains become more circular in cross-section. Such changes would also be expected for a thermotropic transition to a so-called "rotator phase" (see Chapter 8), a structural similarity that has often led to an erroneous interpretation of the damage mechanism. (Reprinted from D. L. Dorset and F. Zemlin (1985) "Structural changes in electron-irradiated paraffin crystals at <15 K and their relevance to lattice imaging experiments," *Ultramicroscopy* **17**, 229–236; with kind permission of Elsevier Science B.V.)

resolution images taken at room temperature from epitaxially oriented paraffin crystals—a row of reflections which also turns out to be the most radiation-sensitive of the whole diffraction pattern (Dorset et al., 1984a)! Images from solution-crystallized monolamellae at 2.5-Å resolution were obtained with a liquid helium-cooled cryoelectron microscope (Zemlin et al., 1985) and this was later extended to 2.1 Å with a liquid nitrogen-cooled sample stage (Brink and Chiu, 1991). However, 3.7-Å images of polyethylene had been obtained about the same time as the cryomicroscopy of paraffin, but at room temperature (Revol and Manley, 1986). There also has been considerable discussion about the merits of "spot-scan," instead of "flood-beam," illumination (Downing and Glaeser, 1986) to reduce the radiation-induced specimen movement. Such specimen movement has been observed experimentally (Dorset and Zemlin, 1985, 1987; Henderson and Glaeser, 1985). Although this illumination mode seems to have been universally adopted by researchers working on thin protein crystals, it is not certain that it is required for high-resolution imaging of organic crystals. For example, the work of Revol (1991) illustrates what can be done at room temperature using great care. In general, the most important requirement for studying radiation-sensitive samples still seems to be to make nearly every electron count toward recording useful structural information.

5.5. Conclusions

It cannot be denied that many potential pitfalls await the unwary when attempts are made to use electron scattering data for ab initio structure analyses. Despite the number of possible data perturbations that can occur, one single criterion can be imposed to improve the likelihood that an analysis will lead to a successful outcome. That is: *Be certain that the data collected from any crystal orientation are at least self-consistent*—i.e., that there is a good match of observations for a large number of individual samples. If there is no such agreement then there is no point in continuing until the source of the problem is identified and corrected. When corrections to the data are attempted, it is also a prerequisite that the physical model correspond to the experimental parameters employed for data collection. For example, a model appropriate for a nanometer-diameter area may be quite inadequate for a selected area diffraction experiment sampling a micrometer diameter. The former experimental condition favors a perfectly flat crystalline slab of uniform thickness whereas the latter test object would comprehend curvilinear distortions and a number of defects. Thus, while dynamical scattering is a rigorous model for the former case, the latter condition may necessitate the inclusion of secondary scattering and the effect of elastic crystal bending. Description of a powder or texture is more complicated, especially if millimeter diameters are irradiated, and, as will be seen, partly explains why the two-beam model has often served as a useful approximation for such data sets.

Part II

Applications

Nuscht macht dem Forscher so gnatzig wie Zeijen die er selbst zerdäppert hat!

SIEGFRIED LENZ, *Heimatmuseum*

Nothing irritates the experimentalist so much as clues (to solving a problem) that he has obscured himself.

6

Molecular Organic Structures

6.1. Background

The first attempt to solve a molecular organic crystal structure quantitatively from electron diffraction intensity data was made by Rigamonti (1936), who characterized a paraffin layer packing. Because the application of electron diffraction techniques to the study of polymethylene compounds is of appreciable historical and practical importance, these results will be discussed in detail in later chapters.

For crystals of other classes of small molecules, the first attempt at a quantitative structure analysis, based on estimated intensities, was made by Karpov (1941) for 1-methyl-7-isopropylphenanthrene. The structure of "retene" was solved by trial-and-error techniques but no crystallographic residual was reported. This was the beginning of the extensive electron crystallographic study of organics in Moscow, much of which will be reconsidered below in light of modern techniques for determination of crystallographic phases.

Other analyses in this early Russian effort, which will not be discussed in detail, include the study of the urotropine structure (Lobachev, 1954), based on powder electron diffraction intensities. A model was constructed from the fit of molecular symmetry to the highly symmetric space group but, again, no agreement to the measured intensity data was given. Most recently, copper salts of amino acids were investigated (Vainshtein et al., 1971; D'yakon et al., 1977), using single-crystal diffraction patterns in combination with oblique texture data or by themselves. The crystal structures were solved by finding heavy-atom positions in Patterson maps. For copper DL alaninate, the final R-factor was 0.19, but the intensity data have not been published. More detailed analyses are described in the following sections.

6.2. Early Data Sets from Moscow

6.2.1. Diketopiperazine

An oblique texture diffraction data set was collected from diketopiperazine by Vainshtein (1955), consisting of 317 intensities, 289 of which were observed (Table 6.1). The space group is $P2_1/a$, with cell constants $a = 5.20$, $b = 11.45$, $c = 3.97$ Å, β

Table 6.1. Observed and Calculated Structure Factors for Diketopiperazine after Least-Squares Refinement

hkl	$\|F_o\|$	F_c	*hkl*	$\|F_o\|$	F_c	*hkl*	$\|F_o\|$	F_c
020	11.94	12.02	021	2.51	−3.94	351	0.00	0.54
040	11.94	−11.87	031	4.30	4.18	361	4.06	−4.71
060	4.18	−4.51	041	1.07	0.73	371	1.07	−1.07
110	7.16	7.60	051	4.78	−4.08	381	0.00	1.44
120	3.70	5.16	061	1.55	2.76	391	2.86	−2.55
(130	3.58	−0.26)	071	3.82	−4.66	3,10,1	5.85	4.08
140	0.60	−1.27	111	22.68	−21.78	31,−1	2.65	−1.63
150	3.34	−3.53	121	6.45	5.93	32,−1	2.98	1.44
160	5.97	−5.46	131	2.03	3.15	33,−1	2.86	−2.55
210	6.09	−4.77	141	2.15	1.67	34,−1	0.84	0.95
(220	10.39	−6.22)	151	8.48	8.19	35,−1	1.19	−0.29
230	1.91	−2.61	161	1.67	−1.69	36,−1	1.67	−1.33
240	2.51	−2.85	171	0.96	−0.13	37,−1	2.03	2.03
(250	9.07	5.54)	181	0.96	0.32	38,−1	3.58	−2.75
260	2.51	1.83	11,−1	7.76	5.52	39,−1	1.43	0.89
270	1.79	1.71	12,−1	5.97	5.73	3,10,−1	2.63	−1.61
280	4.78	4.16	(13,−1	9.55	7.12)	401	0.96	−0.58
290	5.97	−4.78	14,−1	2.51	−1.47	411	0.96	1.29
210,0	4.18	3.01	15,−1	0.48	1.34	421	2.03	2.49
310	0.60	−0.64	16,−1	4.54	−3.67	431	1.43	0.16
320	2.39	−0.86	17,−1	6.68	−5.14	441	3.22	3.32
330	6.21	−6.50	18,−1	4.42	2.91	451	3.34	−3.66
340	1.43	1.70	19,−1	5.97	−4.70	461	2.39	−2.44
350	2.98	1.70	1,10,−1	4.18	4.16	471	2.86	−3.26
360	4.78	3.65	201	6.09	−5.88	481	2.98	−4.33
370	7.04	6.22	211	1.55	−2.02	491	0.00	0.06
380	2.98	2.24	221	1.79	−2.03	40,−1	6.68	−5.68
390	4.78	3.77	231	0.00	−0.75	41,−1	5.01	4.29
400	0.00	0.69	241	2.75	2.73	42,−1	4.54	−3.05
410	0.72	−0.48	251	4.18	3.81	43,−1	5.61	4.92
420	1.79	1.87	261	2.27	2.34	44,−1	0.00	0.90
430	1.79	−1.29	271	2.27	3.08	45,−1	2.27	−1.37
440	2.75	1.98	281	0.60	0.07	46,−1	2.39	1.84
450	1.79	−0.55	291	0.60	0.79	47,−1	0.72	−1.33
460	0.48	−0.62	(20,−1	4.89	−1.10)	48,−1	1.31	1.31
470	2.51	1.79	21,−1	2.27	−0.49	49,−1	1.79	2.28
480	1.79	−1.62	(22,−1	7.28	5.16)	511	0.84	−0.62
490	3.10	2.38	23,−1	0.00	0.80	521	0.48	0.36
510	2.39	2.67	24,−1	8.72	7.16	531	0.48	−1.78
520	4.18	−5.64	25,−1	1.55	−1.14	541	1.07	1.16
530	1.55	1.50	26,−1	4.78	−3.26	51,−1	1.55	−1.84
540	1.91	−1.88	27,−1	2.98	−2.36	55,−1	1.19	1.85
550	0.00	−0.60	28,−1	7.76	−7.24	601	0.60	−0.64
560	2.03	1.88	29,−1	1.91	−2.23	611	2.63	2.86
600	2.98	1.85	2,10,−1	1.19	−1.00	621	1.67	−1.44
640	2.03	−2.02	311	5.49	5.88	(631	1.43	3.51)
650	0.96	−0.57	321	2.98	3.90	641	0.72	−1.32
001	7.70	−6.91	331	2.39	1.86	012	2.03	2.89

(continued)

Table 6.1. (*Continued*)

hkl	$\vert F_o \vert$	F_c	*hkl*	$\vert F_o \vert$	F_c	*hkl*	$\vert F_o \vert$	F_c
011	3.58	4.91	341	2.39	−2.30	022	4.54	−5.08
032	0.84	2.34	332	1.43	−1.05	123	1.91	−1.99
042	4.66	−4.89	342	1.43	−1.67	133	4.06	4.28
052	1.79	−1.74	352	1.43	−2.15	143	0.00	−0.58
062	1.91	2.52	362	1.55	−2.72	153	2.75	1.09
072	1.67	0.66	3,10,2	1.31	−0.69	163	0.00	−0.35
082	4.78	5.76	31,−2	8.12	6.62	173	3.58	−4.29
092	2.98	3.67	32,−2	6.92	−5.15	183	1.91	−2.20
010,2	1.79	1.90	33,−2	0.00	0.08	193	2.63	−3.23
112	1.19	2.49	34,−1	0.00	1.02	1,10,3	0.96	−1.52
(122	2.27	−4.70)	(35,−2	9.31	−2.76)	11,−3	0.96	−0.98
132	2.51	2.50	36,−2	3.34	2.36	12,−3	0.84	0.84
142	0.84	−0.26	37,−2	0.84	−0.91	13,−3	2.03	−1.56
152	0.00	0.35	38,−2	0.60	−0.71	14,−3	0.72	−0.30
162	2.98	4.06	39,−2	0.84	−0.88	15,−3	0.60	0.51
172	1.91	−2.26	402	4.06	−5.26	16,−3	1.43	−1.75
182	2.15	2.53	412	0.84	−1.58	17,−3	0.84	1.95
192	0.00	0.58	422	1.79	−2.28	18,−3	1.19	−1.85
11,−2	0.72	1.14	432	0.00	−0.41	203	0.00	−0.20
12,−2	1.67	2.49	442	0.00	1.03	213	1.19	1.78
13,−2	7.04	−7.03	452	2.75	3.44	223	0.84	−1.75
14,−2	0.00	0.17	462	0.00	0.34	233	1.79	2.02
15,−2	3.58	−2.32	472	1.19	1.70	243	0.84	−2.04
16,−2	1.31	−0.71	482	1.43	0.35	253	0.60	−1.24
17,−2	6.92	5.98	492	1.79	−2.53	263	0.84	0.97
18,−2	3.10	2.42	40,−2	4.42	3.02	273	1.67	−2.52
19,−2	3.58	4.45	41,−2	1.31	0.49	20,−3	7.64	−8.15
1,10,−2	2.15	2.46	42,−2	0.96	0.28	21,−3	2.27	2.11
202	13.01	13.83	43,−2	0.00	0.13	22,−3	4.18	−3.06
(212	1.79	−4.62)	44,−2	3.34	2.47	23,−3	2.03	2.10
222	5.25	5.03	45,−2	0.00	1.65	24,−3	3.22	2.99
232	5.25	−5.39	46,−2	1.55	−0.63	25,−3	1.19	−1.31
242	3.94	−4.78	512	0.00	0.39	26,−3	2.39	1.08
252	0.00	1.06	522	0.60	−1.41	27,−3	0.96	−0.82
262	3.58	−2.36	532	0.60	−1.96	313	3.70	−5.32
272	1.55	1.29	542	0.60	1.75	323	3.82	5.48
282	0.00	0.29	552	0.00	0.04	333	0.00	0.17
292	2.39	−1.81	562	2.98	3.74	343	1.79	1.76
2,10,2	1.79	−2.33	51,−2	1.79	−3.02	353	2.77	2.15
20,−2	1.43	1.51	54,−2	1.19	0.99	363	0.84	−1.91
21,−2	2.63	−0.87	003	1.19	0.57	31,−3	2.51	−2.08
22,−2	1.79	2.22	013	0.96	2.31	35,−3	1.07	1.69
23,−2	2.15	−1.11	023	2.27	3.24	403	0.96	−1.73
24,−2	1.31	1.29	033	0.96	1.60	443	0.72	1.61
25,−2	0.00	0.76	043	4.78	3.92	42,−3	1.19	1.55
26,−2	0.96	−1.66	053	1.79	−2.38	513	2.51	3.14
27,−2	3.10	2.37	063	0.96	−1.65	543	1.43	−1.42
28,−2	1.19	−1.68	073	0.00	−0.39	553	0.60	−1.03
29,−2	3.22	−1.68	083	4.18	−4.60	563	0.96	−2.19

Table 6.1. *(Continued)*

hkl	$\|F_o\|$	F_c	*hkl*	$\|F_o\|$	F_c	*hkl*	$\|F_o\|$	F_c
312	2.63	2.44	093	2.39	3.20	004	0.00	0.66
322	1.19	0.91	113	2.03	1.24	064	0.60	−1.32
074	0.60	−1.46	134	0.96	−2.37	154	0.00	−0.67
164	0.00	2.57	11,–4	4.30	5.52	15,–4	1.55	−2.70
16,–4	0.96	0.85	224	0.72	−1.78	244	1.43	−2.39
264	1.19	1.59	284	1.79	3.38	20,–4	1.79	1.97
21,–4	0.96	1.31	22,–4	0.96	0.60	23,–4	0.00	0.91
24,–4	1.19	−1.12	25,–4	0.84	−1.90	26,–4	0.60	−0.63
(005	1.91	−4.76)						

*(Reflections in parentheses not used in refinement.)

= 81.9°. Since **b** is the texture axis (hence, within the missing cone), some theoretical values were added for the $0k0$ reflections, based on a contemporary x-ray structure. The measured intensity data were corrected for the Lorentz factor generally applied to such oblique texture measurements (see Chapter 3). In addition, some low-angle reflections were corrected for two-beam dynamical scattering (see Chapter 5). The intent of the crystal structure analysis was to provide more accurate hydrogen atom positions than were available from x-ray determinations. Thus the heavy-atom positions from the known x-ray structure (Corey, 1938) were used as an initial source of crystallographic phases, so that the lighter atom positions could be located in subsequent difference potential maps.

After this work was published, there was some question about whether it was really possible to carry out an ab initio structure analysis from the measured electron diffraction intensity data. This is because the most of the phase information had been determined already from the x-ray crystal structure. For the new structure determination (Dorset, 1991d), the published intensity data were accepted without any further change, even though they were obtained at an accelerating voltage near 50 kV. A Wilson (1942) plot indicated that the overall isotropic temperature factor $B = 0.0$ Å^2. Thus, no adjustment was made to the electron scattering factor tables for calculation of $|E_h|$. The distribution of $|E_h|$ was found to agree well with the expected values for a centrosymmetric structure (Karle et al., 1965) (Table 6.2). Next, Σ_1- and Σ_2-triples and quartets were generated for a direct phase determination.

The unit cell origin was defined by setting $\phi_{-312} = 0$, $\phi_{111} = \pi$. In an analysis of phase accuracy, it was found that 146 Σ_2-triples formed for $A_2 \geq 2.0$ contained only two erroneous relationships and that 90/92 quartet invariants with $B \geq 2.0$ also were correct. Evaluation of the individual invariant sums, including one Σ_1-triple ($\phi_{222} = 0$) yielded phases for 133 reflections. A potential map, calculated when these 133 phase terms were combined with measured $|F_o|$, was directly interpretable in terms of the molecular geometry (Figure 6.1), and the measured fractional atomic coordinates were found to be very similar to the ones obtained by Vainshtein and in a more recent x-ray structure analysis (Degeilh and Marsh, 1959) (Table 6.3). The structure was improved

Table 6.2. Distributions of $|E_h|$ for Diketopiperazine

		Theory	
	Experimental	Centrosymmetric	Noncentrosymmetric
$<\|E_h\|^2>$	1.000	1.000	1.000
$<\|E_h^2 - 1\|>$	0.995	0.968	0.736
$<\|E_h\|>$	0.787	0.798	0.886
% $\|E_h\| > 1.0$	29.9	32.2	36.8
% $\|E_h\| > 2.0$	4.7	5.0	1.8
% $\|E_h\| > 3.0$	0.3	0.3	0.01

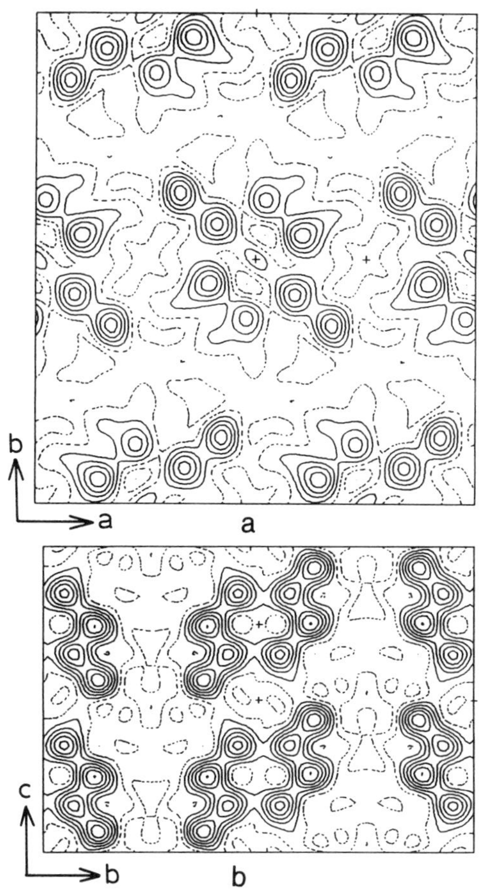

FIGURE 6.1. Electrostatic potential maps for diketopiperazine calculated from zonal phase information obtained by symbolic addition: (a) [001] projection; (b) [100] projection. (Reprinted from D. L. Dorset (1991) "Electron diffraction structure analysis of diketopiperazine—a direct phase determination," *Acta Crystallographica* **A47**, 510–515; with kind permission of the International Union of Crystallography.)

Table 6.3. Fractional Coordinates of Unique Atoms in Diketopiperazine

	Symbolic addition (e.d.)			X-ray structure		
atom	*x*	y	*z*	*x*	*y*	*z*
C 1	–0.175	0.072	0.708	–0.1820	0.0697	0.7170
C 2	0.051	0.120	0.502	0.0515	0.1233	0.5150
N	0.212	0.049	0.308	0.2198	0.0432	0.3098
O	–0.345	0.130	–0.138	–0.3311	0.1328	–0.0956

	After least-squares refinement (e.d.)		
atom	x	*y*	*z*
C 1	–0.181	0.073	0.708
C 2	0.046	0.121	0.514
N	0.223	0.047	0.306
O	–0.343	0.132	–0.106

after the initially obtained atomic coordinates were used to calculate a complete set of phased structure factors so that the new map was based on all phase values.

It was proposed that the theoretical I_{0k0} values used by Vainshtein might bias the direct phasing analysis. These were removed to leave an observed data set with 286 nonzero intensities. After generation of triples and quartets, the direct phase determination was again attempted, using the same reflections as before for origin definition. Again the analysis was successful, producing 93 phase values. When these were combined with the experimental $|F_o|$, the resulting potential maps were again directly interpretable (Figure 6.2), the only difference from the initial maps being some slight elongation of atomic density profiles along **b**.

An automated phasing procedure was also able to determine this structure successfully (Dorset and McCourt, 1994a). A basis set comprised of $\phi_{-312} = 0$, $\phi_{111} = \pi$, $\phi_{3,10,1} = 0$, $\phi_{-281} = a$, $\phi_{370} = b$, $\phi_{-172} = c$, $\phi_{-322} = d$ was constructed and used to generate 2^4 possible solutions with the QTAN version of the tangent formula. Each solution consisted of 239 defined phase values. The correct structure appeared within the four most negative values of NQEST but not at its lowest value, which, in fact, corresponded to a false solution (Figure 4.10). If the correct map was chosen on the basis of chemical knowledge of the molecular geometry, the atomic coordinates again were observed to be close to the values listed in Table 6.3. A successful analysis of these published data also has been reported by Gilmore et al. (1993a), who had utilized the maximum entropy approach. The resultant map again resembled the ones obtained by evaluation of phase invariant sums by more traditional procedures. McMillan (1994) has found a correct solution when these data were phased with another version of the tangent formula. Using the combined figure of merit described by Cascarano et al. (1987), the correct solution corresponded to the most likely phase set.

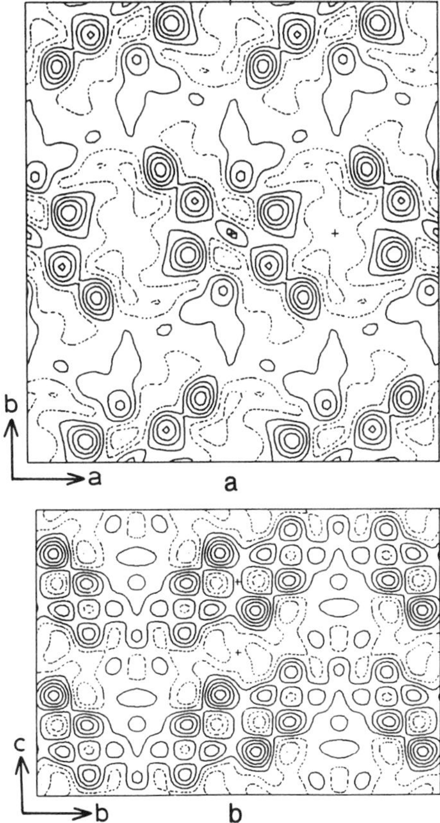

FIGURE 6.2. Electrostatic potential maps calculated from diketopiperazine zonal data from phases determined after unmeasured (0*k*0) reflections were removed from the data set: (a) [001] projection; (b) [100] projection. (Reprinted from D. L. Dorset (1991) "Electron diffraction structure analysis of diketopiperazine—a direct phase determination," *Acta Crystallographica* **A47**, 510–515; with kind permission of the International Union of Crystallography.)

Although Fourier refinement was found to lead to slightly improved atomic peak positions in the experimental potential maps, there was enough data to attempt least-squares refinement of the structure, representing the first practical application of a nearly unconstrained refinement in electron crystallography (Dorset and McCourt, 1994a). The refinement had to be carried out cautiously, however. First it was realized that inclusion of thermal parameters was useless, given the overall intensity distortions imposed by dynamical scattering. Therefore, only atomic positions were refined and, in the use of appropriately revised computer programs from x-ray crystallography (i.e., the appropriate change of electron scattering factor tables), it was soon found that the allowed translational shifts were too large and that a dampening factor of 0.1 to 0.2 had to be applied to constrain the optimization to a local R-factor minimum. Nevertheless, a stable structure was found after 10 cycles minimizing $\Sigma w(F_o - F_c)^2$ and, as shown in Figure 4.13, the molecular geometry was more similar to the one found in a more modern x-ray crystal structure analysis. On the other hand, there was no convincing evidence of hydrogen atoms in experimental potential maps so that these positions could not be included in this analysis. The final $R = 0.25$ for the calculated structure factors in Table 6.1.

These determinations have revealed that the direct phasing relationships are unexpectedly robust in face of the dynamical scattering distortions to the intensity data. The Σ_2-triples (also used in the various versions of the tangent formula) were particularly reliable as were the positive quartet invariants. On the other hand, the evaluation of negative quartets—especially in terms of the NQEST figure of merit—was found to be very sensitive to the nonkinematical intensity values. Dynamical scattering can be regarded approximately as a convolutional smearing of all intensity in the diffraction pattern. Therefore, the kinematically weak reflections to have, on average, larger $|E_h|$ values than expected. Only 79% of experimental and kinematical reflections had the same Miller index value when $|E_h| \leq 0.5$. On the other hand, the QTAN analysis of simulated kinematical data set from this crystal structure actually found the correct model at the lowest NQEST value. (The difficulties experienced in identifying the correct structure seemed to be overcome when a combined figure of merit was used instead, as indicated above.)

Dynamical scattering also can hinder the progress of structure refinement. In general, the refinement should be constrained to a local minimum if the kinematical R-factor is calculated. Search for a "global" minimum, on the other hand, can lead to a badly distorted molecular geometry.

6.2.2. Urea

A single crystal intensity data set from urea (Table 6.4) was collected by Lobachev and Vainshtein (1961) at about 50 kV. The space group is tetragonal $P\bar{4}2_1m$, where $a = b = 5.66$, $c = 4.71$ Å, and the 60 observed data were from a (010) centrosymmetric zone (plane group pmg) from crystals tilted 45° around the needle axis. The intensities had been corrected by a Lorentz factor, accounting for a mosaic distribution around the crystal length and the structure was determined by starting from the coordinates of a contemporary x-ray analysis, again with the intent to find hydrogen positions in the difference potential maps.

Lorentz-corrected intensities from the earlier work were used without any adjustments for direct structure analysis (Dorset, 1991b). A Wilson (1942) plot for these zonal data indicated an isotropic $B = 1.8$ Å^2, but no thermal adjustment was made to the scattering factors for calculation of $|E_h|$. These normalized structure factors were distributed as expected for a centrosymmetric zone (Table 6.5).

For the 15 reflections where $|E_h| \geq 1.0$, 15 Σ_2-triples were generated with $A_2 \geq 2.6$ (one being incorrect) and 8 positive quartets with $B \geq 19.5$. The origin was defined by setting $\phi_{401} = \pi$, $\phi_{103} = 0$, to find new phases from the invariant sums, yielding only four new values. An algebraic value was then given to $\phi_{002} = a$, which permitted access

Table 6.4. Electron Diffraction Data for Urea

$h0l$	$\lvert F_o\rvert$	$\lvert F_c\rvert$	ϕ	ϕ_{LV}*	$h0l$	$\lvert F_o\rvert$	$\lvert F_c\rvert$	ϕ	ϕ_{LV}*
001	0.20	0.90	π	π	306	0.31	0.05	$-\pi/2$	$-\pi/2$
002	3.75	2.74	π	π	307	0.11	0.06	$-\pi/2$	$\pi/2$
003	0.25	1.85	π	π	308	0.06	0.05	$-\pi/2$	$-\pi/2$
004	1.20	0.68	π	π	400	0.80	0.22	0	0
005	0.60	0.37	0	0	401	2.05	2.06	π	π
006	0.28	0.26	0	0	402	0.55	0.72	0	0
007	0.17	0.20	0	π	403	1.45	1.13	0	0
008	0.06	0.13	π	π	404	0.48	0.18	0	π
101	4.75	5.22	$\pi/2$	$\pi/2$	405	0.18	0.20	π	π
102	1.40	2.43	$\pi/2$	$\pi/2$	406	0.12	0.22	π	π
103	1.68	1.84	$-\pi/2$	$-\pi/2$	407	0.08	0.05	π	0
104	0.42	0.54	$\pi/2$	$\pi/2$	501	0.30	0.03	$\pi/2$	$-\pi/2$
105	0.50	0.71	$-\pi/2$	$-\pi/2$	502	0.55	0.02	$\pi/2$	$\pi/2$
106	0.20	0.00	$-\pi/2$	$-\pi/2$	503	0.37	0.48	$-\pi/2$	$-\pi/2$
107	0.28	0.15	$\pi/2$	$\pi/2$	504	0.37	0.46	$\pi/2$	$\pi/2$
108	0.07	0.02	$\pi/2$	$-\pi/2$	505	0.11	0.11	$-\pi/2$	$-\pi/2$
200	3.90	4.01	0	0	506	0.03	0.01	$-\pi/2$	$-\pi/2$
201	4.20	3.49	π	π	600	0.89	1.19	0	0
202	0.82	0.33	0	0	601	0.13	0.22	π	π
203	1.47	1.04	0	0	602	0.32	0.18	π	π
204	0.74	0.04	0	π	603	0.09	0.18	π	π
205	0.28	0.12	π	0	604	0.16	0.10	π	π
206	0.17	0.16	π	π	605	0.16	0.06	0	0
207	0.09	0.11	0	0	701	0.42	0.38	$\pi/2$	$\pi/2$
208	0.04	0.06	π	0	702	0.21	0.32	$\pi/2$	$\pi/2$
301	1.78	1.84	$-\pi/2$	$-\pi/2$	703	0.21	0.19	$-\pi/2$	$-\pi/2$
302	1.02	1.54	$-\pi/2$	$-\pi/2$	704	0.17	0.01	$\pi/2$	$-\pi/2$
303	0.45	0.99	$-\pi/2$	$-\pi/2$	800	0.46	0.34	0	0
304	1.55	1.54	$\pi/2$	$\pi/2$	801	0.23	0.09	π	π
305	0.40	0.09	$\pi/2$	$\pi/2$	802	0.06	0.06	π	π

*ϕ_{LV} is the phase value calculated from coordinates in the paper by Lobachev and Vainshtein (1961). To convert phases to determination described in text, add $\pi h/2$, where h is the first index of $h0l$.

Table 6.5. Distributions of $\lvert E_h\rvert$ for Urea

		Theory	
	Experimental	Centrosymmetric	Noncentrosymmetric
$\langle\lvert E_h\rvert^2\rangle$	1.000	1.000	1.000
$\langle\lvert E_h^2 - 1\rvert\rangle$	0.997	0.968	0.736
$\langle\lvert E_h\rvert$	0.812	0.798	0.886
% $\lvert E_h\rvert > 1.0$	25.0	32.2	36.8
% $\lvert E_h\rvert > 2.0$	10.0	5.0	1.8
% $\lvert E_h\rvert > 3.0$	0.0	0.3	0.01

Table 6.6. Atomic Coordinates for Urea

	Direct phasing (electron diffraction)		X-ray structure	
Atom	*x,y*	z	*x,y*	*z*
C	0	0.308	0	0.3308
O	0	0.571	0	0.5987
N	0.145	0.170	0.1429	0.1848

to nine further phases. Permuting $a = 0,\pi$ lead to two trial potential maps (Figure 4.9), one of which clearly was recognized to be the structure of urea. Atomic coordinates taken from this map were then used to generate phases for all 60 reflections. A map based on these phase values depicted the atomic positions more clearly (Figure 4.9c) and the new atomic coordinates (Table 6.6) were then used to calculate bond distances and angles (since the molecular plane is parallel to the projection plane). Although the final $R = 0.34$, the agreement of these bond parameters to a more recent x-ray crystal structure (Vaughan and Donohue, 1952) is very good (Figure 6.3). On the other hand, there was no convincing evidence for hydrogen atom positions in the experimental potential maps.

a
O
1.24Å
119.2° C
1.33Å
121.3°
N N

b
O
1.28Å
120.9° C
1.35Å
118.2°
N N

c

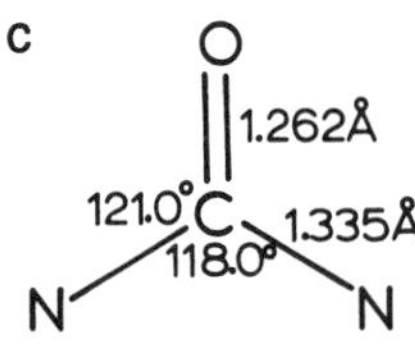

FIGURE 6.3. Bond distances and angles for urea after crystal structure analysis: (a) direct phase determination and Fourier refinement with electron diffraction data; (b) previous results of Lobachev and Vainshtein (1961); (c) x-ray crystal structure. (Reprinted from D. L. Dorset (1991) "Is electron crystallography possible? The direct determination of organic crystal structures," *Ultramicroscopy* **38**, 23–40; with kind permission of Elsevier Science B.V.)

6.2.3. Thiourea, Paraelectric Form

Oblique texture diffraction intensity data (187 independent three-dimensional reflections) were collected at room temperature from thiourea samples at about 50 kV (Dvoryankin and Vainshtein, 1960). The orthorhombic space group is centrosymmetric, Pnma, with unit cell constants a = 7.66, b = 8.54, c = 5.52 Å. The measured intensities were corrected for a phenomenological Lorentz factor and a Blackman two-beam dynamical correction was also applied to some low-angle data. The original structure had been determined using coordinates from a contemporary x-ray analysis.

The overall temperature factor from a Wilson plot, $B = 1.0$ Å^2, was small enough that no correction was made to the scattering factors for calculation of $|E_h|$ (Dorset, 1991b). Since all dynamically affected reflections were not identified in the earlier paper, two possible structure factor relationships were considered for the intensity data. Hence the calculation of normalized structure factors was based on $|F_{obs}| \alpha\ I_{obs}^{1/2}$ or $|F_{obs}| \alpha\ I_{obs}$.

For $|E_h|$ based on the former, "kinematical" distribution in Table 6.7, 177 Σ_2-triples were generated for 127 reflections where $|E_h| \geq 0.5$ and only one relationship was found to disagree with the original phase assignments, down to the threshold value: $A_2 = 2.0$. Only two errors were found for 82 quartets above $B_{min} = 8.0$. After defining the origin by setting $\phi_{136} = \pi$, $\phi_{141} = 0$, $\phi_{272} = \pi$, and accepting phase values from seven Σ_1-triples, a total phase set of 76 reflections was generated to produce three-dimensional potential maps from which coordinates of the atomic positions could be measured. The resultant phase list for all 187 reflections was used to generate final three-dimensional maps as shown in Figure 6.4. Although the R-value obtained from these coordinates (Table 6.5) is 0.25, the bond distances and angles are significantly distorted when compared to a modern x-ray crystal structure (Truter, 1967) (Figure 6.8). However, the sulfur atom position agrees well with that seen in the x-ray structure. Such a result is anticipated when significant dynamical scattering effects are present. In early discussions of the influence of multiple beam scattering on the effective Patterson function for such structures, Cowley and Moodie (1959) predicted that the light atom–heavy atom vectors would be somewhat broadened and flattened, but that light atom–light atom vectors would not be appreciably affected. Heavy atom–heavy atom vectors would be sharpened and even reduced in magnitude by dynamical scattering. Multislice simulations of a polymer structure with similar atomic components (poly(ethylene sulfide)) revealed that, as the crystal thickness was increased, the sulfur atom position could be

Table 6.7. Atomic Coordinates for Thiourea, Paraelectric Form

	E.d. ("kinematic" distribution)			X-ray structure		
Atom	x	y	z	x	y	z
S	–0.004	0.250	0.114	–0.008	0.250	0.114
C	0.088	0.250	–0.134	0.092	0.250	–0.163
N	0.138	0.116	–0.284	0.131	0.117	–0.274

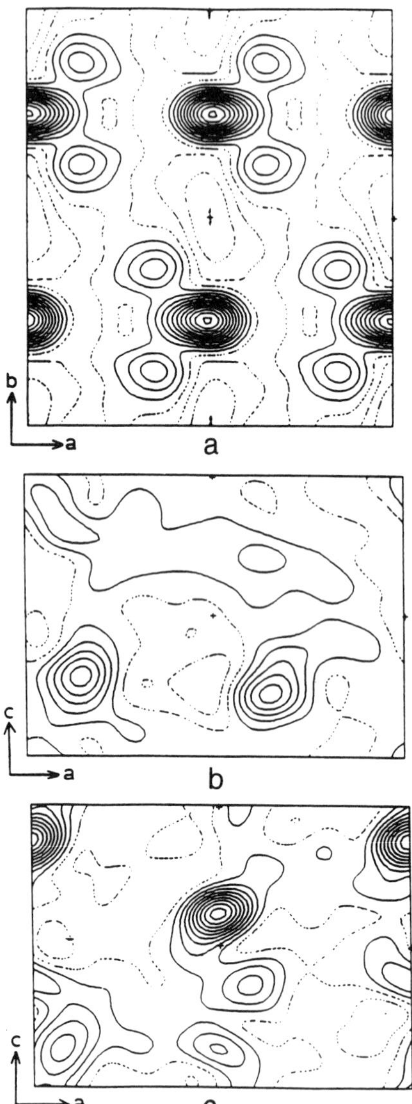

FIGURE 6.4. Potential maps for paraelectric thiourea after crystallographic phase determination by symbolic addition: (a) [001] projection (the C–S bond is not resolved into individual atomic positions); (b) slice at y = 0.12 with N atomic position; (c) Slice at y = 0.25 with C and S positions resolved. (Reprinted from D. L. Dorset (1991) "Is electron crystallography possible? The direct determination of organic crystal structures," *Ultramicroscopy* **38**, 23–40; with kind permission of Elsevier Science B.V.)

accurately located in the potential maps after direct phase determination, but the density representing the lighter atom was increasingly distorted (Dorset and McCourt, 1992).

Equivalent results were obtained when the phase determination was made with the tangent formula (QTAN) (Dorset, 1992b). For this analysis, the origin was defined as before and the seven Σ_1-triple phases ($\phi_{060} = \phi_{260} = \phi_{020} = \phi_{002} = \pi$; $\phi_{402} = \phi_{080} = \phi_{240} = 0$) were used to expand the basis set. Five additional phases: ϕ_{201}, ϕ_{411}, ϕ_{121}, ϕ_{352},

Table 6.8. Phases for Paraelectric Thiourea after QTAN Compared to 1960 Determination by Dvoryankin and Vainshtein (DV)

hkl	$\phi_{\text{QTAN, kin}}$	$\phi_{\text{QTAN, dyn}}$	ϕ_{DV}	*hkl*	$\phi_{\text{QTAN, kin}}$	$\phi_{\text{QTAN, dyn}}$	ϕ_{DV}
002	π	π	π	134	0	0	0
004	0	0	0	135	π	π	π
006	π	π	π	136	π	π	π
008	0	0	0	137	0	0	π
00,10	π	π	π	138	0	0	0
021	π	π	π	13,10	π	π	π
022	π	π	π	140	0	π	0
023	π	π	π	141	π	π	π
024	0	0	0	142	π	π	π
025	0	0	π	143	0	0	π
026	π	π	π	144	0	0	0
027	0	0	0	145	π	π	π
028	0	0	0	146	π	0	π
029	π	π	0	147	0	0	0
042	π	π	π	149	π	π	π
043	0	0	0	150	0	0	0
044	0	0	0	151	π	π	π
045	π	π	π	152	π	π	π
046	π	π	π	154	0	0	0
061	0	0	0	161	π	π	π
062	π	π	π	164	0	0	0
101	0	0	0	165	π	π	π
103	0	0	0	200	π	π	π
105	π	π	π	202	π	π	π
107	π	π	π	204	0	0	0
110	0	0	0	206	0	0	0
111	π	π	0	208	π	π	π
112	π	π	π	20,10	π	π	π
113	0	0	0	211	π	π	π
114	0	0	0	212	π	π	π
115	π	π	π	213	0	0	0
116	π	π	π	214	0	0	0
118	0	0	0	215	π	π	π
119	π	π	π	216	π	π	0
11,10	π	π	π	217	0	0	0
120	0	0	0	220	0	0	0
121	π	π	π	221	0	0	0
122	π	π	π	222	π	π	0
123	0	0	0	223	0	0	0
124	π	π	π	224	π	π	0
125	π	π	π	226	π	π	π
126	π	π	π	227	π	π	π
127	0	0	0	228	0	0	0
128	0	0	0	230	0	0	0
130	0	0	0	231	π	π	π
131	π	π	π	232	π	π	π
132	π	π	π	233	0	0	0
133	0	0	0	234	0	0	0

(continued)

Table 6.8. (*Continued*)

hkl	$\phi_{QTAN, kin}$	$\phi_{QTAN, dyn}$	ϕ_{DV}	*hkl*	$\phi_{QTAN, kin}$	$\phi_{QTAN, dyn}$	ϕ_{DV}
235	π	π	π	350	π	π	π
236	π	π	π	351	0	0	0
240	0	0	0	354	0	0	π
241	π	π	π	360	0	0	0
242	π	π	π	361	π	π	π
243	π	π	π	363	0	0	0
244	π	π	π	400	π	π	π
245	0	0	0	402	0	0	0
246	π	π	π	404	π	π	π
250	0	0	0	406	0	0	0
251	π	π	π	408	π	π	π
252	π	0	0	410	0	0	0
253	0	0	0	411	π	π	π
255	π	π	π	413	0	0	0
301	π	π	π	414	π	π	0
303	π	0	0	415	π	π	π
305	π	π	π	420	π	π	π
307	0	0	0	422	0	0	0
310	π	0	π	423	π	π	π
312	0	0	0	424	π	π	π
313	π	π	π	426	0	0	0
314	π	π	π	430	0	0	0
315	0	0	0	431	π	π	π
316	0	0	0	435	0	0	0
318	π	π	π	440	π	π	π
320	π	π	π	441	0	0	0
321	π	π	π	442	0	0	0
322	π	π	π	444	0	π	π
323	0	0	0	446	0	0	0
324	0	0	0	503	0	0	0
325	π	π	π	510	π	π	π
330	π	π	π	511	0	0	0
331	0	0	0	512	0	0	0
332	0	0	0	514	π	π	π
333	π	π	π	515	π	π	0
334	0	0	π	516	0	0	0
335	0	0	0	530	π	0	π
336	0	0	0	532	0	0	0
338	π	π	π	541	0	0	0
340	0	0	0	610	π	π	π
341	π	π	π	611	0	0	0
342	π	0	π	613	π	π	π
343	0	0	0	630	π	π	π
344	0	π	π	631	0	0	0
345	π	π	π	632	0	0	π
347	0	0	0				

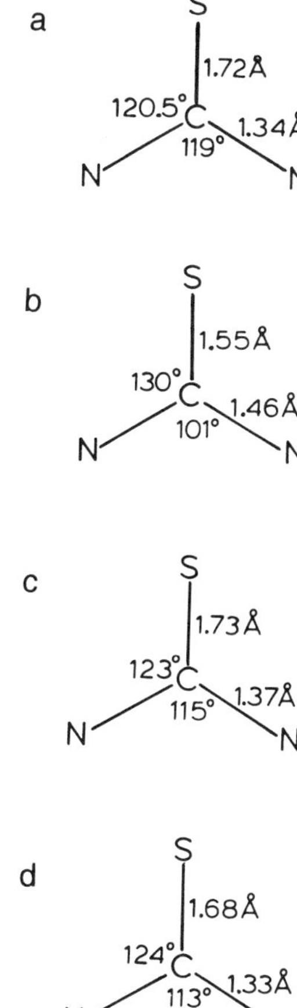

FIGURE 6.5. Bond distances and angles for paraelectric thiourea after phase determination: (a) x-ray crystal structure; (b) after symbolic addition with an electron diffraction data set assuming a "kinematical" distribution, $|F_h| = kI_h^{1/2}$; (c) after symbolic addition on data assuming $|F_h| = kI_h$, a "dynamical" distribution; (d) values from a previous electronographic determination by Dvoryankin and Vainshtein (1960). (Reprinted from D. L. Dorset (1991) "Is electron crystallography possible? The direct determination of organic crystal structures," *Ultramicroscopy* **38**, 23–40; with kind permission of Elsevier Science B.V.)

ϕ_{112}, were assigned symbolic values which, after permutation, led to 32 solutions, each with 134 phase terms. A map equivalent to Figure 6.4 was found at the lowest NQEST value and the phase list was found to contain only 9 disagreements with the original determination (Table 6.8).

If a "dynamical" relationship was used to calculate normalized structure factors, 142 reflections, where $|E_h| \geq 0$ were used to form triple and quartet invariants, as before. The origin was defined by setting: $\phi_{136} = \pi$, $\phi_{1,10,1} = \pi$, $\phi_{272} = \pi$ and Σ_1-triples were found to add 10 more phase terms. After evaluation of Σ_2-triples and positive quartets, 82 phases were obtained and the $|F_o|$ map produced from this set led to atomic coordinates (Table 6.8) more in accord with the x-ray crystal structure (Truter, 1967). Even though the R-factor was higher than before (0.28), the structure was found to be geometrically correct (Figure 6.5).

Similar results were again obtained with QTAN. The basis set was defined using the same phase origin, accepting Σ_1-terms: $\phi_{060} = \phi_{260} = \phi_{404} = \phi_{0,10,0} = \phi_{204} = \phi_{004} = \phi_{220} = \phi_{020} = \pi$; $\phi_{080} = \phi_{402} = 0$. Allowing ϕ_{464}, ϕ_{352}, and ϕ_{433} to be permuted, 8 solutions were generated and the best phase set was found to occur again at the lowest NQEST value. Of the 179 phases found, 33 disagreed with the assignments of Dvoryankin and Vainshtein (1960), as shown in Table 6.8. However, when more reflections were used to calculate the potential map, the bond distances and angles were again distorted.

The results of these analyses are very interesting. It is clear that dynamical scattering did not affect the phase determination much. Unlike the diketopiperazine example discussed above, it also did not lead to any problems with NQEST as a figure of merit when the tangent formula was used. On the other hand, if a "kinematical" relationship between observed intensities and structure factors was assumed, a geometrically correct structure could not be determined from the potential maps. Emphasis of higher angle reflections for the phase determination, on the other hand, by assuming a "dynamical" relationship between intensity and structure factor magnitude, led to a chemically reasonable structure but it could not be refined against the whole data set. Clearly, the global R-factor minimum corresponded to a distorted structure in either case.

6.2.4. Thiourea, Ferroelectric Form

A set of 240 unique three-dimensional oblique texture intensities was collected from thiourea in its ferroelectric phase at 133 K by Dvoryankin and Vainshtein (1962). The orthorhombic space group was found to be noncentrosymmetric, $Pmc2_1$, with unit cell constants $a = 8.52$, $b = 5.49$, $c = 7.54$ Å. Dynamical scattering contributions were identified in the original work and the structure analysis was again based on a known x-ray structure.

For generation of $|E_h|$, only the relationship $|F_h| \alpha I_h^{1/2}$ was considered (Dorset, 1992b). Because the space group has only one centrosymmetric zone, the tangent formula was used for phase determination. From origin definition: $\phi_{310} = \phi_{411} = \phi_{340} = 0$ comprised the known basis set and values of ϕ_{334}, ϕ_{012}, ϕ_{013}, ϕ_{211}, ϕ_{114} were permuted to produce 62 unique phase solutions. The unrestricted reflections were permuted in phase by $\pi/4 + n\,\pi/2$ through four quadrants. Each possible solution listed values for 183/240 terms. The correct structure was found again at the lowest value of NQEST, with an origin shifted (0,0.5,–0.218) from the original choice made by Dvoryankin and Vainshtein. (The latter arbitrary origin shift value is, of course, possible for the 2_1 axis.) After compensating for the origin shift, the initial phase set differed from a previous x-ray determination by an mean value of 33.4° (Table 6.9).

Atomic coordinates were then obtained from the three-dimensional potential map to calculate phase values for all reflections (Figure 6.6). Refinement by Fourier techniques lowered the crystallographic residual from 0.39 to 0.30 (applying an isotropic temperature factor $B = 4.0$ Å^2 for all atoms). Final refined atomic coordinates (Table 6.10) resembled those found by x-ray crystallography but the C–S bond in one of the two half molecules of the asymmetric unit was somewhat short (Figure 6.7).

Table 6.9. Comparison of QTAN Phases to 1962 Determination by Dvoryankin and Vainshtein (DV): Determination of the Ferroelectric Thiourea Structure

hkl	ϕ_{DV}	ϕ_{QTAN}	*hkl*	ϕ_{DV}	ϕ_{QTAN}
200	0	0	840	180	180
400	0	0	050	0	0
600	0	0	350	180	0
800	0	0	450	180	0
10,00	0	0	550	180	0
110	180	180	060	0	0
210	180	180	260	0	0
310	0	0	360	0	0
410	180	180	011	0.3	7.6
510	0	0	111	−90.7	−80.8
610	180	180	211	−0.1	−0.1
710	180	180	311	78.6	103.6
910	180	180	411	15.8	−1.2
11,10	0	0	511	98.9	101.5
020	0	0	611	5.5	7.8
120	180	0	711	189.3	109.6
220	0	180	811	−2.0	7.3
320	0	0	911	−87.9	−80.3
420	180	180	10,11	5.2	−17.6
520	0	0	11,11	15.0	101.8
620	0	0	021	60.7	103.7
720	180	180	121	42.4	18.4
820	0	0	221	127.5	102.8
10,20	180	180	321	15.3	12.8
030	0	180	421	−101.7	−80.3
130	180	180	521	0.2	19.1
230	180	180	621	−55.2	−80.5
330	180	0	721	69.4	186.0
430	180	180	821	79.5	102.0
530	180	180	921	32.4	13.7
730	180	180	10,21	163.8	105.4
930	180	180	031	−13.0	−0.2
040	180	0	131	−81.8	−81.3
240	180	180	231	−9.6	−61.9
340	0	0	331	−20.5	−84.7
440	180	180	431	−10.9	−63.8
540	0	0	412	−87.6	−61.6
640	180	180	512	210.0	202.2
531	98.7	109.2	612	93.6	106.9
631	−14.2	0.3	712	212.4	201.9
731	−122.1	−83.5	812	93.6	109.1
831	−11.6	−2.1	912	211.7	201.8
931	−71.0	−81.6	022	40.5	−3.3
041	−47.4	−124.8	122	−81.6	−63.1
141	196.0	167.1	222	−72.7	−68.0
241	−150.9	−135.1	322	83.9	103.9
341	169.5	38.6	422	268.1	230.1

(continued)

Table 6.9. (*Continued*)

hkl	ϕ_{DV}	ϕ_{QTAN}	*hkl*	ϕ_{DV}	ϕ_{QTAN}
441	−113.1	−93.8	522	4.4	97.2
541	129.5	167.7	622	−34.2	−27.5
641	−66.9	−95.0	722	−83.2	−64.4
051	156.0	90.6	922	−77.4	−59.5
151	−62.2	−65.6	032	66.7	96.3
251	137.3	180.5	132	26.7	44.9
351	2.3	8.3	232	−16.1	−13.5
451	211.0	180.5	332	55.5	35.4
551	84.0	87.3	432	−68.0	−55.4
651	192.1	94.5	532	51.7	36.5
061	47.9	111.1	632	34.8	64.6
161	236.1	176.5	042	44.9	13.1
361	180.1	37.9	142	−128.4	−87.2
102	−86.4	−68.1	242	7.3	10.1
202	205.3	204.8	342	112.7	127.4
302	93.6	111.0	442	−28.8	5.7
402	−156.1	−155.8	542	125.2	135.6
502	93.3	110.4	052	260.0	126.8
602	206.0	205.1	152	135.6	110.0
702	−86.4	−68.2	252	252.6	260.7
802	207.3	205.9	013	194.2	210.7
10,02	204.9	204.5	113	−75.8	−58.9
012	93.7	108.9	213	204.4	211.2
112	212.1	201.9	313	8.1	127.0
212	94.6	101.6	413	268.0	208.6
513	−54.9	−60.5	804	184.2	42.4
613	212.3	205.8	014	−57.7	108.9
713	−77.6	−58.4	114	28.1	45.9
023	88.6	117.2	214	−60.7	−45.8
123	−29.6	26.9	314	113.8	82.6
223	95.2	93.0	414	−61.8	−46.3
323	180.4	217.2	514	83.2	54.5
423	128.2	−19.1	614	−59.7	−45.3
523	204.8	213.4	714	24.8	45.5
623	98.2	116.2	814	−57.2	114.6
723	264.6	214.4	024	184.2	215.8
033	165.1	207.3	124	204.7	257.7
133	167.3	176.8	224	182.1	184.3
233	225.3	209.6	324	162.6	151.5
333	146.6	126.5	424	5.3	56.3
433	−38.1	48.7	524	164.8	156.8
533	145.3	131.2	624	184.1	208.4
633	70.5	39.5	724	−107.4	−82.6
043	107.2	110.3	034	166.3	134.3
143	22.6	33.4	134	−86.5	−71.2
243	111.8	87.6	234	88.3	139.8
343	97.0	34.4	334	224.0	234.2
443	93.0	57.2	434	2.3	−49.2
543	178.9	211.0	534	230.7	236.4
153	10.1	−79.5	634	156.4	136.4

(continued)

Table 6.9. (*Continued*)

hkl	ϕ_{DV}	ϕ_{QTAN}	*hkl*	ϕ_{DV}	ϕ_{QTAN}
253	−42.0	−41.5	044	207.5	218.8
104	59.2	−45.1	144	−2.0	−61.2
204	65.7	50.7	244	232.1	212.5
304	84.4	132.6	344	182.7	140.2
404	33.4	46.9	154	−101.5	−15.8
504	82.1	132.4	015	120.8	209.2
604	98.2	56.8	115	111.5	143.1
704	33.0	−45.7	215	56.1	57.6
312	210.5	202.4	315	143.7	142.9
415	39.1	56.0	535	−49.3	−34.3
515	−105.7	−27.4	045	−88.2	−46.0
615	111.6	61.3	145	−38.5	30.6
715	107.6	140.8	245	−26.4	−40.6
025	−85.7	−43.7	016	−91.8	−30.4
125	34.9	51.9	116	255.4	267.1
225	−54.0	−39.7	216	166.6	161.5
325	72.7	28.2	316	237.6	250.6
425	49.9	−48.1	416	112.0	183.1
035	139.5	226.3	026	−5.0	246.8
135	56.3	125.0	126	83.1	121.6
235	160.7	133.6	226	49.9	62.8
335	−91.1	−11.0	326	−60.9	0.9
435	101.2	64.2	426	51.9	64.0

After refinement, the final mean phase error, when compared to the x-ray determination of Goldsmith and White (1959), was 16.3°. Compared to the original electron crystallographic analysis this figure was 11.5°. No evidence for hydrogen atom positions was found in the experimental potential maps.

Again, dynamical scattering was found not to compromise direct phase determination seriously, although it did lead to some geometrical inaccuracies in the final refined structure. In all of the determinations discussed above, based on oblique texture data collected at a rather low accelerating voltage, the average sample thickness was

Table 6.10. Atomic Coordinates for Thiourea, Ferroelectric Form

	Electron diffraction			X-ray structure		
Atom	*x*	y	*z*	*x*	*y*	*z*
S 1	0.500	0.123	0.012	0.500	0.125	0.010
C 1	0.500	0.381	−0.095	0.500	0.399	−0.103
N 1	0.367	−0.483	−0.151	0.365	−0.491	−0.149
S 2	0.000	0.353	−0.002	0.000	0.349	−0.003
C 2	0.000	0.052	0.052	0.000	0.049	0.075
N 2	0.133	−0.058	0.125	0.135	−0.071	0.105

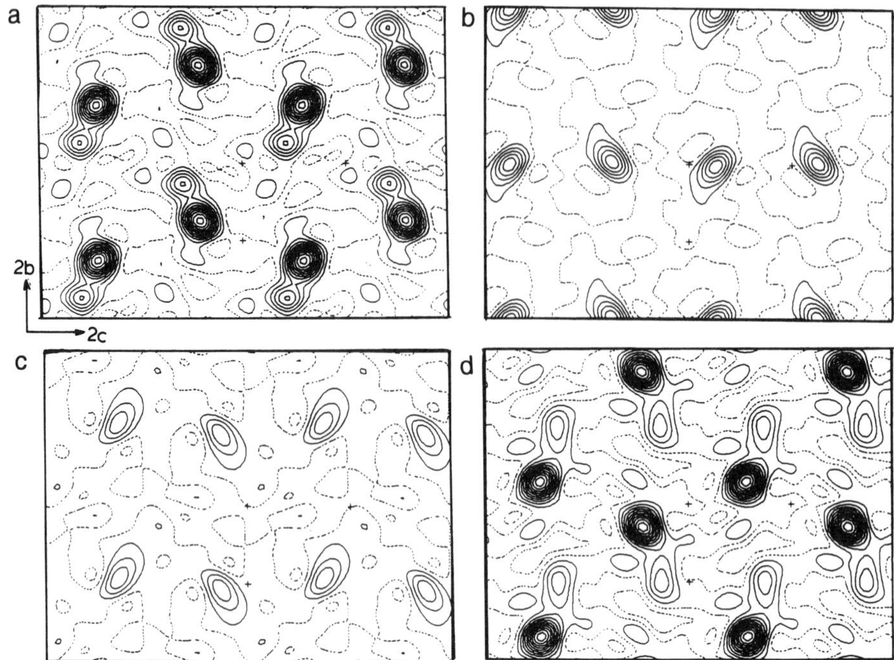

FIGURE 6.6. Potential map for ferroelectric thiourea after phasing with the tangent formula: (a) slice at $x = 0.5$, showing sulfur and carbon positions for molecule 1; (b) slice at $x = 0.37$, showing nitrogen position of molecule 1; (c) slice at $x = 0.13$, showing nitrogen position for molecule 2; (d) slice at $x = 0$, showing sulfur and carbon positions for molecule 2. (Reprinted from D. L. Dorset (1992) "Automated phase determination in electron crystallography: thermotropic phases of thiourea," *Ultramicroscopy* **45**, 357–364; with kind permission of Elsevier Science B.V.)

an unknown quantity. Thus, even though the orientational distribution of the crystallites in the texture can be beneficial for collection of useful intensity data, the presence of a heavy atom and use of large wavelength electrons for data collection can still hamper the final structure refinement.

6.3. Recent Analyses Based upon Selected Area Diffraction Data

6.3.1. Copper Perchlorophthalocyanine

Copper perchlorophthalocyanine has been a standard material for study in the electron microscope. Thin crystals (approximately 100-Å layers) were oriented epitaxially on freshly cleaved and outgassed (001) KCl crystal faces by sublimation from a heated source (Dorset et al., 1991) as described originally by Uyeda et al., (1972). Initially, selected area diffraction experiments were carried out at 100 kV with a JEOL JEM-100B7 electron microscope equipped with a side-entry goniometer stage tilted at 26.5°. Rotation of the grid allowed the crystal packing to be projected down the

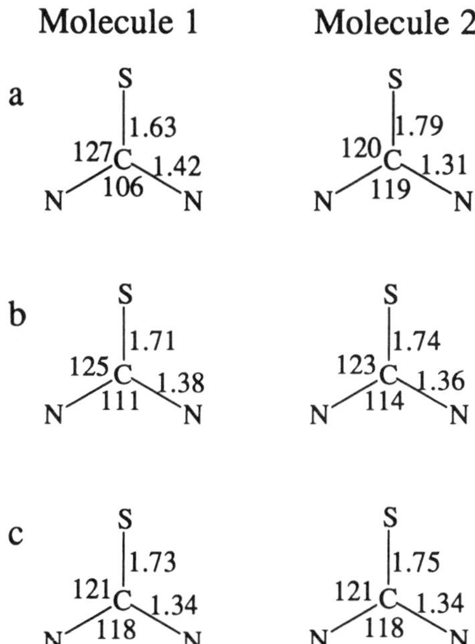

FIGURE 6.7. Bond distances and angles for ferroelectric thiourea: (a) tangent formula analysis of electron diffraction data; (b) results of Dvoryankin and Vainshtein (1962); (c) x-ray crystal structure. (Reprinted from D. L. Dorset (1992) "Automated phase determination in electron crystallography: thermotropic phases of thiourea," *Ultramicroscopy* **45**, 357–364; with kind permission of Elsevier Science B.V.)

molecular columns. It was readily apparent, after evaluation of the measured intensities, that these data were insufficient for ab initio structure analysis, since features of the unit cell transform could not be easily discerned in the diffraction pattern (Figure 1.7b). The experiments were then repeated at 1200 kV using an AEI EM7 high-voltage electron microscope (Figure 6.8).

Intensity data were extracted from experimental diffraction patterns after scanning the films with a flat-bed microdensitometer, yielding, after integration of peaks, 198 unique $hk0$ values (Table 6.11). The measured cell constants for the projection in plane group cmm were d_{100}= 17.56, b = 26.08 Å, consistent with the $C2/m$ unit cell constants: a = 19.62, b = 26.08, c = 3.76 Å, β = 116.5°. (Thus, d_{100}= a sin 116.5°.) From a Wilson plot it was seen that no thermal adjustment to scattering factor tables was necessary for calculation of $|E_h|$, since B = 0.0 Å^2. This indicates that the data are perturbed somewhat by multiple scattering. The distribution of $|E_h|$ values (Table 6.12) also does not correspond to a centrosymmetric data set.

Reflections, where $|E_h| \geq 0.6$, were used to generate 31 Σ_2-triple invariants in plane group *cmm*. Because the projection is centered, only one reflection could be used for origin definition, i.e., with index parity $uu0$ (ϕ_{930}). Three additional symbolic phases

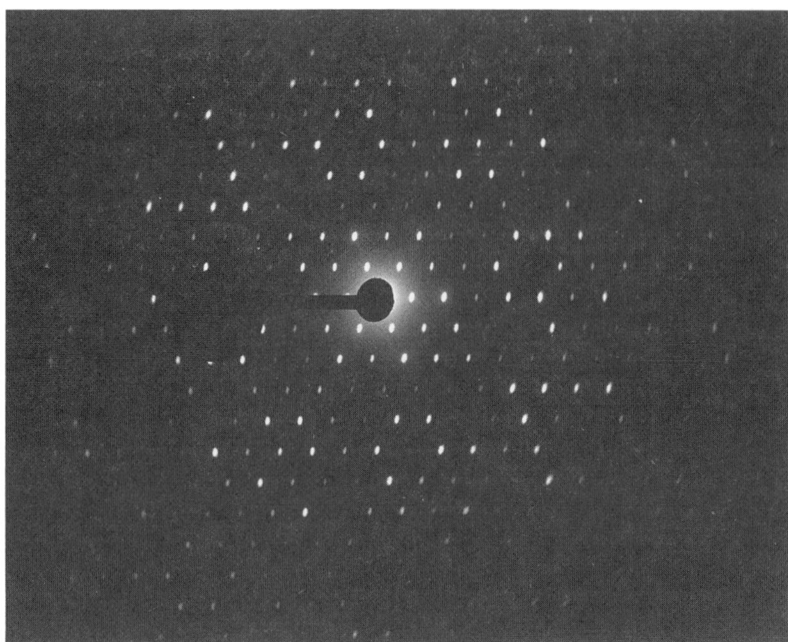

FIGURE 6.8. Electron diffraction pattern (*hk*0) from copper perchlorophthalocyanine obtained at 1200 kV. (Reprinted from D. L. Dorset, W. F. Tivol, and J. N. Turner (1991) "Electron crystallography at atomic resolution: ab initio structure analysis of copper perchlorophthalocyanine," *Ultramicroscopy* **38**, 41–45; with kind permission of Elsevier Science B.V.)

Table 6.11. Final Observed and Calculated Structure Factors for Copper Perchlorophthalocyanine Solved from 1200-kV Electron Diffraction Intensities

hkl	$\lvert F_o \rvert$	F_c	*hkl*	$\lvert F_o \rvert$	F_c
0 2 0	1.78	1.83	8 10 0	0.77	–0.85
0 4 0	1.70	1.99	8 12 0	0.74	0.41
0 6 0	1.02	–0.29	8 14 0	0.41	0.35
0 8 0	1.74	–2.23	8 16 0	0.28	0.41
0 10 0	1.24	1.74	8 18 0	0.37	0.62
0 12 0	1.62	–2.64	8 20 0	0.40	–0.10
0 14 0	0.77	–0.21	8 22 0	0.55	0.57
0 16 0	0.79	1.22	8 24 0	0.28	0.31
0 18 0	0.41	–0.18	9 1 0	0.63	–0.52
0 20 0	1.10	1.44	9 3 0	2.10	2.05
0 22 0	0.92	–1.04	9 5 0	0.81	0.56
1 1 0	2.29	2.12	9 7 0	1.02	0.84
1 3 0	1.41	–1.01	9 9 0	0.42	–0.06
1 5 0	0.67	0.30	9 11 0	0.98	0.92
1 7 0	0.87	0.93	9 13 0	0.98	1.10
1 9 0	1.76	1.82	9 15 0	0.57	–0.50

(continued)

Table 6.11. (*Continued*)

hkl	$\|F_o\|$	F_c	*hkl*	$\|F_o\|$	F_c
1 11 0	0.93	1.26	9 17 0	0.50	–0.53
1 13 0	0.89	1.07	9 19 0	0.98	–0.61
1 15 0	0.73	–0.70	9 21 0	0.40	0.12
1 17 0	0.91	–0.52	9 23 0	0.26	0.32
1 19 0	0.92	0.42	9 25 0	0.20	–0.14
1 21 0	0.33	0.16	10 0 0	0.48	–0.05
2 0 0	1.90	1.08	10 2 0	1.25	1.65
2 2 0	2.57	–2.60	10 4 0	0.48	0.64
2 4 0	0.88	0.96	10 6 0	0.39	0.14
2 6 0	0.69	–0.30	10 8 0	0.50	–0.30
2 8 0	1.02	–0.95	10 10 0	0.46	–0.08
2 10 0	0.84	0.22	10 12 0	0.48	0.56
2 12 0	1.89	2.26	10 14 0	0.63	0.58
2 14 0	0.67	0.73	10 16 0	0.60	0.34
2 16 0	0.54	0.22	10 18 0	0.53	0.16
2 18 0	0.37	0.07	11 1 0	0.60	–0.40
2 20 0	0.87	1.12	11 3 0	0.63	0.18
2 22 0	0.46	0.05	11 5 0	0.57	–0.58
3 1 0	1.69	1.32	11 7 0	0.61	0.94
3 3 0	0.82	0.50	11 9 0	1.24	0.89
3 5 0	1.82	–1.86	11 11 0	0.41	0.24
3 7 0	1.82	1.93	11 13 0	0.24	0.33
3 9 0	0.69	–0.18	11 15 0	0.77	0.43
3 11 0	0.57	–0.27	11 17 0	0.36	0.54
3 13 0	0.68	1.05	11 19 0	0.10	–0.12
3 15 0	0.42	–0.22	11 21 0	0.20	–0.25
3 17 0	0.97	1.29	11 23 0	0.20	0.32
3 19 0	0.28	0.10	11 25 0	0.17	0.13
3 21 0	0.64	–0.28	12 0 0	0.96	0.99
4 0 0	1.51	–0.99	12 2 0	0.42	–0.16
4 2 0	0.66	0.26	12 4 0	0.85	0.74
4 4 0	1.95	–1.87	12 6 0	0.37	0.52
4 6 0	0.55	–0.10	12 8 0	1.24	–1.09
4 8 0	2.29	3.40	12 10 0	0.74	0.88
4 10 0	1.11	–1.29	12 12 0	0.62	–0.73
4 12 0	0.81	0.98	12 14 0	0.76	0.77
4 14 0	0.60	–0.35	12 16 0	0.62	0.70
4 16 0	0.62	0.64	12 18 0	0.10	0.00
4 18 0	0.54	0.75	12 20 0	0.39	0.36
4 20 0	0.60	–0.76	13 1 0	1.07	–0.67
4 22 0	0.53	–0.27	13 3 0	0.46	0.38
5 1 0	0.92	0.67	13 5 0	0.67	0.77
5 3 0	0.52	0.16	13 7 0	0.33	–0.21
5 5 0	2.04	2.15	13 9 0	0.58	0.08
5 7 0	1.38	–1.42	13 11 0	0.42	0.23
5 9 0	1.52	–1.62	13 13 0	0.36	0.25
5 11 0	0.44	–0.18	13 15 0	0.36	0.40
5 13 0	0.37	0.29	13 17 0	0.41	–0.17
5 15 0	0.85	0.90	14 0 0	0.66	0.25
5 17 0	0.40	0.11	14 2 0	0.69	–0.17

(*continued*)

Table 6.11. (*Continued*)

hkl	$\|F_o\|$	F_c	*hkl*	$\|F_o\|$	F_c
5 19 0	0.46	0.27	14 4 0	0.10	0.14
5 21 0	0.33	0.15	14 6 0	0.20	0.33
5 23 0	0.47	−0.53	14 8 0	0.35	−0.26
6 0 0	2.33	−2.34	14 10 0	0.24	0.73
6 2 0	0.81	0.01	14 12 0	0.62	−0.55
6 4 0	1.91	1.69	14 14 0	0.10	0.18
6 6 0	1.02	1.20	14 16 0	0.33	0.18
6 8 0	1.04	0.99	15 1 0	0.90	1.06
6 10 0	0.52	−0.21	15 3 0	0.48	0.65
6 12 0	0.60	1.07	15 5 0	0.52	−0.35
6 14 0	0.57	0.05	15 7 0	0.37	−0.08
6 16 0	1.12	−1.09	15 9 0	0.10	−0.10
6 18 0	0.50	−0.66	15 11 0	0.30	−0.37
6 20 0	0.51	−0.47	15 13 0	0.30	−0.35
6 22 0	0.10	−0.05	16 0 0	0.73	−0.54
6 24 0	0.50	0.12	16 2 0	0.10	0.38
7 1 0	2.19	2.28	16 4 0	0.67	−0.11
7 3 0	1.13	−1.14	16 6 0	0.33	−0.07
7 5 0	0.85	0.35	16 8 0	0.10	0.19
7 7 0	1.70	−1.76	16 10 0	0.36	0.17
7 9 0	1.18	−0.41	16 12 0	0.33	0.27
7 11 0	0.45	−0.44	17 1 0	0.26	−0.05
7 13 0	0.57	−0.84	17 3 0	0.26	−0.02
7 15 0	1.14	1.32	17 5 0	0.41	−0.19
7 17 0	0.87	0.39	17 7 0	0.51	−0.36
7 19 0	0.39	0.39	17 9 0	0.28	−0.05
7 21 0	0.49	0.26	17 11 0	0.28	0.29
7 23 0	0.39	0.04	18 0 0	0.20	0.06
8 0 0	0.97	−0.14	18 2 0	0.10	0.24
8 2 0	1.77	1.61	18 4 0	0.10	−0.28
8 4 0	0.82	−0.78	18 6 0	0.10	−0.16
8 6 0	0.71	−1.02	18 8 0	0.37	0.29
8 8 0	0.76	0.10	18 10 0	0.39	0.22

a, b, c were assigned requiring the generation of 8 potential maps from the 27 phased reflections (only 8 of which were assigned phases unequivocally, including a Σ_1 estimate for $\phi_{12,0,0} = 0$). In one map, likely positions of the copper and some chlorine atoms could be located (Figure 6.9) and these were used to generate trial phases for all 198 unique reflections. These positions were reinforced in the next map and probable locations for additional atoms also were found. The process was repeated until no new atomic positions could be found, whereupon difference Fourier syntheses based on $2|F_o| - |F_c|$ and $|F_o| - |F_c|$ were used to find the remaining lighter atoms, as outlined in Figure 6.14. After optimization of the structure by further Fourier refinement, the final crystallographic residual was 0.32, or 0.28 for 150 reflections corresponding to $|E_h| > 0.6$. Multislice calculations could lower the R-factor to 0.18 if a sin

Table 6.12. Distributions of $|E_h|$ for Copper Perchlorophthalocyanine

	Experimental	Theory: Centrosymmetric	Theory: Noncentrosymmetric
$<|E_h|^2>$	1.000	1.000	1.000
$<|E_h^2-1|>$	0.676	0.968	0.736
$<|E_h|>$	0.909	0.798	0.886
% $|E_h| > 1.0$	34.8	32.2	36.8
% $|E_h| > 2.0$	1.0	5.0	1.8
% $|E_h| > 3.0$	0.0	0.3	0.01

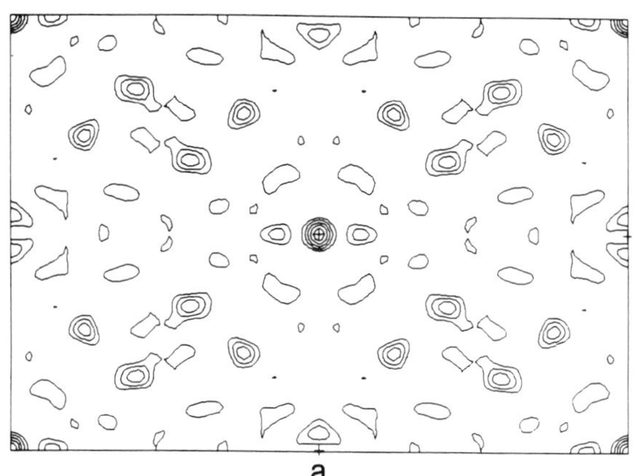

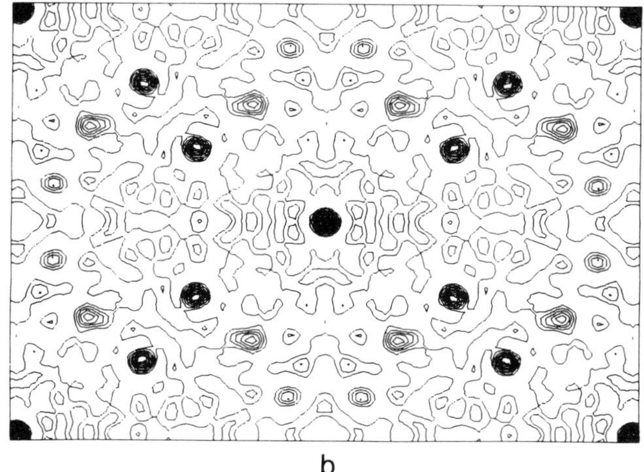

FIGURE 6.9. (a) Initial atomic positions for copper perchlorophthalocyanine after symbolic addition phase assignments for 27 reflections. (b) Map obtained after phasing 137 reflections by QTAN, for solution where NQEST = –0.176. (Reprinted from D. L. Dorset, M. P. McCourt, J. R. Fryer, W. F. Tivol, and J. N. Turner (1994) "The tangent formula in electron crystallography: phase determination of copper perchlorophthalocyanine," *Microscopy Society of America Bulletin* **24**, 398–404; with kind permission of the Microscopy Society of America.)

Table 6.13. Zonal Atomic Coordinates for Copper Perchlorophthalocyanine

Atom	x	y
Cu	0.000	0.000
Cl 1	0.081	0.302
Cl 2	0.157	0.202
Cl 3	0.271	0.120
Cl 4	0.405	0.057
N 1	0.000	0.070
N 2	0.093	0.000
N 3	0.117	0.090
C 1	0.136	0.042
C 2	0.203	0.025
C 3	0.265	0.055
C 4	0.327	0.025
C 5	0.055	0.106
C 6	0.036	0.158
C 7	0.074	0.202
C 8	0.036	0.246

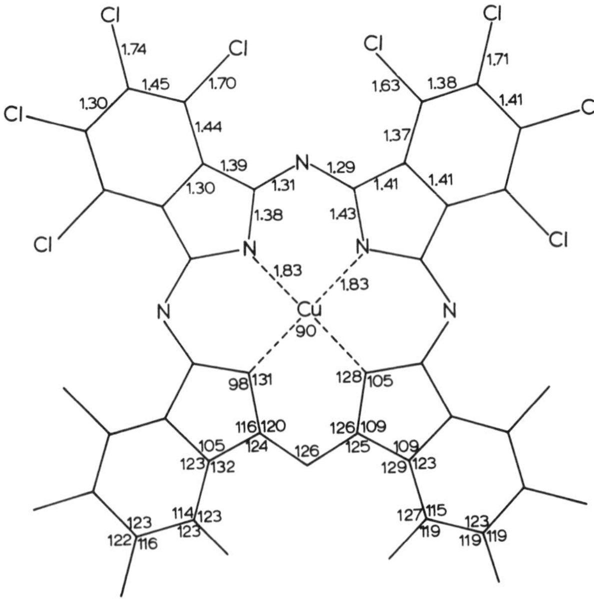

FIGURE 6.10. Bond distances and angles for copper perchlorophthalocyanine after Fourier refinement. (Reprinted from D. L. Dorset, W. F. Tivol, and J. N. Turner (1991) "Electron crystallography at atomic resolution: ab initio structure analysis of copper perchlorophthalocyanine," *Ultramicroscopy* **38**, 41–45; with kind permission of Elsevier Science B.V.)

θ/λ cutoff of 0.27 was imposed to avoid the underestimation of high-angle data due to Ewald sphere curvature (see Chapter 5). The final atomic coordinates (Table 6.13) corresponded to a chemically reasonable structure (Figure 6.10).

If the 27 reflections used to generate the first map were also used as a basis phase set for the tangent formula (Dorset et al., 1993), it was possible to determine the structure directly (Figure 6.11) from phase values assigned to 137 reflections. Data from the Fourier transform of a 2.4-Å-resolution image taken at 500 kV by Dr. J. R. Fryer (Figure 6.12) could also be used as a basis set for the tangent formula but the correct solution did not, in this case, correspond to the lowest NQEST value. The Sayre equation is more successful for carrying out this phase refinement (Figure 6.13). The image was averaged again with the CRISP programs (instead of another, less convenient software package used earlier) and the resultant phases expanded successfully with the Sayre equation, even though the starting phase set contained 10/40 phases in error, due to uncorrected objective lens astigmatism.

At about the same time that this structure was solved by evaluation of individual triple invariant sums, followed by Fourier refinement, Fan et al. (1991) used the Fourier transform of the 500 kV electron micrograph, obtained by Uyeda et al. (1978–1979) to 2-Å resolution. This image is a significant exception to Spence's (1980) rather pessimistic appraisal of the field in Appendix 4 of his book, i.e., "at the time of writing the author is not aware of the existence of any experimental atomic

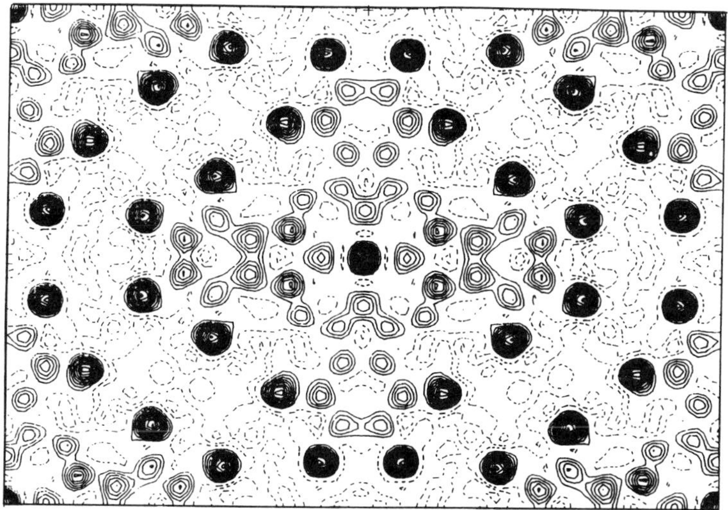

FIGURE 6.11. Potential map for copper perchlorophthalocyanine after tangent formula expansion of phases obtained by symbolic addition. (Reprinted from D. L. Dorset, M. P. McCourt, J. R. Fryer, W. F. Tivol, and J. N. Turner (1994) "The tangent formula in electron crystallography: phase determination of copper perchlorophthalocyanine," *Microscopy Society of America Bulletin* **24**, 398–404; with kind permission of the Microscopy Society of America.)

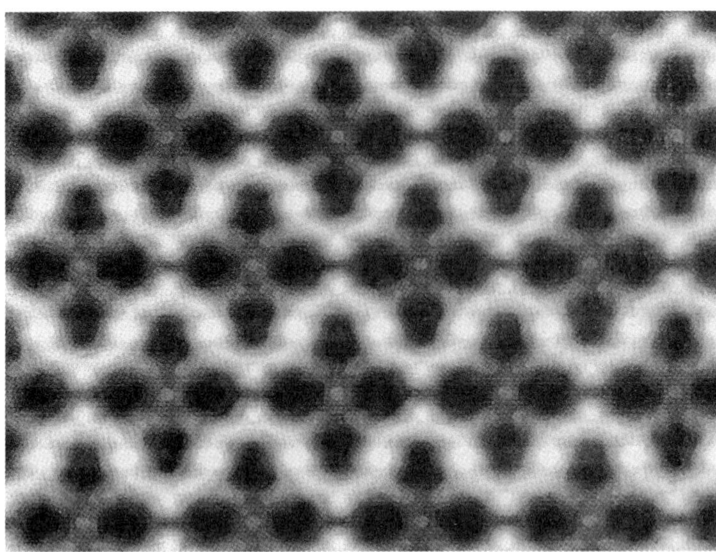

FIGURE 6.12. Electron micrograph of copper perchlorophthalocyanine taken at 500 kV after image averaging with CRISP. (Original micrograph courtesy of Dr. J. R. Fryer.)

resolution structure images." The heavy atoms of this structure were clearly discerned at their correct positions in the averaged images. (Other high-resolution images of organic crystals, in which the shapes of the molecules were observed, are listed in Table 6.14.) When the image phases were used as a basis set for RANTAN, the resolution enhancement to the 1-Å limit seen in the electron diffraction pattern was readily accomplished. Methods for image deconvolution to remove the transfer

Table 6.14. High-Resolution Images of Molecular Organic Crystals

Compound	Resolution	Reference
copper perchlorophthalocyanine	2.0 Å	Uyeda et al. (1978–1979)
copper perbromophthalocyanine	3.5 Å	Fryer and Holland (1983)
anthracene	4.8 Å	Fryer (1978)
quaterrylene	3.5 Å	Smith and Fryer (1981)
p-hexaphenyl	3.2 Å	Kawaguchi et al. (1986)
zinc phthalocyanine	2.8 Å	Kobayashi et al. (1981)
copper phthalocyanine	ca. 3 Å	Murata et al. (1976)
lanthanide phthalocyanines	ca. 3 Å	Zhang et al. (1989b)
phthalocyanine	ca. 3 Å	Fryer (1979b)
silver 7,7,8,8 tetracyanoquindodimethane (TCNQ)	2.4 Å	Uyeda et al. (1980)
perchlorocoronene	3.2 Å	Dong et al. (1992)
octacyanophthalocyanine/metal complexes	4.4 Å	Ashida (1991)
paraffin	2.5 Å	Zemlin et al. (1985)
paraffin	2.1 Å	Brink and Chiu (1991)

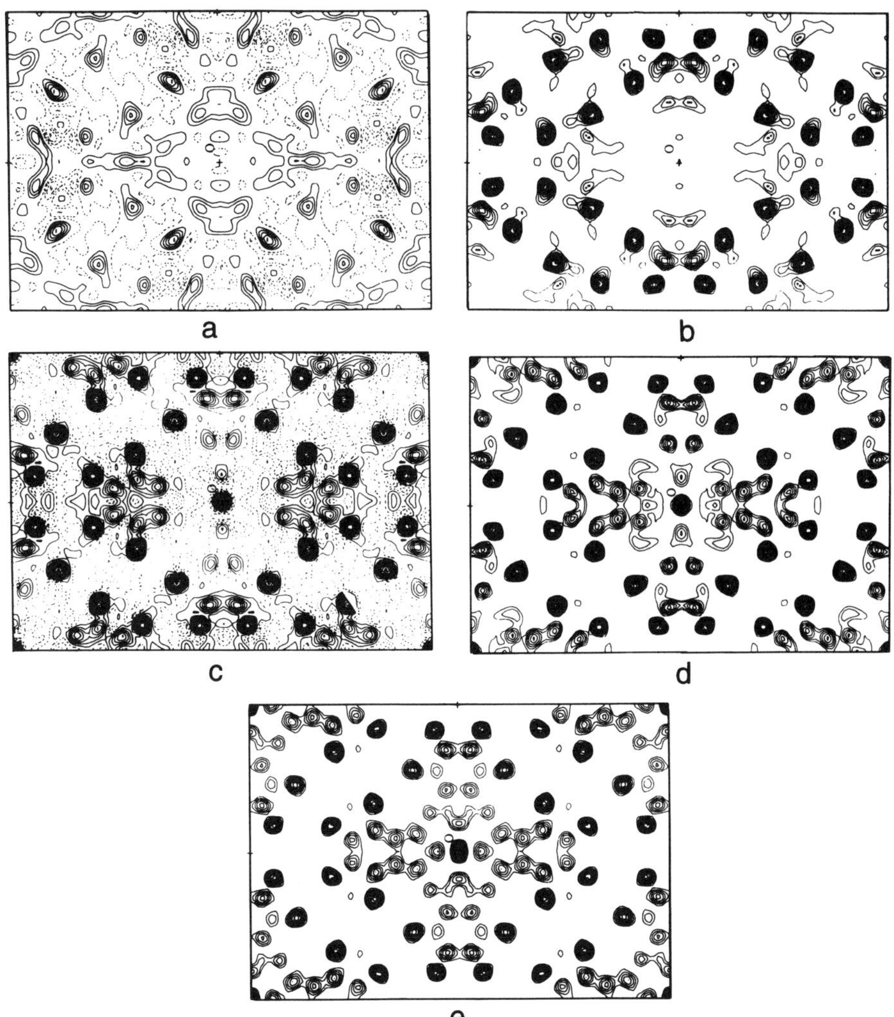

FIGURE 6.13. Progress of a phase refinement for copper perchlorophthalocyanine via the Sayre equation starting with the Fourier transform of the image in Figure 6.12: (a) after initial resolution enhancement; (b) first map after selecting atomic coordinates from (a) for a structure factor calculation; (c, d, e) subsequent Fourier refinement steps.

function phase shifts, discussed by Li (1961) (Chapter 3), greatly facilitated this process. More recently, image data at somewhat poorer resolution (e.g., 2.5 Å) was successfully extended to the electron diffraction resolution limit by use of the Sayre equation (H. F. Fan and F. H. Li, personal communication), in agreement with our own work (Dorset et al., 1995).

An interesting result was found when data were recorded from these crystals at different accelerating voltages. By contrast to the theoretical prediction of an optimal accelerating voltage for electron diffraction (Jap and Glaeser, 1980), it appears that the measured intensity data continued to be improved as the electron wavelength decreased (Tivol et al., 1993). Even though errors in the intensity measurements, due to several sources, including slight misorientations of the crystal and local variations of crystal thickness, were seen, a trend could be observed. The tangent formula would readily find structural solutions for data obtained, e.g., at 700 kV but not at 400 kV (Dorset et al., 1993). (However, the Sayre equation was not so sensitive to this specific voltage change.) There appeared to be at least two reasons for this discrepancy between theory and experiment. First, the crystals were not flat over the selected area used for data collection so that the resolution limit itself was not defined by the Ewald sphere curvature (i.e., through the shape transform of the crystal), although this curvature has very little effect at 1200 kV. Second, and most important, the actual resolution limit for diffraction from molecular organic crystals at room temperature was defined by the thermal motion of the molecules, so the higher resolution data necessary for justification of an "optimal" voltage made no significant contribution to the total diffracted intensity. (These points are discussed in Chapter 5.) Although the influence of n-beam dynamical scattering has been experimentally observed, it will be shown in a future publication that the major perturbation of the 1200-kV data from this compound is actually secondary scattering and that a simple correction dramatically lowers the R-value from 0.36 to 0.21 for all 198 measured 1200-kV data (based on scaling data such that $\Sigma F_{\text{obs}} = k\Sigma F_{\text{calc}}$ rather than the $\Sigma I_{\text{obs}} = k\Sigma I_{\text{calc}}$ criterion used above).

6.3.2. Copper Perbromophthalocyanine

After the successful copper perchlorophthalocyanine structure determination, it was interesting to attempt an analysis of its perbromo- analog (Dorset et al., 1992b). Thin (about 100-Å) microcrystals were grown by vapor deposition onto freshly cleaved and outgassed KCl crystal plates to achieve the same epitaxial orientation obtained for the perchloro- compound. After depositing a carbon film onto this oriented film, the organic layer was floated onto a clean water surface and then picked up with bare copper grids. After drying, the preparation was examined in the electron microscope.

Electron diffraction patterns were obtained at 1200 kV from a specimen tilted 26° to the electron beam, allowing 168 unique intensities to be observed (Figure 6.14). Unit cell constants measured from the zone with *cmm* symmetry were $d_{100} = 17.88(9)$ Å (corresponding to $a = 19.89$ Å), $b = 26.46(15)$ Å. The tilting experiment implied that $\beta = 116°$. It is also assumed that the projected cell length is near 3.76 Å for later multislice calculations.

Triple and quartet phase invariant sums were evaluated in an attempt to determine the crystal structure. The best solution revealed probable positions for the Cu and two Br atoms (Figure 4.11) and these positions were used for a structure factor calculation

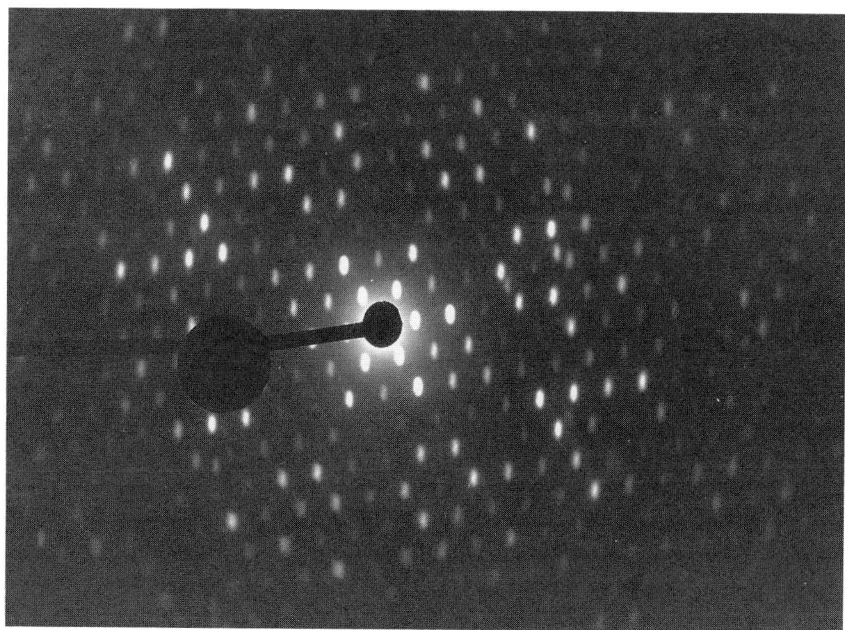

FIGURE 6.14. Electron diffraction pattern of copper perbromophthalocyanine obtained at 1200 kV. (Reprinted from D. L. Dorset, W. F. Tivol, and J. N. Turner (1992) "Dynamical scattering and electron crystallography—ab initio structure analysis of copper perbromophthalocyanine," *Acta Crystallographica* **A48**, 562–568; with kind permission of the International Union of Crystallography.)

to generate trial phase values for all 168 reflections. The heavy-atom positions were reinforced in the subsequent map and iterative uses of the procedure could locate all heavy atoms and two light atoms. Difference Fourier syntheses were then employed to find all but one of the remaining light atoms, a benzene ring carbon, that is inserted finally from known structural geometry (Figure 4.12). When Fourier refinement was continued, it was possible to find a "reasonable" potential map (Figure 4.12a) at $R = 0.36$. However, the molecular geometry was distorted. A geometrically more reasonable model (Figure 4.12c) corresponded to $R = 0.42$ and was very close to the structure of the perchloro- derivative. If coordinates from the perchloro- derivative were used ($R = 0.41$), the bond distances and angles in Figure 6.15, were calculated. Isotropic temperature factors assigned to the atoms were $B_{Cu} = 2.0$, $B_N = 4.0$, $B_C = 4.0$, and $B_{Br} = 6.0$ Å^2. The geometry of the organic residue corresponded to expected values (Brown, 1968) for the phthalocyanines but the carbon-halogen bond distances were about 0.1 Å too short (Sutton, 1968). Attempts to lengthen these distances by Fourier refinement were not successful.

Obviously, even though the geometry of the organic residue agrees with the expected molecular architecture for such compounds, a crystallographic residual of 0.41 is unacceptable as proof of a structure solution. For this reason, a multislice

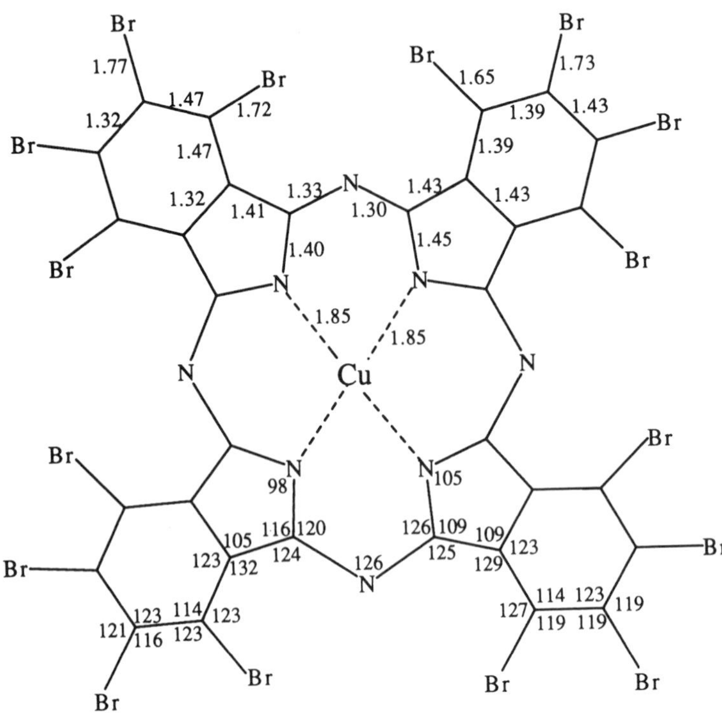

FIGURE 6.15. Final bond distances and angles for copper perbromophthalocyanine. (Reprinted from D. L. Dorset, W. F. Tivol, and J. N. Turner (1992) "Dynamical scattering and electron crystallography—ab initio structure analysis of copper perbromophthalocyanine," *Acta Crystallographica* **A48**, 562–568; with kind permission of the International Union of Crystallography.)

dynamical calculation was carried out, again based on the structure of the perchloro-derivative, to seek a convergence to the experimental structure factor magnitudes. Initially, this calculation was carried out at a resolution of 1.36 $Å^{-1}$, but reflections, specifically within the Miller index range, $h \leq 10$, $k \leq 9$ were monitored. Dynamical intensities were monitored at 10 slice intervals up to 50 slices. At $t = 188$ Å, the total summed intensity retained 98.58% of the incident beam within this aperture limit.

Table 6.15 and Figure 6.16 demonstrate that an optimal agreement of observed and calculated structure factors occurred between crystal thicknesses of 113 and 150 Å, lowering the crystallographic residual for the data subset from 0.39 to 0.27. This thickness range is in good agreement with the experimentally estimated value. Most importantly, the multislice calculation predicted which reflections were most affected by dynamical scattering, viz., 020, 040, 110, 200, and 480, i.e., most deviations occurred at $d^* \leq 0.15$ $Å^{-1}$. That the agreement did not improve further probably means that other data perturbations, such as secondary scattering, are also present, as found for the perchloro- analog. This structure analysis, no doubt, represents an extreme case for accepting data to be used for ab initio determinations.

Table 6.15. Observed and Calculated Structure Factors for Copper Perbromophthalocyanine, Including a Multislice Correction for Dynamical Scattering

hk	$\|F_o\|$	$\|F_c\|$ (kinematical)	$\|F_c\|$, (dynamical, t = 113 Å)
02	1.86	0.71	1.29
04	1.66	1.24	1.35
06	0.70	0.24	0.34
08	1.47	2.13	1.79
11	2.02	0.78	1.53
13	0.98	0.72	0.93
15	0.29	0.20	0.19
17	0.57	0.99	0.84
19	1.27	1.50	1.28
20	1.28	0.44	0.68
22	2.40	1.73	2.09
24	0.47	1.20	1.04
26	0.41	0.16	0.44
28	0.61	1.16	1.04
31	1.24	1.01	0.87
33	0.42	0.69	0.71
35	1.41	1.90	1.56
37	1.33	1.67	1.38
39	0.36	0.19	0.19
40	1.12	1.17	1.00
42	0.44	0.40	0.48
44	1.46	1.84	1.53
46	0.35	0.09	0.16
48	1.51	3.04	2.61
51	0.64	0.81	0.81
53	0.27	0.13	0.19
55	1.65	1.93	1.62
57	0.96	1.50	1.29
59	0.93	1.35	1.18
60	1.95	2.16	1.81
62	0.51	0.51	0.28
64	1.37	0.54	1.16
66	0.67	0.95	0.82
68	0.65	0.76	0.68
71	1.81	2.00	1.67
73	0.82	1.18	1.13
75	0.49	0.48	0.82
77	1.17	1.42	1.20
79	0.76	0.13	0.76
80	0.82	0.09	0.22
82	1.36	1.22	1.04
84	0.55	0.63	0.54
86	0.45	0.72	0.72
88	0.38	0.23	0.68
91	0.33	0.58	0.72
93	1.62	1.52	1.29
95	0.52	0.42	0.38

(continued)

Table 6.15. (*Continued*)

hk	$\lvert F_o \rvert$	$\lvert F_c \rvert$ (kinematical)	$\lvert F_c \rvert$, (dynamical, t = 113 Å)
97	0.66	0.65	0.56
99	0.48	0.07	0.12
10,0	0.28	0.01	0.31
10,2	0.86	1.25	1.06
10,4	0.17	0.46	0.38
10,6	0.15	0.06	0.24
10,8	0.28	0.24	0.31
R		0.39	0.27

6.3.3. C_{60} Buckminsterfullerene

Thin crystals of C_{60} buckminsterfullerene were grown by evaporation of a dilute benzene solution on a carbon film-covered electron microscope grid. Electron diffraction patterns from various crystal projections are depicted in Figure 2.16. The pattern from the untilted crystals (i.e., the [111] projection) contained weak forbidden reflections originating from layer stacking faults, as discussed by many researchers. Aside from the weak reflections in this projection, the indexed zonal reflections from six

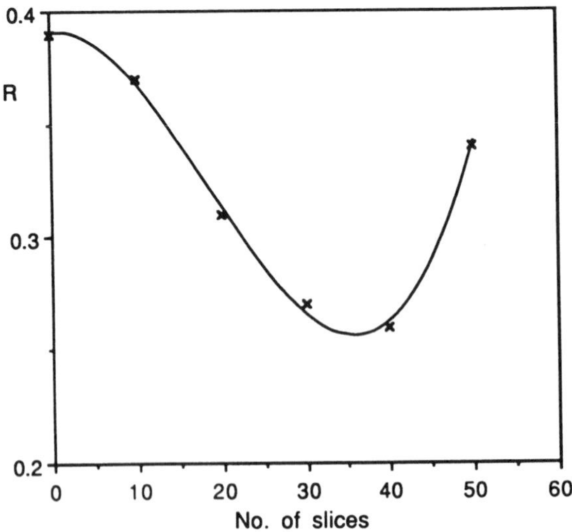

FIGURE 6.16. Improvement of crystallographic residual *R* as a multislice calculation progresses through an increasing number of crystal slices with thickness 3.8 Å. (Reprinted from D. L. Dorset, W. F. Tivol, and J. N. Turner (1992) "Dynamical scattering and electron crystallography—ab initio structure analysis of copper perbromophthalocyanine," *Acta Crystallographica* **A48**, 562–568; with kind permission of the International Union of Crystallography.)

orientations corresponded to the cubic space groups *Fm*3*m* or *Fm*3, with measured cell axis a = 14.26(23) Å, in good agreement with earlier measurements (Fleming et al., 1991). Intensity data (42 unique values) were collected at 100 kV from the various zonal projection after integration of densitometer scans of the diffraction films. It was important to establish first that the intensity data from individual projections were internally consistent, this found by comparison of several diffraction patterns from a given crystal orientation. Data from the separate zones were placed on the same relative scale by equating the sum of reflection intensities shared between any two sets.

Direct phase determination involved the evaluation of Σ_2-triples and positive quartets based on *hhl* reflections and their permutations allowed by the cubic symmetry in space group *Fm*3 (Dorset and McCourt, 1994b). This space group was chosen because it shares a common projection along [110] with *Fm*3*m*, so the former could be used to halve the number of symmetry operations needed to generate the structure invariant sums. Only one reflection could be used to define the origin (parity uuu), so ϕ_{333} = 0 was specified. An algebraic value was also given to ϕ_{113} to evaluate 17 Σ_2-triples and 16 positive quartets above specified threshold values for A_2 and B. The resultant phase set is given in Table 6.16. After permutation of each Miller index *hhl* to equivalent *hlh* and *lhh*, the phases were combined with observed structure factor magnitudes to generate a potential map based on 17 unique reflections (Figure 6.17).

The appearance of the initial potential map was surprising since it seemed to indicate a somewhat static density distribution, disagreeing with the result of other experimental work, i.e., that the molecules must rotate freely in the crystal at room temperature. Nevertheless, a regular carbon icosahedron was fit to this experimental density map and a good fit could be found with a molecular radius similar to the one proposed by André et al. (1992) (Table 6.17). This corresponded to a crystallographic residual R = 0.30 for data restricted to sin $\theta/\lambda \leq 0.30$ Å^{-1}. A multislice calculation (assuming a static model) improved this agreement, i.e., R = 0.23 for a crystal thickness of t = 100 Å. Thus, there appeared to be merely a two-fold disorder around unit cell axial directions, but a good fit of other holes in the density by the centers of six membered rings was a very compelling argument for a somewhat static structure, reminiscent of the original model of Fleming et al. (1991).

As enticing as this result was, there were also significant problems. For example, the distorted bond geometry indicated in the first x-ray analysis (Fleming et al., 1991)

Table 6.16. Crystallographic Phases for C_{60} Buckminsterfullerene by Direct Methods

0kl	ϕ	*1kl*	ϕ	*2kl*	ϕ	*3kl*	ϕ	*4kl*	ϕ	*5kl*	ϕ
022	*a*	113	*a*	224	0	333	0	444	*a*	553	*a*
		111	0	222	*a*	331	0	400	0	555	0
		115	0			335	*a*	440	*a**		
		117	*a*					442	0		
		119	0								
	a = π										

*Disagrees with free rotor model.

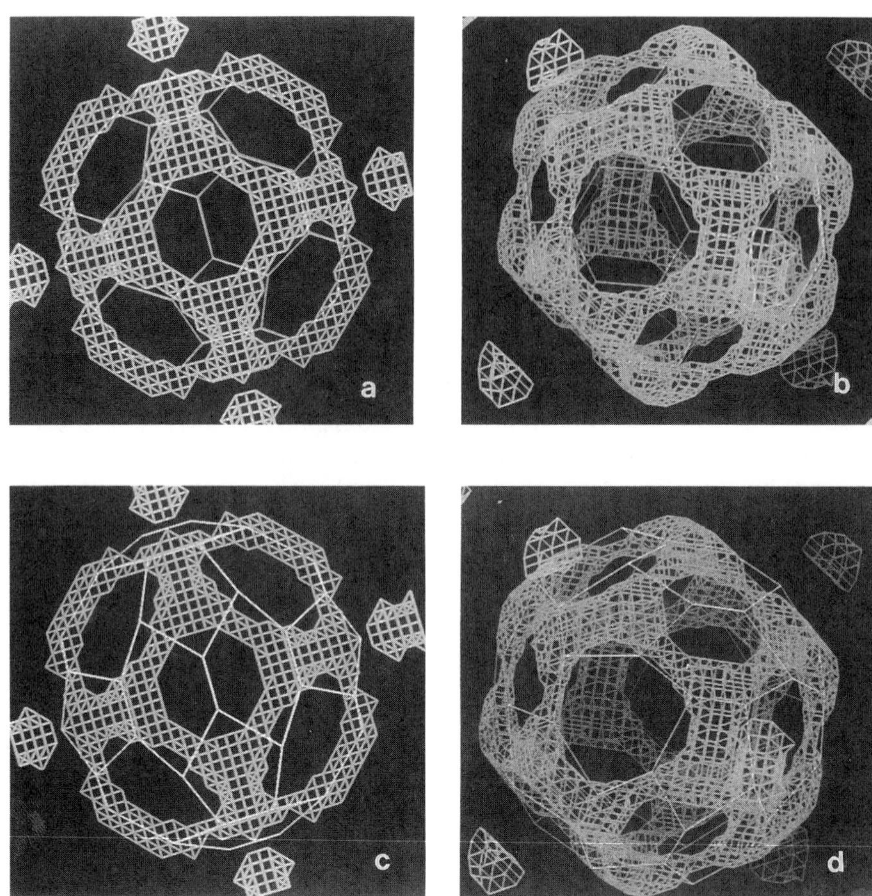

FIGURE 6.17. Initial potential map for C_{60} buckminsterfullerene after phasing of 17 unique reflections. (a, b) Fit with an ideal carbon icosahedron; (c, d) fit with the distorted structure of Fleming et al. (1991). (Reprinted from D. L. Dorset and M. P. McCourt (1994) "Disorder and the molecular packing of C_{60} buckminsterfullerene: a direct electron crystallographic analysis," *Acta Crystallographica* **A50**, 344–351; with kind permission of the International Union of Crystallography.)

Table 6.17. Model Coordinates for C60 Buckminsterfullerene

	Electron diffraction			Model of André et al. (1992)*		
	x	*y*	*z*	*x*	*y*	*z*
C 1	0.052	0	0.249	0.049	0	0.246
C 2	0.105	0.085	0.220	0.101	0.083	0.214
C 3	0.185	0.052	0.165	0.184	0.051	0.163

*As discussed in this work, these coordinates can be derived from the two unique C-C bond lengths (taken to be 1.45 Å and 1.40 Å) and the spherical radius *R* (defined as 3.548 Å).

Table 6.18. Observed and Calculated Structure Factors for C_{60} Buckminsterfullerene

hkl	$\lvert F_o \rvert$	F_c	*hkl*	$\lvert F_o \rvert$	F_c
002	1.10	-1.02^*	440	1.31	1.24
004	1.02	1.03^*	442	0.87	0.81
006	0.72	0.76^*	444	1.25	−1.14
008	0.66	-0.52^*	446	0.45	0.50
0010	0.43	0.29^*	448	0.42	0.31
111	2.59	3.54	551	1.05	−0.98
113	2.75	−3.38	553	0.81	−0.60
115	1.21	1.82	555	0.74	0.58
117	0.84	−0.98	557	0.51	0.76
119	0.65	0.51	660	0.88	0.56
220	4.87	−4.58	662	0.60	0.60
222	2.16	−2.77	664	0.61	0.45
224	1.80	2.09	666	0.57	−0.30
226	1.01	−1.11	771	0.41	0.28
228	0.64	0.57	773	0.40	−0.26
2210	0.34	−0.27	260	0.84	−1.06
331	1.57	1.56	264	0.64	−0.69
331	1.99	1.92	042	1.82	1.65
335	1.00	−1.08	064	0.88	−0.99
337	0.54	0.51	028	0.51	0.35
339	0.41	0.78	048	0.86	0.52

*Reflections most affected by secondary scattering; correction averaged over three separate zones.

actually was in better agreement to the observed *hhl* data ($R = 0.25$) than the model used for the multislice calculation, despite the ample crystallographic data that support a regular icosahedron. The source of this problem was found in the measured I_{h00} values, which, for electron diffraction, were observed to be very large. In a qualitative study of this material by Van Tendeloo et al. (1992), however, these strong reflections were found to abruptly disappear when the row was rotated away from a zone axis—a characteristic of secondary scattering. Thus, the *h*00 reflections must be weak, as they are for the x-ray data from such crystals (André et al., 1992). If they are weak, then a freely rotating structure can be justified.

To calculate the structure factors from the rotationally disordered packing, it was required only to consider the Fourier transform of a uniform disc in any projection (André et al., 1992):

$$\mathrm{F}(\mathbf{s}) = \sum_j f_j' \, (\sin(2\pi \mathbf{r}_j \cdot \mathbf{s})/2\pi \mathbf{r}_j \cdot \mathbf{s}) \exp 2\pi i \mathbf{r}_j \cdot \mathbf{s}$$

Such a calculation produced a very good fit to the measured intensities if the *h*00 reflections were omitted. To model secondary scattering, the relationship $I_h' = I_h + mI_h{*}I_h$ was used for three zones containing *h*00 reflections ([110], [310], and

Table 6.19. Observed and Calculated Structure Factors for Graphite

hk	$\lvert F_o \rvert$	F_c	hk	$\lvert F_o \rvert$	F_c
10	2.30	3.23	50	0.18	0.21
11	7.91	5.58	33	0.41	0.42
20	1.28	1.21	42	0.23	0.19
21	0.87	0.73	51	0.11	0.17
30	2.01	1.77	60	0.28	0.26
22	1.54	1.27	43	0.10	0.13
31	0.45	0.41	52	0.17	0.23
40	0.37	0.34	61	0.12	0.11
32	0.25	0.28	62	0.11	0.08
41	0.59	0.61			

[100]). The best adjusted weight m for the intensity convolution was nearly the same for each projection and, thus, was fixed to a constant value. When the I_{h00} were averaged, the fit to all 42 measured data, assuming $B = 3.0$ Å^2, was $R = 0.17$ (Table 6.18). Comparison of the calculated phases to the ones determined directly detected only one discrepancy. Use of all reflections to calculate a potential map then resulted in a more uniform density profile. The best analysis, therefore, agreed with the results of previous crystallographic and spectroscopic determinations of the room temperature structure, i.e., the molecules are rotationally disordered in the lattice.

Secondary scattering is a commonly occurring perturbation to intensity data from organic crystals, but generally appears as weak violations of space-group forbidden reflections, e.g., axial reflections with odd index. The secondary scattering phenomenon, observed here as strong even-order intensities, was unexpected initially. On the other hand, the incoherent multiple scattering did not compromise the success of direct phase determination. Only the interpretation of the structure was initially misleading, limited also by the series termination effect resulting from a small number of phased $\lvert F_o \rvert$ to calculate the initial potential map.

6.3.4. Graphite

The crystal structure of graphite was determined by Ogawa et al. (1994) using 800-kV selected area diffraction intensities from thin crystals obtained from patterns recorded on imaging plates. Careful convergent beam diffraction experiments had been carried out earlier by Goodman (1976) to determine that a perfect hexagonal structure actually does exist and that the alternative trigonal structure was a result of stacking faults. The hexagonal space group is $P6_3/mmc$, with cell constants $a = b = 2.46$, $c =$ 6.70 Å. In projection down c (plane group $p6m$), carbon atoms were found to lie on special positions (0,0), (1/3,2/3), (2/3,1/3), resulting in calculated structure factor values, all with zero phase angles. The phasing model corresponded to an R-value of 0.23 for 19 unique reflections (see Table 6.19).

6.4. Conclusions

Results of early structure analyses based on texture electron diffraction intensities carried out in Moscow have been unfairly criticized by the crystallographic community. From the examples given above, it is clear that the correct structures can be determined directly from the measured data and, therefore, do not depend on any existing x-ray determinations to find crystallographic phases. Despite the obvious perturbations due to dynamical scattering (low electron accelerating voltage, unknown average crystal thickness), the direct phase determination often leads to chemically reasonable molecular architectures and crystal packings. There are also obvious benefits to using the texture diffraction geometry for data collection and this procedure should be improved greatly when higher-voltage electron sources are employed. In general, the minimization of strong nonsystematic effects due to the distribution of crystal orientations over a wide sampling area seems to be beneficial for collection of directly analyzable data.

Single crystal diffraction patterns also contain useful intensity data and direct phasing of these again leads to reasonable results. Although the phase determination based on Σ_2-triples (either by symbolic addition of via the tangent formula) is somewhat "robust," in spite of multiple scattering contributions to the intensities, there is no doubt that the highest voltage electron source possible is the best one for data collection. For example, there seems to be no convincing experimental evidence for an "optimal" accelerating voltage in the 400-kV range. High-voltage diffraction intensities are particularly important for structure refinement so that the structure model with the best bonding geometry will also correspond to the lowest crystallographic residual.

7

Inorganic Structures

7.1. Background

With procedures for electron diffraction structure analysis established in Moscow, intensity data, mainly from oblique texture patterns, were often used for quantitative crystal structure analyses of inorganic compounds, in addition to the organic materials discussed in Chapter 6. While favorable results might be expected for light-atom structures such as boric acid (Cowley, 1953a), ammonium sulfate (Udalova and Pinsker, 1964), and ammonium chloride (Kuwabara, 1969), as originally claimed, it might be more difficult to accept the validity of heavier-atom structures described in the books of Pinsker (1953), and Zvyagin (1967), and the recent chapter by Vainshtein, Avilov, and Zvyagin (1992). However, a typical analysis could be cited, that of the semiconductor $AgTlSe_2$, for which Imamov and Pinsker (1965) collected 200 independent oblique texture intensities. Agreement of the final model to the observed data was $R \approx 0.20$, and it was decided that the bonds are covalent. Difficulties experienced in determining the structures of basic lead carbonate (Cowley, 1956) and the λ-phase of alumina (Cowley, 1953b) from single-crystal patterns, on the other hand, lead one to appreciate why Cowley and Moodie (1957) would have been motivated to develop a more accurate, multiple-beam model for dynamical electron diffraction. It was obvious that the Blackman two-beam model accepted by other laboratories did not adequately account for the measured deviations from the kinematical theory, especially when selected area diffraction intensities from single crystals were used for structure analysis.

Because of the complexity of the dynamical diffraction calculations, described in Chapter 5, the prevailing viewpoint nowadays is that ab initio structure analyses, based on electron diffraction intensity data or the interpretation of high-resolution electron micrographs, is absolutely pointless for inorganic substances, again because of the strong scattering cross-section of the heavy-atom components. A very recent statement of this viewpoint has been given by Eades (1994), viz.: "Electron diffraction intensities are not simply related to the corresponding structure factors except in the case of 'dynamical extinctions' . . . If it was possible to find conditions when 'dynamical extinction' occurs for reflections that are not forbidden, it would be possible to use electron diffraction intensities for the determination of structure factors in a simple way. Unfortunately, it can be shown that this is not possible."

As a rigorous statement, this is correct. In fact, faced with this problem, especially given the unique opportunities provided by convergent beam techniques, suggestions of combined electron diffraction and powder x-ray diffraction structure analyses have been made (Gjønnes et al., 1989). The x-ray data would be best employed for model refinement after the unit cell symmetry is correctly determined by convergent beam electron diffraction. Again, no criticism can be made of this approach. Using only electron scattering information, the accepted procedure for structure analysis, currently, is to start with a preconceived atomic model for the crystal structure. Based on the atomic coordinates, a multiple-beam scattering calculation is tested for convergence to the observed diffraction or image data. The main problem, however, i.e., to find a suitable approximation to the actual crystal structure, remains unsolved. With trial-and-error methods, it is difficult to arrive at such a solution at random, even given "reasonable" packing parameters (see Chapter 4). While the rigor of Eades' statement above cannot be disputed, the real goal of the experimental work should be to measure "quasi-kinematical" diffraction intensities where enough of the direct unit cell transform is preserved to permit an ab initio structure determination to be carried out. The approach taken by Russian investigators to employ texture diffraction data for this purpose, therefore, offers some advantages over the use of single-crystal data. As described by Cowley (1967): "However, one must take into account that in patterns from polycrystalline materials there will be an averaging over thickness which will smooth out the oscillations of the single-crystal curves. Also, more importantly, the averaging over orientations will reduce the effect of many-beam interactions. Of all the crystal orientations for which a particular reflection takes place, a relative small proportion will be the special orientations for which there are strong non-systematic interactions and available evidence suggests that for these special orientations the contribution to the particular reflection, if it is a strong one, will be considerably reduced. Most of the intensity of the reflection can be considered to arise in the presence of dynamical effects from only the systematic interactions."

Certainly dynamical effects were thought to become very important as the atomic number of the scattering species became large and/or the unit cell was also about the same size as for the materials used in tests of dynamical theory. (The importance of such interactions could be detected by the need to make a large two-beam correction to a particular intensity.) However, Cowley arrived at the following conclusion: "For most of the structure analyses which have been made on the basis of arc or ring patterns, the interpretation of the intensities would appear to be justified on the basis that the deviations from the kinematical approximation are no greater than, for example, the 'extinction effects' present in the data for many contemporary X-ray diffraction structure analyses. Also, particularly for light-atom structures where some correction has been made for dynamical effects on the two-beam basis, the number of reflections seriously affected by n-beam systematic interactions has not been high."

The following examples will demonstrate that a more direct approach to electron crystallographic structure analysis of inorganics may, indeed, be reasonable. Certainly the existence of higher voltage sources than used in pioneering work allows the quasi-kinematical approximation to be satisfied for samples that would have caused problems earlier.

7.2. Structures Solved from Electron Diffraction Data

7.2.1. Boric Acid

Cowley (1953a) had published a single-crystal *hk*0 electron diffraction data set from boric acid collected at room temperature with 133 nonzero intensities. The unit cell parameters:

$$a = b = 7.04,\ c = 6.56\ \text{Å}$$

$$\alpha = 92.5°,\ \beta = 101.2°,\ \gamma = 120.0°$$

were taken from the x-ray analysis of Zachariasen (1934). (Note that a hexagonal layer packing is implied.) Subsequently a low-temperature (108 K) set had been collected at 100 kV from preparations crystallized by the evaporation of a dilute aqueous solution to give 83 unique intensities (Dorset, 1992c). After calculation of $|E_h|$, their distribution (Table 7.1) for the room temperature data was found to correspond to a noncentrosymmetric structure, despite the fact that $P\bar{1}$ was assumed to be the space group. From a Wilson plot, the overall temperature factor was estimated to be $B = 0.0$ Å^2. From this value, it was apparent that some perturbation of the data was taking place.

The progress of the direct phase determination (Dorset, 1992c) from a few Σ_1-triples, 16 Σ_2-triples, and 35 quartets is outlined in Table 7.2. The initial map, based on 28 phased reflections shown in Figure 7.1a, was found when the algebraic unknown was assigned the value $a = \pi$. If trial atomic positions were taken from this map, assuming the threefold axis through the borate residue, the initial residual for the initial set of calculated structure factors was 0.39. If this total phase set was then used to calculate another potential map, the structure was more clearly defined (Figure 7.1b). The new coordinates correspond to $R = 0.29$. Hydrogen atom positions were found by

Table 7.1. Distributions of $|E_h|$ for Boric Acid

		Theory	
	Experimental	Centrosymmetric	Noncentrosymmetric
$<\lvert E_h\rvert^2>$	1.00	1.000	1.000
$<\lvert E_h^2 - 1\rvert>$	0.71	0.968	0.736
$<\lvert E_h\rvert>$	0.89	0.798	0.886
% $\lvert E_h\rvert > 1.0$	24.8	32.2	36.8
% $\lvert E_h\rvert > 2.0$	2.3	5.0	1.8
% $\lvert E_h\rvert > 3.0$	0.8	0.3	0.01

Table 7.2. Phase Determination for Boric Acid (Room Temperature)

origin definition: $\phi_{-630} = \phi_{-360} = 0$
Σ_1-triples: $\phi_{-820} = \phi_{620} = \phi_{260} = \phi_{-680} = 0$; $\phi_{460} = \pi$
Σ_2-triples:
  $\phi_{330} = 0$; $\phi_{900} = 0$; $\phi_{-990} = 0$; $\phi_{090} = 0$
  Let $\phi_{-110} = a$; this leads to $\phi_{-520} = a$
  Let $\phi_{-140} = b$; this leads to $\phi_{-410} = b$, $\phi_{510} = b$, $\phi_{-740} = a$
    $\phi_{150} = b$, $\phi_{130} = \pi$, $\phi_{-250} = a$
    $\phi_{-430} = \pi$, $\phi_{-560} = \pi$, $\phi_{700} = \pi$
positive quartets: $\phi_{-290} = \phi_{200} = \phi_{420} = \phi_{-160} = \pi$; $\phi_{-220} = b$
correct map calculated when $a = b = \pi$

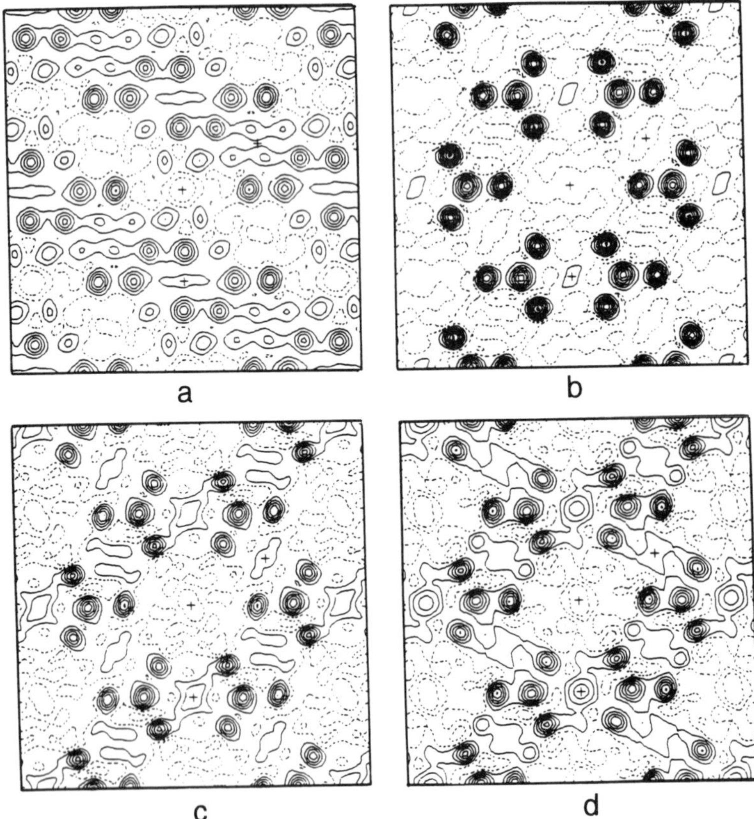

FIGURE 7.1. Structure analysis of boric acid, room temperature data: (a) potential map calculated from initial phase set; (b) map from structure factor phases based on atomic positions in (a); (c) attempt to locate hydrogen positions based on difference map ($2|F_o| - |F_c|$), with all measured data; (d) hydrogen positions found when only high $|E_h|$ data are used. (Reprinted from D. L. Dorset (1992) " Direct methods in electron crystallography—structure analysis of boric acid," *Acta Crystallographica* **A48**, 568–574; with kind permission of the International Union of Crystallography.)

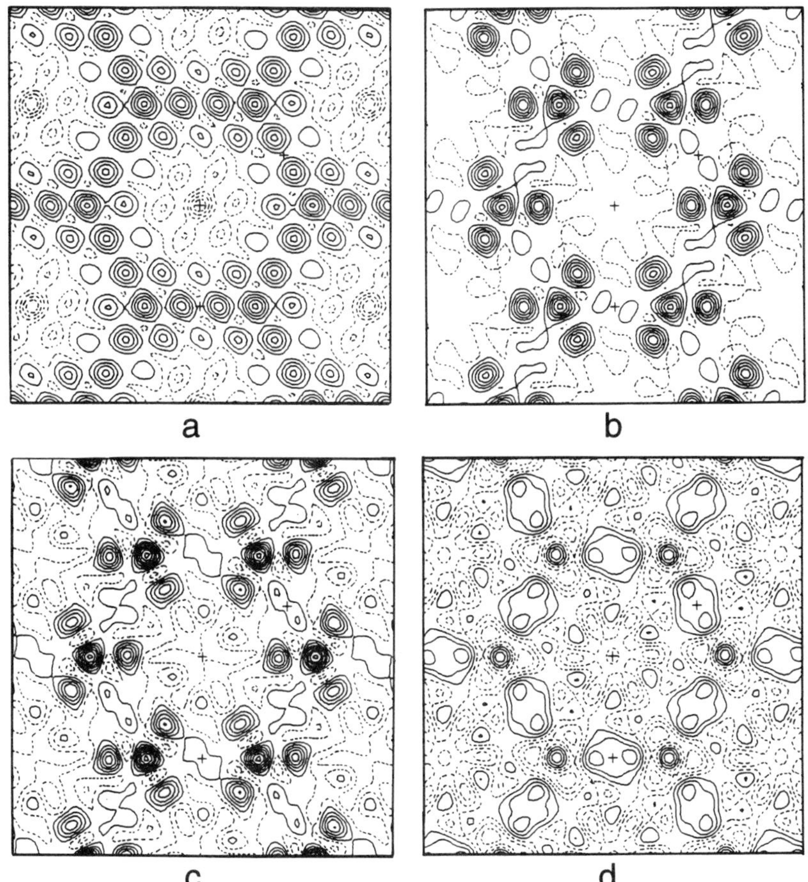

FIGURE 7.2. Structure analysis of boric acid, low temperature data: (a) potential map based on initial phase set; (b) map based on phases from atomic positions located in (a); (c) difference map $(2|F_o| - |F_c|)$; (d) difference map based on $(|F_o| - |F_c|)$. (Reprinted from D. L. Dorset (1992) " Direct methods in electron crystallography—structure analysis of boric acid," *Acta Crystallographica* **A48**, 568–574; with kind permission of the International Union of Crystallography.)

generating a potential map based on $2|F_o| - |F_c|$ (Figure 7.1c) but their locations were not well defined. If another difference map based on $|F_o| - |F_c|$ was calculated for reflections corresponding to $|E_h| \geq 0.85$ (Figure 7.1d), these lighter-atom positions were more clearly discerned. The resulting bond distances and angles could then be compared to values determined in a later neutron diffraction analysis (Craven and Sabine, 1966): <B–O> = 1.36 (1) Å (1.367 Å), <O–B–O> = 120.1(2)° (120°), <O–H> = 1.11 (11) Å (0.97 Å), <B–O–H> = 111.3 (60)° (113.3°).

The structure analysis with the low-temperature data proceeded in a way similar to that for room temperature analysis. An initial potential map based on the phase

Table 7.3. Observed and Calculated Structure Factors for Boric Acid (Low Temperature)

hk	$\|F_o\|$	F_c	*hk*	$\|F_o\|$	F_c
1 0	1.61	−1.73	−4 3	0.94	−0.98
2 0	0.77	−1.26	−5 3	0.65	−0.13
3 0	0.40	−0.41	−6 3	1.51	1.28
4 0	0.69	0.81	−7 3	0.47	−0.27
5 0	0.56	0.38	0 4	0.69	0.81
6 0	0.37	−0.27	1 4	0.57	0.44
7 0	0.35	−0.36	2 4	0.52	−0.40
0 1	1.61	−1.73	3 4	0.47	−0.28
1 1	0.75	1.10	−1 4	0.94	−0.91
2 1	0.80	0.73	−2 4	0.66	0.28
3 1	0.94	−0.93	−3 4	0.92	−1.13
4 1	0.52	0.24	−4 4	0.69	0.83
5 1	0.57	−0.46	−5 4	0.52	0.25
6 1	0.42	0.47	−6 4	0.49	−0.40
−1 1	1.61	−1.74	−7 4	0.45	−0.37
−2 1	0.75	1.10	0 5	0.56	0.38
−3 1	0.73	1.20	1 5	0.57	−0.43
−4 1	0.92	−1.06	2 5	0.35	0.21
−5 1	0.57	0.43	3 5	0.35	−0.28
−6 1	0.57	−0.43	−1 5	0.52	0.24
−7 1	0.37	0.42	−2 5	0.65	−0.11
0 2	0.77	−1.26	−3 5	0.85	−0.42
1 2	0.73	1.22	−4 5	0.57	0.44
2 2	0.66	0.28	−5 5	0.56	0.35
3 2	0.65	−0.12	−6 5	0.52	−0.43
4 2	0.49	−0.42	−7 5	0.41	0.27
5 2	0.53	0.28	0 6	0.37	−0.27
6 2	0.42	0.16	1 6	0.37	0.41
−1 2	0.75	1.10	−1 6	0.57	−0.47
−2 2	0.77	−1.31	−2 6	0.49	−0.43
−3 2	0.80	0.75	−3 6	1.51	1.28
−4 2	0.60	0.28	−4 6	0.52	−0.38
−5 2	0.86	−0.39	−5 6	0.57	−0.39
−6 2	0.52	−0.41	−6 6	0.37	−0.32
−7 2	0.35	0.21	−7 6	0.44	0.46
0 3	0.40	−0.41	0 7	0.35	−0.36
1 3	0.92	−1.07	−1 7	0.42	0.47
2 3	0.86	−0.40	−2 7	0.53	0.28
3 3	1.51	1.28	−3 7	0.45	−0.34
4 3	0.45	−0.35	−4 7	0.47	−0.30
−1 3	0.80	0.72	−5 7	0.35	0.20
−2 3	0.73	1.24	−6 7	0.37	0.40
−3 3	0.40	−0.33	−7 7	0.35	−0.31

Table 7.4. Final Atomic Positions for Boric Acid (Low Temperature)

	Electron diffraction		Neutron diffraction*	
	x	*y*	*x*	*y*
B	0.666	0.334	0.670	0.332
O 1	0.454	0.226	0.451	0.208
O 2	0.774	0.226	0.795	0.232
O 3	0.774	0.546	0.769	0.555
H 1	0.443	0.371	0.372	0.291
H 2	0.629	0.072	0.710	0.073
H 3	0.928	0.557	0.922	0.627

*After origin shift (0.024,–0.094) for a single layer.

estimate of 23 reflections is shown in Figure 7.2a. It was more difficult to pick out the boric acid residues unequivocally, but if this model was correctly deduced from the peak positions, the first map based on a complete phase set again shows a clear representation of the structure (Figure 7.2b). Fourier refinement resulted in a structure that agreed with the observed structure factors by $R = 0.29$ (Table 7.3). The final coordinates are compared to the results of other analyses in Table 7.4.

It is clear that the observed data represent a hexagonal layer packing and not the triclinic unit cell described by Zachariasen (1934). Cowley (1953a) proposed that the crystals used for data collection contained a stacking disorder parallel to [001]. Although the later neutron diffraction analysis (Craven and Sabine, 1966) did not indicate that layers exist where the atoms are coplanar, the observation of continuous reciprocal lattice rods in tilt series (Figure 7.3) supports Cowley's thesis. An incoherent diffraction phenomenon due to crystal bending might be an alternative way to explain the diffraction from isolated hexagonal layers, although the rather small *c*-spacing given above would imply that this bend component must be rather large. However, sharp bend contours were not observed in bright field images. If the stacking disorder exists, secondary scattering of the type considered above in the analysis of C_{60} buckminsterfullerene also could be quite significant. Correction of either analysis above for this incoherent multiple scattering was found to lower the residual to $R = 0.22$.

7.2.2. Celadonite

The crystal structure of celadonite, $K_{0.8}(Mg_{0.7}Fe_{1.4}(Al_{0.4}Si_{3.6}O_{10}))(OH)_2$, as originally determined by B. B. Zvyagin (1957) from 82 oblique texture diffraction data collected at about 50 kV. The monoclinic unit cell, space group *C*2/*m*, has dimensions $a = 5.20$, $b = 9.00$, $c = 10.25$ Å, $\beta = 100.1°$. Direct phasing based on the evaluation of phase invariant sums treated the $h0l$ and $0kl$ data sets separately (Dorset, 1992d).

For the 41 $h0l$ data, two phases were accepted from Σ_1-triples and a partial origin definition was made by setting $\phi_{201} = 0$. Phase values were found for a total of 19

a

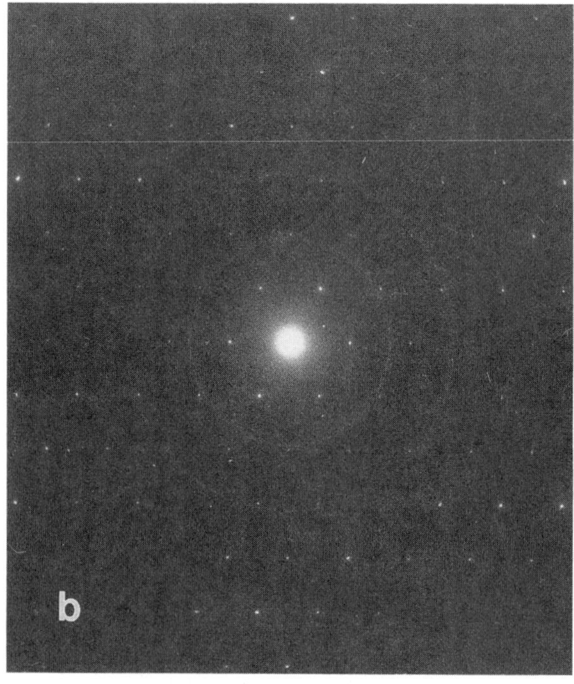
b

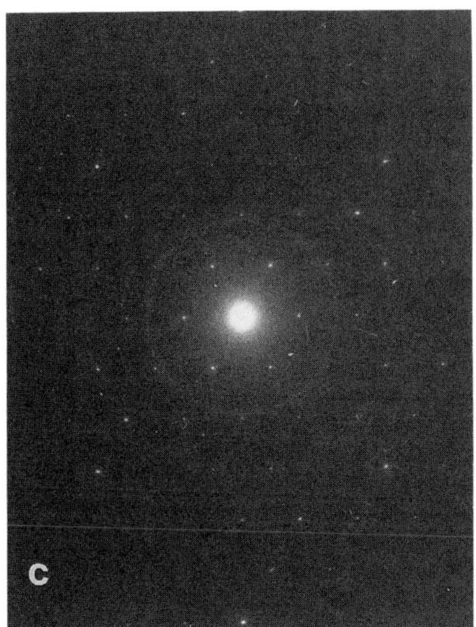

FIGURE 7.3. Electron diffraction patterns from tilted crystal of boric acid (common tilt axis for one crystal): (a) 0°; (b) 5°; (c) 10°; (d) 15°. (Reprinted from D. L. Dorset (1992) "Direct methods in electron crystallography—structure analysis of boric acid," *Acta Crystallographica* **A48**, 568–574; with kind permission of the International Union of Crystallography.)

Table 7.5. Initial Phase Assignments for Celadonite, [010] Projection

h0l	|*E*|	ϕ
2 0 0	2.55	0
2 0 1	1.86	0
2 0 2	1.64	0
2 0 4	2.18	0
2 0 6	0.99	a^*
–2 0 1	1.60	a^*
–2 0 3	0.88	0
–2 0 4	1.02	0
–2 0 5	0.90	a^*
–2 0 6	1.74	0
–2 0 7	1.85	0
4 0 4	1.15	0
–4 0 1	1.76	a^*
–4 0 2	2.17	0
–4 0 3	1.36	0
–4 0 5	0.96	0
–4 0 6	1.16	0
–6 0 2	1.38	0
–6 0 9	0.73	0

$^*a = \pi$.

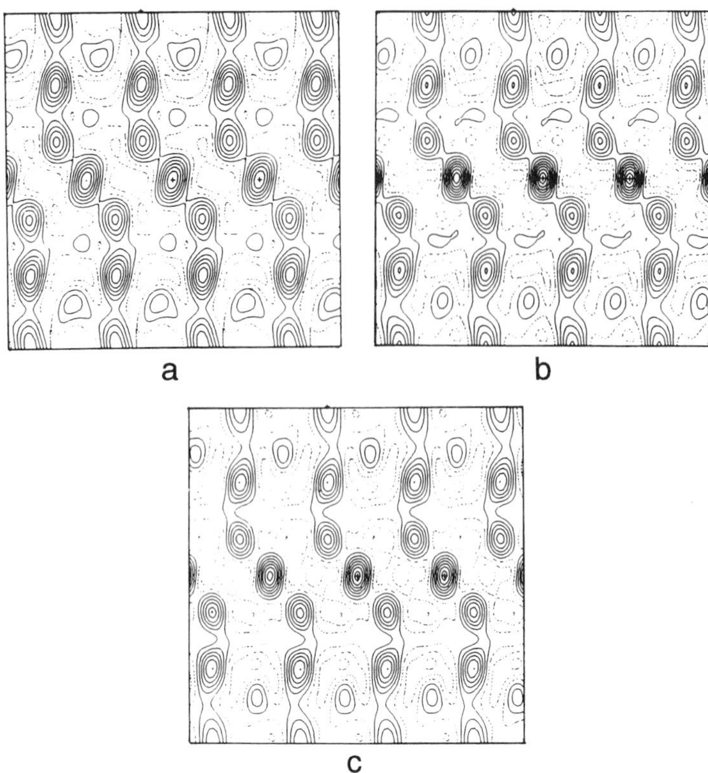

FIGURE 7.4. Potential maps for celadonite, [010] projection: (a) from initial phase set; (b) from structure factor phases based on atomic positions in (a); (c) Zvyagin's structure. (Reprinted from D. L. Dorset (1992) "Direct phasing in electron crystallography: determination of layer silicate structures," *Ultramicroscopy* **45**, 5–14; with kind permission of Elsevier Science B.V.)

reflections (Table 7.5) after including an algebraic unknown. Two assignments were rejected for the $(\bar{6},0,1)$ and $(\bar{6},0,3)$ reflections due to inconsistencies in the evaluation of Σ_2-triples. The initial potential map, calculated when $a = \pi$ (Figure 7.4a), revealed the atom positions. After x,z coordinates were determined for Mg, Fe, Si, Al, K, and two oxygens, to be used for a structure factor calculation, a map based on all of the zonal data was calculated from the new phase estimates (Figure 7.4b). This agrees well with the map based on the phase set in Zvyagin's original paper. There are only 4/41 discrepant phases in the final list for this zone.

Next, the 41 $0kl$ data were analyzed as a separate set. After setting the origin partly by letting $\phi_{061} = 0$, phases were accepted from three Σ_1-triples to allow assignments to be made for 14 reflections via two algebraic unknowns, after evaluating 16 Σ_2-triples and one quartet (Table 7.6). After permuting these unknowns to find a likely map (Figure 7.5a), y,z positions were found for the major atomic sites. The resulting phase

Table 7.6. Initial Phase Assignments for Celadonite, [100] Projection

$0kl$	$\|E\|$	ϕ^*
0 2 0	2.29	b
0 2 2	1.00	$a + b$
0 2 3	1.63	a
0 4 1	1.00	b
0 4 4	1.21	a
0 4 6	0.60	0
0 4 7	0.92	b
0 6 1	1.54	0
0 6 4	0.99	$a + b$
0 6 7	1.41	0
0 10 0	0.88	b
0 12 0	1.21	0
0 12 2	0.65	0
0 12 3	0.73	$a + b$

*$a = b = \pi$.

set from the structure factor calculation disagreed with Zvyagin's values for only 5/41 reflections. The maps (Figure 7.5b,c), are in close agreement.

7.2.3. Muscovite

Muscovite, $KAl_2(AlSi_3O_{10})(OH)_2$, crystallizes in the monoclinic space group $C2/c$ with cell constants $a = 5.18$, $b = 8.96$, $c = 20.10$ Å, $\beta = 95.67°$. Zvyagin and Mischenko (1961) had collected 128 $h0l$ and $0kl$ texture diffraction data for structure analysis. (Note, however, that selected area diffraction data from this material could not be used for crystal structure determination because of the elastic bend distortion (Cowley and Goswami, 1961), with the resultant diffraction incoherence discussed in Chapter 5.) Again the direct phase analyses treated each zone separately (Dorset, 1992d).

Starting with the $h0l$ set, a Σ_1-triple was used to find $\phi_{-4,0,12} = 0$, and an additional algebraic unknown was used to evaluate 50 Σ_2-triples, resulting in a total set of 23 assignments for this zone (Table 7.7). (An origin definition is not permitted for this projection because all of the measured reflections are phase invariants.) The map calculated when $a = \pi$ is shown in Figure 7.6a. From this x,z positions of the Al, Si, K metals and three oxygens could be discerned, but the assumption that an Al could be substituted for a Si on one site was not made. After calculating structure factors, the map (Figure 7.6b,c) resembles the one calculated with phase set of Zvyagin and Mischenko (1961), although there are 11/51 phases that disagree.

Analysis of the other zone was more difficult. An origin-defining reflection was given by $\phi_{0,8,13} = 0$ and the value $\phi_{0,12,0} = 0$ was taken from a number of Σ_1-triples. After evaluating Σ_2-triples, phase values are assigned to 18 reflections, as listed in Table 7.8. Two algebraic unknowns were needed for these determinations. The best

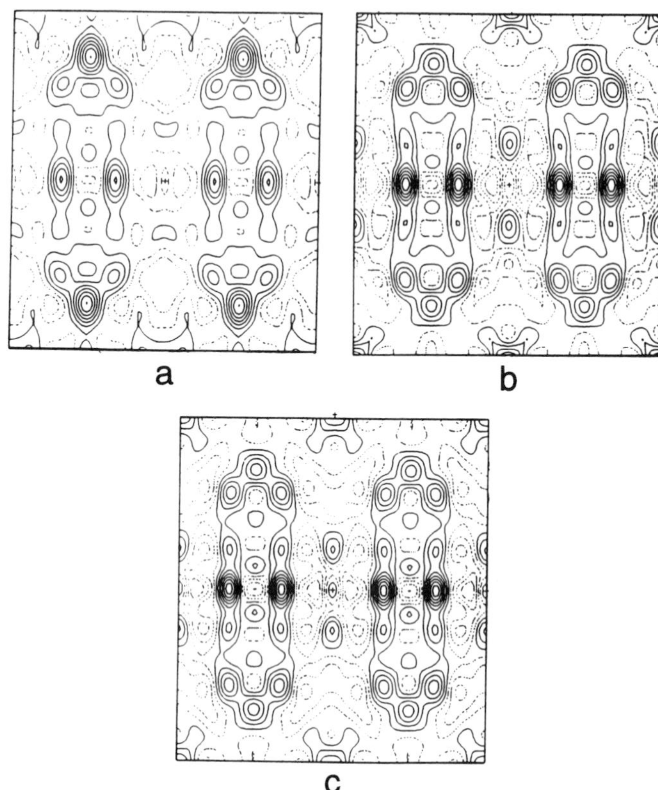

FIGURE 7.5. Potential maps for celadonite, [100] projection: (a) from initial phase set; (b) from structure factor phases based on atomic positions in (a); (c) Zvyagin's structure. (Reprinted from D. L. Dorset (1992) "Direct phasing in electron crystallography: determination of layer silicate structures," *Ultramicroscopy* **45**, 5–14; with kind permission of Elsevier Science B.V.)

map (Figure 7.7) could be used to find all heavy-atom positions but some of the oxygen positions were not visualized (compare with original assignments given by Zvyagin (1967) in his book, p. 271). If the identified atomic positions were used to calculate structure factors, the ensuing map more closely resembled the one calculated from the earlier phase set (Figure 7.7b,c) but there were still 21 phase discrepancies with the 77 0*kl* values listed in the earlier determination.

7.2.4. Phlogopite-Biotite

The structure of phlogopite-biotite, $KMg_3AlSi_3O_{10}(OH)_2$, is somewhat similar to that of celadonite (Zvyagin and Mischenko, 1963). The monoclinic space group is again *C*2/*m* with cell constants $a = 5.28$, $b = 9.16$, $c = 10.30$ Å, $\beta = 99.83°$. Most of the intensity of the oblique texture patterns is contained in the *h*0*l* set. The first set of

Table 7.7. Initial Phase Assignments for Muscovite, [010] Projection

$h0l$	$\lvert E\rvert$	ϕ
200	1.10	0
206	1.02	a
20,10	1.45	a
20,14	0.84	a
20,18	1.28	a
–202	1.32	a
–206	1.55	a
–20,10	1.77	a
–20,16	1.92	0
–20,18	1.77	a
400	1.64	0
408	1.91	0
40,14	1.12	a
–402	2.32	a
–404	0.72	0^*
–40,12	1.77	0
–40,16	1.28	0
608	0.97	0
60,10	1.48	a^*
–602	1.72	a
–606	1.41	a
–60,14	2.51	a^*
–60,18	0.79	a^*

$^*a = \pi$; disagrees with assignment of Zvyagin and Mischenko (1961).

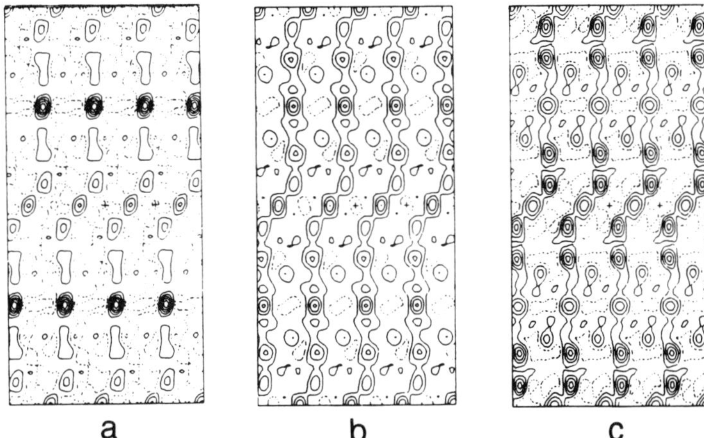

FIGURE 7.6. Potential maps for muscovite, [010] projection: (a) calculated from initial phase set; (b) calculated from structure factors based on atomic positions in (a); (c) Zvyagin's structure. (Reprinted from D. L. Dorset (1992) "Direct methods in electron crystallography: determination of layer silicate structures," *Ultramicroscopy* **45**, 5–15, with kind permission of Elsevier Science B.V.)

Table 7.8. Initial Phase Assignments for Muscovite, [100] Projection

0*kl*	|*E*|	ϕ
023	0.98	*a*
024	1.24	*a* + *b*
025	1.45	*a*
040	0.90	π
041	0.82	*b*
044	0.79	*b*
04,11	0.72	0
04,12	2.38	*b*
04,13	1.00	0*
060	2.89	*a*
068	1.08	*a*
06,10	1.10	*a**
06,16	2.94	*a*
08,11	1.08	0*
08,13	1.04	0
0,10,1	0.72	*a* + *b*
0,10,2	1.26	*a* + *b*
0,12,0	1.16	0

*$a = b = \pi$; disagrees with assignment of Zvyagin and Mischenko (1961).

Table 7.9. Initial Phases for Phlogopite-Biotite, [010] Projection

*h*0*l*	|*E*|	ϕ^*
200	2.16	*a*
201	2.21	0
202	1.99	*a*
204	2.68	*a*
206	0.93	*b*
207	1.03	*b*
208	1.14	*a*
–201	2.01	*a* + *b*
–204	1.48	*a*
–206	1.61	*a*
–207	2.84	0
–208	0.89	*a*
404	2.49	0
405	1.34	*a*
–401	2.21	*b*
–402	2.02	0
–403	1.65	*a*
–405	0.72	*a*
–406	2.66	0
600	0.74	*a*
606	1.05	*a*
–602	1.58	*a*
–603	1.02	0
–606	0.84	*a*
–609	1.07	0
–60,10	0.94	*a*
–60,12	1.05	*a*
–60,13	0.76	0

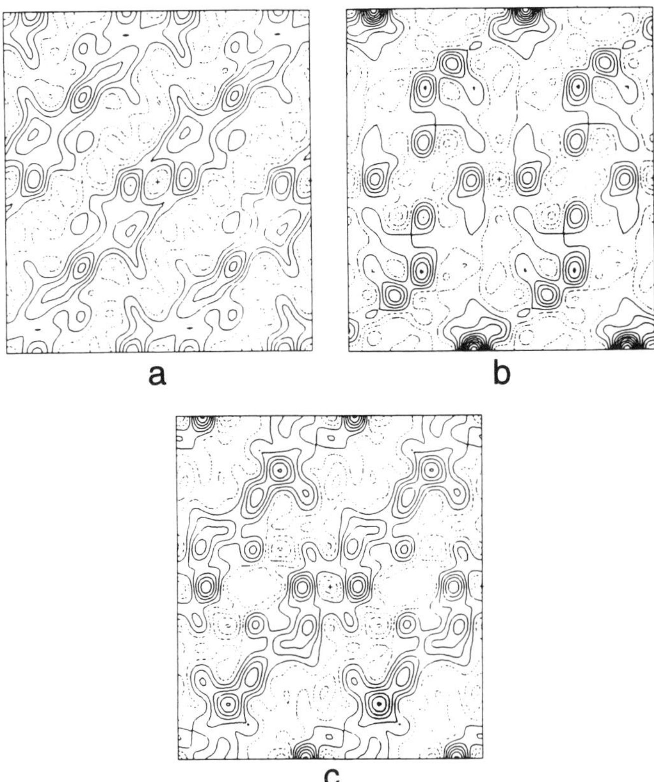

FIGURE 7.7. Potential maps for muscovite, [100] projection: (a) calculated from initial phase set; (b) calculated from structure factors based on atomic positions in (a); (c) Zvyagin's structure. (Reprinted from D. L. Dorset (1992) "Direct methods in electron crystallography: determination of layer silicate structures," *Ultramicroscopy* **45**, 5–14; with kind permission of Elsevier Science B.V.)

$|E_h|$ was calculated using all intensity data to find the overall scale factor and then a separate set was generated for just the $0kl$ intensities alone (Dorset, 1992d).

Phase determination with the $h0l$ set began with an origin-defining reflection, $\phi_{-2,0,7} = 0$, and acceptance of two more phases from Σ_1-triples. Two algebraic unknowns were required to assign values to 28 reflections (Table 7.9) but permutation of these phases generated one of four maps resembling the [010] projection for celadonite (Figure 7.8a). From this map, x,z coordinates could be assigned to two Mg atoms, individual Si, K, and Al, as well as three oxygens. When these atomic positions were used to calculate structure factors for all 59 $h0l$ reflections, the ensuing map was almost indistinguishable from the one calculated from the earlier assignment (Figure 7.8b,c). There were only two phase errors.

There was very little difficulty in assigning phase values to the $0kl$ reflections from the rescaled $|E_h|$ list, even though they are relatively weak. One origin reflection, $\phi_{067} = 0$, was defined as well as a single Σ_1-assignment (0,12,0). With three algebraic

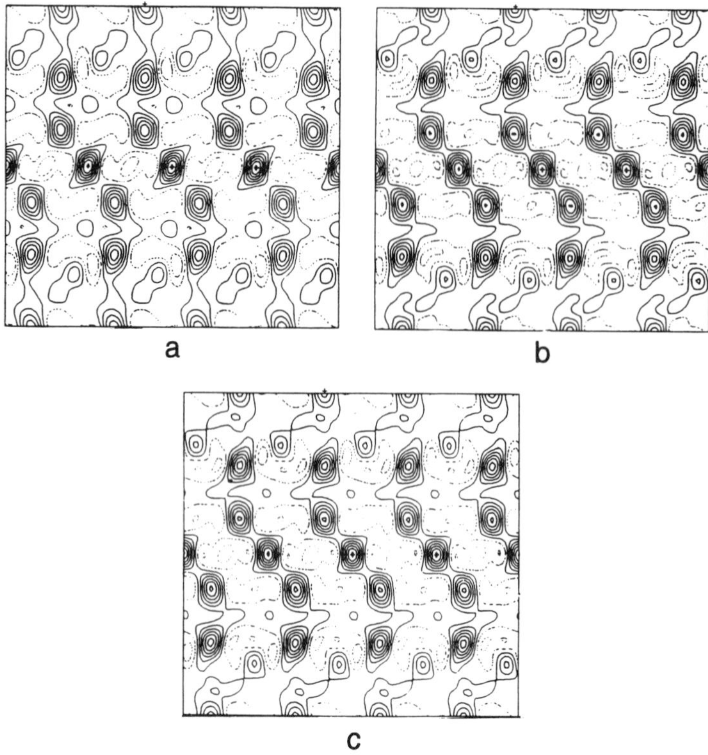

FIGURE 7.8. Potential maps for phlogopite-biotite, [010] projection: (a) calculated from initial phase set; (b) calculated from structure factors based on atomic positions in (a); (c) Zvyagin's structure. (Reprinted from D. L. Dorset (1992) "Direct methods in electron crystallography: determination of layer silicate structures," *Ultramicroscopy* **45**, 5–14; with kind permission of Elsevier Science B.V.)

unknown, phases were found for 17 of the 40 $0kl$ data (Table 7.10), requiring the calculation of eight maps. Structure factor calculations based on the atomic positions from the best map (Figure 7.9a) resulted in only seven disagreements with the original solution. Again, the maps are very similar (Figure 7.9b,c).

7.2.5. λ-Alumina

Single crystal electron diffraction data were obtained from the λ-phase of alumina by Cowley (1953b). The formula is $3NiO{\cdot}5Al_2O_3$. The structure was taken to be pseudotetragonal, actually orthorhombic, with cell constants $a = b = 7.63$, $c = 2.89$ Å. From evaluation of systematic absences in the uncorrected 81 $hk0$ data as well as the nonfourfold distribution of intensities, the plane group *pgm* was assumed for the direct phase determination. (Cowley had chosen the space group *Pmma*, which has *pmm* symmetry in the [001] projection, but there is still a row of systematic absences along one of the reciprocal axes that also would allow *pgm* to be chosen.) From these a set

Table 7.10. Initial Phases for Phlogopite-Biotite, [100] Projection

$0kl$	$\|E\|$	ϕ^*
022	2.14	0
023	1.46	a
041	0.75	a
044	0.97	a
045	0.74	a
046	1.51	0
047	0.91	b
060	2.78	c
064	0.73	0
065	1.18	b
067	1.02	0
068	1.46	0
06,10	1.35	$a+b$
086	0.89	0
0,12,0	1.99	0
0,12,5	0.79	$b+c$
0,12,7	0.86	c

* $a = b = \pi$; $c = 0$.

of $|E_h|$ was generated assuming no thermal motion. The origin was defined by setting $\phi_{730} = \phi_{320} = \pi$, to maintain an agreement with Cowley's assignment, and five more phases were obtained from Σ_1-triples. After evaluation of Σ_2-triples, nondiscrepant assignments were made to 48 reflections, with nine potentially false values (Table 7.11). It was possible to find many of the atomic positions in the first potential map, when $a = \pi$ (Figure 7.10) at locations corresponding to those assigned by Cowley. Ensuing Fourier refinement appeared to be successful, but it was not possible to obtain an R-value lower than 0.48. No dynamical scattering correction was attempted since the third dimension of the atomic positions remained unknown. As in the case of copper perbromophthalocyanine, discussed in the previous chapter, there was enough structural information left in the diffraction intensities to allow a large phase set to be determined but the structure could not be refined kinematically to a reasonable crystallographic residual.

7.2.6. Basic Copper Chloride

Texture electron diffraction data from basic copper chloride, $CuCl_2 \cdot 3Cu(OH)_2$, have been published by Voronova and Vainshtein (1958). The space group is $P2_1/m$ with unit cell parameters $a = 5.73$, $b = 6.12$, $c = 5.63$ Å, $\beta = 93.75°$ in agreement with a powder x-ray determination of Aebi (1948). The three-dimensional electron diffraction data set contains 120 unique intensities, from which $|E_h|$ values were calculated.

In the direct phase analysis based on the evaluation of phase invariant sums (Dorset, 1994a), the origin was defined by setting $\phi_{063} = \phi_{-1,0,2} = 0$. A third reflection

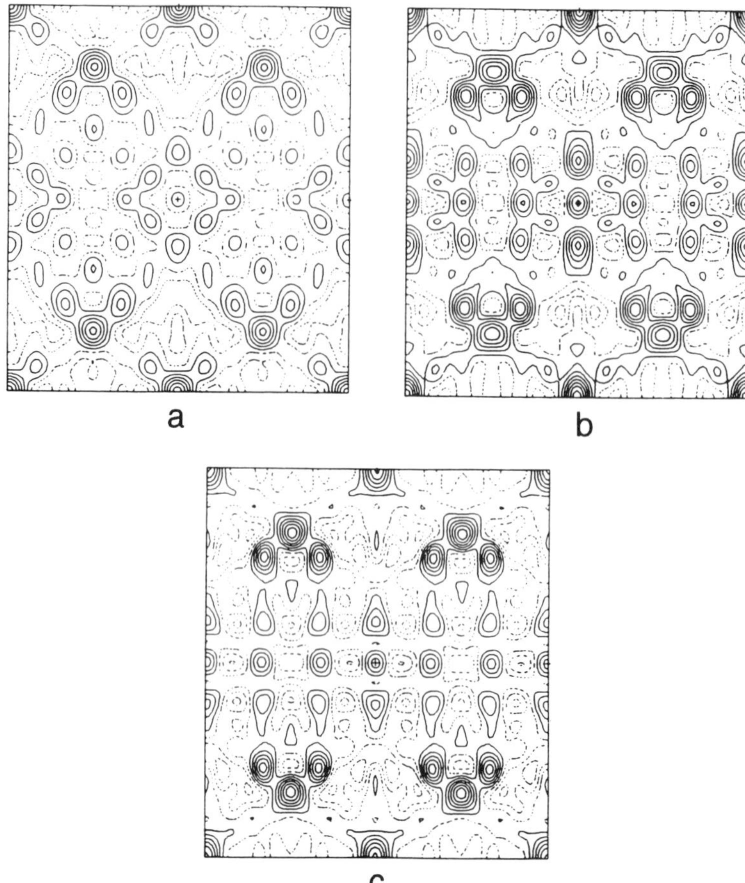

FIGURE 7.9. Potential maps for phlogopite-biotite, [100] projection: (a) calculated from initial phase set; (b) calculated from structure factors based on atomic positions in (a); (c) Zvyagin's structure. (Reprinted from D. L. Dorset (1992) "Direct methods in electron crystallography: determination of layer silicate structures," *Ultramicroscopy* **45**, 5–14; with kind permission of Elsevier Science B.V.)

could not be specified because of the restrictive index parity distribution in the data set, but this additional phase was not required. Three phases were accepted from high probability Σ_1-triples and, after evaluation of 225 Σ_2-triples and 47 quartets, 58 unique reflections were assigned phase values (Table 7.12), all of which were equal to 0°. Nevertheless, only one assignment (–2,6,1) was incorrect. From the ensuing three-dimensional potential maps (Figure 7.11), it was possible to find all atomic positions. From these, structure factors could be calculated. In this list, there were only 7 phase discrepancies from the original solution, including terms with a value of π (Table 7.13). Unusual as this solution may seem, the same result was found in a separate determi-

Table 7.11. Phase Assignments for λ-Alumina ($a = \pi$)

hkl	Direct methods	Cowley model
400	0	0
600	0	0
700	0	0
800	π	π
110	π	π
310	0	π
410	0	0
610	0	π
120	π	π
220	π	π
320	π	π
420	π	0
520	π	π
620	π	0
720	π	π
820	π	π
130	0	0
230	π	π
330	π	0
430	0	0
630	π	π
730	π	π
040	0	0
240	0	0
340	0	0
440	π	π
540	0	0
740	0	0
250	π	π
350	0	π
450	0	0
550	0	0
750	0	0
850	0	0
060	π	π
160	π	0
360	π	π
460	π	π
560	π	0
660	π	π
760	0	π
170	π	π
470	π	π
570	π	π
080	0	π
180	0	0
280	0	0
380	0	0

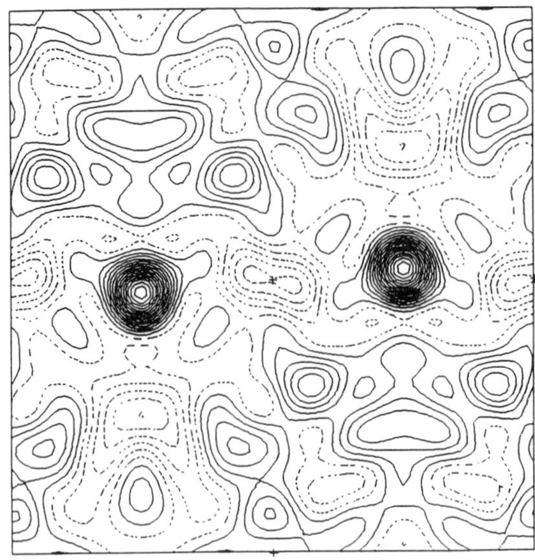

FIGURE 7.10. Potential map for λ-alumina calculated from direct phase determination when ambiguity is defined $a = \pi$.

nation with maximum entropy techniques (Gilmore et al., 1993a), and later by the use of the minimal principle (M. P. McCourt, unpublished data).

The ensuing Fourier refinement to find better atomic positions (Table 7.14) led to an improved fit of the model to the observed data. The final residual was 0.25 if neutral atoms were used and 0.24 if charged atoms were assumed. After the refinement there were only four phase discrepancies to the original determination, three of which

Table 7.12. Phase Assignments for Basic Copper Chloride, by Direct Methods

hkl	ϕ	*hkl*	ϕ	*hkl*	ϕ	*hkl*	ϕ	*hkl*	ϕ	*hkl*	ϕ
040	0	121	0	10,–2	0	063	0	144	0	12,–5	0
140	0	161	0	14,–2	0	023	0	104	0	165	0
240	0	22,–1	0	30,–2	0	16,–3	0	24,–4	0	32,–5	0
340	0	321	0	402	0	163	0	44,–4	0	025	0
		361	0	34,–2	0	263	0	34,–4	0	42,–5	0
		42,–1	0	20,–2	0	263	0	30,–4	0	125	0
		26,–1	0*	442	0	32,–3	0	204	0	006	0
		261	0	042	0	323	0	044	0	046	0
		421	0	202	0	22,–3	0	004	0	10,–6	0
		221	0	40,–2	0			404	0	027	0
		32,–1	0	302	0						
		061	0	102	0						
		021	0								

*Incorrect.

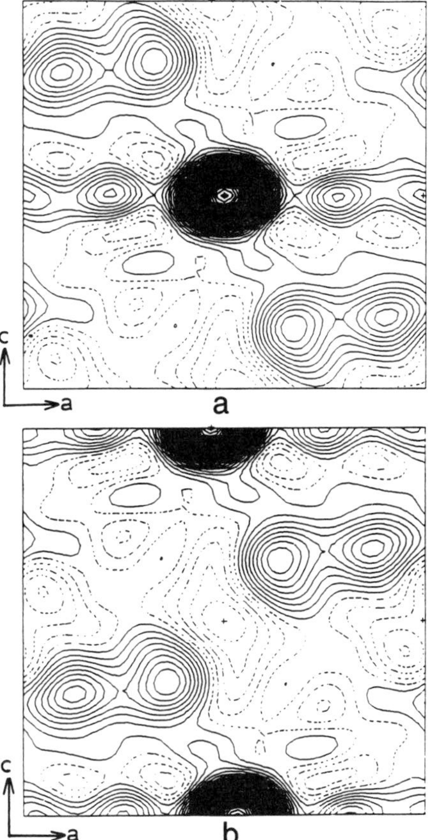

FIGURE 7.11. Potential maps for basic copper chloride after direct phase determination: (a) section at $y = 0.0$; (b) section at $y = 0.25$. (Reprinted from D. L. Dorset (1994) "Electron crystallography of inorganic compounds. Direct determination of the basic copper chloride structure $CuCl_2 \cdot 3Cu(OH)_2$," *Journal of Chemical Crystallography* **24**, 219–224; with kind permission of Plenum Publishing Corp.)

Table 7.13. Observed and Calculated Structure Factors, Basic Copper Chloride

hkl	$\|F_o\|$	F_c	*hkl*	$\|F_o\|$	F_c
001	1.96	0.94	212	3.27	–2.40
110	1.78	–1.49	131	2.92	–2.16
011	3.17	–2.72	013	2.75	1.67
101	3.40	–2.60	11,–3	1.10	0.78
11,–1	3.07	–3.30	22,–2	2.05	2.36
111	3.24	3.27	230	2.06	–2.68
020	2.74	–2.16	032	1.00	0.25
120	3.76	3.72	222	1.77	–1.41
021	6.38	4.47	30,–2	6.80	7.11
20,–1	3.93	–4.42	20,–3	2.84	1.45

(continued)

Table 7.13. (*Continued*)

hkl	$|F_o|$	F_c	*hkl*	$|F_o|$	F_c
10,–2	10.10	12.21	23,–1	2.13	–1.88
012	1.13	–0.71	023	7.80	8.37
201	2.64	4.31	32,–1	4.00	2.59
12,–1	2.46	2.18	132	1.62	–2.13
102	3.73	3.70	21,–3	1.10	–2.03
21,–1	1.45	2.99	321	6.85	7.61
121	8.51	10.61	302	2.91	2.37
211	1.35	0.79	040	8.00	8.31
112	3.26	3.48	203	1.86	–1.76
20,–2	7.35	6.95	312	1.04	–0.15
022	1.44	0.47	140	4.77	4.30
22,–1	6.35	4.97	041	1.47	0.37
12,–2	3.66	–2.93	213	1.65	–1.88
202	3.94	3.65	23,–2	1.28	0.53
221	5.63	6.29	32,–2	1.77	0.85
031	1.59	1.32	22,–3	2.02	2.91
122	2.17	–1.73	232	1.62	1.53
004	3.73	2.60	33,–3	1.25	–0.82
033	2.16	–1.10	224	1.60	1.58
33,–1	1.51	–0.65	005	0.88	–0.21
240	2.23	3.42	134	1.38	1.14
31,–3	1.34	1.16	34,–2	2.20	3.80
042	2.53	1.60	11,–5	1.40	–0.36
104	4.65	4.40	115	1.04	–1.02
223	3.09	4.27	025	2.02	1.17
24,–1	2.09	–1.88	12,–5	3.04	3.26
14,–2	5.32	5.28	044	2.50	1.58
40,–2	3.11	3.88	125	2.03	1.62
142	1.27	1.58	144	3.07	2.55
42,–1	4.50	5.35	061	2.54	1.18
33,–2	1.15	–1.66	24,–4	2.51	1.97
32,–3	2.47	2.57	161	3.12	2.78
421	3.34	3.28	433	1.98	0.48
402	4.98	5.27	404	2.79	1.87
124	1.31	0.35	442	1.80	3.08
204	3.30	3.39	26,–1	2.32	1.52
051	1.49	–0.51	16,–2	0.75	–0.70
340	2.07	1.78	32,–5	2.04	2.18
323	2.02	1.43	261	2.32	2.08
22,–4	2.06	–2.18	006	3.04	2.63
043	0.71	–0.18	10,–6	1.86	2.04
14,–3	2.14	1.33	34,–4	1.77	1.26
30,–4	2.79	1.96	063	4.07	2.69
034	0.68	–0.86	026	1.27	0.21
12,–6	0.36	0.67	16,–3	2.67	1.71
361	2.32	2.51	163	2.42	1.57
126	0.44	0.40	42,–5	1.76	1.84
44,–4	1.77	1.91	263	1.74	1.57
046	2.32	1.70	027	1.86	1.02
081	0.29	0.07	065	2.04	0.55

Table 7.14. Final Coordinates, Basic Copper Chloride

	Direct analysis			X-ray structure (bromide)		
	x	y	z	x	y	z
Cu(I)	0	0	0	0	0	0
Cu(II)	0	0.25	0.50	0	0.25	0.50
Cl(*Br*)	0.396	0.25	0.392	*0.375*	*0.25*	*0.208*
OH(I)	0.845	0.25	0.872	0.83	0.25	0.87
OH(II)	0.871	0	0.324	0.87	0	0.32

corresponded to weak reflections. The resulting structure agrees well with the x-ray determination for the isostructural basic copper bromide (Aebi, 1950), especially if differences in ionic radii are taken into account (see Table 7.14).

7.2.7. High T_c Superconductor

The Pb-doped Bi-2223 ($Bi_2Sr_2Ca_2O_x$) phase is found to have an incommensurate modulated structure with lattice constants a = 5.49, b = 5.41, c = 37.1 Å in the four-dimensional space group $P^{Bbmb}{}_{111}$, with a modulation vector $q = 0.117\ b^*$ (Mo et al., 1992). A high-resolution electron micrograph was taken of this structure, projecting down the **a**-axis, and its Fourier transform was found to provide phase information to $(1.7\ \text{Å})^{-1}$ (Li, 1993). If phases from the average structure were used to assign values to the most intense reflections of the electron diffraction pattern, the phases of the weaker superlattice reflections, which were due to the incommensurately modulated superstructure, could be assigned via the Sayre equation, expressed in four dimensions (Fan, 1993). Such a four-dimensional convention is a convenient way to treat such materials as periodic objects (Amelinčkx and Van Dyck, 1993b) and thus the phase assignment utilizes the same hyperspatial model.

7.2.8. Potassium Niobium Oxide

Crystals of a potassium- niobium oxide, $K_2O \cdot 7Nb_2O_5$, were used for high-resolution electron microscopy and electron diffraction data collection (Hu et al., 1992). The material crystallizes in a tetragonal unit cell with dimensions $a = b = 27.5$, c = 3.94 Å. The phase contrast transfer function was deconvoluted from the image using the maximum entropy technique discussed in Chapter 3, to produce a Scherzer focus representation of the structure to 1.9-Å resolution, from which crystallographic phase information was derived. These phases could then be extended to the 1-Å resolution of the electron diffraction pattern via the Sayre equation, resulting in an atomic resolution model of the crystal structure. The final coordinates corresponded to a crystallographic R-factor of 0.28.

7.2.9. Aluminum–Germanium Alloys

In a highly innovative approach to solving inorganic structures from dynamical electron diffraction data, Vincent and Exelby (1991, 1993) have investigated the use of higher-order Laue zone (HOLZ) reflections from alloy microcrystals in various polymorphic phases. It was reasoned by these researchers that reflections in this region should be amenable to ab initio structure analysis if most of the data originated from crystals thinner than the HOLZ extinction distances for these reflections. The data collection from nanometer-size areas was undertaken with great care to ensure that reflections (often within the first-order Laue zone (FOLZ)) were correctly indexed. The resulting data were found to correspond, e.g., to $hk1$ or $hk2$ layers (depending also on the zone axis used for the crystal projection). Although direct methods were attempted in one case (Vincent and Exelby, 1991), Patterson maps were employed most often. Due to the restricted data sets used, vector interactions within individual crystal layers could be isolated. Useful structure results were obtained for the Al-Ge alloys in rhombohedral and monoclinic phases (Vincent and Exelby, 1991) and, later, in a tetragonal phase (Vincent and Exelby, 1993). Despite the somewhat large computed R-factors (e.g., 0.42), the packing geometry appeared to be reasonable. Although dynamical scattering was somewhat problematic, spurious peaks in the potential maps were found most often to be due to the undersampling of diffraction data. More recent work (Vincent and Midgley, 1994) has added a double conical beam rocking system for recording FOLZ electron diffraction intensities so that the effects of the shape transform of a flat crystal (via the excitation error) are eliminated and the nonsystematic dynamical perturbations are, in addition, minimized. It was argued that this improved method for data collection would enable sampling of very small, flat crystalline areas to approach the conditions already experienced when much larger areas are used for data collection (e.g., in texture diffraction patterns). Preliminary structural analyses of $Er_2Ge_2O_7$ by these methods have been very promising.

7.3. Structures from High-Resolution Electron Micrographs

7.3.1. Potassium Niobium Tungsten Oxide

A high resolution electron micrograph (Figure 7.12) was obtained at 200 kV from a potassium niobium tungsten oxide with composition $K_{8-x}Nb_{16-x}W_{12+x}O_{80}$, with $x \approx 1$ (Hovmöller et al., 1984). The unit cell constants in projection are $a = 22.1$, $b = 17.8$ Å, and the projected plane group symmetry is pgg. After digitization of the micrograph, the limits of the phase contrast transfer function envelope were defined (with the contrast reversal at 2.4-Å resolution). The Fourier transform of the image sampled at reflection centers was used to derive crystallographic phases. It was important to impose the symmetry operations of the plane group before calculation of the final map (Figure 7.13). In this map, the positions of all the heavy atoms were clearly visible and were found to compare favorably to the x-ray structure of an isostructural sodium compound (Table 7.15). Oxygen atoms were added at stereochemically reasonable sites to give the final set of coordinates.

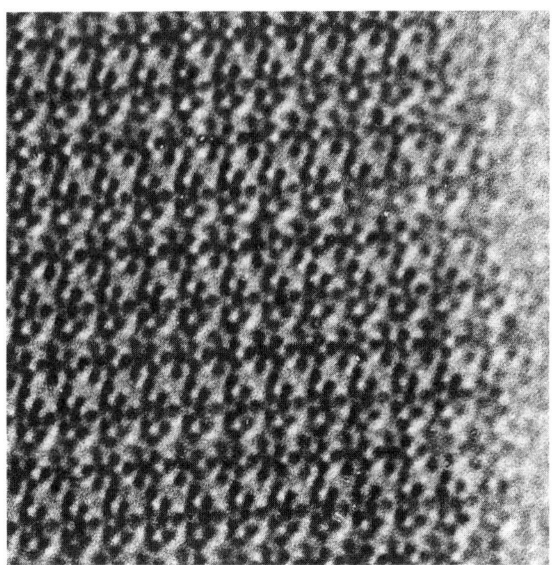

FIGURE 7.12. High-resolution electron micrograph from a potassium niobium tungsten oxide. (Thanks to Dr. S. Hovmöller for permission to use an unpublished image from his work.)

FIGURE 7.13. (left) Final potential map for potassium niobium tungsten oxide after image analysis and structure interpretation. (right) Structural model for potassium niobium tungsten oxide. (Reprinted from S. Hovmöller, A. Sjögren, G. Farrants, M. Sundberg, and B. O. Marinder (1984) "Accurate atomic positions from electron microscopy," *Nature* **311**, 238–241; with kind permission of Macmillan Journals Ltd. and Dr. S. Hovmöller.)

Table 7.15. Metal Atom Coordinates for Potassium Niobium Tungsten Oxide

	Electron microscopy		X-ray crystallography	
	x	y	*x*	*y*
Me(1)	0.1069	0.4077	0.1060	0.4137
Me(2)	0.1578	0.2014	0.1616	0.2059
Me(3)	0	0.25	0	0.25
Me(4)	0.25	0.3588	0.25	0.3508
Me(5)	0.25	0.0364	0.25	0.0338

*See Hovmöller et al. (1984).

7.3.2. Staurolite

Crystals of the mineral staurolite, $HFe_2Al_9Si_4O_{24}$, were sectioned to produce projections along [100], [010], and [001] and were thinned to about 30-Å thicknesses, electron transparent sheets by an ion-beam method (Downing et al., 1990; Wenk et al., 1992). With the goniometer stage of the high-voltage electron microscope operated at 800 kV, it was also possible to project down [101] and [310]. The space group is described as being *C2/m* but not deviating significantly from *Ccmm*, with unit cell constants $a = 7.87$, $b = 16.61$, $c = 5.66$ Å.

Although the experimental envelope of the transfer function was found to have its first minimum near 1.4 Å, only data to 1.6-Å resolution were accepted. Using the Fourier transform of the image projections to obtain crystallographic phases, a three-dimensional image of the crystal structure was reconstructed, resolving all atoms including the oxygens (Figure 7.14). The experimental image was well matched by multislice dynamical calculations based on the known crystal structure Hence, in general, there is a good agreement with the x-ray crystal structure (Smith, 1968).

Although no electron diffraction intensities were used for the image reconstruction, an attempt was made to solve the crystal structure by direct phasing as if the amplitudes from the Fourier transform of the image (Wenk et al., 1992) were electron diffraction data. (However, as shown by the standard deviations published in the original work, there were large errors in these image amplitudes.) After origin definition and acceptance of three algebraic phase ambiguities (requiring the calculation of eight maps), phase values were assigned for 37 of the 59 *hkl* data (Table 7.16). The best map (Figure 7.14) closely resembles the one calculated from image phases in the section at $z = 0$ but the match is less satisfactory for the section at $z = 0.25$. It may be possible to use this map to begin a Fourier refinement of the structure but better diffraction amplitudes are necessary for this analysis to continue.

7.3.3. Sodium Niobium Fluoroxide

Crystals of the material $Na_3Nb_{12}O_{31}F$ were used (Li and Hovmöller, 1988) to obtain high-resolution electron micrographs (Figure 7.15) at 200 kV corresponding to 2.5-Å resolution. From an x-ray structure, the cell constants in space group *P*4 were

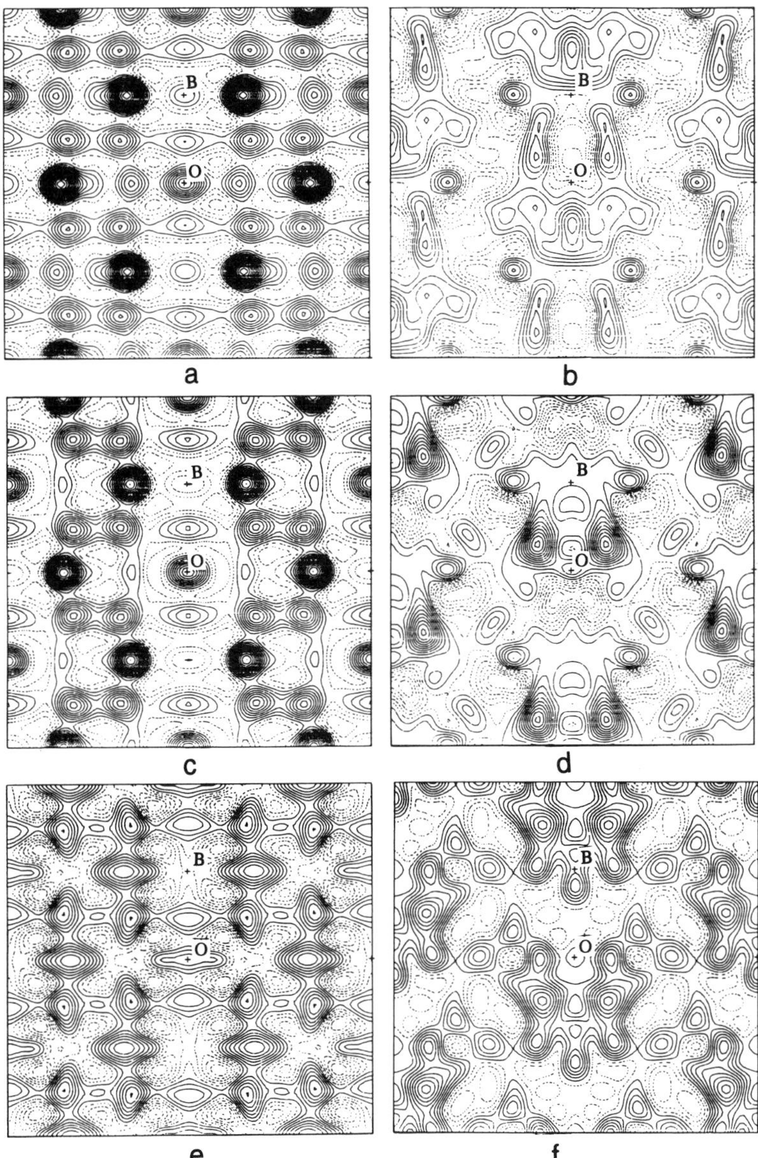

FIGURE 7.14. Potential maps for staurolite: (a) after symbolic addition, section at $z = 0$; (b) after symbolic addition, section at $z = 0.25$; (c) after Fourier refinement, section at $z = 0$; (d) after Fourier refinement, section at $z = 0.25$; (e) image phases of Wenk et al. (1992), section at $z = 0$; (f) image phases of Wenk et al. (1992), section at $z = 0.25$.

Table 7.16. Direct Phase Determination of Staurolite Based on Published Amplitudes*

0*kl*	ϕ	1*kl*	ϕ	2*kl*	ϕ	3*kl*	ϕ	4*kl*	ϕ	5*kl*	ϕ
004	0	111	b	202	a	330	0	400	a	511	b†
040	0	130	a	240	a	310	0	440	a	510	c
021	π					312	0	420	a†	530	c
023	π					350	0	402	a†		
042	0					313	π†				
002	0†										
022	0†										

6*kl*	ϕ	7*kl*	ϕ	8*kl*	ϕ	9*kl*	ϕ	10*kl*	ϕ	12*kl*	ϕ
600	0	711	b	822	0+	910	0+	10,02	a	12,00	0
640	0†	730	a	820	0	930	0	10,20	a†		
602	0										
620	0†										
623	π										

*Structure determination made in space group Cmcm, thus reversing *a* and *b* axes of the original study and hence *h and k* for the reflections. $a = c = \pi$, $b = 0$.
†Erroneous.

found to be $a = b = 17.49$, $c = 3.94$ Å. After digitization of the image, the reverse Fourier transform was used to obtain crystallographic phases, imposing the *p*4 projection symmetry. The average map (Figure 7.16) accurately provided the positions of the heavy atoms in the crystal which could be compared favorably to the x-ray crystallographic results (Table 7.17).

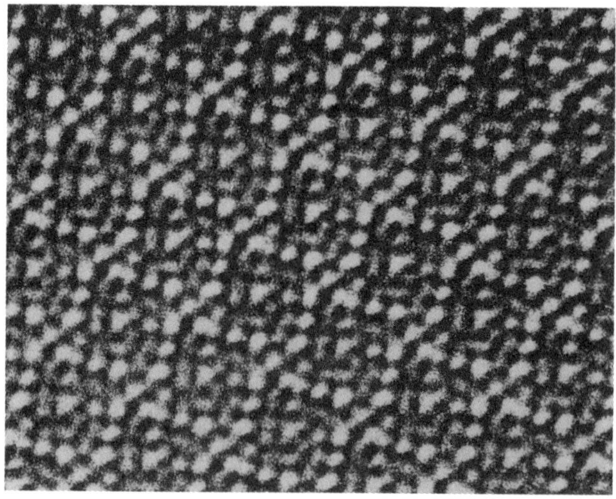

FIGURE 7.15. High-resolution electron micrograph of a sodium niobium fluoroxide. (Thanks to Dr. S. Hovmöller for permission to use an unpublished image from his work.)

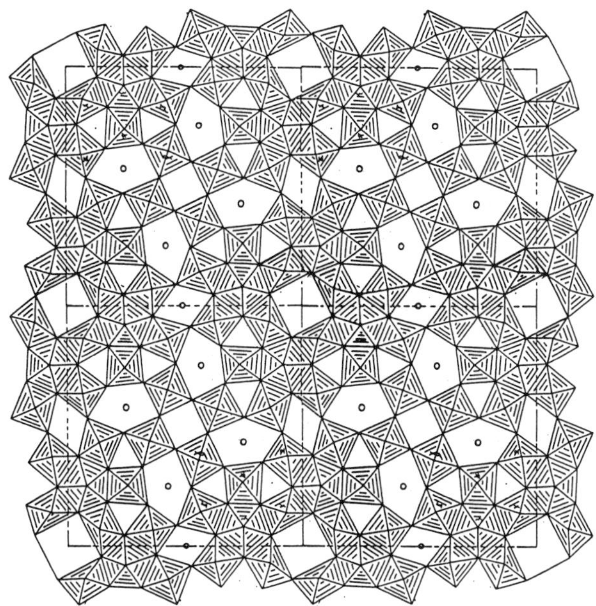

FIGURE 7.16. Final model for sodium niobium fluoroxide after image analysis and structure interpretation. (Reprinted from D. X. Li and S. Hovmöller (1988) "The crystal structure of $Na_3Nb_{12}O_{31}F$ determined by HREM and image processing," *Journal of Solid State Chemistry* **73**, 5–10; with kind permission of Academic Press, Inc. and Dr. S. Hovmöller.)

7.3.4. Zeolites

Recently, an extensive study of two zeolite structures was made, based on high-resolution images and electron diffraction intensities collected on an intermediate voltage electron microscope (Pan and Crozier, 1993). After real space averaging of the 2-Å resolution images from suitably thin specimen areas, it was found that the average unit cell potential, as seen on the maps, could be directly interpreted in terms of the known crystal structures, i.e., the weak phase object approximation was valid for data collected at 400 kV. While the observed diffraction intensities could be matched well by a dynamical scattering model, it was clear that data from a thinner specimen area

Table 7.17. Niobium Coordinates for Sodium Niobium Fluoroxide*

	Electron microscopy		X-ray crystallography	
	x	y	*x*	*y*
Nb(1)	0.0654	0.1357	0.0704	0.1393
Nb(2)	0.0731	0.3740	0.0676	0.3689
Nb(3)	0.2478	0.0674	0.2506	0.0785
Nb(4)	0.2509	0.2601	0.2496	0.2663
Nb(5)	0.4299	0.1421	0.4321	0.1450
Nb(6)	0.4478	0.3559	0.4382	0.3588

*See Li and Hovmöller (1988).

also would correspond adequately to the kinematical approximation to permit a structure analysis to be carried out.

The possibilities for ab initio structure analysis of zeolites from measured electron diffraction intensities have been brilliantly demonstrated recently in the direct determination of the MCM-22 structure by Nicopoulos and his coworkers (S. Nicopoulos, private communication). Using three-phase invariants formed from the normalized electron diffraction intensities a model was obtained that was found subsequently to agree well with the independent Rietveld determination by Leonowicz et al. (1994) with powder x-ray data. Nevertheless, the x-ray model was not needed for the electron diffraction analysis, since the laboratories were not in communication before the independent studies were made.

7.4. Conclusions

The success of ab initio structure analyses of inorganic crystals, based on electron diffraction intensities and, especially, on electron microscope images, is a surprising result, given the extreme pessimism expressed by many theoreticians (e.g., Heidenreich, 1964; Spence, 1980; Humphreys and Bithell, 1992; Eades, 1994). As shown in this chapter, direct phase determination leads to reasonable results for a variety of materials, even when intensity data were collected at rather low voltages (however, in this case, sampling rather large areas, including a distribution of crystal orientations, to minimize strong nonsystematic diffraction contributions). It may not be possible, however, to refine these structures without some recourse to dynamical scattering calculations, but, at the very least, a useful structural model is obtained directly from the phase determination. Therefore, one is not forced to use trial-and-error procedures (or even results from previous x-ray crystallographic analyses) to find a starting model for the multislice calculation.

Most surprising is the success of high-resolution microscopy for structure determination. It is certainly true that dynamical scattering distorts the density distribution of high-resolution micrographs so that there will be some point where the intuitive interpretation of that image will not be possible (i.e., when the weak phase object approximation is no longer valid). On the other hand, when the measured intensity distribution of an experimental image is Fourier-transformed to find estimated crystallographic phases (imposing further constraints due to choice of unit cell origin and the space-group symmetry operators), it is not yet clear what amount of dynamical scattering can take place before these derived phase terms are completely meaningless. It was demonstrated effectively that, in some cases, accurate metal atom coordinates can be found for oxide structures that agree well with x-ray results, even though dynamical images are being analyzed. More work is needed in this area to estimate the practical experimental limitations to the use of data from such materials for ab initio structure determinations. Obviously, careful control of parameters such as electron wavelength and crystal thickness must be imposed to ensure that the image-derived phases are useful for finding a correct crystal structure.

8

The Alkanes

8.1. Background

Crystalline *n*-alkanes were among the first materials studied by quantitative electron crystallographic techniques. Rigamonti (1936) had obtained *hk*0 electron diffraction intensity data from thin films of a paraffin fraction melting in the range 49–50 °C, as well as pure *n*-$C_{31}H_{64}$ obtained from the synthetic reduction of palmitone. Structure factors were computed from electron form factors derived from the Mott formula via x-ray scattering factor tables. No Lorentz correction was made to the observed intensity data. A chain model was constructed from the earlier x-ray model of Müller (1930) and the rotational parameter $90° - \varphi$ of the projected chain plane (Figure 8.1), where φ is the setting angle, was varied from 20 to 50° with most likely values found between 35 and 40°. The C-H distance was estimated to lie in the range 1.1 to 1.4 Å.

Much of the electron diffraction information from *n*-paraffins in this early period actually came from (Bragg) reflection diffraction (i.e., RHEED) studies of chain layers on flat support surfaces. For example, in single-crystal RHEED experiments, Thiessen and Schoon (1937) and, later, Schoon (1938) described the polymorphism of *n*-$C_{30}H_{62}$: one form packs with inclined chains while the other forms a rectangular layer. The single, rectangular-layer polymorph of *n*-$C_{31}H_{64}$ was also described, all of this information still consistent with modern knowledge of preferred paraffin chain packing motifs (Nyburg and Potworowski, 1973). (Other RHEED work on alkane derivatives will be discussed in Chapter 9.) More recently, Sutula and Bartell (1962) have published a compendium of reflection diffraction results from *n*-paraffins in terms of crystal polymorph.

The next attempt to carry out a quantitative electron diffraction structure analysis of a paraffin layer packing was made by Vainshtein and Pinsker (1950), who examined a fraction melting at 53.5 °C. Spot and texture diffraction patterns were used to determine this structure, comprised of 44 *hk*0 and 47 *hkl* intensities. The initial crystallographic phases were derived from a known chain model. The resultant structure for the rectangular layer had the following bonding parameters: $d_{C\text{-}C} = 1.52$ Å, $d_{C\text{-}H} = 1.17$ Å, C-C-C angle: 110°, H-C-H angle: 105°, results that were imagined to be more accurate than any possible from contemporary x-ray determinations.

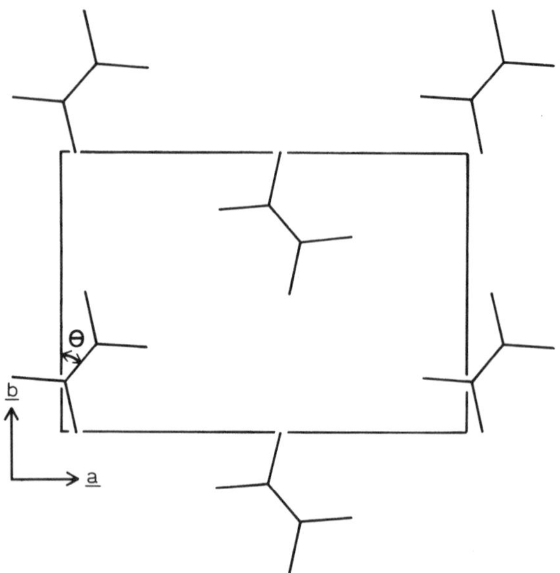

FIGURE 8.1. Setting angle of paraffin chains in an orthorhombic layer, viewing down the chain axes. (Reprinted from B. Moss, D. L. Dorset, J. C. Wittmann, and B. Lotz (1984) "Electron crystallography of epitaxially grown paraffin," *Journal of Polymer Science, Polymer Physics Edition* **22**, 1919–1929; with kind permission of John Wiley & Sons., Inc.)

Vainshtein, Lobachev, and Stasova (1958) continued this work with other paraffin layers, including the mixture of chain components examined earlier by Vainshtein and Pinsker, as well as *hk*0 powder diffraction intensities from n-$C_{18}H_{38}$ and n-$C_{30}H_{62}$, both in an orthorhombic form. From these analyses, the C-H bond distance was found to be about 1.12 Å. Secondary scattering was found to be an important perturbation to these diffraction patterns by Cowley, Rees, and Spink (1951). It was claimed that a correction for incoherent multiple scattering would improve the detectability of the hydrogen positions in the ensuing potential maps.

8.2. Contemporary Structure Analyses

8.2.1. Even-Chain Paraffins

In Chapter 5, it was mentioned how intensity data from paraffin mono- and multilayers crystallized from solution have been used as standards to detect various diffraction perturbations. For example, multiple-beam dynamical scattering was found for all crystal thicknesses measured and at all electron accelerating voltages. Spurious increments of data resolution can result as a consequence of secondary scattering, in addition to the appearance of space-group forbidden reflections. One of the most significant observations, however, was the dramatic effect of elastic bending on the diffraction intensity from multilayer crystals (Dorset, 1980). Because of the projection down a very large unit cell axis, only the features of the subcell scattering (see Chapter 10) are noted and not the true bilayer unit cell scattering, which must account for the translational offset of two monolayers. Given the number of times *n*-paraffin structures have been studied by electron diffraction techniques, it is surprising that this important detail remained unnoticed for so long.

Monolayer structures for the even-chain paraffins have been determined quantitatively for examples from *n*-$C_{24}H_{50}$ to *n*-$C_{94}H_{190}$ (Dorset, 1980; Dorset et al., 1992). The single crystal diffraction pattern is always the same, as illustrated in Figure 8.2. A paradigm for the series can be found in the orthorhombic form of *n*-$C_{36}H_{74}$ (Dorset, 1976b). Beginning with 42 measured *hk*0 intensity data, a starting atomic model based on the known layer packing was refined by Fourier techniques to find hydrogen positions. The resulting values (corresponding to $R = 0.24$) derived from experimental maps (Figure 8.3) were, on average, $d_{C\text{-}H} = 1.04$ Å, and the H-C-H angle: 98°. In other words, the bond length is similar to the result found from x-ray crystal structures but

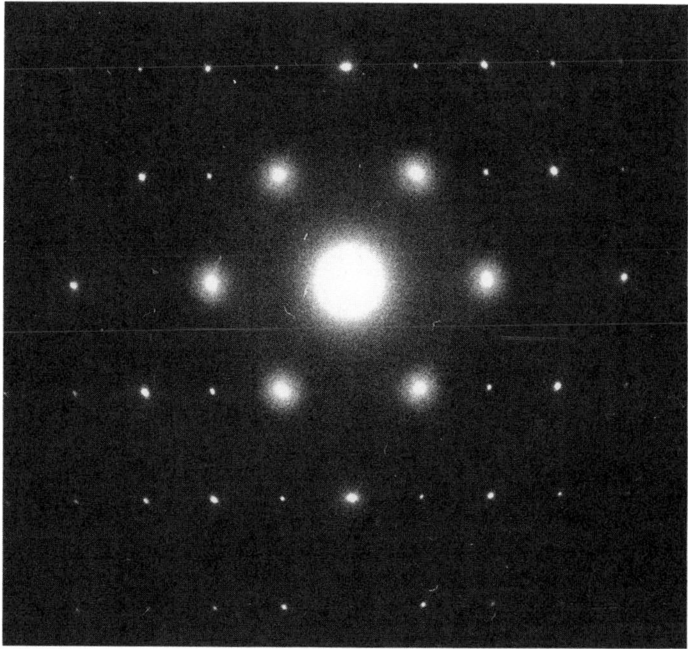

FIGURE 8.2. Selected area electron diffraction pattern (*hk*0) from an orthorhombic paraffin layer.

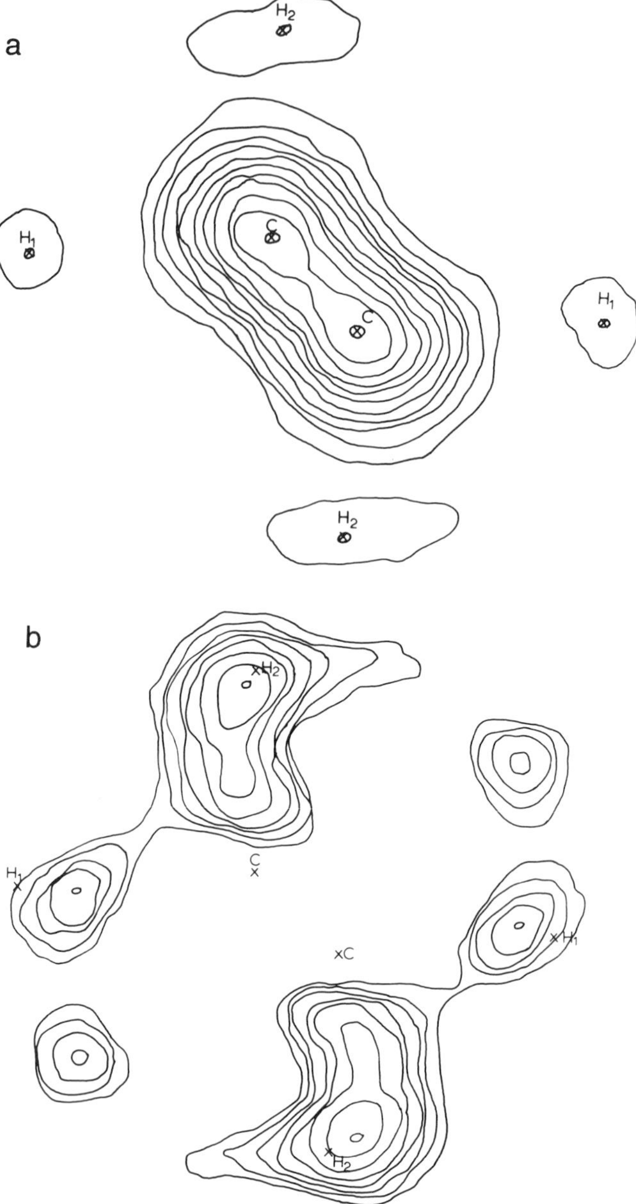

FIGURE 8.3. Potential maps for a paraffin layer: (a) based on phased F_h(obs); (b) difference map. (Reprinted from D. L. Dorset (1976) "The interpretation of quasi-kinematical single-crystal electron diffraction intensity data from paraffins," *Acta Crystallographica* **A32**, 207–215; with kind permission of the International Union of Crystallography.)

the bond angle is about 10° too small. The experimental setting angle, 42.1°, agrees well with the value found in the x-ray crystal structure of this polymorph (Teare, 1959). More recently, structural determinations have also been carried out with intensity data from perdeuterated analogs of the *n*-paraffins, e.g., *n*-$C_{30}D_{62}$ and *n*-$C_{36}D_{74}$, demonstrating that the layer packings are exactly the same as found for the perprotonated parents (Dorset, 1991e). As a benchmark experiment on the use of high-resolution electron microscopy to image lattices of aliphatic materials, correlation-averaged images of *n*-$C_{44}H_{90}$ obtained from a cryoelectron microscope (Figure 8.4) revealed details to 2.5-Å resolution, consistent with the measured objective lens phase contrast transfer function (Zemlin et al., 1985). This resolution has been recently improved to 2.1 Å (Brink and Chiu, 1991). It was originally thought that edge dislocations could be observed directly in these images, a presumption reinforced by model multislice image calculations incorporating the defect in a superlattice (Dorset et al., 1986). However, as was pointed out later as a caveat by Pradere et al. (1988b), such interpretations of low-contrast images can easily go astray. For example, phase noise included in the lattice averaging by Fourier peak filtration will cause the appearance of spurious "defects". To date, the only reliable way of detecting edge dislocations in such crystals by the use of moiré fringe patterns in diffraction contrast images taken at low magnifications.

It is significant to note here that the observed *hk*0 electron diffraction intensities from *n*-$C_{36}H_{74}$ were the first ever used to solve a crystal structure quantitatively by direct methods (Dorset and Hauptman, 1976). For this determination, only Σ_2-triples and quartets were utilized and the phase estimate for the former was made through the so-called MDKS formula, i.e.:

$$\cos(\phi_{h(1)} + \phi_{h(2)} + \phi_{h(3)}) = M(D - KS)$$

where

$$D = \sum_i m_i D_i / \sum_i m_i$$

and

$$D_i = <(|E_{-h(i)+k}|^2 - 1)\ |E_k| \geq t,\ |E_{h(j)+k}| \geq t>_k$$

$$S_i = <2(|E_{-h(i)+k}|^2 - 1)\ |E_k| \geq t >_k$$

The indices $i = 1, 2, 3$ when $j = 2, 3, 1$, respectively. The terms M and K are scaling parameters. The origin was defined for two reflections in the list of $|E_h|$ by setting $\phi_{310} = \phi_{720} = 0$. The sequence of the phase determination, which assigned values to 24 reflections, is reviewed in Table 8.1. The positions of the carbon atoms were readily apparent in the first potential map (Figure 8.5).

Another part of the electron diffraction pattern that should not be ignored is the non-Bragg continuous signal that can be observed to join the discrete diffraction maxima (Dorset, 1977b). Such continuous streaks were noted, for example, in the early

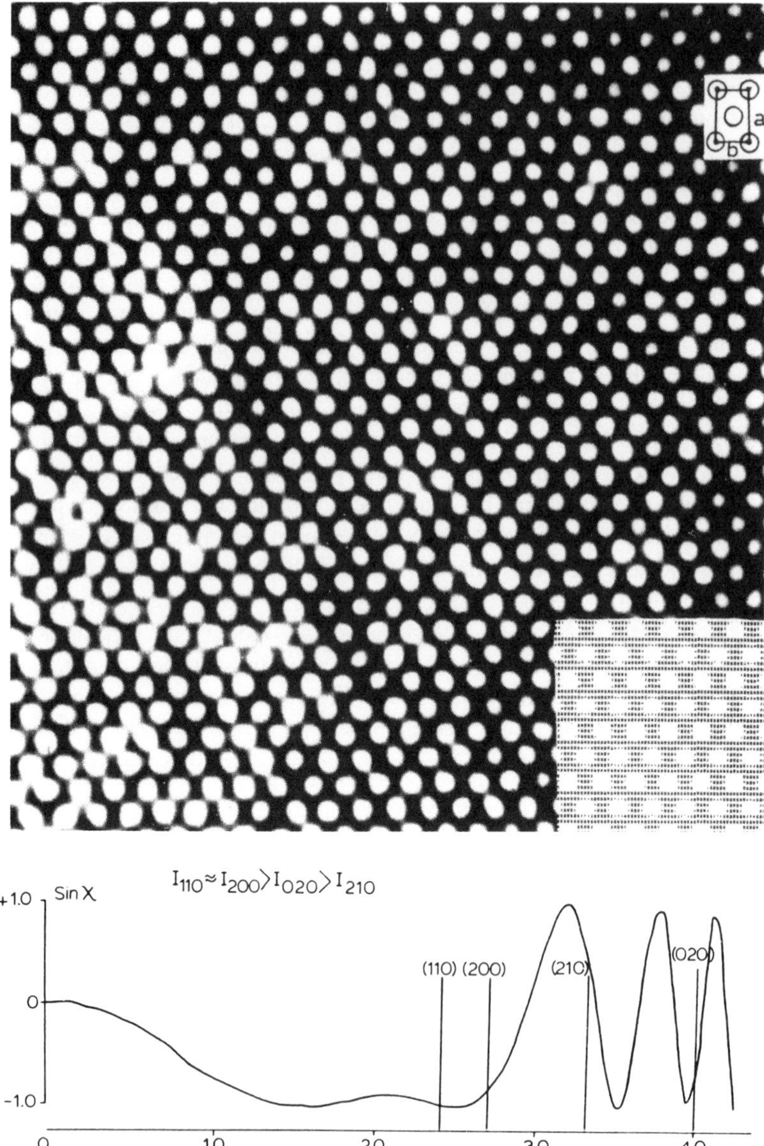

FIGURE 8.4. High-resolution image of n-$C_{44}H_{90}$ reconstructed after correlation alignment of crystal subareas (see Chapter 4). The experimental transfer function of the cryoelectron microscope is indicated below and the results of an image calculation based on the known crystal structure is shown as an inset. (Reprinted from F. Zemlin, E. Reuber, E. Beckmann, E. Zeitler, and D. L. Dorset (1985) "Molecular resolution electron micrographs of monolamellar paraffin crystals," *Science* **229**, 461–462; with kind permission of the American Association for the Advancement of Science.)

Table 8.1. Direct Phase Determination for n-$C_{36}H_{74}$, $hk0$ Data

origin definition:	
$\phi_{310} = \phi_{720} = 0$	
quartet relationships:	
$\phi_{540} = 0$	$\phi_{110} = 0$
$\phi_{520} = 0$	$\phi_{320} = 0$
$\phi_{740} = 0$	$\phi_{340} = 0$
$\phi_{910} = \pi$	$\phi_{710} = \pi$
also: if $\phi_{830} = a$, then $\phi_{430} = a$; $\phi_{820} = a + \pi$; $\phi_{630} = a$	
Σ_2-triples:	
$\phi_{630} = 0$, thus $a = 0$	
also:	
$\phi_{230} = 0$	$\phi_{800} = \pi$
$\phi_{220} = 0$	$\phi_{150} = \pi$
$\phi_{610} = 0$	$\phi_{450} = 0$
$\phi_{020} = 0$	$\phi_{810} = 0$
$\phi_{200} = 0$	$\phi_{410} = 0$

electron diffraction studies of anthracene (Charlesby et al., 1939) and are often a prominent part of the pattern from n-paraffin layers (Figure 8.6). Two of the most common causes of such diffuse scattering are thermal motion and packing disorder, and these have been discussed extensively in two treatises (Guinier, 1963; Amoros and Amoros, 1968), as well as a recent review (Welberry and Butler, 1994). As pointed out by Guinier (1963), when considering the diffraction from a crystal, scattering always must be conserved. Therefore, any signal that does not contribute to the Bragg reflections (due to the ordered part of a structure) must be found in a continuous signal corresponding to the disorder in the crystalline lattice. In the case of thermal motion, an expression for the diffuse intensity is written:

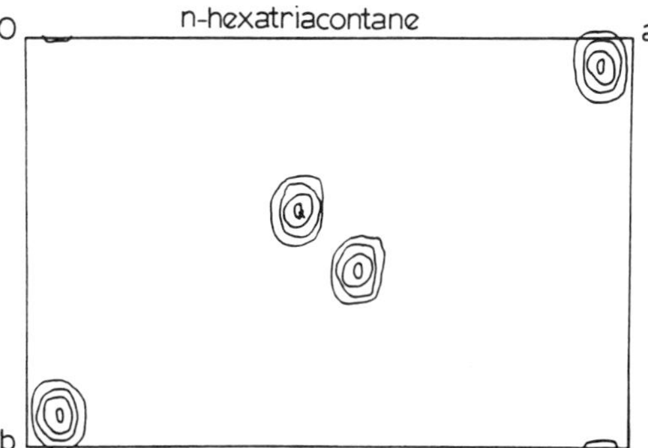

FIGURE 8.5. Results of direct phase determination with ($hk0$) electron diffraction intensities from an n-$C_{36}H_{74}$ layer. (Reprinted from D. L. Dorset and H. A. Hauptman (1976) "Direct phase determination for quasi-kinematical electron diffraction intensity data from organic microcrystals," *Ultramicroscopy* **1**, 195–201; with kind permission of Elsevier Science B.V.)

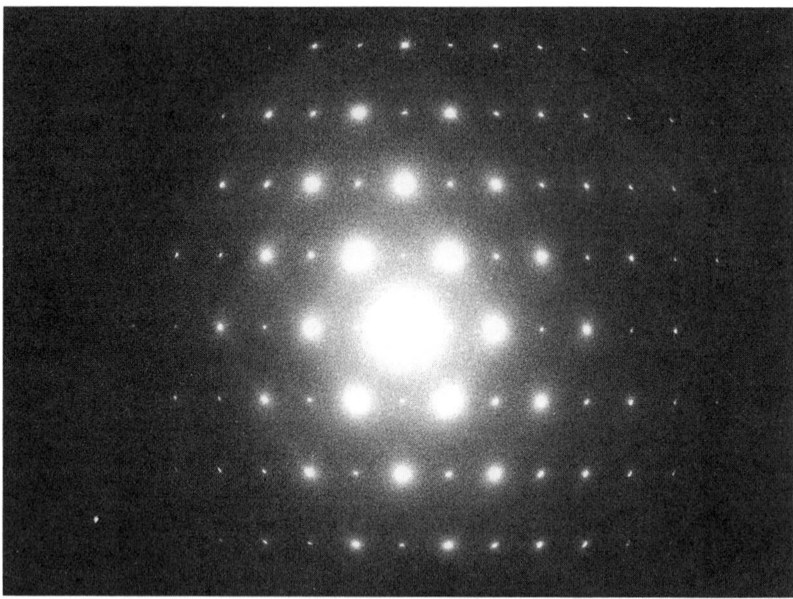

FIGURE 8.6. Continuous diffuse scattering observed in an $(hk0)$ electron diffraction pattern from an n-paraffin multilayer.

$$I_{\mathrm{DFT}} = |F_h|^2\{1 - \exp(-(1/2)B|\mathbf{s}|^2\}$$

where B is the isotropic thermal parameter and $|\mathbf{s}|$ is the reciprocal lattice vector length. For molecular crystals, it has often been stated that the structure factor terms $|F_h|$ should be calculated for isolated single molecules or groups of atoms in the unit cell, i.e., assuming that the motions of these groups are uncorrelated with those of the others. Thus their separate contributions would be added incoherently. For example, in x-ray studies, Hoppe and his coworkers (Hoppe, 1956; Hoppe and Baumgärtner, 1957; Hoppe et al., 1957; Hoppe and Ranck, 1960) used this assumption to find crystallographic phase information not present in the Patterson function.

In the analysis of the continuous diffuse electron scattering signal from solution-crystallized n-paraffins, it was found (Dorset et al., 1991) that the intensity was quite sensitive to temperature, establishing that the probable origin was due to the thermal vibration of the atoms in the crystal. However, if the proposed incoherent contribution by individual molecules in the unit cell was assumed, the model, while correctly finding the region of reciprocal space where the continuous scattering occurred, did not account for the observed narrowness of the diffuse streak (Dorset, 1977b). Only when correlations involving the total unit cell contents were considered did the calculated diffuse scattering adequately match the experimental signal (Figure 8.7).

Analysis of electron diffraction patterns alone does not provide a complete picture of n-paraffin layer structures, however. As discussed in Chapter 5, chain monolayers

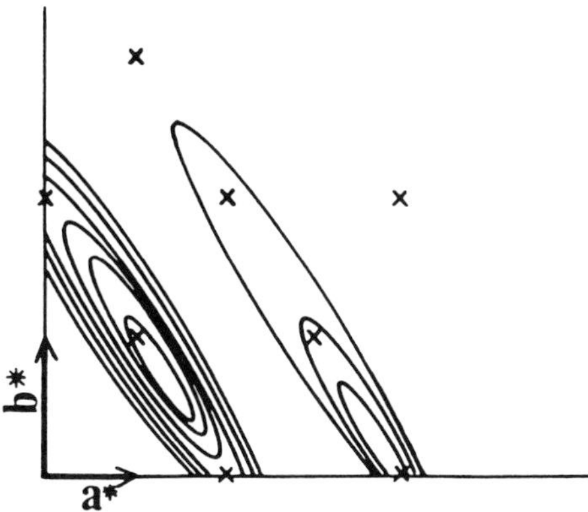

FIGURE 8.7. Calculated diffuse scattering from a paraffin (one *hk*0 quadrant represented), assuming correlations between the two chains of the unit cell layer. (Reprinted from D. L. Dorset, H. Hu, and J. Jäger (1991) "Continuous diffuse scattering from polymethylene chains—an electron diffraction study of crystalline disorder," *Acta Crystallographica* **A47**, 543–549; with kind permission of the International Union of Crystallography.)

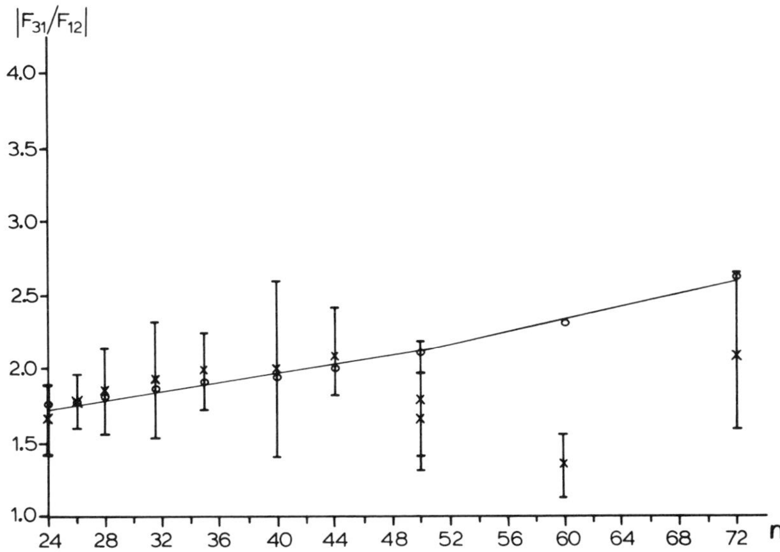

FIGURE 8.8. Plot of mean dynamical structure factor magnitudes for orthorhombic paraffin monolayers as a function of layer thickness (expressed as the number of chain carbons). Compare with Figure 5.5. Beyond a limiting chain length, there is a large discrepancy between theory and experiment. (Reprinted from D. L. Dorset (1986) "Sectorization of n-paraffin crystals," *Journal of Macromolecular Science—Physics* **B25**, 1–20; with kind permission of Marcel Dekker, Inc.)

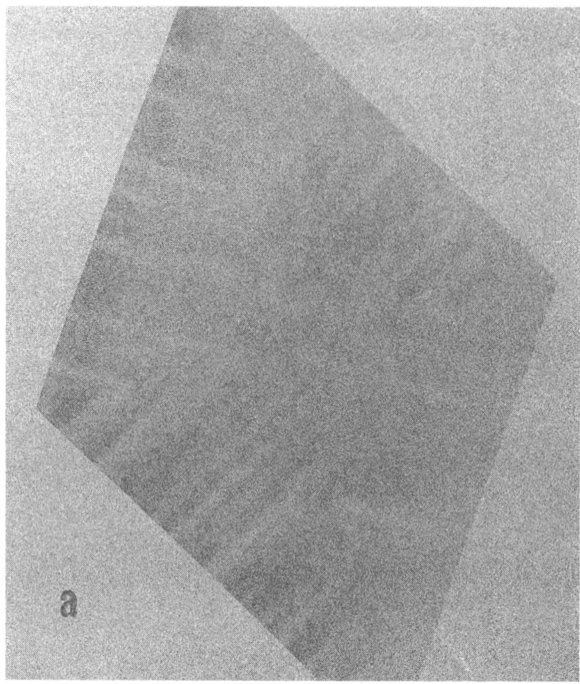

FIGURE 8.9. (a) Bright field electron micrograph of a monolayer n-$C_{82}H_{166}$ crystal showing sectorization behavior. (Reprinted from D. L. Dorset (1986) "Sectorization of n-paraffin crystals," *Journal of Macromolecular Science—Physics* **B25**, 1–20; with kind permission of Marcel Dekker, Inc.) (b) High-resolution electron microscopy demonstrates that the contrast change is due to alternation of untilted and tilted chain layers, resulting after collapse of a three-dimensional habit. (Reprinted from D. L. Dorset (1994) "Electron crystallography of linear polymers," in *Characterization of Solid Polymers*, S. J. Spells, ed., Chapman and Hall, London, pp. 1–17; with kind permission of Chapman and Hall, Ltd.) (c) When the chain length is great enough to permit chain-folding (e.g., n-$C_{168}H_{338}$), then the sectorization behavior seen in bright-field micrographs resembles that of polyethylene. (Reprinted from D. L. Dorset, R. G. Alamo, and L. Mandelkern (1993) "Surface order and the sectorization of polymethylene lamellae," *Macromolecules* **26**, 3143–3146; with kind permission of the American Chemical Society.)

were used as crystals with exactly defined thicknesses to follow the progress of n-beam dynamical scattering effects. However, beyond a certain chain length, e.g., for n-$C_{50}H_{102}$, the relative experimental structure factor magnitude was found to deviate significantly from the value predicted by dynamical scattering theory (Dorset, 1986b) (Figure 8.8). Only after observation of bright-field electron micrographs was it revealed that the crystal habit was now sectorized in a way very similar to that found for polyethylene (Revol and Manley, 1986). Sharp contours in diffraction contrast images were found to lie along the <130> directions in {110} sectors instead of the <1 $\bar{3}$ 0> direction found for the polymer. Although such paraffin crystals do not contain chain folds, this morphology was found to persist up to n-$C_{94}H_{190}$. Beyond this chain

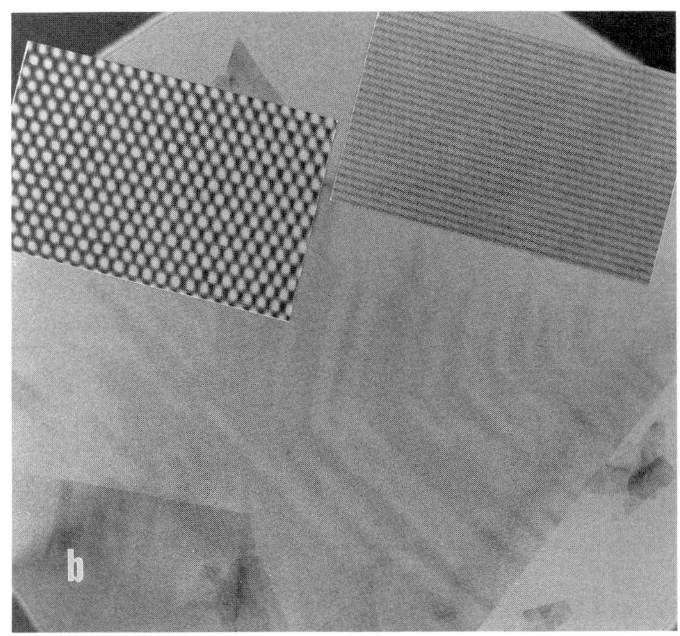
b

c

length, e.g., for n-$C_{168}H_{338}$, where chain folding is permitted, the sectorization behavior resembled that of the infinite polymer (Ungar et al., 1985; Organ and Keller, 1987; Dorset et al., 1993). It was obvious that the newly discovered sectorized layer structure, which occurs for nonfolded long paraffins, was the result of a collapse of a three-dimensional habit formed in suspension as the crystals were grown from a poor solvent by a process reminiscent of self-seeding. High-resolution electron micrographs of n-$C_{82}H_{166}$ (Figure 8.9) demonstrated that the alternation of dark and light bands in the bright field images was due to rectangular and oblique chain layers, respectively, as they had also been observed for polyethylene (Dorset et al., 1990). These experiments demonstrated, therefore, that many of the characteristics of the crystal morphology in the polymer are determined by the stem packing and not the chain folding. On the other hand, surface decoration (Wittmann and Lotz, 1985) of this lamellar structure by polymethylene chains revealed the overall bidirectional nucleation found for the lower alkanes, with only weak indications of the layer sectorization behavior found at the crystal surface.

Diffraction data from solution-crystallized paraffins can only provide information about the methylene subcell packing. Because of elastic crystal bending, a shorter unit cell length must be oriented parallel to the incident beam if the electron diffraction intensities are to represent the total unit cell contents. As discussed in Chapter 3, certain organic substrates such as naphthalene or benzoic acid can be used to nucleate the [100] projection of a paraffin crystal, from which $0kl$ and hhl electron diffraction patterns can be obtained (Moss et al., 1984) (Figure 8.10), as well as high-resolution electron microscope lattice images (Dorset and Zemlin, 1990) (Figure 8.11). The nucleation procedure can be understood in terms of the eutectic phase diagram (Dorset et al., 1989) and the lattice matching of interacting crystal faces for the two eutectic components (see Chapter 3).

Moss et al. (1984) analyzed electron diffraction intensity data from an epitaxially crystallized n-paraffin (n-$C_{36}H_{74}$) to demonstrate that these agree well with the x-ray crystal structure of the orthorhombic form solved by Teare (1959). For the combined data set, the agreement of the model to experimental values is $R = 0.25$ when $B = 3.0$ Å^2 (Table 8.2) (although there is some disparity in the temperature factors for the two data projections). An agreement of I_{00l} data for the $0kl$ and hhl data sets indicated that the diffraction intensities were not seriously perturbed by dynamical scattering. A low temperature data set was also obtained to observe an experimental lowering of the Debye-Waller factor for the $0kl$ data set. Somewhat later, using a liquid helium cooled cryoelectron microscope, Dorset and Zemlin (1990) published 5-Å resolution electron micrographs of the lamellar repeat (Figure 8.11). The computed transform (Figure 8.12) could be used to obtain crystallographic phase information (e.g., via CRISP) after the envelope of the phase contrast transfer function was found (Figure 3.12) using the procedure of Unwin and Henderson (1975).

With the crystallographic phases determined from the images coupled with the use of traditional direct phasing procedures (i.e., generation of Σ_1- and Σ_2- three-phase invariants), it was possible to solve the structure from the electron diffraction intensity data. The orthorhombic space group is noncentrosymmetric $Pca2_1$ with cell constants

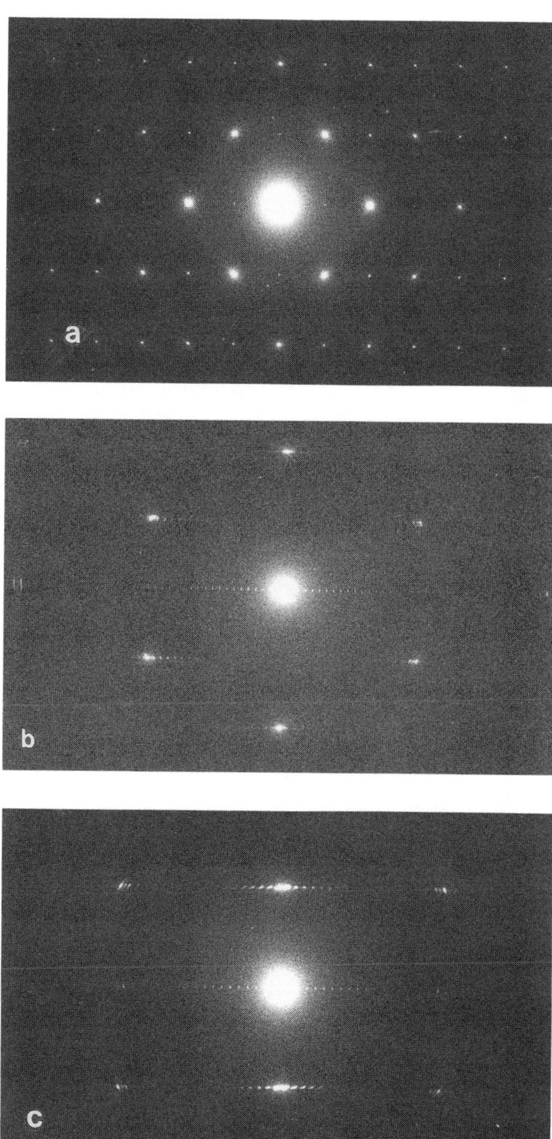

FIGURE 8.10. Electron diffraction pattern from *n*-hexatriacontane: (a) *hk*0 from a solution-crystallized sample; (b) 0*kl* from an epitaxially oriented sample; (c) *hhl* from an epitaxially oriented sample tilted 33° around $\mathbf{c}^*$. (Reprinted from D. L. Dorset, J. Hanlon, and G. Karet (1989) "Epitaxy and structure of paraffin-diluent eutectics," *Macromolecules* **22**, 2169–2176; with kind permission of the American Chemical Society.)

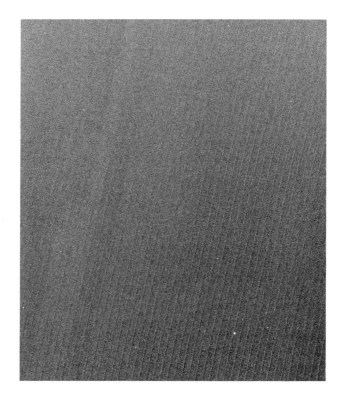

FIGURE 8.11. High-resolution electron micrograph of n-$C_{36}H_{74}$, epitaxially crystallized on benzoic acid (obtained in a cryoelectron microscope). (Reprinted from D. L. Dorset and F. Zemlin (1990) "Direct phase determination in electron crystallography: the crystal structure of an n-paraffin," *Ultramicroscopy* **33**, 227–236; with kind permission of Elsevier Science B.V.)

$a = 7.42$, $b = 4.96$, $c = 95.14$ Å. Since the structure is dominated by a centrosymmetric subcell packing of methylene groups, centrosymmetric estimates were given to the phases, following the procedure outlined in Table 8.3. (Due to symmetry constraints, the lateral subcell origin must lie at $b/4$ in the unit cell. This means that odd k-index reflections will have phase values near $\pm\pi/2$, whereas even-index reflections will have phases near $0,\pi$.) As shown in the phase comparison in Table 8.3, this approximation did not lead to serious error, so that the ensuing potential maps clearly revealed the characteristic packing motif of the paraffin chains (Figure 8.13) in the [100] projection.

Epitaxial growth was also possible for a number of other even-chain paraffins up to n-$C_{60}H_{122}$ and the measured lamellar spacings were found to correspond well to their literature values (Dorset, 1986b). For the longer chain paraffins, it becomes more and more difficult to orient the molecules epitaxially from a dilute co-melt with, say, benzoic acid. Other techniques, such as deposition from the vapor phase onto sub-

Table 8.2. Electron Diffraction Data from Epitaxially Oriented n-$C_{36}H_{74}$

hkl	$\|F_o\|$	ϕ (deg)	hkl	$\|F_o\|$	ϕ (deg)
002	0.61	180.16	01 32	0.45	91.60
004	0.60	180.25	01 34	0.55	91.95
006	0.59	180.18	01 36	1.08	92.19
008	0.58	180.65	01 38	2.41	−87.73
00 10	0.56	180.35	01 40	0.50	−87.80
00 12	0.50	180.80	020	3.12	180.00
00 14	0.42	181.06	022	0.34	0.16
00 16	0.46	180.49	024	0.36	0.25
00 18	0.54	181.55	02 74	0.58	184.43
00 20	0.41	181.03	02 76	0.55	4.58
00 22	0.47	181.13	03 36	0.90	−87.81
00 24	0.31	182.15	03 38	1.01	92.27
00 26	0.31	180.73	110	1.29	180.00
00 28	0.18	182.05	111	3.73	179.85
00 30	0.18	182.07	113	0.53	179.54
00 32	0.24	180.92	115	0.51	170.20
00 70	0.25	3.68	117	0.48	178.99
00 72	0.28	4.06	119	0.39	178.46
00 74	1.11	4.43	11 11	0.31	178.35
00 76	0.72	184.58	11 34	0.35	80.64
00 78	0.15	184.42	11 36	0.72	87.54
01 22	0.18	90.83	11 38	1.34	−85.77
01 24	0.13	91.63	11 78	1.14	184.44
01 26	0.18	91.78	220	1.69	180.00
01 28	0.21	91.30	22 37	1.06	−89.25
01 30	0.25	92.09	22 38	1.02	88.89

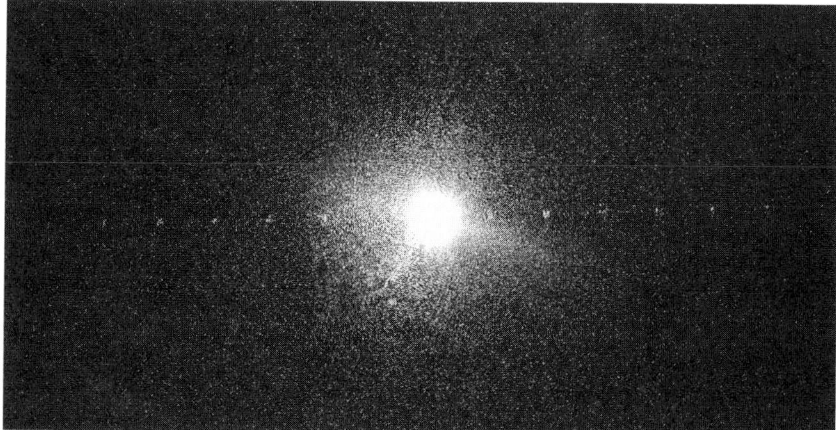

FIGURE 8.12. Computed transform of the image in Figure 8.11—the diffraction resolution extends to 5 Å. (Reprinted from D. L. Dorset and F. Zemlin (1990) "Direct phase determination in electron crystallography: the crystal structure of an n-paraffin," *Ultramicroscopy* **33**, 227–236; with kind permission of Elsevier Science B.V.)

Table 8.3. Direct Phase Determination for n-$C_{36}H_{74}$ $0kl$ and hhl Data

electron microscopy:
within first envelope of phase contrast transfer function:
$\phi_{002} = \phi_{004} = \phi_{006} = \phi_{008} = \phi_{0010} = \pi$
structure invariants:
$0kl$ data: origin definition: $\phi_{01,36} = \pi/2$
Σ_1-triple: $\phi_{020} = \pi$
Σ_2-triples: $\phi_{00,74} = 0$
$\phi_{00,76} = \pi$
$\phi_{01,38} = -\pi/2$
$\phi_{02,74} = \pi$
$\phi_{02,76} = 0$
$\phi_{03,36} = -\pi/2$
$\phi_{03,38} = \pi/2$
hhl data: origin definition: $\phi_{111} = \pi$, $\phi_{11,36} = \pi/2$
Σ_1-triple: $\phi_{220} = \pi$
Σ_2-triples: $\phi_{11,38} = -\pi/2$
$\phi_{11,75} = \pi$
$\phi_{22,37} = -\pi/2$
00l data:
Σ_1-triples: $\phi_{004} = \phi_{008} = \phi_{00,12} = \phi_{00,16} = \phi_{00,20} = \phi_{00,28} = \phi_{00,32} = \pi$

strates such as potassium hydrogen phthalate, are easier (Zhang and Dorset, 1990). However, in this case, a metastable form of the paraffin structure with "nematic"-like disorder was noted after the material was initially deposited onto the nucleating substrate. At first, the $0kl$ diffraction pattern contained only subcell (or "polyethylene") reflections. Annealing the sample in the presence of the substrate eventually caused the appearance of the first lamellar reflections to appear. At higher temperatures, the most intense subcell reflections were split as the final, stable lamellar structure was reached. A model for this change involving longitudinal chain diffusion has been proposed (Figure 8.14). The same crystallization phenomenon has been noted by

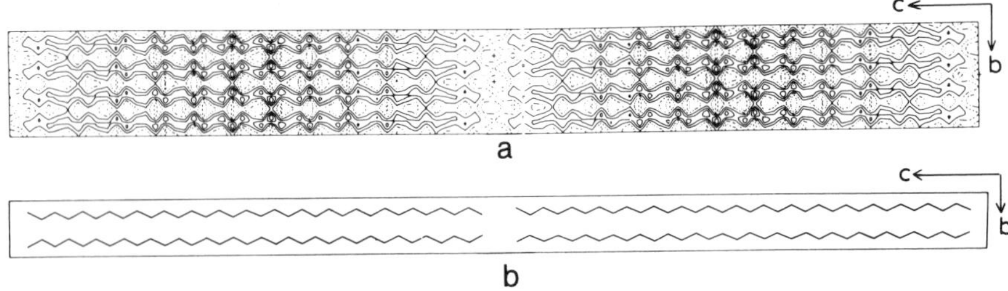

FIGURE 8.13. (a) Potential map for epitaxially oriented n-hexatriacontane based on phases from the image transform and direct methods. (b) X-ray crystal structure. (Reprinted from D. L. Dorset and F. Zemlin (1990) "Direct phase determination in electron crystallography: the crystal structure of an n-paraffin," *Ultramicroscopy* **33**, 227–236; with kind permission of Elsevier Science B.V.)

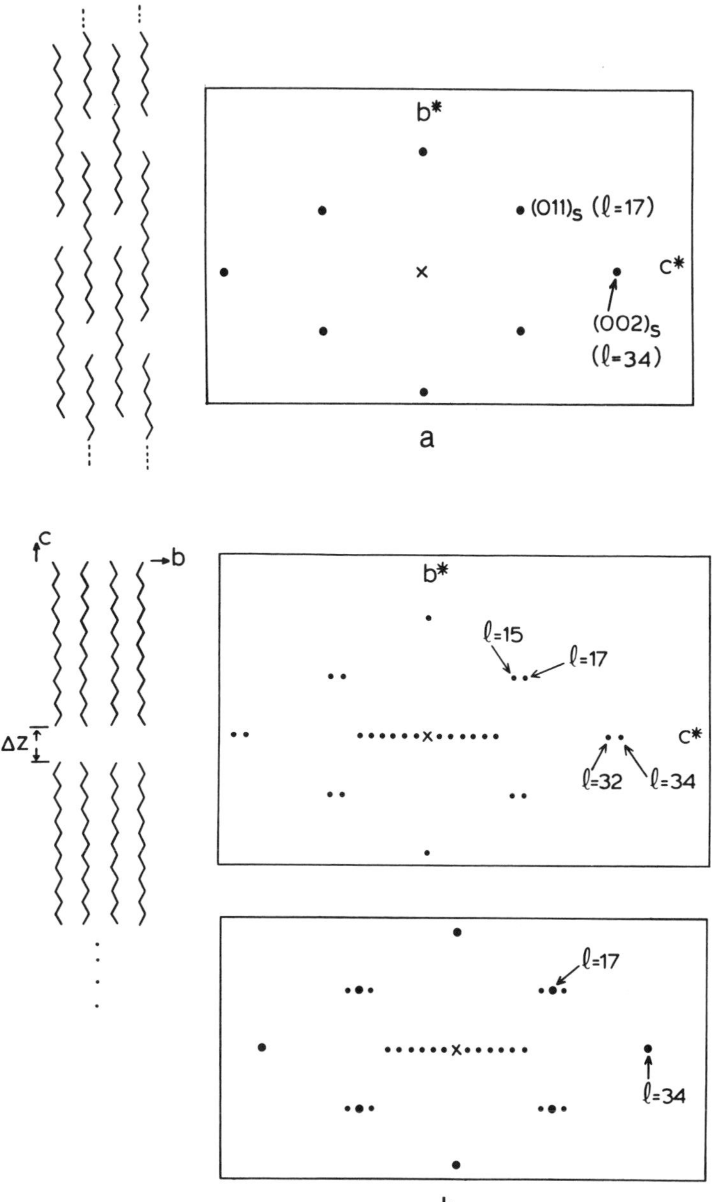

FIGURE 8.14. Sequence of structural changes that occur when a very long paraffin is epitaxially oriented on a substrate at room temperature from the vapor phase and then annealed. (a) Initially, the diffraction pattern resembles that of polyethylene—i.e., only the polymethylene lattice is ordered so that there is a nematiclike displacement of chains. (b) As the sample is heated in the presence of the substrate, lamellar reflections begin to appear, followed by some splitting of the (01*l*) polyethylene reflections, but not the intense (00*l*) reflections. Eventually all of the polyethylene reflections are split, indicating that the interlayer spacing Δz is no longer a simple multiple of the methylene subcell repeat $\mathbf{c}_s$. (Reprinted from W. P. Zhang and D. L. Dorset (1990) "Phase transformation and structure of n-$C_{50}H_{102}$/n-$C_{60}H_{122}$ solid solutions formed from the vapor phase," *Journal of Polymer Science B Polymer Physics* **28**, 1223–1232; with kind permission of John Wiley & Sons, Inc.)

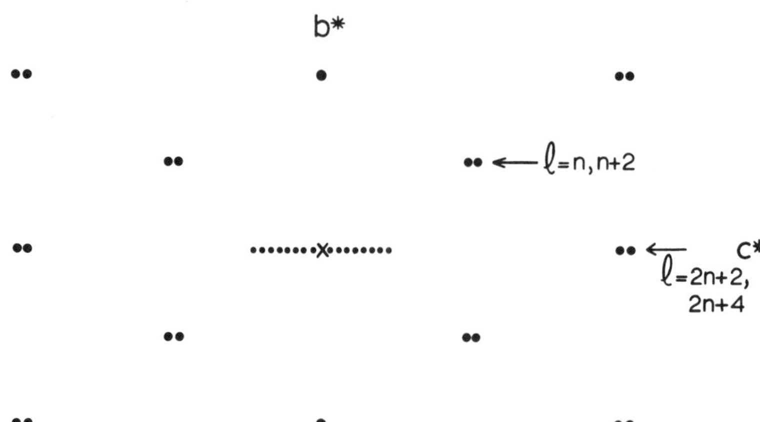

FIGURE 8.15. Relationship between indices of the most intense (0*kl*) reflections and chain carbon number for orthorhombic *n*-paraffins packing either in $Pna2_1$ (if even) or $A2_1am$ (if odd). (Reprinted from D. L. Dorset (1987) "Role of symmetry in the formation of n-paraffin solid solutions," *Macromolecules* **20**, 2782–2788; with kind permission of the American Chemical Society.)

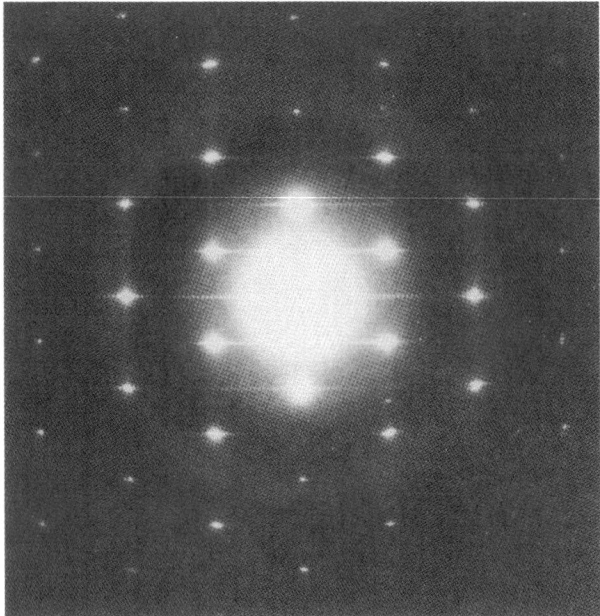

FIGURE 8.16. Electron diffraction pattern from epitaxially oriented *n*-$C_{36}H_{74}$ held at 15 K. Strong diffuse streaks are observed perpendicular to the lamellar reflections. (Reprinted from D. L. Dorset, B. Moss, and F. Zemlin (1985–86) "Kink defects in linear chain molecules—structure analyses based on spot and continuous diffuse electron diffraction intensities," *Journal of Macromolecular Science—Physics* **B24**, 87–97; with kind permission of Marcel Dekker, Inc.)

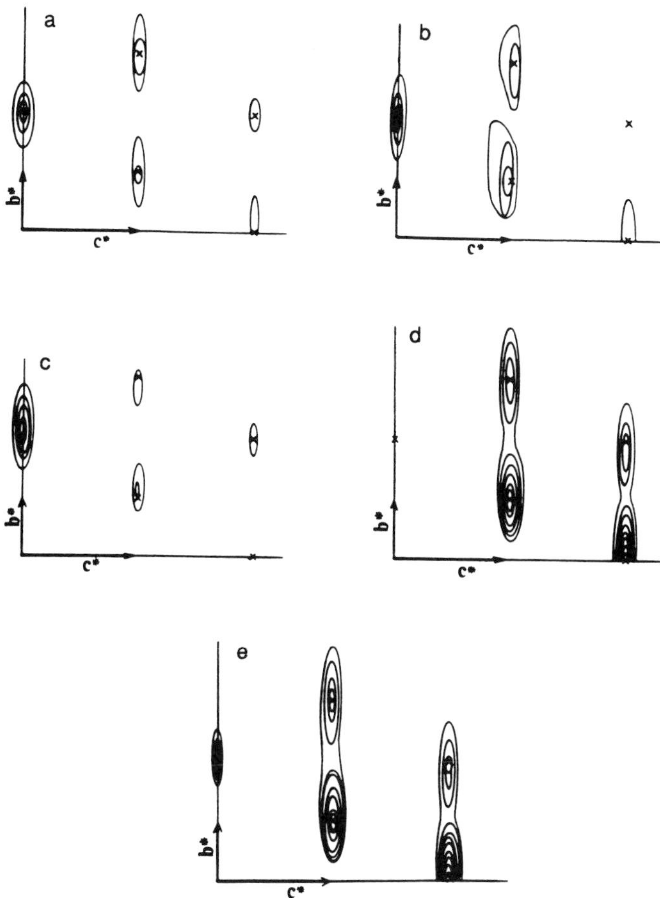

FIGURE 8.17. Models for continuous diffuse scattering in Figure 8.16: (a) thermal diffuse scattering model; (b) model incorporating chain kink defects; (c) model with torsional disorder; (d, e) longitudinal chain positional disorder at low or room temperature, respectively. Only the latter explains the experimental observations. (Reprinted from D. L. Dorset, H. Hu, and J. Jäger (1991) "Continuous diffuse scattering from polymethylene chains—an electron diffraction study of crystalline disorder," *Acta Crystallographica* **A47**, 543–549; with kind permission of the International Union of Crystallography.)

vibrational spectroscopy for shorter-chain-length paraffins deposited onto a cold substrate (Hagemann et al., 1987).

For epitaxially oriented samples of even paraffins, in space group $Pca2_1$, indices of the most intense $0kl$ reflections correspond to $l = n, n + 2$ for the $01l$ row, and $l = 2n + 2, 2n + 4$ for the $00l$ row, where n is the carbon number for $n\text{-}C_nH_{2n+2}$ (see Figure 8.15). The crystallization of the thermodynamically stable monoclinic polymorph (Shearer and Vand, 1956) (space group $P2_1/a$) has never been observed for epitaxially oriented samples and only very rarely for solution-crystallized layers.

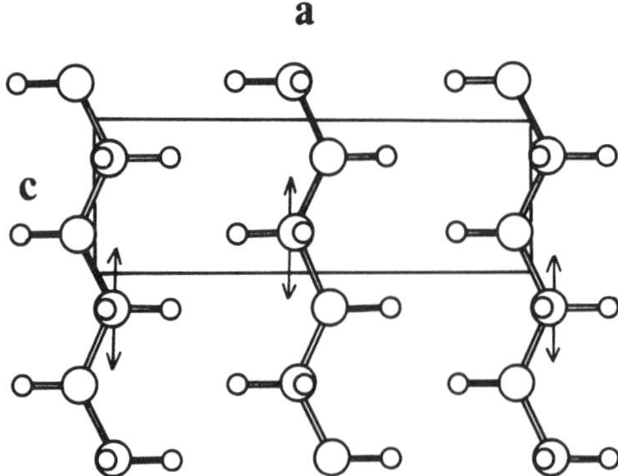

FIGURE 8.18. Longitudinal displacement model explaining diffuse scattering in Figure 8.16. (Reprinted from D. L. Dorset, H. Hu, and J. Jäger (1991) "Continuous diffuse scattering from polymethylene chains—an electron diffraction study of crystalline disorder," *Acta Crystallographica* **A47**, 543–549; with kind permission of the International Union of Crystallography.)

As is the case for solution-crystallized paraffins, patterns from the epitaxially oriented samples also contained intense continuous diffuse scattering streaks connecting some of the strong Bragg peaks (Figure 8.16). Apart from the line along $k = 0$, however, this diffuse signal was not sensitive to lowered temperature (Dorset et al., 1985; Dorset et al., 1991), indicating that its origin was related to a static disorder rather than to thermal vibration. This was underscored by the failure of a thermal vibration model to account for the relative intensity of the diffuse signal (Figure 8.17), even though the positions of the continuous streaks were correctly predicted. The problem was solved with a model for which small longitudinal translations were incorporated into the chain packing (Figure 8.18), giving an expression in the form:

$$I_{\text{long}}(\mathbf{s}) = |F|^2 \{1 - \exp[-4\pi^2 l^2 \sigma_L^2]\}$$

where l is the index of the methylene subcell and σ_L is the average molecular shift. As shown in Figure 8.17, the match of intensity was now satisfied for the $0kl$ pattern, as well as other reciprocal lattice projections. The significance of these findings will be discussed further below when thermotropic phase behavior is considered.

8.2.2. Odd-Chain Paraffins

Odd-chain *n*-paraffins can crystallize in two polymorphs, as described by Piesczek and his coworkers (1974). The lowest temperature form is orthorhombic, space group *Pcam*, a structure originally reported by Smith (1953) and revised by Nyburg and Potworowski (1973). The chain monolayer structure is much the same as for the even-chain paraffins examined above and the $hk0$ diffraction patterns, accordingly, resemble Figure 8.2. For multilamellar crystals grown from solution, it is impossible to assign the space group from the observed $hk0$ zonal data. For example, if the low-energy orthorhombic form is formed, systematic space group absences would be expected to occur only for the *h*00 reflections, when $h = 2n + 1$. If the bilayer packing is the slightly higher energy form, then, as described by Piesczek et al. (1974), a monoclinic space group *Aa* with nearly orthorhombic cell angles would be expected. Here the extinctions for *hk*0 require all reflections where $h = 2n + 1$ to be absent. Obviously the electron diffraction patterns from multilamellar crystals satisfied neither condition (since the observed diffraction pattern was identical to Figure 8.2) so that the influences of the elastic crystal bending discussed in Chapter 5 were again observed. That is to say, in the projection down a long unit cell axis, only the monolayer packing (specifically the methylene subcell) was expressed by the *hk*0 electron diffraction intensities and agreed with the assignment of *pgg* plane group symmetry (Dorset, 1987c).

Epitaxial orientation of odd-chain paraffins on substrates such as benzoic acid clearly demonstrates that the crystal structure of the crystallized polymorph was not that of the lowest energy form solved by Smith. Systematic absences for the 0*kl* diffraction pattern were in accord with space group *Aa*, if the unit cell is monoclinic, or $A2_1am$, if it is orthorhombic. The possibility of an *Aa* packing model was justified with electron diffraction intensities from *n*-$C_{33}H_{68}$, using cell constants $a = 7.57$, $b = 4.98$, $c = 87.94$ Å (Dorset, 1986b). (The β-angle, deviating only slightly from 90°, would not expressed in this rectangular projection, except, possibly, to slightly shorten the measured long spacing.) With theoretical carbon ($B = 6$ Å^2) and hydrogen ($B = 8$ Å^2) positions, the final residual was $R = 0.23$ (see Table 8.4). On the other hand, convincing arguments for the orthorhombic unit cell were given in later work involv-

Table 8.4. Electron Diffraction Data for *n*-$C_{33}H_{68}$, 0*kl* Reflections*

hkl	$\lvert F_o \rvert$	ϕ(deg)
00 2	0.25	179.7
00 4	0.24	180.3
00 8	0.31	179.3
00 12	0.28	177.0
00 16	0.28	182.2
00 20	0.21	176.4
00 24	0.12	175.6
00 28	0.10	185.2
00 32	0.09	168.9
00 68	0.48	–5.0
00 70	0.46	175.0
01 33	0.38	–92.5
01 35	0.75	87.5
02 0	1.33	180.0
02 68	0.31	175.1
02 70	0.26	–5.0
03 35	0.35	–92.5

*Only reflections accessed by direct methods are listed.

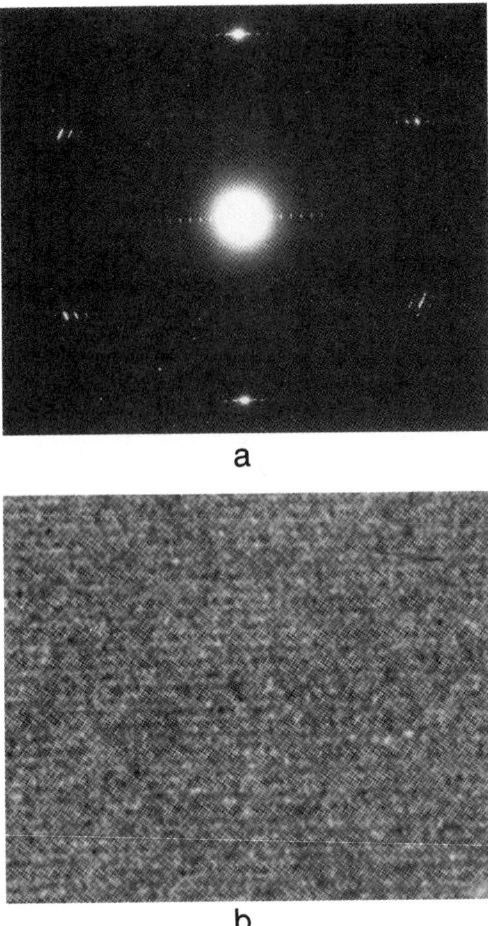

FIGURE 8.19. Epitaxially crystallized *n*-tritriacontane: (a) electron diffraction pattern (0*kl*); (b) electron micrograph. (Reprinted from D. L. Dorset and W. P. Zhang (1991) "Electron crystallography at atomic resolution: the structure of the odd chain paraffin n-tritriacontane," *Journal of Electron Microscope Technique* **18**, 142–147; with kind permission of John Wiley & Sons, Inc.)

ing tilts around various reciprocal axes (Hu et al., 1989). In general, the 0*kl* electron diffraction patterns (Figure 8.19) satisfied the same index rules for the strong 00*l* and 01*l* reflections as do the orthorhombic even-chain paraffins, in terms of the carbon number *n*.

Starting with lower-resolution lattice images, easily obtained from a conventional electron microscope, it has been possible to combine image analysis and direct phase determination to solve the crystal structure of n-$C_{33}H_{68}$ from the structure factor

Table 8.5. Direct Phase Determination for n-$C_{33}H_{68}$, $0kl$ Data

electron microscopy:

$\phi_{002} = \phi_{004} = \pi$

phase invariant sums:

lamellar reflections:

Σ_1-triples: $\phi_{004} = \phi_{008} = \phi_{00,12} = \phi_{00,16} = \phi_{00,20} = \phi_{00,24} = \phi_{00,28} = \phi_{00,32} = \pi$

Σ_2-triples: $\phi_{00,68} + \phi_{00,70} = \pi$

zonal reflections: origin definition: $\phi_{01,35} = \pi/2$

Σ_1-triple: $\phi_{020} = \pi$

Σ_2-triples: $\phi_{03,35} = -\pi/2$

$\phi_{02,70} = 0$

$\phi_{00,70} = \pi$, then from above, $\phi_{00,68} = 0$

$\phi_{02,68} = \pi$

$\phi_{01,33} = -\pi/2$

magnitudes listed in Table 8.4 (Dorset and Zhang, 1991). The procedure used is reviewed in Table 8.5. Although the space group is again noncentrosymmetric, the presence of the centrosymmetric methylene subcell causes the phases of most intense reflections were found to lie near centrosymmetric values so that the correct structure was readily visualized in the potential map (Figure 8.20). (Again, the shift of the chain—and, hence, the methylene subcell—origin by $b/4$ required that odd k-index reflections would have phases near $\pm\pi/2$ while even-index reflections would have values near $0,\pi$.)

Continuous diffuse diffraction from epitaxially crystallized odd-chain paraffins were also analyzed in the way discussed above for the even-chain compounds. The results were identical to those found for the even-chain n-paraffins, indicating that the diffuse signal is caused by small longitudinal displacements along the chain axes (Dorset et al., 1991).

8.2.3. Thermotropic Phase Transitions of Linear Paraffins

It is well known (Asbach and Kilian, 1970) that n-paraffins in the chain length regime 11 to 43, for odd-chain members, and 26 to 38, for even-chain members,

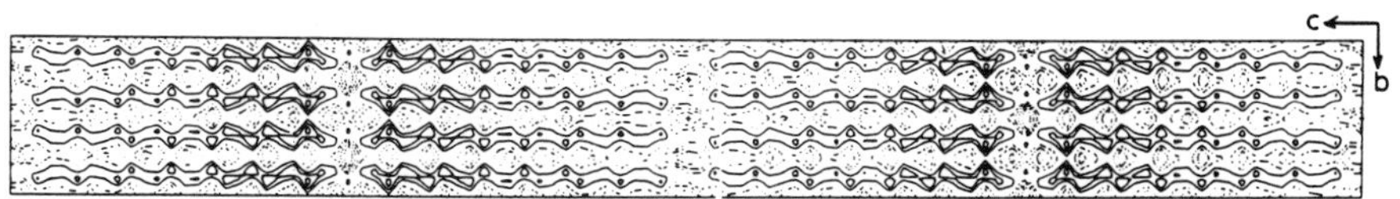

FIGURE 8.20. Potential map for n-$C_{33}H_{68}$ calculated from phases obtained by direct methods and from the Fourier transform of the electron micrograph in Figure 8.19b. (Reprinted from D. L. Dorset and W. P. Zhang (1991) "Electron crystallography at atomic resolution: the structure of the odd chain paraffin n-tritriacontane," *Journal of Electron Microscope Technique* **18**, 142–147; with kind permission of John Wiley & Sons, Inc.)

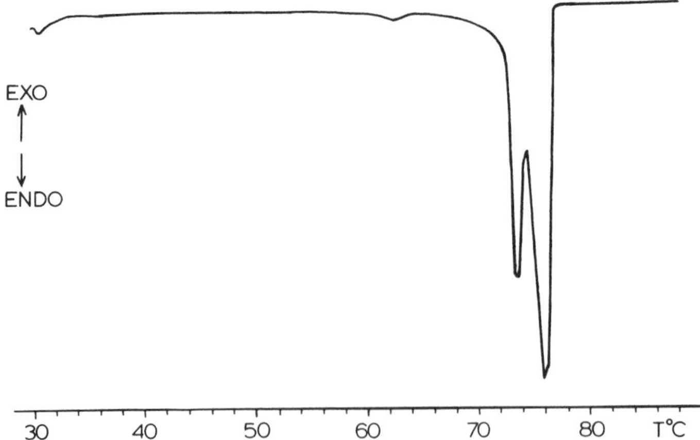

FIGURE 8.21. DSC scan of n-$C_{36}H_{74}$. With increasing temperature, small endotherm is found first for the monoclinic to orthorhombic crystal–crystal transition, followed by a larger one for the orthorhombic to hexagonal ("rotator") transition and then the largest for the melt transition.

undergo a crystal–crystal transition to the so-called "rotator" phase before the melt, seen as an endotherm in a heating scan with a differential scanning calorimeter (Figure 8.21). The electron diffraction pattern from the rotator phase is well known (Figure 8.22), and, for these paraffins, can be visualized in heating experiments on crystalline

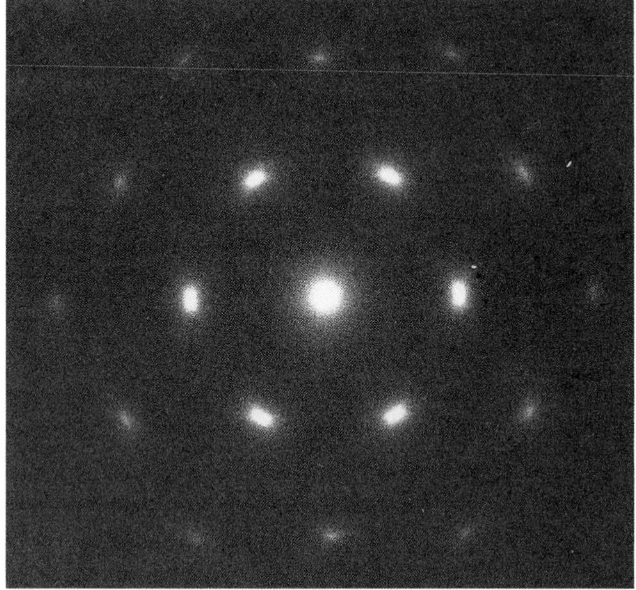

FIGURE 8.22. Electron diffraction pattern from chains in the hexagonal "rotator" phase.

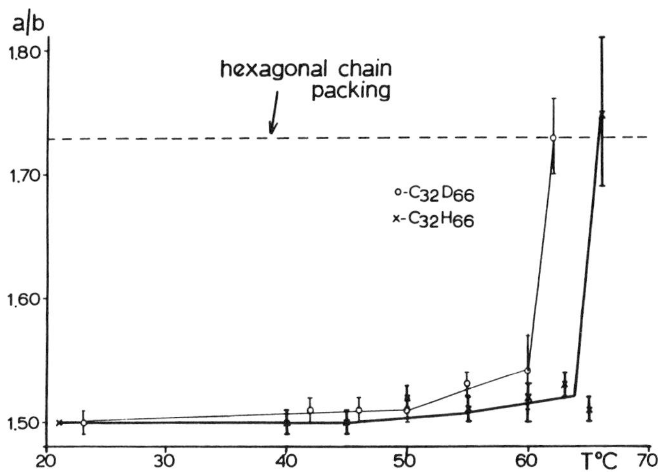

FIGURE 8.23. Comparison (in terms of unit cell *a/b* ratio) of two isotopically related paraffins that transform to the "rotator" phase. Perdeuteration lowers the temperature of this transition. (Reprinted from D. L. Dorset (1991) "Structure interactions between n-paraffins and their perdeuterated analogs; binary compositions with identical chain lengths," *Macromolecules* **24**, 6521–6526; with kind permission of the American Chemical Society.)

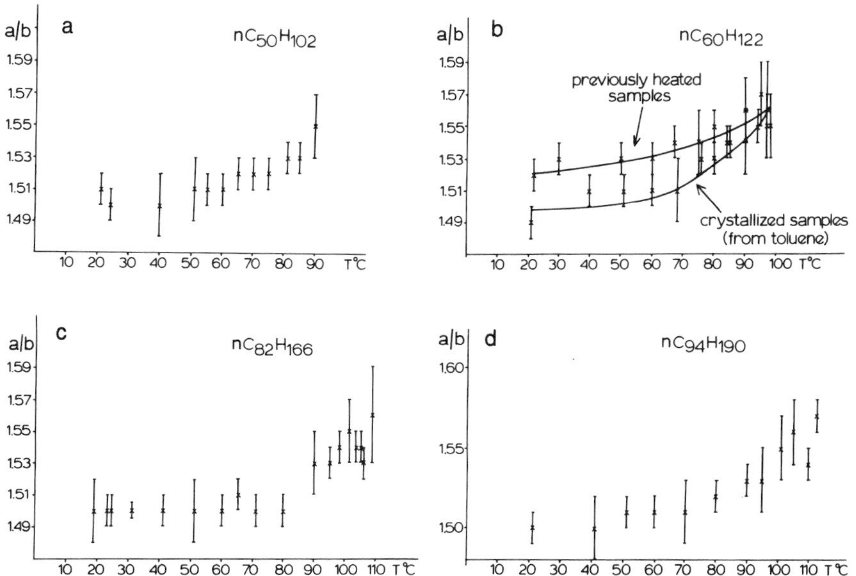

FIGURE 8.24. Change of the *a/b* ratio for paraffins that do not undergo the transition to the rotator phase. (Reprinted from D. L. Dorset, R. G. Alamo, and L. Mandelkern (1992) "Premelting of long n-paraffins in chain-extended lamellae: an electron diffraction study," *Macromolecules* **25**, 6284–6288; with kind permission of the American Chemical Society.)

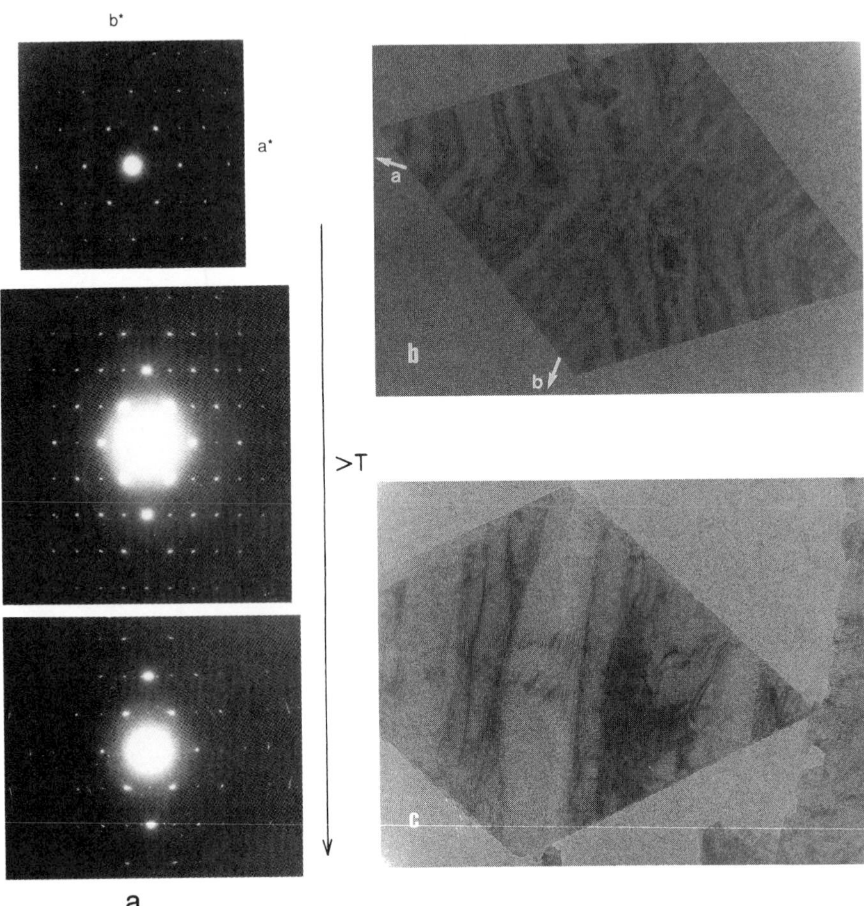

Figure 8.25. Change on heating for a paraffin that has no "rotator" phase. The layer begins to tilt around the subcell **b**-axis as observed (a) in the electron diffraction pattern, and the characteristic sectorization behavior (b) is found to transform to a banded structure (c) with striations along **b**. (Reprinted from D. L. Dorset, R. G. Alamo, and L. Mandelkern (1992) "Premelting of long n-paraffins in chain-extended lamellae: an electron diffraction study," *Macromolecules* **25**, 6284–6288; with kind permission of the American Chemical Society.)

layers in the electron microscope. The rotator phase was originally characterized by Müller (1932) in his x-ray studies of the paraffins and, when it exists, the resultant orthorhombic lattice expansion is found to reach an axial ratio $a/b = \sqrt{3}$. This corresponds to a hexagonal layer with a new unit cell constant: $a = 4.80$ Å.

Using solution-crystallized paraffin layers, it is informative to follow the progress of the *a/b* ratio by electron diffraction patterns as the sample is heated toward the rotator transition temperature. As shown in Figure 8.23, there is a very small increase of this value until this transition is reached, whereupon the abrupt expansion to the given hexagonal packing limit (Dorset, 1991e) occurs. Axial ratios for lateral packings of chains, which are beyond the length domain allowing the rotator phase to appear, continue to expand gradually until the melt is reached (Figure 8.24), but, as shown by bright-field electron micrographs, as well as the diffraction patterns, a transition to an oblique packing occurs just below the melt (Figure 8.25) (Dorset et al., 1992), responsible, no doubt, for the "roof structures" seen in heated alkane crystals (Keller, 1961; Fischer, 1971; Takimizawa et al., 1982). Such longitudinal translations, which are also responsible for the screw dislocation growth of the multilamellar crystals (Dawson and Vand, 1951; Dawson, 1952), are preserved on a miniature scale as positional disorder when the paraffins are crystallized from the melt (e.g., by epitaxial orientation on a substrate) and would account for the residual diffuse scattering signal described above (Dorset et al., 1991). Perdeuteration of the chain is known also to lower the pretransition temperature as well as the melting temperature. The isotopic substitution also lowers the chain length where the rotator disappears to *n*-$C_{36}D_{74}$ (Stehling et al., 1971; Dorset, 1991e). In this case, the rotator transition has been found only when the melt was recrystallized, but heating samples directly only caused an expansion to an axial ratio of about 1.68 before melting occurs (Dorset, 1991e) (Figure 8.26).

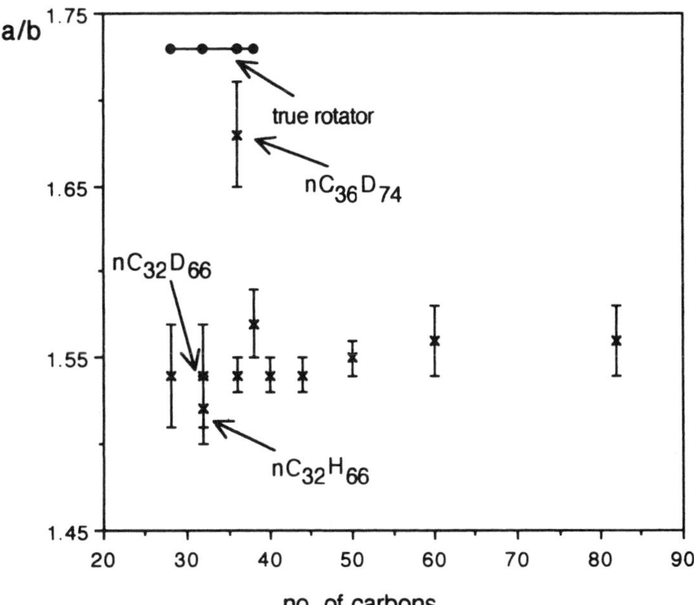

FIGURE 8.26. Length domain of even chain paraffins in terms of ultimate *a/b* ratio upon heating. For perdeuterated members, the boundary for the "rotator" phase is found at *n*-$C_{36}D_{74}$, which does not transform to this phase unless cooled from the melt. (Reprinted from D. L. Dorset (1991) "Structure interactions between n-paraffins and their perdeuterated analogs; binary compositions with identical chain lengths," *Macromolecules* **24**, 6521–6526; with kind permission of the American Chemical Society.)

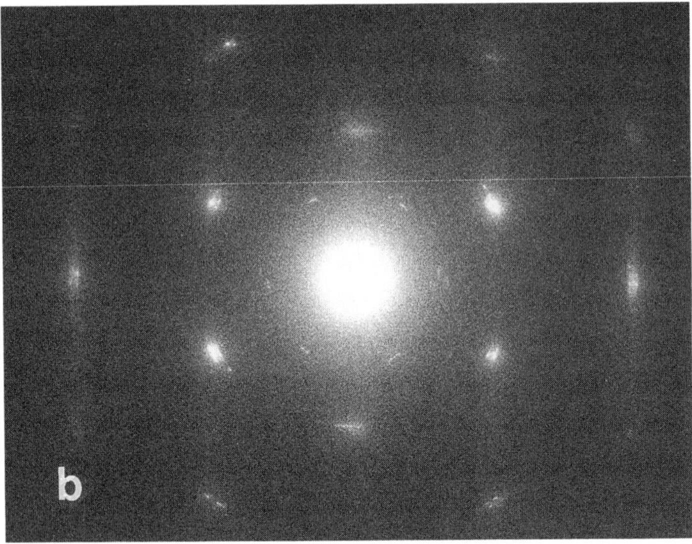

FIGURE 8.27. When epitaxially oriented crystals of $n\text{-}C_{36}H_{74}$ (a) are heated toward the rotator phase, the resolution of the lamellar reflections (b) greatly decreases, corresponding to a disordered lamellar interface. (Reprinted from D. L. Dorset, B. Moss, J. C. Wittmann, and B. Lotz (1984) "The pre-melt phase of n-alkanes: crystallographic evidence for a kinked chain structure," *Proceedings of the National Academy of Sciences USA* **81**, 1913–1917; copyright held by the author.)

The term "rotator phase" for the layer structure appearing after the premelt transition is actually the result of historical convention. Various packing models for this average structure have been proposed, including an chain helix, similar to poly(tetrafluoroethylene) (D'Ilario and Giglio, 1974), or an accumulation of chain defects (Maroncelli et al., 1982), in addition to the true rotor itself. All of these cause the average chain projection to have a more or less circular cross section, and thus cannot be distinguished from one another in the projection down the chain axes. Disordered layer packings with average hexagonal symmetry have been proposed as a further alternative by Ungar and Masic (1985) and by Scaringe (1991).

More specific information about the premelt changes in the paraffin crystal structure can be obtained from the electron diffraction of samples that had been epitaxially oriented (Dorset et al., 1984b). For example, in $0kl$ diffraction patterns from n-$C_{36}H_{74}$ crystals (Figure 8.27) major changes were found, upon heating, to occur in the so-called "lamellar reflections," which become more and more limited in resolution as the sample is heated, whereas the intensities of "polyethylene reflections," which are related to the methylene subcell packing, were much less affected. Such a change can only be explained in terms of a disordered lamellar interface of the type also seen in vibrational spectroscopic measurements. Similar changes to the lamellar reflection row had also been noted in x-ray diffraction experiments (Craievich et al., 1984). However, the electron diffraction measurements were the first observation of such changes in terms of a single-crystal pattern from a complete zone. The "rotator" phase, therefore, cannot involve an actual chain rotor, nor can it be explained by a helical geometry, since either one or the other would leave the interfacial interactions of methyl groups virtually intact. Conformational disorders have to be built into the chain, and, according to vibrational spectroscopic measurement, these have the highest density at the chain ends with decreasing concentration as the chain center is approached (Maroncelli et al., 1985).

As already mentioned, when the chain length is too long to permit formation of the hexagonal layer packing, another sequence of events can be followed with epitaxially oriented samples via heating experiments. For example, if the indexing rules given above to determine the apparent carbon number of a lamellar layer were used to follow the changes in heated n-$C_{50}H_{102}$, not only was the resolution of the lamellar row shown to decrease with higher temperature (Figure 8.28), but a structure similar to n-$C_{51}H_{104}$ was found, on average (Dorset et al., 1992a), indicating that the lamellar disorder is being caused mainly by slight longitudinal shifts. As long as the final temperature was not too close to the melting point, the transition was observed to be more or less reversible, although the total resolution of the lamellar row was not fully recovered (Figure 8.28). Similar loss of lamellar resolution had also been observed by x-ray diffraction (Fischer, 1971) and vibrational spectroscopy indicates chain disorder at high temperature similar to that found for the shorter paraffins (Kim et al., 1989).

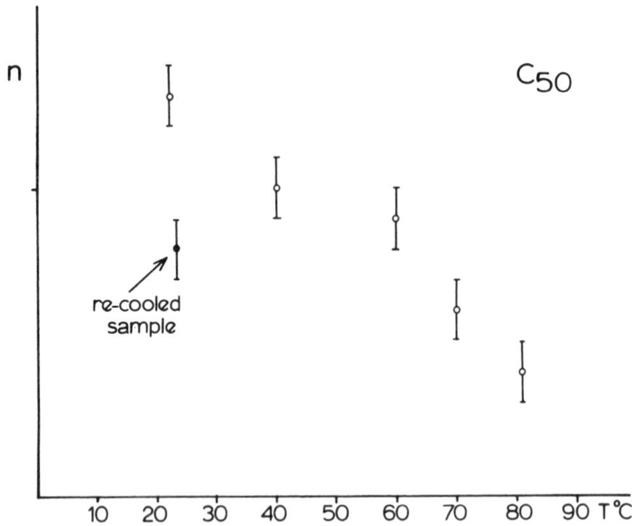

FIGURE 8.28. Longer paraffins that do not transform to the "rotator" phase, e.g., n-$C_{50}H_{102}$, also have attenuated lamellar resolution in their electron diffraction patterns taken at higher temperature. There is also a longitudinal transition to an average structure very much like n-$C_{51}H_{104}$. Upon cooling the sample, the original resolution of the lamellar row is not totally restored. (Reprinted from D. L. Dorset, R. G. Alamo, and L. Mandelkern (1992) "Premelting of long n-paraffins in chain-extended lamellae: an electron diffraction study," *Macromolecules* **25**, 6284–6288; with kind permission of the American Chemical Society.)

8.2.4. Binary (and Multicomponent) Phase Behavior in Paraffins

8.2.4.1. Solid Solutions. Binary combinations of n-paraffins have been often studied in order to determine the crystallographic and volumetric requirements important for the stabilization of solid solutions. Based on a review of early powder x-ray measurements (Mnyukh, 1960), Kitaigorodskii (1961) was able to propose that the nonoverlap volume Δ for the two chains in a solid solution must be minimal. If the overlap volume is r, then one can define $\varepsilon = 1 - \Delta/r$. For complete solubility, then $\varepsilon \geq 0.8$ should hold. More importantly, he reasoned that, if the pure molecules are to form a solid solution, they should also have symmetrically compatible crystal structures. If a molecule A is added to a lattice of B, then either the symmetry of B is unaltered or the symmetry of B is lowered by the addition of A for a stable solid solution to be formed.

The principles can be tested in terms of localized microstructure by electron diffraction measurements on epitaxially oriented crystals. Intensity data were recorded (Figure 8.29) from a nearly 1:1 solid solution of n-$C_{32}H_{66}$ with n-$C_{36}H_{74}$ (Dorset, 1990b). Using the Fourier transform of a lattice image and direct phasing procedures (see outline in Table 8.6), it was possible to solve the average crystal structure. Indices of most intense reflections in the $0kl$ diffraction pattern, following the rules outlined above, indicated that the average structure must be n-$C_{34}H_{70}$, corresponding also to

FIGURE 8.29. (a) Electron diffraction pattern from a 1:1 solid solution of n-$C_{32}H_{66}$/n-$C_{36}H_{74}$ epitaxially crystallized on benzoic acid. (b) Potential map after structure determination. (Reprinted from D. L. Dorset (1990) "Direct structure analysis of a paraffin solid solution," *Proceedings of the National Academy of Sciences USA* **87**, 8541–8544; copyright held by the author.)

Table 8.6. Direct Phase Determination for 1:1 n-$C_{32}H_{66}$/n-$C_{36}H_{74}$ Solid Solution: $0kl$ Data

electron microscopy:	
$\phi_{002} = \phi_{004} = \pi$	
structure invariant sums:	
zonal reflections:	
origin definition:	$\phi_{03,36} = \pi/2$
Σ_1-triple:	$\phi_{020} = \pi$
Σ_2-triples:	$\phi_{00,70} = a$
	$\phi_{00,72} = \pi$
	$\phi_{01,34} = \pi/2$
	$\phi_{01,36} = -\pi/2$
	$\phi_{02,70} = a + \pi$
	$\phi_{02,72} = 0$
	$\phi_{03,34} = -\pi/2$
lamellar reflections:	
Σ_1-triples:	$\phi_{004} = \phi_{008} = \pi$
Σ_2-triple:	$\phi_{00,70} = 0 = a$

Table 8.7. Electron Diffraction Data for the Paraffin Solid Solution

hkl	$\|F_o\|$	$\|F_c\|$	ϕ
00 2	0.67	0.85	180.16
00 4	0.55	0.38	180.23
00 6	0.37	0.30	180.32
00 8	0.27	0.30	180.73
00 68	0.27	0.26	4.30
00 70	0.86	1.14	4.87
00 72	0.83	0.84	185.03
00 74	0.21	0.24	184.74
01 30	0.25	0.24	93.69
01 32	0.42	0.33	94.06
01 34	0.81	0.90	91.65
01 36	1.57	1.77	–87.34
01 38	0.37	0.23	–90.83
02 0	1.93	2.42	180.00
02 2	0.44	0.21	–10.50
02 70	0.47	0.41	185.45
02 72	0.51	0.31	4.12
03 32	0.37	0.20	–90.83
03 34	0.60	0.53	–86.48
03 36	1.17	1.06	92.75

the cell constants $a = 7.42$, $b = 4.96$, $c = 90.05$ Å. The space group is $Pca2_1$, as found also for the pure even-chain components. If the average structure was used as a model (Figure 8.29) and the occupancies of outer chain carbon atoms were refined (Figure 8.30), the match to the experimental data was rather good (Table 8.7) with $R = 0.23$. The partial occupancy was indicated by the limited resolution of the (00*l*) reflections at low angle, seen earlier in x-ray diffraction studies (Asbach et al., 1979; Craievich et al., 1984; Denicolo et al., 1984). These results are consistent with Kitaigorodskii's rules and represent the first quantitative crystallographic analysis of a paraffin solid solution.

On the other hand, a rather good qualitative analysis of an n-$C_{20}H_{42}$/n-$C_{22}H_{46}$ solution had been made earlier by Lüth et al. (1974), based on single-crystal x-ray data. Here, the pure components pack in a triclinic unit cell (space group $P\bar{1}$), while the solid solution adopts an average orthorhombic packing in space group $Bb2_1m$, i.e., with *higher* symmetry than that of either component! (Recently, a more quantitative determination has been made for a similar n-$C_{24}H_{50}$/n-$C_{26}H_{54}$ solid solution (Gerson and Nyburg, 1994) yielding the same general result.) There were also claims that n-$C_{35}H_{72}$ could form a stable solution with n-$C_{36}H_{74}$ (Mazee, 1958), an observation also in disagreement with Kitaigorodskii's symmetry rules.

The apparent discrepancies with Kitaigorodskii's symmetry criteria for solid solution formation could be explained only after the earlier diffraction observations on bulk systems were improved by electron diffraction measurements of the micro-

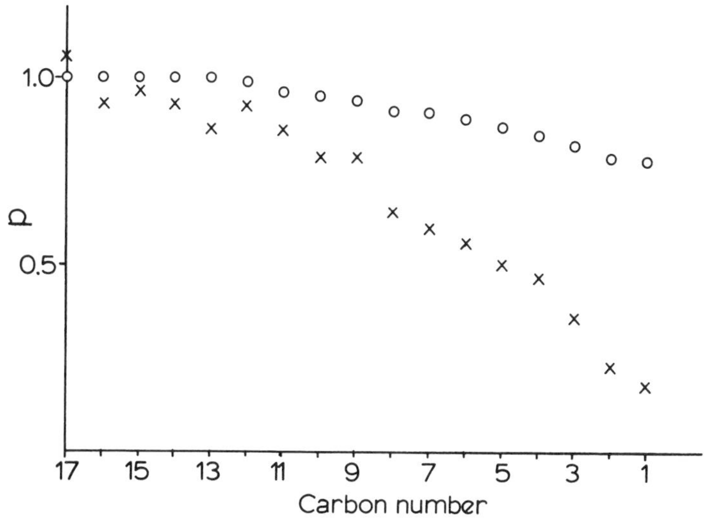

FIGURE 8.30. Occupancy of outer chain carbons. When the relative atomic peak heights (p) in the potential map "x" are used as a direct indication of chain occupancy ($1-p$), the agreement to the observed diffraction data is not very good. It is improved when these peak heights are dampened by 0.25. (Reprinted from D. L. Dorset (1990) "Direct structure analysis of a paraffin solid solution," *Proceedings of the National Academy of Sciences USA* **87**, 8541–8544; copyright held by the author.)

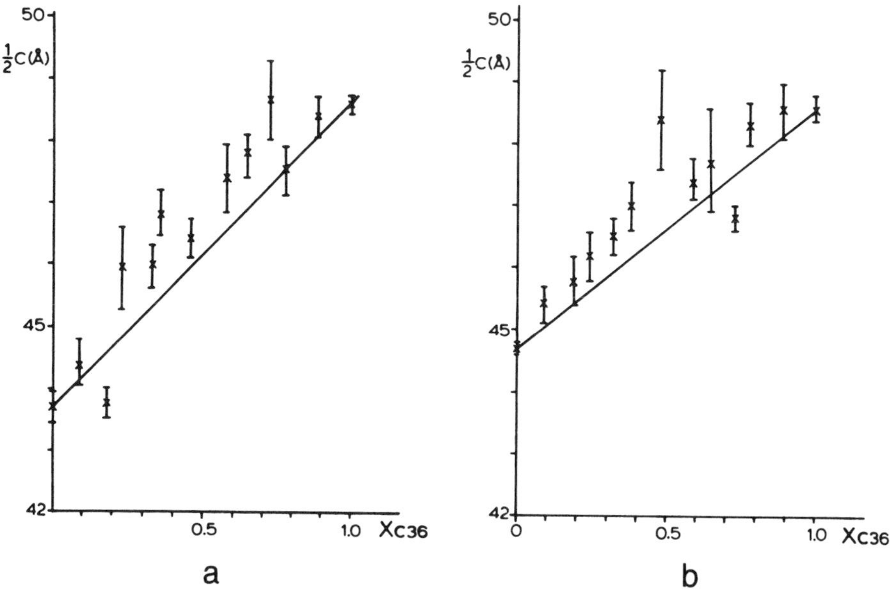

FIGURE 8.31. Average lamellar thickness for solid solutions over a complete range of concentrations: (a) n-$C_{32}H_{66}$/n-$C_{36}H_{74}$; (b) n-$C_{33}H_{68}$/n-$C_{36}H_{74}$. (Reprinted from D. L. Dorset (1985) "Crystal structure of n-paraffin solid solutions—an electron diffraction study," *Macromolecules* **18**, 2158–2163; with kind permission of the American Chemical Society.)

state. Particularly important is the description of a layer packing for a binary combination in terms of a local crystal structure (Dorset, 1987d), expressed as the apparent carbon number of the layer from the indexing rules given in Figure 8.15. For example, both n-$C_{33}H_{68}$ and n-$C_{32}H_{66}$ had been found to form stable solid solutions with n-$C_{36}H_{74}$ as shown in Figure 8.31, evidenced by measurement of their electron diffraction patterns over all concentrations (Dorset, 1985a, 1987d). If the average lamellar spacings are plotted against concentration of the longer component, then values lying somewhat above the line defined by Vegard's law are observed (Dorset, 1985a) (Figure 8.32), consistent with earlier x-ray measurements (Mnyukh, 1960). However, if each electron diffraction pattern is measured individually in terms of a local crystal structures, another picture begins to emerge (Dorset, 1987d). That is, the space-group symmetry of a solid solution is not found to remain invariant with changes in concentration, despite Kitaigorodskii's predictions. In fact, for a nominal concentration, local domains may pack as an average odd-chain structure while contiguous areas can also pack as the longer or shorter even-chain structure (Figure 8.32). The small chain movements needed to realize this difference in unit cell symmetry, require only a longitudinal shift by one methylene unit (Figure 8.33). The lamellar spacings of these localized structures also match closely the values expected for the pure paraffin corresponding to the measured average chain number. It is clear, therefore, that an

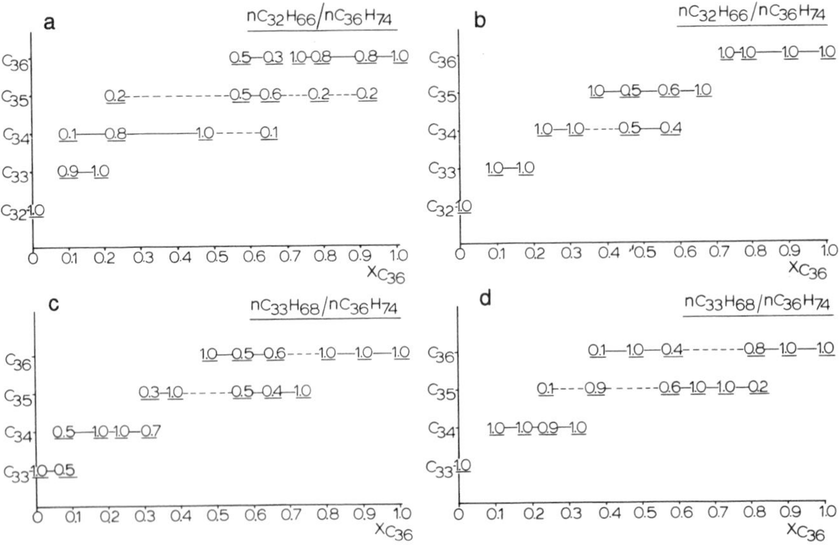

FIGURE 8.32. Local crystal structures for concentration series of n-$C_{32}H_{66}$/n-$C_{36}H_{74}$ and n-$C_{33}H_{68}$/n-$C_{36}H_{74}$. The local structure is obtained from the indices of individual $0kl$ electron diffraction patterns according to the rule outlined in Figure 8.15. As can been seen, several crystal structures can coexist for any nominal bulk concentration and these structures can mimic odd– or even–paraffin lamellar packings. (Reprinted from D. L. Dorset (1987) "Role of symmetry in the formation of n-paraffin solid solutions," *Macromolecules* **20**, 2782–2788; with kind permission of the American Chemical Society.)

FIGURE 8.33. How an average odd-chain paraffin layer can be formed by even-chain components. Perfect odd and even chain packings are represented in (a) and (b). In (c), an even chain translates by one methylene to form an average odd layer. (Reprinted from D. L. Dorset (1987) "Role of symmetry in the formation of n-paraffin solid solutions," *Macromolecules* **20**, 2782–2788; with kind permission of the American Chemical Society.)

average layer spacing larger than the value predicted by Vegard's law is not entirely due to the dominance of the crystal packing by the longer component of the binary solution, as originally claimed (Kitaigorodskii et al., 1958). Rather, the deviation from Vegard's law was actually found to be an ensemble average over a number of step functions defined by the local layer structures (Figure 8.34). In addition, some shortening of the longer component by a kink defect was required to accommodate it into smaller lamellar repeats.

The odd-chain-type lamellar packing from two even chains (0.23 n-$C_{24}H_{50}$/0.77 n-$C_{26}H_{54}$) was again demonstrated recently in the first quantitative x-ray crystal structure of a paraffin solid solution (Gerson and Nyburg, 1994). This determination arrived at many of the conclusions reached earlier in our electron diffraction analyses. One difference, however, was that the average layer structure in the x-ray determination was actually larger (n-$C_{27}H_{56}$) than that of the longest chain component, something never observed in electron diffraction studies of binaries grown from the melt.

In summary, it is found that the layer symmetry (including the methylene subcell and its orientation in the layer) is a more important consideration for the stability of paraffin solid solutions than is unit cell symmetry. On the other hand, the volume rules formulated by Kitaigorodskii are useful, although relative values change somewhat with average chain length (Dorset, 1990c), approximately according to the line given by Mathieson and Smith (1985) (Figure 8.35). For example, the experimentally characterized (Zhang and Dorset, 1990) stable solid solution n-$C_{50}H_{102}$/n-$C_{60}H_{122}$ would correspond to a parameter $\varepsilon = 0.80$.

The stability of n-paraffin solid solutions has been observed to persist for multicomponent "wax" mixtures. For example, single-crystal electron diffraction patterns have been obtained (Figure 8.36a) from epitaxially oriented "Gulfwax," a commercial paraffin wax extract from petroleum (Dorset, 1987d). Evaluation of average lamellar structures from these microcrystals arrives at average chain lengths

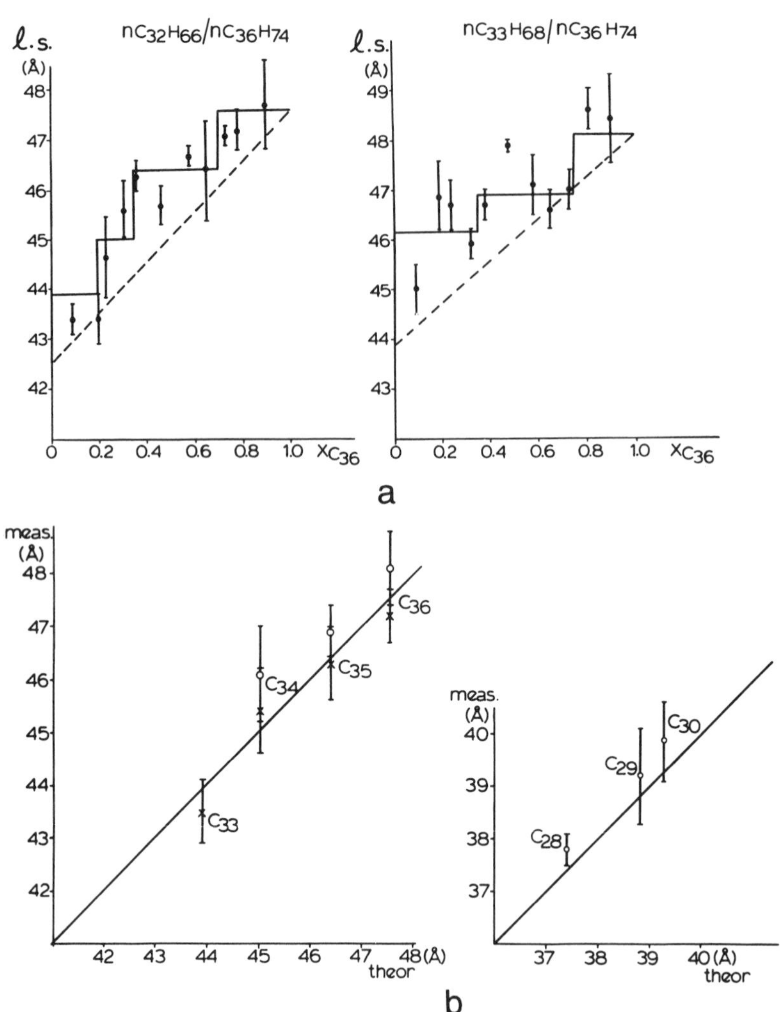

FIGURE 8.34. (a) Plot of individual lamellar spacing for paraffin solid solutions at nominal concentrations. (b) When the solid solution layer packing appears to mimic a certain pure paraffin, its lamellar thickness is also close to that of the pure alkane. On the left are apparent structures from n-$C_{32}H_{66}$/n-$C_{36}H_{74}$ solid solutions and on the right are those from localized crystallites of a "Gulfwax" preparation. (Reprinted from D. L. Dorset (1987) "Role of symmetry in the formation of n-paraffin solid solutions," *Macromolecules* **20**, 2782–2788; with kind permission of the American Chemical Society.)

consistent with an earlier powder x-ray analysis (Chichakli and Jessen, 1967). A quantitative structure analysis has been carried out with similar intensity data from microcrystals of a diesel wax. Excellent diffraction patterns have also been obtained from a "flat" (i.e., equimolar) distribution of even alkanes from C_{26} to C_{36} (Figure 8.36b). Several artificial Gaussian distributions of chain components have also been

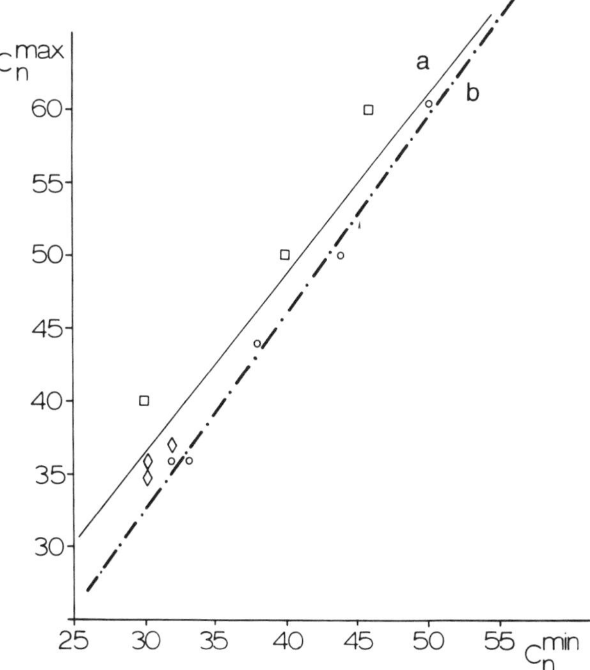

FIGURE 8.35. Experimental stability domain for solid solution in terms of C_n^{min}/C_n^{max}. A fitted line of $C_n^{max} = 1.33\ C_n^{min} - 7.32$ is found to lie somewhat to the right of the line determined earlier by Matheson and Smith (1985) in their study of shorter chain alkane binaries. (Reprinted from D. L. Dorset (1990) "Chain length and the co-solubility of n-paraffins in the solid state," *Macromolecules* **23**, 623–633; with kind permission of the American Chemical Society.)

examined. Single crystals of a wax from a child's birthday candle have been grown by epitaxial orientation (Figure 8.36c). In the latter examples, the amount of overall crystal ordering, despite the interlamellar disorder (expressed again by the attenuated $00l$ reflections near the unscattered beam), is quite striking. This means that there is a strong correlation between layers.

Crystal structure determinations have been carried out for the waxes depicted in Figures 8.36b,c. Both crystallize in space group $A2_1am$, characteristic of an overall odd-chain orthorhombic structure (see above). The "flat" wax distribution of even-chain components crystallizes most frequently as an average n-$C_{33}H_{68}$ structure. The direct phase determination is outlined in Table 8.8. From the first potential map, it was clear that fractional chain occupancies were required for the outer carbons. Using a Gaussian distribution similar to the one employed for the binary solid, n-$C_{32}H_{66}$/n-$C_{36}H_{74}$, except for a smaller weight at the terminal methyl, a reasonable match was found to the observed data (Table 8.9). It will be shown elsewhere (Dorset, 1995) that a correction for secondary scattering improves this fit. Analysis of the birthday candle wax structure (most frequently expressed as an average n-$C_{29}H_{60}$ lamella) is outlined

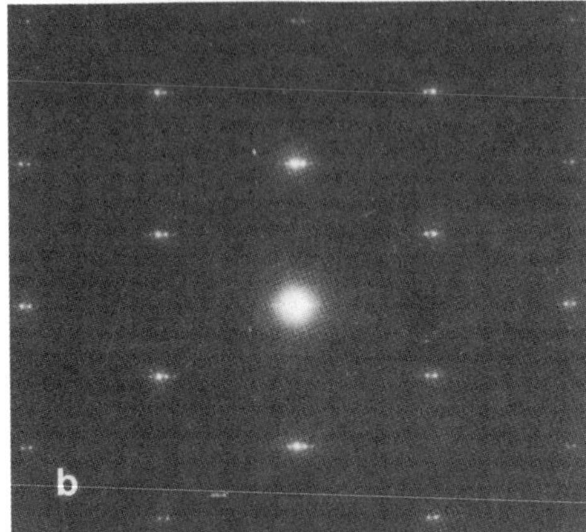

FIGURE 8.36. Electron diffraction pattern from epitaxially oriented multicomponent waxes: (a) "Gulfwax"; (reprinted from D. L. Dorset (1987) "Role of symmetry in the formation of n-paraffin solid solutions," *Macromolecules* **20**, 2782–2788; with kind permission of the American Chemical Society); (b) "flat" wax composed of equimolar amounts of all even chain paraffins from C_{26} to C_{36}; (c) child's birthday candle wax.

Table 8.8. Direct Phase Determination for a "Flat" Multicomponent Paraffin Solid Solution

origin definition

let $\phi_{01,35} = \pi/2$

ambiguities:

let $\phi_{00,70} = a$, $\phi_{00,68} = b$

Σ_1-triples:

$\phi_{020} = \pi$, $\phi_{040} = 0$

Σ_2-triples:

$\phi_{01,103} = b + \pi/2$, $\phi_{01,105} = -\pi/2$

$\phi_{02,68} = b + \pi$, $\phi_{02,70} = \pi$

$\phi_{03,33} = b + \pi/2$, $\phi_{03,35} = -\pi/2$, $\phi_{03,103} = b - \pi/2$, $\phi_{03,105} = -\pi/2$

$\phi_{04,68} = b$, $\phi_{04,70} = 0$

$\phi_{05,35} = \pi/2$

solution at $a = \pi$, $b = 0$, for example

Table 8.9. Observed and Calculated Structure Factors for Multicomponent "Flat" Wax

0*kl*	$\lvert F_o \rvert$	$\lvert F_c \rvert$
002	0.81	0.85
004	0.47	0.45
006	0.25	0.37
00,66	0.24	0.23
00,68	0.57	0.99
00,70	0.69	0.79
00,72	0.20	0.21
01,31	0.25	0.27
01,33	0.47	0.69
01,35	0.87	1.33
01,37	0.25	0.14
01,103	0.18	0.15
01,105	0.18	0.04
020	1.92	2.08
022	0.35	0.23
024	0.29	0.12
02,66	0.22	0.09
02,68	0.46	0.42
70	0.44	0.34
03,31	0.18	0.16
03,33	0.47	0.43
03,35	0.80	0.86
03,37	0.22	0.09
03,103	0.18	0.18
03,105	0.18	0.06
040	0.25	0.31
042	0.20	0.03
044	0.18	0.02
04,68	0.26	0.08
04,70	0.30	0.06
05,33	0.16	0.21
05,35	0.22	0.43

Table 8.10. Direct Phase Determination for a Wax from a Child's Birthday Candle

origin definition:
let $\phi_{01,31} = \pi/2$

ambiguities:
let $\phi_{00,60} = a$, $\phi_{00,58} = b$

Σ_1-triples:
$\phi_{020} = \pi$, $\phi_{040} = 0$

Σ_2-triples:
$\phi_{002} = a - b$, $\phi_{01,29} = a - \pi/2$
$\phi_{02,58} = b + \pi$, $\phi_{02,60} = a + \pi$
$\phi_{03,29} = a + \pi/2$, $\phi_{03,31} = -\pi/2$
$\phi_{05,29} = a - \pi/2$, $\phi_{05,31} = \pi/2$

solution at, e.g., $a = \pi$, $b = 0$

in Table 8.10. Here, only the final three carbons at either end of the molecule were assigned nonunitary occupancies. The agreement to the measured data is again reasonably good (Table 8.11).

8.2.4.2. Binodal Phase Boundary. When the volume increment between the two chains was increased, it was possible to form a metastable solid solution from the melt which, in time, phase-separated in the solid state. Electron diffraction patterns

Table 8.11. Final Observed and Calculated Structure Factor Magnitudes for Candle Wax Determination

$0kl$	$\lvert F_o \rvert$	$\lvert F_c \rvert$
002	0.85	1.25
004	0.79	1.03
006	0.68	0.73
008	0.46	0.42
00,58	0.83	0.20
00,60	0.75	0.97
01,25	0.26	0.18
01,27	0.37	0.34
01,29	0.71	0.79
01,31	1.13	1.36
020	2.39	2.13
022	0.30	0.33
024	0.20	0.27
02,58	0.35	0.08
02,60	0.36	0.38
03,27	0.18	0.19
03,29	0.44	0.45
03,31	0.68	0.79
040	0.36	0.32
042	0.19	0.05
05,29	0.23	0.20
05,31	0.30	0.36

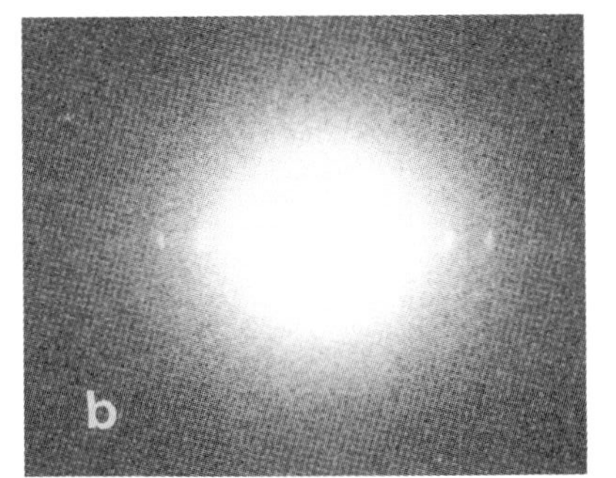

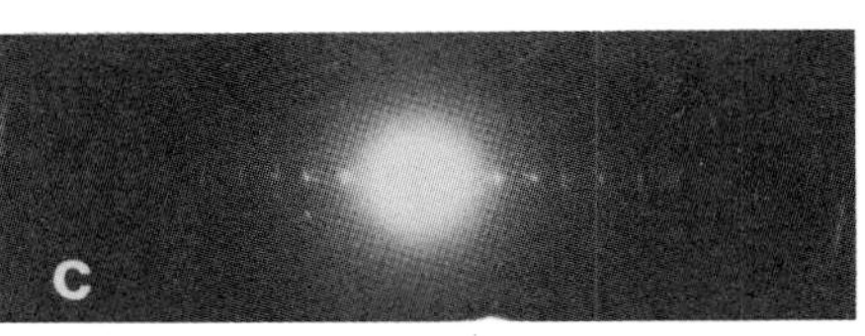

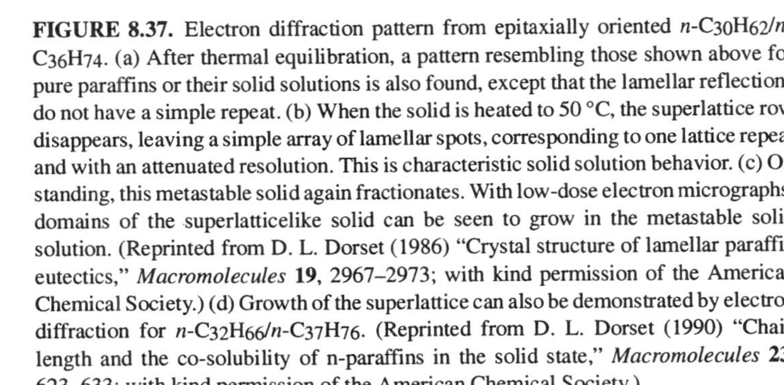

FIGURE 8.37. Electron diffraction pattern from epitaxially oriented n-$C_{30}H_{62}$/n-$C_{36}H_{74}$. (a) After thermal equilibration, a pattern resembling those shown above for pure paraffins or their solid solutions is also found, except that the lamellar reflections do not have a simple repeat. (b) When the solid is heated to 50 °C, the superlattice row disappears, leaving a simple array of lamellar spots, corresponding to one lattice repeat and with an attenuated resolution. This is characteristic solid solution behavior. (c) On standing, this metastable solid again fractionates. With low-dose electron micrographs, domains of the superlatticelike solid can be seen to grow in the metastable solid solution. (Reprinted from D. L. Dorset (1986) "Crystal structure of lamellar paraffin eutectics," *Macromolecules* **19**, 2967–2973; with kind permission of the American Chemical Society.) (d) Growth of the superlattice can also be demonstrated by electron diffraction for n-$C_{32}H_{66}$/n-$C_{37}H_{76}$. (Reprinted from D. L. Dorset (1990) "Chain length and the co-solubility of n-paraffins in the solid state," *Macromolecules* **23**, 623–633; with kind permission of the American Chemical Society.)

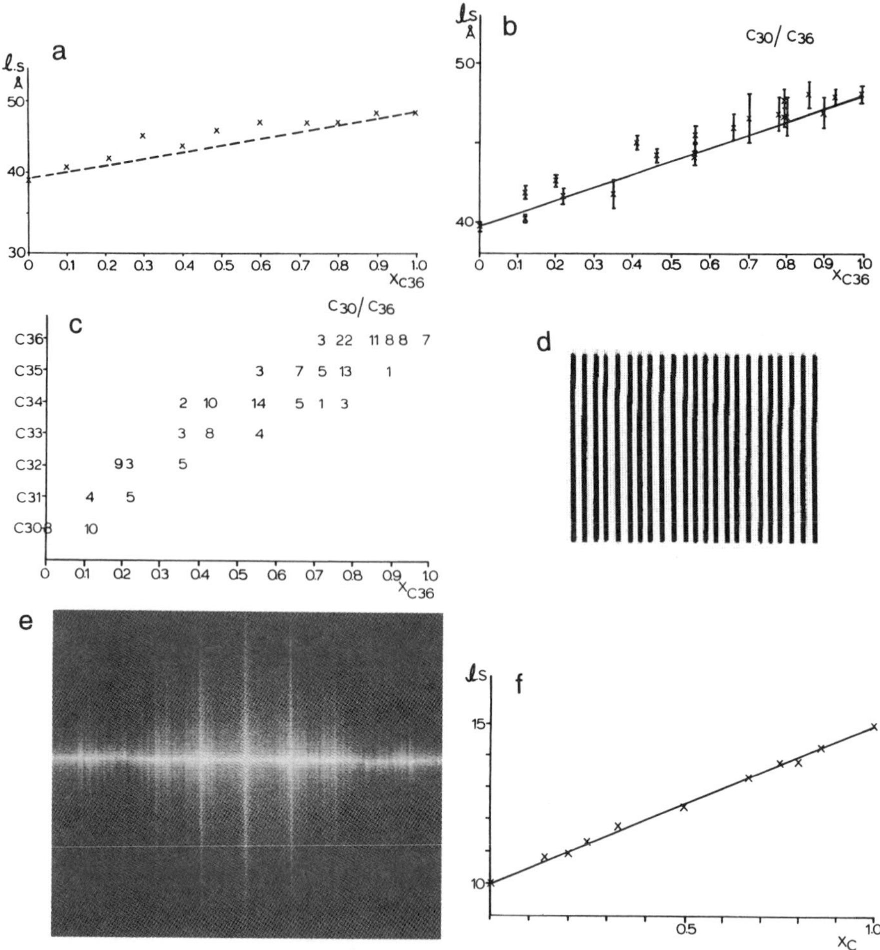

FIGURE 8.38. A plot of average lamellar spacing for n-$C_{30}H_{62}$/n-$C_{36}H_{74}$ (based on the most intense lamellar spots) lies closer to the Vegard's law line than was found for stable solid solutions: (a) x-ray diffraction spacings; (b) electron diffraction spacings; (c) Local crystal structures found from indices of electron diffraction patterns. (d) A random alternation of long and short lamellae accounts for (e) the superlatticelike pattern and (f) the spacings based on the crude lamellar model conform to the experimental observations. (Reprinted from D. L. Dorset (1990) "Chain length and the co-solubility of n-paraffins in the solid state," *Macromolecules* **23**, 623–633; with kind permission of the American Chemical Society.)

(Dorset, 1986b), e.g., from n-$C_{30}H_{62}$/n-$C_{36}H_{74}$ binaries (Figure 8.37) initially resembled those from stable solid solutions. The lamellar reflections corresponded to a single average lattice repeat and were attenuated in resolution, again indicating a disordered lamellar interface. The average spacings often lay closer to the line defined by Vegard's law than was found for true solid solutions (Figure 8.38). Eventually, the lamellar reflections increased in resolution and formed an apparent superlattice repeat (Figure 8.37). Indeed, electron microscope lattice images could be used to monitor the growth of a new phase in the aged solid solution (Dorset, 1990c) (Figure 8.37d). While a model for the sequence of nearly pure lamellae has been formulated to explain this phase separation (Dorset, 1986b), it was clear that the process strongly resembles the formation of incommensurate phase structures, as described for certain inorganic materials (Amelinckx and Van Dyck, 1993b). Additional work to monitor the progress of the phase separation has been undertaken with perdeuterated probes via vibrational spectroscopy (Snyder et al., 1992, 1993) and neutron scattering (White et al., 1990). Curiously enough, studies including the perdeuterated compounds have shown that small adjustments to molecular volume can lead to marked phase separation effects. For example, electron diffraction measurements on epitaxially oriented binaries of n-$C_{30}H_{62}$/n-$C_{36}D_{74}$ or n-$C_{30}D_{62}$/n-$C_{36}D_{74}$ resulted in average lamellar spacings which strongly resembled those from the all-protonated binary. However, plots of n-$C_{30}D_{62}$/n-$C_{36}H_{74}$ lamellar spacings showed that this combination of chains formed a eutectic of solid solutions, similar to the case of n-$C_{30}H_{62}$/n-$C_{40}H_{82}$, and consistent with the experimental phase diagram. This striking isotope effect had nothing to do with any imagined changes in the atom–atom potential function induced by the substitution of hydrogen with deuterium on a paraffin chain. For example, DSC and electron diffraction measurements of binary-phase behavior of deuterocarbon/hydrocarbon binaries, consisting of components with the same chain length, revealed that the solid solution interaction could be expressed by a nearly ideal Raoult's law relationship, even for the transition of the orthorhombic layer packing to a less ordered "rotator" phase (Dorset, 1991e). The isotope effect, therefore, seemed to be merely a consequence of small volumetric differences in the component molecules.

8.2.4.3. Eutectics. Eutectics are formed when the chain length difference becomes too great. First one observes a phase separation of solid solutions, as exemplified by the n-$C_{30}D_{62}$/n-$C_{36}H_{74}$ example given above and others, including n-$C_{30}H_{62}$/n-$C_{40}H_{82}$. The electron diffraction pattern (Dorset, 1986b) from such a solid contains a "superlattice" lamellar spacing along with a spacing from a nearly pure

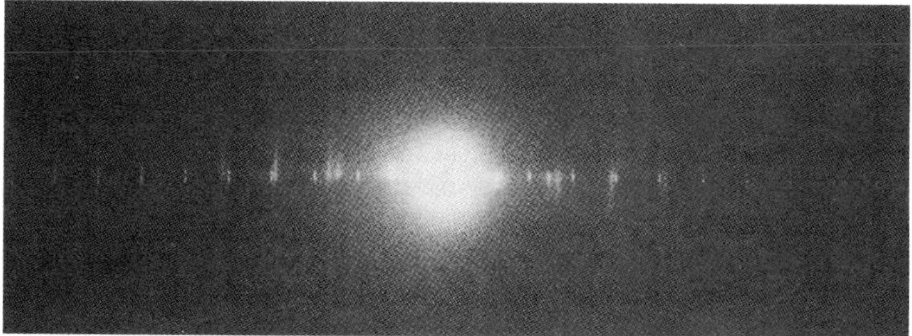

FIGURE 8.39. Lamellar row from a 1:2 combination of n-$C_{30}H_{62}$/n-$C_{40}H_{82}$. A superlattice (arced spots) coexists with a fraction of the pure longer chain component (sharp spots). (Reprinted from D. L. Dorset (1986) "Crystal structure of lamellar paraffin eutectics," *Macromolecules* **19**, 2967–2973; with kind permission of the American Chemical Society.)

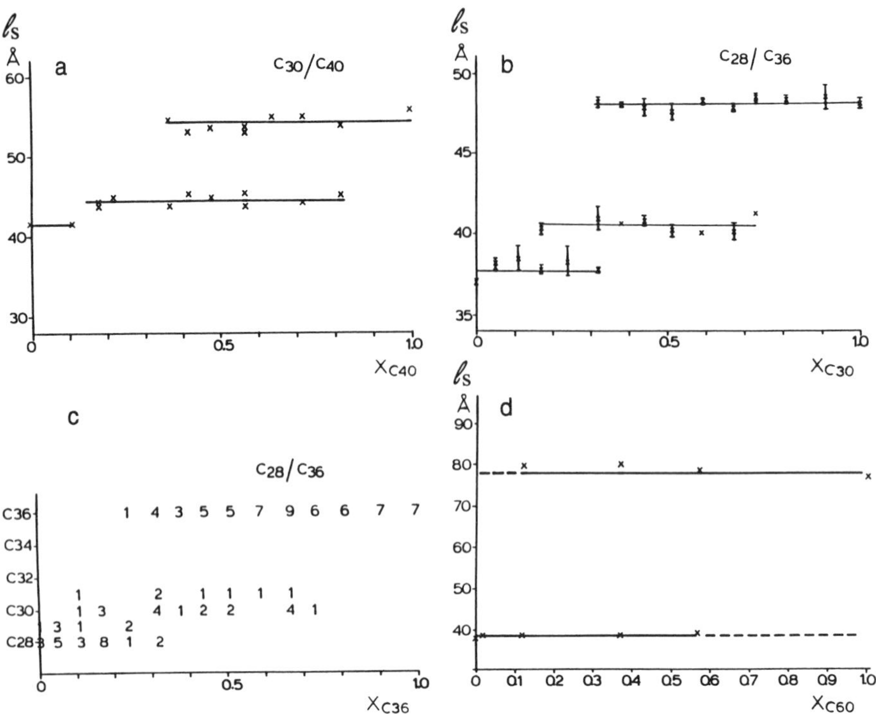

FIGURE 8.40. Plot of electron diffraction lamellar spacings for two fractionated paraffin binaries wherein some cosolubility is retained: (a) n-$C_{30}H_{62}$/n-$C_{40}H_{82}$; (b) n-$C_{28}H_{58}$/n-$C_{36}H_{74}$. In both cases, an intermediate lamellar row is found due to the superlattice repeat in Figure 8.39; (c) the apparent crystal structures of the combination in *b* as measured from electron diffraction patterns; (d) for totally fractionated combinations, e.g., n-$C_{28}H_{58}$/n-$C_{60}H_{122}$ eutectics, no intermediate spacings are found in diffraction patterns. (Reprinted from D. L. Dorset (1990) "Chain length and the co-solubility of n-paraffins in the solid state," *Macromolecules* **23**, 623–633; with kind permission of the American Chemical Society.)

component (Figure 8.39). The apparent superlattice does not change with concentration, unlike the case of the metastable solid solutions, as can be demonstrated by the plot in Figure 8.40. Electron micrographs (Figure 8.41) reveal a sharp interface between the two solid domains, indicating that an exact epitaxy between methyl group surfaces occurs. Evidence for a similar crystallographic boundary for fully phase-separated eutectics has also been seen in light-microscopic observation of sequential crystallization behavior.

To summarize, electron crystallographic studies of paraffin binary combinations have revealed for the first time that local microcrystalline structures are very important for understanding the growth and stabilization of the solid. In general, the progression from a stable solid solution to a fully phase-separated eutectic solid is subtle, involving a continuous sequence of crystal structures (Dorset, 1990d). Hence, the concept of "mechanical mixtures" is often incorrect when dealing with organic eutectics. This is

FIGURE 8.41. Low-dose electron micrograph of an n-$C_{30}H_{62}$/n-$C_{40}H_{82}$ solid. A sharp boundary is found between the superlattice and the pure n-$C_{40}H_{82}$ domain. (Reprinted from W. P. Zhang and D. L. Dorset (1989) "Direct lattice imaging of domain boundaries in n-paraffin binary eutectics formed from the vapor phase," *Proceedings of the 47th Annual Meeting of the Electron Microscopy Society of America*, San Francisco Press, San Francisco, pp. 702–703; with kind permission of the San Francisco Press.)

also exemplified by the mechanism of epitaxial orientation of linear molecules by crystals of other organics—a process understood by a cooling a co-melt into a eutectic solid.

8.2.5. Cycloalkanes

Interest in the crystal structure of cycloalkanes arises largely because they can be models for the chain-folding of polymers such as polyethylene (Newman and Kay, 1967). X-ray structures exist for two members with rather small perimeters, viz.: cyclotetratriacontane, c-$(CH_2)_{34}$, that packs in a unit cell with triclinic symmetry (Kay and Newman, 1968) and also has a methylene subcell with triclinic symmetry (see Chapter 10), and cyclohexatriacontane, c-$(CH_2)_{36}$, that packs in a monoclinic unit cell with methylene groups in a deformed triclinic subcell (Trzebiatowski et al., 1982). Although the structures are, in themselves, interesting for demonstrating how different perimeters can support a regular or distorted fold geometry, neither is particularly relevant as a model for polyethylene, since neither packs in the orthorhombic subcell

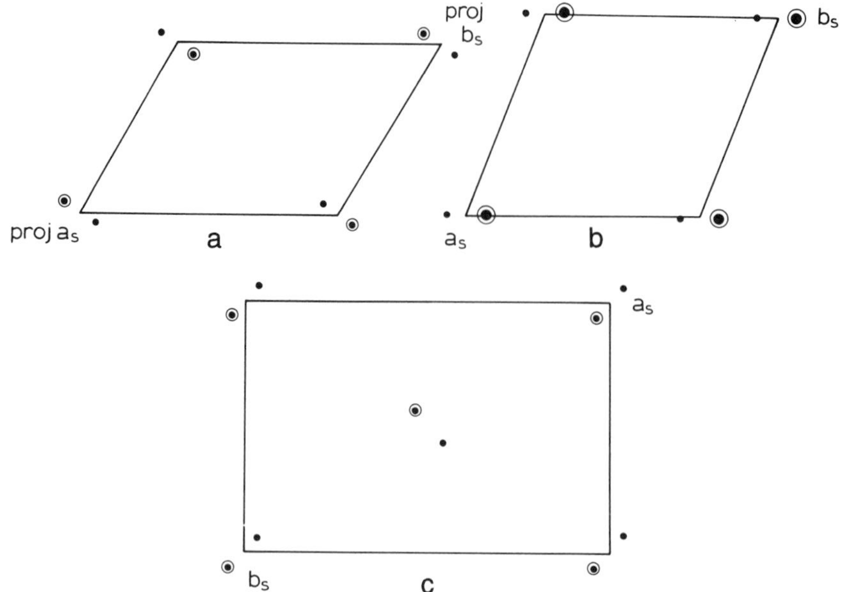

FIGURE 8.42. Polymethylene chain layer packings observed for cycloalkanes: (a) triclinic; (b) monoclinic; (c) orthorhombic. (Reprinted from D. L. Dorset and S. L. Hsu (1989) "Polymethylene chain packing in epitaxially crystallized cycloalkanes: an electron diffraction study," *Polymer* **30**, 1596–1602; with kind permission of Butterworth-Heinemann, Ltd.)

found for the polymer. For this reason, longer-perimeter cycloalkanes were investigated by electron diffraction and electron microscopy.

A qualitative description of polymorphism in higher-molecular-weight cycloalkanes, based on electron diffraction measurements, was reported by Lieser et al. (1988). Later, quantitative studies of cyclooctatetracontane, c-$(CH_2)_{48}$ (Dorset and Hsu, 1989) confirmed the existence of some polymorphic forms found in the earlier study, in particular one with an untilted chain structure packing in a subcell with

Table 8.12. Structure Factors for c-$(CH_2)_{48}$ Based on the $M_{\parallel}$ Subcell

hkl	$\lvert F_o \rvert$	F_c
100	3.11	+3.20
200	1.40	+0.92
010	2.42	+3.49
020	0.70	+0.41
110	1.40	+1.85
1,–1,0	3.11	+3.03
1,–2,0	1.40	+0.27
2,–1,0	1.37	+1.34
2,–2,0	0.84	+0.73

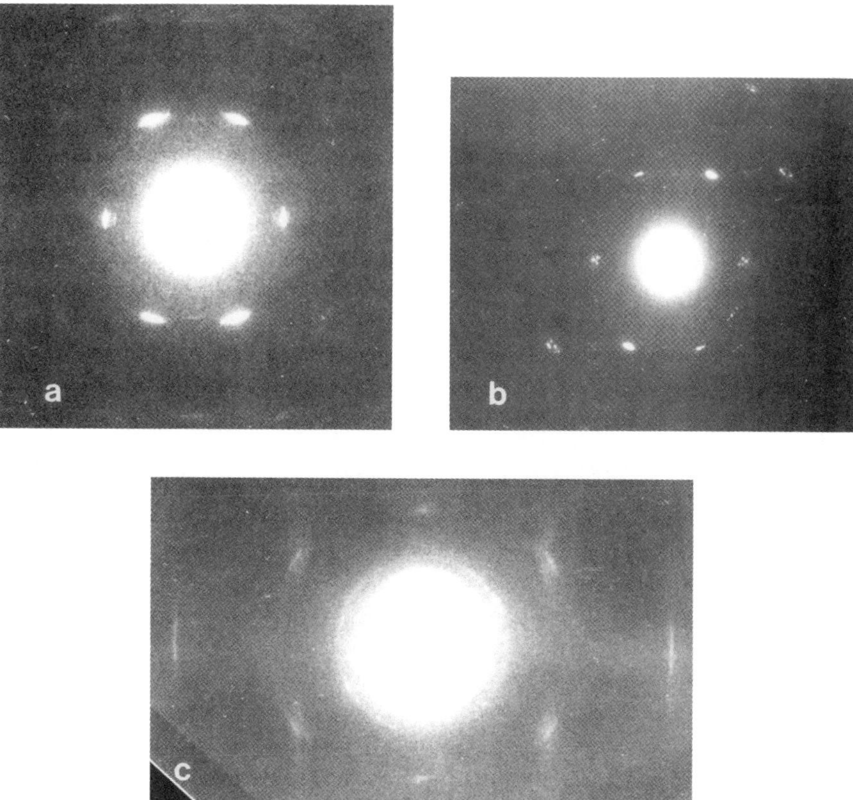

FIGURE 8.43. Electron diffraction patterns from c-$(CH_2)_{48}$: (a, b) from hot xylene solution ($M_{\|}$ subcell); (c) epitaxially oriented ($O_{\perp}$ subcell). (Reprinted from D. L. Dorset and S. L. Hsu (1989) "Polymethylene chain packing in epitaxially crystallized cycloalkanes: an electron diffraction study," *Polymer* **30**, 1596–1602; with kind permission of Butterworth-Heinemann, Ltd.)

parallel chains (Figure 8.42). Since the intensity data (Figure 8.43), by themselves (Table 8.12), could not be used to distinguish between $T_{\|}$ and $M_{\|}$, the latter option was chosen since, according to Kitaigorodskii (1961), it was the only one that could correspond to a rectangular layer packing (see Chapter 10). This choice was corroborated by examination of electron diffraction patterns from epitaxially oriented samples. The larger perimeter cyclodoheptacontane, c-$(CH_2)_{72}$ was found to crystallize in three forms. Initially from solution, the $T_{\|}$ was found in an inclined chain packing. When recrystallized from the melt, either the $M_{\|}$ or a rotator-like phase were seen. Epitaxial crystallization, on the other hand, formed either the $M_{\|}$ structure, or one packing in the $O_{\perp}$ form, as in polyethylene (Figure 8.44). Cycloalkanes with a larger perimeter, e.g., cyclohexanonacontane, c-$(CH_2)_{96}$, could be grown in the $O_{\perp}$ form from the melt (Table 8.13), or, alternatively, the rotator form. However, a nonorthorhombic subcell was again observed when the material was grown from solution. The inference made

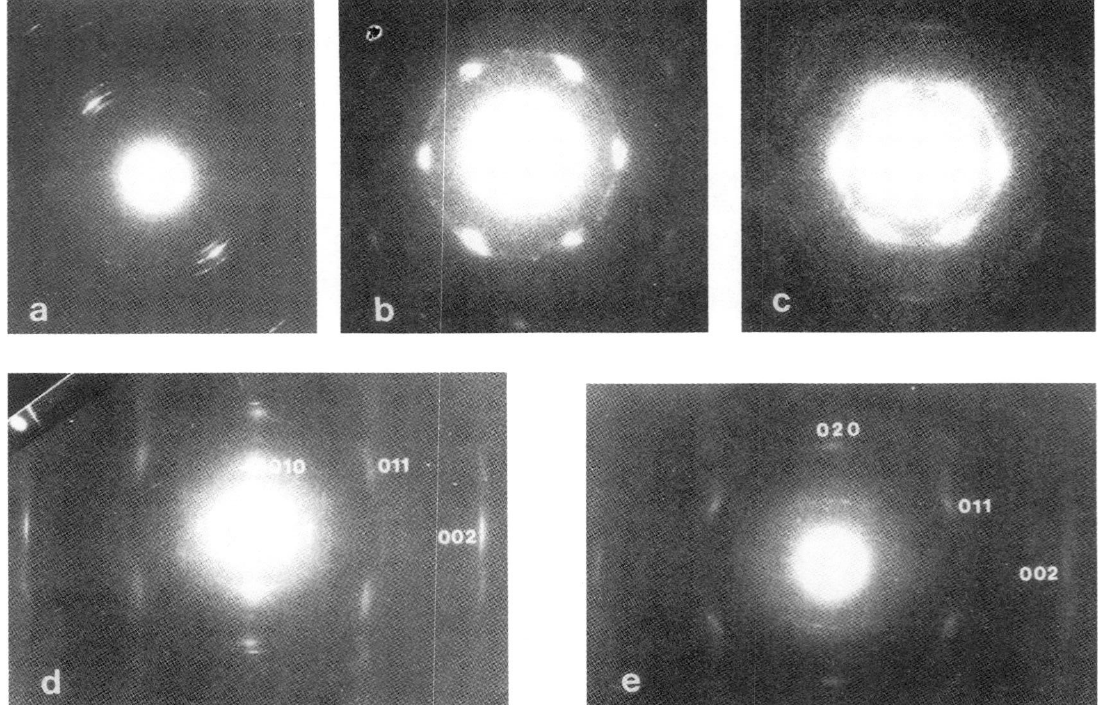

FIGURE 8.44. Electron diffraction patterns from c-$(CH_2)_{72}$: (a) solution-crystallized ($T_{\parallel}$-tilted chains); (b) melt-crystallized ($M_{\parallel}$); (c) heated crystals in rotator phase; (d) epitaxially crystallized ($M_{\parallel}$); (e) epitaxially crystallized ($O_{\perp}$). (Reprinted from D. L. Dorset and S. L. Hsu (1989) "Polymethylene chain packing in epitaxially crystallized cycloalkanes: an electron diffraction study," *Polymer* **30**, 1596–1602; with kind permission of Butterworth-Heinemann, Ltd.)

Table 8.13. Structure Factors for Epitaxially Oriented c-$(CH_2)_{96}$ Compared to the Orthorhombic Perpendicular Subcell

$0kl$	$\|F_o\|$	F_c
002	0.87	−0.73
011	1.12	−1.45
020	1.55	−1.68
031	0.75	−0.68
040	0.32	−0.06

from these studies was that there is an interplay of stem and fold packing for these structures so that, when the folds are regular, a stem methylene subcell packing is required with parallel chain planes. When the fold geometry is loosened, on the other hand, the perpendicular arrangement of adjacent chain planes is permitted in the methylene subcell found for polyethylene. Eventually, with a large enough chain perimeter, e.g., as in c-$(CH_2)_{120}$, the orthorhombic chain packing was found to form characteristic lozenges by solution crystallization (Ihn et al., 1990). Decoration of the crystal surface with polymethylene chain fragments revealed the same sectorization behavior found for polyethylene, and would imply that all folds of the polymer should be oriented in nearly the same direction for each fold sector (see Chapter 11).

8.2.6. Perfluoroalkanes

Preliminary x-ray studies have been reported for the perfluoroalkanes n-$C_{16}F_{34}$ and n-$C_{20}F_{42}$ (Bunn and Howells, 1954; Starkweather, 1986; Schwickert et al., 1991) and the results were compared to what is known about the infinite polymer (Clark and Muus, 1962a,b). Microcrystals had been prepared of the shorter compound (Figure 8.45) and $hk0$ electron diffraction intensity data (Fig. 8.46) were well fit by rigid perfluoromethylene rotor (Table 8.14) (Dorset, 1977b). The hexagonal unit cell axis is $a = 4.96$ Å, with resultant rotor distances, $r_C = 0.408$ Å, $r_F = 1.65$ Å. The structure factor calculation was carried out with the following expression (Vainshtein, 1963):

$$F(\mathbf{s}) = \sum_j f'_j J_0(2\pi\, \mathbf{r}_j \cdot \mathbf{s})$$

where J_0 is a Bessel function and r_j are radii for the various atoms from the chain center. It was possible to improve the fit to the experimental data with an n-beam dynamical scattering calculation (phase grating approximation).

The orthogonal view onto the chain axis was more difficult to crystallize, since typical organic substrates are not wetted by fluorocarbons (Dorset, 1990e). Lathlike crystals of n-$C_{24}F_{50}$ were grown by deposition of the fluorocarbon from the vapor phase onto KCl or NaCl substrates in high vacuum (Zhang and Dorset, 1990). Because of the cubic symmetry of the nucleating surfaces on the salt crystals, two equivalent

FIGURE 8.45. Solution-crystallized sample of n-$C_{16}F_{34}$. (Reprinted from D. L. Dorset (1977) "Perfluoroalkanes: a model for the hexagonal methylene subcell?," *Chemistry and Physics of Lipids* **20**, 13–19; with kind permission of Elsevier Science Ireland Ltd.)

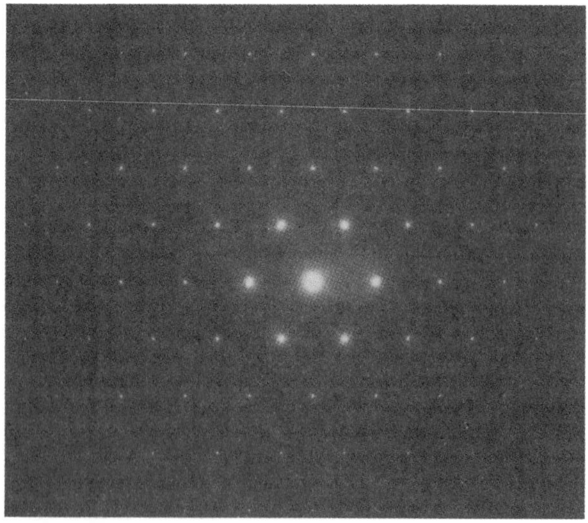

FIGURE 8.46. Perfluoroalkane electron diffraction pattern in a projection down the chain axes. (Reprinted from W. P. Zhang and D. L. Dorset (1990) "Epitaxial growth and crystal structure analysis of perfluorotetracosane," *Macromolecules* **23**, 4322–4326; with kind permission of the American Chemical Society.)

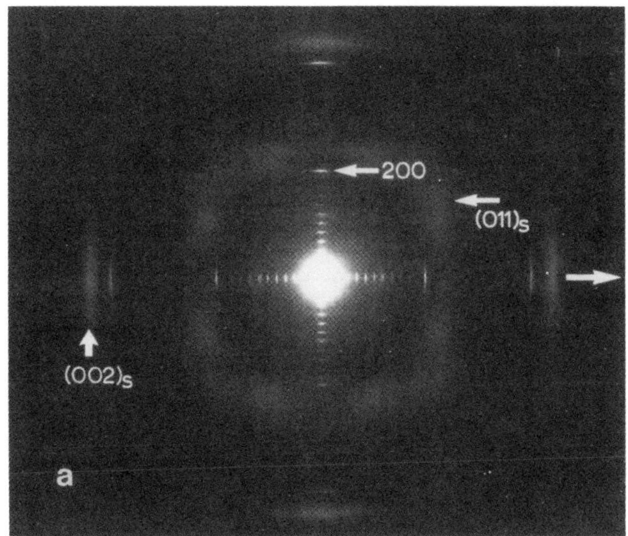

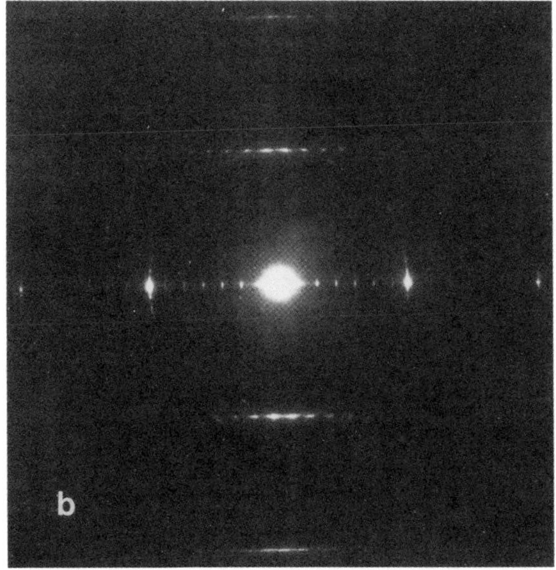

FIGURE 8.47. (a) *h*0*l* (doubled arrangement due to symmetry of epitaxial substrate); (b) *hhl* electron diffraction patterns from *n*-$C_{24}F_{50}$. (Reprinted from W. P. Zhang and D. L. Dorset (1990) "Epitaxial growth and crystal structure analysis of perfluorotetracosane," *Macromolecules* **23**, 4322–4326; with kind permission of the American Chemical Society.)

Table 8.14. Observed and Calculated Structure Factors for Perfluorocetane

$hk0$	$\|F_o\|$	$\|F_{kin}\|$	$\|F_{dyn}\|$
100	2.60	2.53	2.44
200	0.36	0.11	0.31
300	0.49	0.64	0.60
110	0.31	0.23	0.38
210	0.54	0.49	0.47
310	0.31	0.40	0.38
220	0.43	0.48	0.46

crystalline directions were grown, as is evident in the $h0l$ diffraction pattern (Figure 8.47). Tilt of the substrate by 30°, on the other hand, isolated a single hhl pattern (Figure 8.47). Measured unit cell constants were $a = 5.68$, $b = 9.84$, $c = 67.4$ Å (the latter corresponding to a lamellar spacing equivalent to the value given by Starkweather (1986)). The lamellar repeat could also be observed directly in electron micrographs (Figure 8.48). From the average molecular geometry and the systematic absences in the electron diffraction pattern, the space group was reckoned to be *Fmmm*. Using a rotor model for each perfluoromethylene unit, the agreement to the measured data was

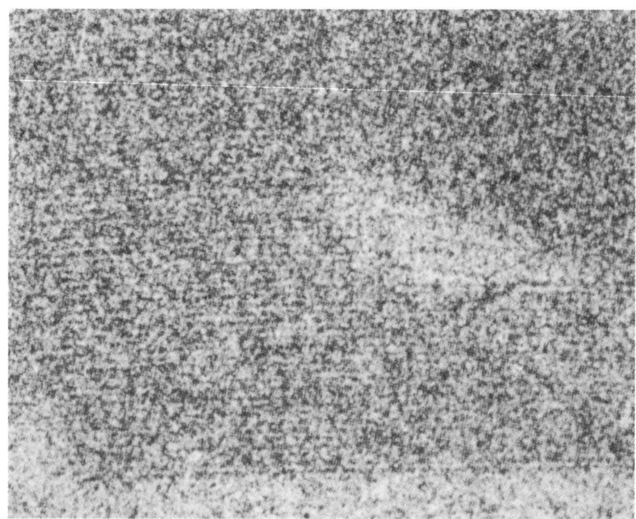

FIGURE 8.48. Low-dose electron micrograph of epitaxially oriented n-$C_{24}F_{50}$ showing lamellar repeat. (Reprinted from W. P. Zhang and D. L. Dorset (1990) "Epitaxial growth and crystal structure analysis of perfluorotetracosane," *Macromolecules* **23**, 4322–4326; with kind permission of the American Chemical Society.)

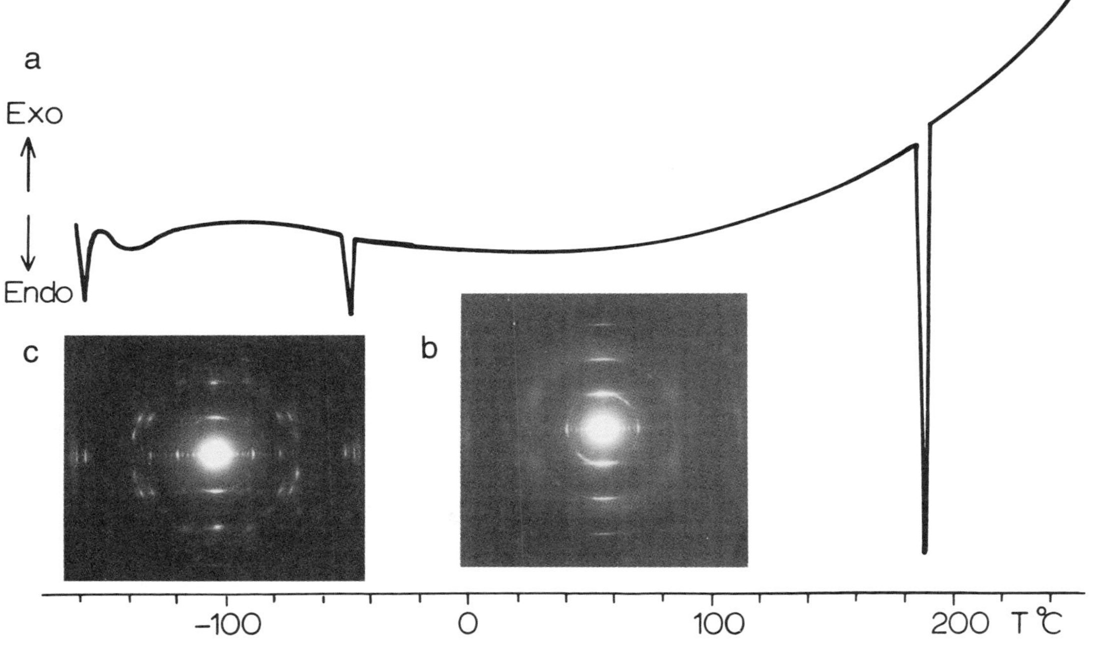

FIGURE 8.49. Subambient phase transition of *n*-$C_{24}F_{50}$ as revealed by electron diffraction of epitaxially oriented samples and correlated to a DSC scan. (Reprinted from D. L. Dorset (1994) "Electron crystallography of organic molecules," *Advances in Electronics and Electron Physics* **88**, 111–197; with kind permission of Academic Press, Inc.)

Table 8.15. Observed and Calculated Structure Factors for Epitaxially-Oriented n-$C_{24}F_{50}$

$0kl$	$\|F_o\|$	F_c
00 2	0.89	0.64
00 4	0.87	–0.63
00 6	0.83	0.60
00 8	0.71	–0.59
00 10	0.62	0.55
00 12	0.51	–0.50
00 14	0.46	0.46
00 16	0.37	–0.41
00 18	0.35	0.35
00 20	0.30	–0.32
00 22	0.27	0.26
00 24	0.25	–0.22
11 1	2.59	4.29
11 3	1.63	–1.42
11 5	1.05	0.84
11 7	0.91	–0.59
11 9	0.48	0.44

$R = 0.27$ (Table 8.15). The structure is obviously disordered and a more highly ordered pattern is observed when the specimen is cooled below –50 °C (Figure 8.49).

8.3. Conclusions

Alkanes were among the first organic materials studied by electron diffraction techniques, also serving as good test samples for detection of many nonideal data perturbations. Before anything new could be discovered about these compounds in the electron microscope, new crystallization methods, such as the epitaxial orientation techniques developed by Wittmann and Lotz (1990), had to be employed. These allowed collection of useful electron diffraction intensity data from the most informative projection onto the molecular chains. Low-dose electron microscopy has increased the detail visualized from all crystal forms to resolutions well beyond the limits "predicted" from the Rose equation (see Chapter 5). Because of these advances, considerable progress has been made in the study of disordered layer packing, as well as the structure of more ordered phases, when single microcrystals are examined in the electron beam. For example, the study of the binary solid state uncovers details not obtainable from the x-ray study of bulk specimens. This is also true for multicomponent paraffin waxes. Single-crystal characterizations of thermotropic changes also lead to a more accurate model for the crystal–crystal transitions of these materials than is possible from earlier x-ray studies.

9

Alkane Derivatives

9.1. Background

The use of electron diffraction to study simple alkane chain derivatives is intimately tied with the early history of the technique, beginning in the 1930s and 1940s. A major motivation for this early work was to understand the nature of boundary lubricant films on metal bearings, as well as the nature of Langmuir–Blodgett films. With lubricants, for example, in situ experiments were carried out with RHEED (Thomson and Murison, 1933), as well as transmission diffraction experiments using thin foils (or other films) as the support surface (Lebedeff, 1931; Garrido and Hengstenberg, 1932; Hengstenberg and Garrido, 1932). In some of this work, provisions were also made to heat the sample to study their phase transitions.

In the study of boundary lubricants, e.g., as described by Andrew (1936), various oily substances were rubbed onto metal plates to determine which would leave the most stable, ordered chain packing in the surface monolayer. Reactive metals were found to give the most satisfactory results, especially if the hydrocarbon was slightly oxidized. Advancing beyond the investigation of impure fatty substances (Rupp, 1934; Storks and Germer, 1937), linear molecules with different functional groups (Trillat and Motz, 1935) were considered, to study their relative activities. This work included heating experiments (Tanaka, 1938, 1940), and the best results were obtained from fatty acids on, e.g., copper surfaces. It was soon realized that heavy metal soaps were being created at the interface, since the thermal transitions observed in electron diffraction experiments occurred at a point much higher than expected for the isotropic organic solid (Tanaka, 1939). If nonreactive metals such as platinum were used, on the other hand, the thermal behavior was not changed (Cowley, 1948). Other RHEED experiments (Spink, 1950; Menter and Tabor, 1951) were accompanied by transmission diffraction experiments (Menter, 1950). Other polar substances such as the fatty alcohols (Saunders and Tabor, 1951) were shown to produce similar results if reactive metals were used as a substrate.

In studies of Langmuir–Blodgett films (Storks and Germer, 1936; Germer and Storks, 1938) carried out in this period, it was possible to distinguish between two projections of a tilted chain packing corresponding to the directions parallel or perpendicular to the dipping direction through the monolayer. Later RHEED and transmission experiments (Germer and Storks, 1938; Havinga and De Wael, 1937a,b; DeWael and Havinga, 1940) established that the first layer picked up from the trough could have a different structure than the layers in a built-up multilayer—i.e., the chain axes in the first layer could be untilted while they would be tilted in subsequent layers. Epstein (1951) investigated the effect of extracting free fatty acid from mixed monolayer of the acid with its barium salt, inducing surface diffusion and to form micellelike surface structures. Similar work on Langmuir–Blodgett films has been revived recently, especially involving unsaturated chains that can be chemically cross-linked into a surface π-electron system (Peterson and Russell, 1984; Peterson, 1987; Peterson et al. 1988).

Particularly important in this early work was the study of polymorphism for these fatty materials. For example, Trillat and Hirsch (1933) characterized a form of stearic acid not yet found in x-ray experiments. Other investigations include the work of Thiessen and Schoon (1937). These studies were also extended to the wax ethers and esters (Natta and Rigamonti, 1935; Natta et al., 1935; Thiessen and Schoon, 1937; Schoon, 1938; Coumoulos and Rideal, 1941). It soon became apparent that only a limited number of alkyl chain packing motifs were possible, leading eventually to the concept of methylene chain "subcells" (to be discussed in Chapter 10), as well as the permissible layer packings (i.e., chain tilts) that will accommodate them. The average chain tilt ψ, if not immediately apparent from the RHEED patterns of single crystals, can be measured by comparing the distance r between the second and fourth orders of the reflection diffraction pattern (see Figure 1.14) to that of an untilted chain layer, as shown by Karle and Brockway (1947). Thus:

$$\psi = \cos^{-1}\,(r/2.55\ \text{Å})$$

More recent application of electron diffraction data for quantitative structure analysis of these alkane derivatives will be discussed in the following section.

9.2. Contemporary Structure Analyses

9.2.1. Fatty Alcohols

In recent studies of fatty alcohols single crystals formed by evaporation of dilute solutions in various solvents, Precht (1976a,b) was able to characterize two principal packing forms, both of which adopt the methylene subcell structure found also for paraffins. One of these forms packs with chain axes normal to the crystal plate. Phased electron diffraction amplitudes from these crystals were used to calculate a potential map that strongly resembles the ones shown in Chapter 8 for paraffins. This form also has been observed in RHEED experiments by Sutula and Bartell (1962). An orthorhombic polymorph of n-tricosanol has also been studied by Li (1963), who corrected the electron diffraction intensity data for dynamical scattering with a two-beam model. The resulting potential map from this analysis again resembles those from orthorhombic paraffins. The second crystal form has a tilted chain structure, where the tilt axis

FIGURE 9.1. Electron diffraction pattern (*hk*0) from a multilamellar crystal of 1-octadecanol. (Reprinted from D. L. Dorset (1979) "Orthorhombic *n*-octadecanol: an electron diffraction study," *Chemistry and Physics of Lipids* **23**, 337–347; with kind permission of Elsevier Science Ireland, Ltd.)

corresponds to the $b = 4.94$ Å axis of the orthorhombic subcell. The chain packing would correspond to the crystal structure determined by Abrahamsson et al. (1960). A form packing in the hexagonal layer characteristic of the "rotator" phase on *n*-paraffins was also described by Frede and Precht (1974), but its appearance may actually be the result of radiation damage, as discussed in an Chapter 5.

A closer look at the untilted chain structure of *n*-alcohols was based on selected area diffraction patterns from *n*-octadecanol (Dorset, 1979). The cell constants measured from the single crystal patterns from multilamellar samples (Figure 9.1) were $a = 7.33$, $b = 5.01$ Å. Note that the intensity of the (110) reflection is lower than that of the (200). This is due to the $\sin \theta/\lambda$ dependence of diffraction from slightly bent crystals (see Chapter 5). As with similar examples from multilamellar paraffin crystals, most of the diffraction pattern appears to originate from a single layer (Table 9.1). As the crystals are thinned by sublimation, the spacing changes to $a = 7.39$ Å and the diffraction intensities appear to originate entirely from a polymethylene monolayer (Table 9.2). It is interesting to note that there is a preferred direction for specimen removal as it sublimes, as seen in the low-magnification electron micrographs. The

Table 9.1. Observed and Calculated Structure Factors for *n*-Octadecanol Multilayer Crystals

hk	$\lvert F_o \rvert$	$\lvert F_c \rvert$ unit cell*	$\lvert F_c \rvert$ subcell
200	4.48	5.52	4.26
400	1.89	0.72	1.96
110	1.96	2.38	4.48
210	0.95	1.00	0.93
310	1.83	1.92	1.37
410	0.84	1.16	0.75
510	0.82	1.31	0.76
020	2.05	3.80	2.60
120	1.25	1.40	0.85
220	1.40	1.07	0.87
320	1.16	0.94	1.05
420	0.75	0.11	0.39
520	0.86	0.14	0.94
130	1.29	0.58	1.34
230	1.27	1.07	1.03
330	0.99	0.51	0.48
430	1.16	1.31	0.91

*Based on the orthorhombic crystal structure of even-chain paraffins.

remaining crystal strands are elongated in a direction corresponding to the unit cell *a*-axis, meaning that this must also be the direction of hydrogen-bonding rows. This is consistent with the findings for the inclined chain crystal structure (Abrahamsson et al., 1960). Thus, the hydrogen-bonding scheme stabilizes the polar group packing of two polymorphs that can interconvert.

Table 9.2. Observed and Calculated Structure Factors for *n*-Octadecanol Monolayer Crystals

hk	$\lvert F_o \rvert$	$\lvert F_c \rvert$ subcell
200	4.24	4.16
400	1.96	1.91
110	3.82	4.37
210	0.79	0.91
310	1.74	1.34
410	0.74	0.73
020	2.20	2.54
120	0.89	0.83
220	1.40	0.85
320	0.93	1.02
420	0.61	0.38
520	0.83	0.92
130	1.18	1.31
230	0.97	1.00
330	0.69	0.47
430	0.64	0.89

9.2.2. Fatty Acids

The fatty acids have often been examined by electron diffraction, as discussed above. RHEED measurements on various linear, even-chain acids by Sutula and Bartell (1962) detected mainly the *C*-polymorph that is inclined around the orthorhombic subcell $b_{sub} \approx 5.0$ Å axis. In our transmission experiments on the fatty acids (Dorset, 1976a, 1983b), we have most often seen patterns from the *B*-form, where the chains are tilted around the subcell $a_{sub} \approx 7.5$ Å axis (Figures 3.7 and 5.13a), although patterns from the *C*-form have also been observed. As mentioned before, this feature can be used to calibrate the tilt-axis position of an electron microscope goniometer stage, when the crystals are tilted by about 27° and then rotated into the proper orientation. The electron diffraction pattern then resembles the ones obtained from untilted orthorhombic paraffins and the observed structure factors agree well with the theoretical values calculated from the untilted polymethylene chain subcell (Table 9.3).

Table 9.3. Structure Factor Calculation for Behenic Acid in Two Orientations, Based on the Orthorhomic Perpendicular Subcell

$(hk0)_{subcell}$	$\|F_o\|$	$\|F_c\|$	$\|F_{dyn}\|$
27° tilt projecting down the chain axes:			
200	2.12	2.86	2.69
400	1.05	1.11	1.00
600	0.31	0.24	0.20
110	2.28	3.03	2.81
210	0.95	0.60	0.58
310	1.10	0.83	1.02
410	0.62	0.41	0.40
510	0.52	0.37	0.36
610	0.27	0.27	0.27
020	1.22	1.63	1.54
120	0.69	0.52	0.50
220	1.01	0.51	0.79
320	0.72	0.58	0.56
420	0.45	0.20	0.38
520	0.42	0.42	0.45
130	0.62	0.71	0.66
230	0.51	0.52	0.51
330	0.23	0.23	0.30
430	0.36	0.39	0.45

hkl	$\|F_o\|$	$\|F_c\|$
untilted crystal:		
020	4.96	5.95
040	1.62	1.91
060	0.39	0.20
120	0.48	0.12
130	0.30	0.27
140	0.37	0.12
150	0.30	0.07
510	1.11	0.97
530	0.66	0.58

Attempts have been made to solve the structure of soaps from observed electron diffraction intensities. For example, Stephens and Tuck-Lee (1969) measured selected area patterns and powder x-ray diffraction films from lead stearate multilayers. From the observed single crystal data, they guessed that the chains would pack in oblique layer, even though the electron diffraction patterns clearly resembled those from untilted orthorhombic paraffins. More recently, a correct analysis of this structure was made by Vainshtein and Klechkovskaya (1993), who interpreted Patterson functions generated from three-dimensional texture electron diffraction intensities to justify a structural model with untilted chain layers.

9.2.3. Ketoalkanes

Samples of palmitone, 15-ketohentriacontane, were crystallized onto potassium hydrogen phthalate or KCl substrates from the vapor phase to obtain an epitaxial orientation (Zhang and Dorset, 1989). If this crystallization was carried out at room temperature, only diffraction patterns with "polyethylene" reflections were seen, as was also found for the growth of n-$C_{50}H_{102}$ and n-$C_{60}H_{122}$ from the vapor phase (see Chapter 8). The layer growth was achieved for this ketone after annealing but, unlike the case of paraffins, the orientation of the crystal was along [110] so that tilts around the c^* axis would access the $0kl$ and $h0l$ diffraction patterns. Representative zonal patterns are shown in Figure 9.2. The unit cell constants are $a = 7.56$, $b = 4.93$, $c = 82.8$ Å. From systematic absences, the space group was found to be $A2_1am$, consistent with the structure of an odd-chain alkane (*not* the $Pca2_1$ claimed in the original paper, which would be found for only the even-chain alkane layers). When cleaved KCl crystals were used as a substrate, the projection was [120]. It was apparent that the keto group had some effect on the nucleation of this molecule, which, otherwise, packed as if it were an n-paraffin, consistent with earlier findings for a shorter-chain homolog (Malta et al., 1974). This conclusion was also supported by the agreement of the layer spacing with that of the unsubstituted hydrocarbon.

9.2.4. Wax Esters

Symmetric esters synthesized from fatty acids and fatty alcohols of the same chain length have also been studied by electron diffraction since the early years. RHEED measurements by Thiessen and Schoon (1937) on cetyl palmitate and those by Sutula and Bartell (1962) on this and other esters, yielded cell constants essentially in agreement with the early x-ray analysis of Kohlhaas (1938). Transmission electron diffraction measurements made at accelerating voltages from 100 kV to 900 kV seemed to indicate that two polymorphic forms occur for cetyl palmitate (Dorset, 1976e). Measured cell constants from untilted monoclinic crystals were $a = 5.59$, $b = 7.46$ Å. Plots of intensities along a reciprocal lattice row obtained from a tilt series also were in accord with a lamellar spacing consistent with Kohlhaas' value of 77.9 Å. Tilt of the crystals around the unit cell b-axis by 30° yielded a pattern similar to the orthorhombic form of alkanes and the intensity data accordingly were well matched by the polymethylene lattice model (Table 9.4). (In terms of chain packing, the electron diffraction study of ethylene di(11-bromoundecanoate) arrived at similar results with an *R*-factor of 0.20 in the projection down the chain axes (Dorset, 1983b). Clearly, the bromine scattering contribution was not seen because of the crystal bending.) Continuous diffuse scattering patterns from the untilted crystals had also been observed and they could be well explained by a thermal motion model, if correlations of all the unit cell contents were considered (Dorset, 1978b) (Figure 9.3).

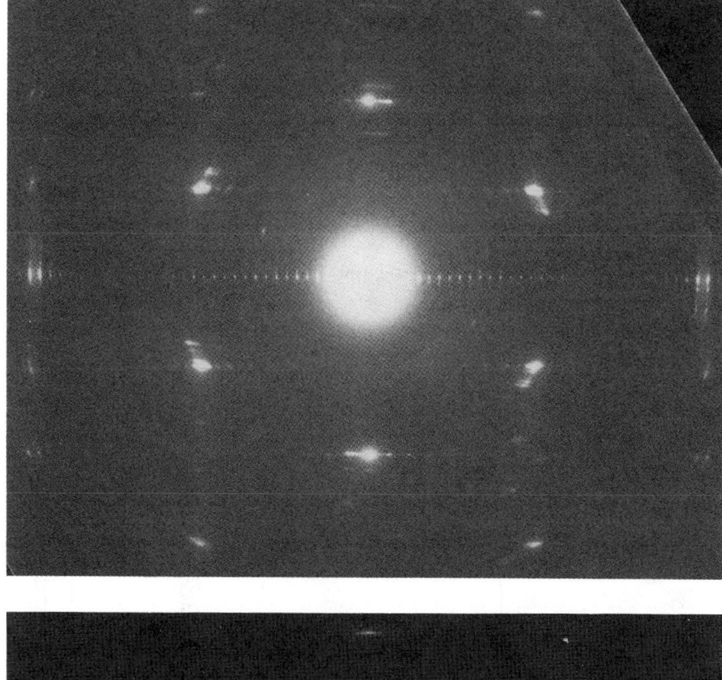

FIGURE 9.2. Electron diffraction patterns from epitaxially oriented 15-ketohentriacontane: (a) 0*kl* data; (b) *hhl* data.

Table 9.4. Observed and Calculated Structure Factors for Cetyl Palmitate Monoclinic Crystals

hk	$\lvert F_o \rvert$	$\lvert F_c \rvert$ subcell	$\lvert F_{dyn} \rvert$
tilted 30° around *b*-axis			
200	4.07	5.44	5.05
400	2.26	2.12	1.89
600	0.99	0.46	0.37
110	4.50	5.77	5.29
210	1.06	1.15	1.08
310	2.10	1.58	1.91
410	0.80	0.79	0.76
510	1.17	0.71	0.67
610	0.47	0.52	0.51
710	0.59	0.27	0.35
020	2.08	3.11	2.89
120	0.57	1.00	0.93
220	1.91	0.98	1.49
320	1.08	1.11	1.05
420	1.13	0.37	0.71
520	0.66	0.80	0.84
620	0.50	0.04	0.19
720	0.42	0.45	0.49
130	1.15	1.36	1.23
230	0.73	1.00	0.95
330	0.76	0.44	0.56
430	0.84	0.50	0.85
530	0.50	0.39	0.29

hk	$\lvert F_o \rvert$	$\lvert F_c \rvert$	$\lvert F_{dyn} \rvert$
untilted crystals:			
020	4.20	6.20	4.12
040	1.98	1.99	1.80
060	0.73	0.21	1.03
110	0.71	0.58	0.36
120	0.00	0.13	0.07
130	0.46	0.28	0.33
140	0.41	0.12	0.10
150	0.53	0.07	0.28
160	0.00	0.07	0.10
510	1.70	1.01	0.71
520	0.00	0.01	0.01
530	0.91	0.60	0.59
540	0.00	0.02	0.03
550	0.68	0.20	0.48

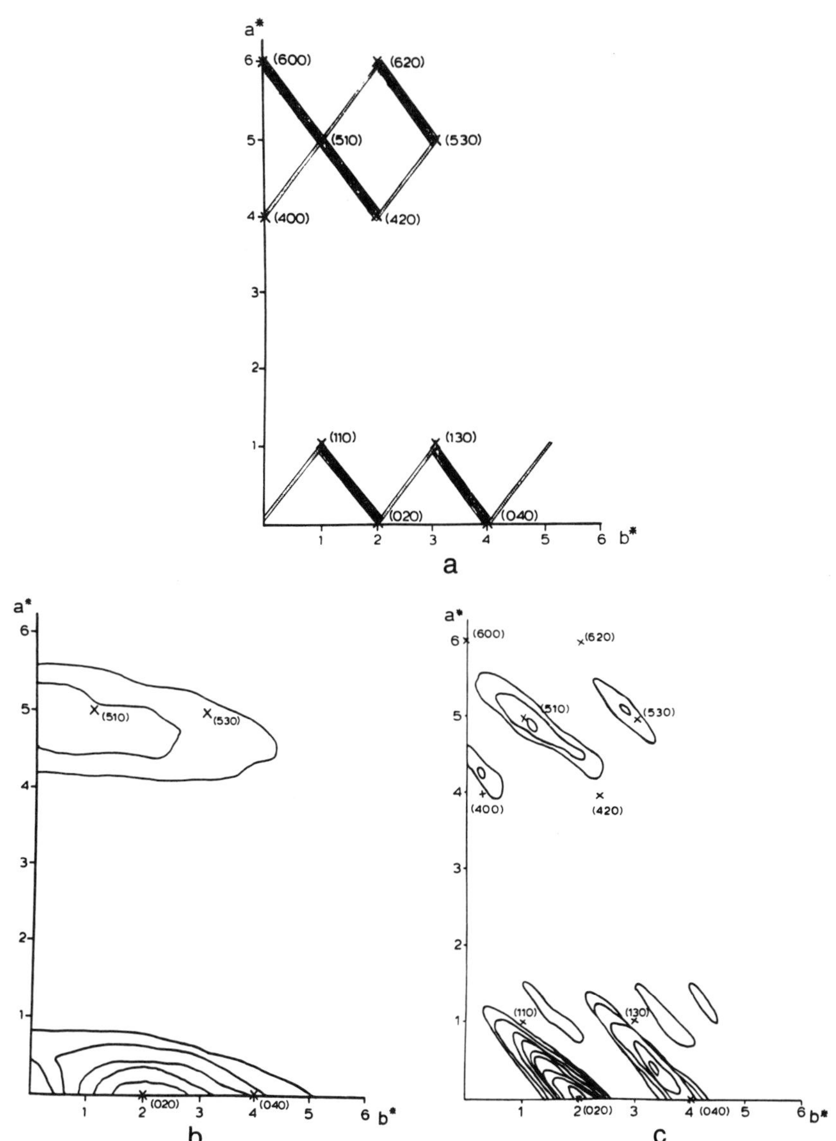

FIGURE 9.3. (a) Schema of observed diffuse diffraction in *hk*0 patterns from monoclinic cetyl palmitate. (b) Diffuse scattering model where all molecules are uncorrelated. (c) Diffuse scattering model based on correlated molecules. (Reprinted from D. L. Dorset (1978) "Continuous diffuse electron scattering from polymethylene compounds. II. Oblique layer crystals," *Zeitschrift für Naturforschung, Teil A* **33a**, 1090–1092; with kind permission of the Verlag der Zeitschrift für Naturforschung.)

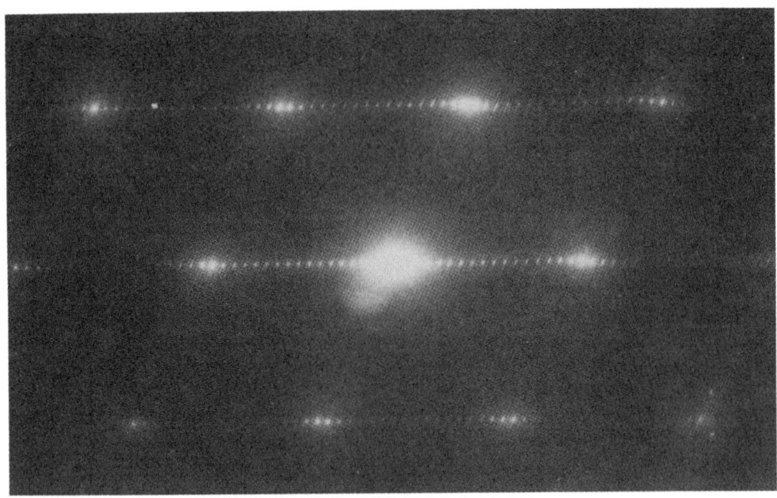

FIGURE 9.4. Electron diffraction pattern from epitaxially oriented stearyl myristate.

Stearyl myristate was also epitaxially oriented on benzoic acid to give the [010] projection of the unit cell with constants $a = 5.60$, $b = 7.40$, $c = 86.20$ Å, $\beta = 115.0°$, consistent with the monoclinic polymorph of the symmetric esters (Figure 9.4). Its crystal structure is currently being determined from the electron diffraction intensities but many features of the chain packing can be seen already in the initial potential maps (Figure 9.5). A preliminary structure analysis has also been carried out for epitaxially crystallized myristyl stearate, which favors an orthorhombic form, with cell constants $a = 7.63$, $b = 4.98$, $c = 87.8$ Å, in space group $A2_1am$ (Zhang et al., 1989a). Before annealing, the as-grown crystals were disordered, evidenced by streaks in the $0kl$ diffraction pattern. These streaks were attributed to stacking faults, as justified by an optical transform of a chain packing model.

A putative second polymorph observed for cetyl palmitate (Dorset, 1976e) appeared to be orthorhombic but was found only for crystals grown from hot solvent.

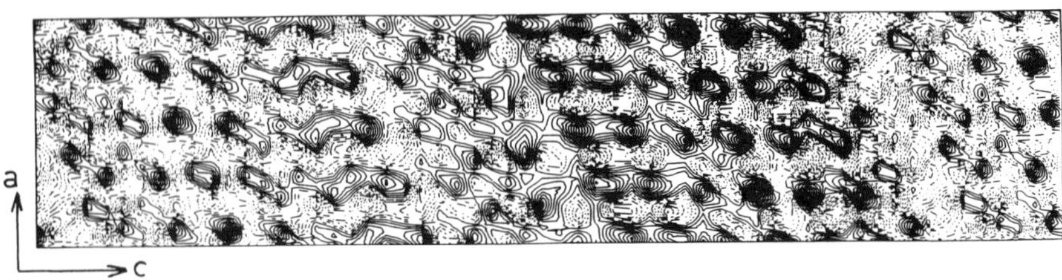

FIGURE 9.5. Initial structure model for stearyl myristate (half unit cell) determined from electron diffraction intensities measured from Figure 9.4.

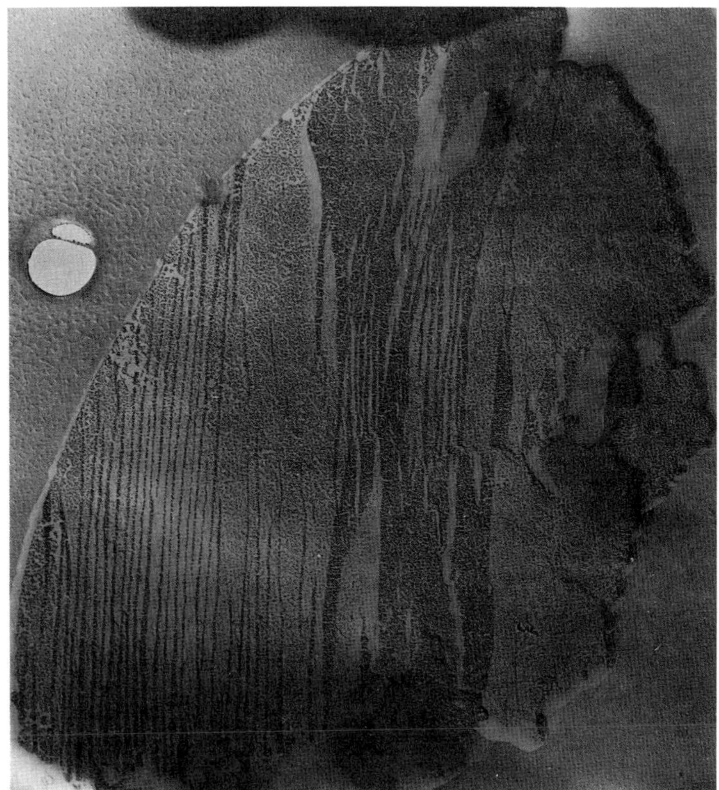

FIGURE 9.6. Gold-shadowed microcrystal of cetyl palmitate grown from hot solvent, showing "rooflike" corrugations. (Reprinted from D. L. Dorset (1978) "Transmission electron diffraction intensities from real organic crystals: thin plate microcrystals of paraffinic compounds," *Zeitschrift für Naturforschung, Teil A* **33a**, 964–982; with kind permission of the Verlag der Zeitschrift für Naturforschung.)

The intensity data again agreed with the untilted polymethylene subcell structure. Although orthorhombic forms are known for unsymmetric wax esters (Aleby et al., 1970), consistent with our single-crystal observations on myristyl stearate, only the oblique packing is predicted for the symmetric members because of the bulk of the ester group in the center of the molecule, the cross section of which must be compensated by crystal tilt. It was therefore decided to test the correctness of this assumption by attempting to form solid solutions of this ester with the paraffin n-$C_{36}H_{74}$. There is very little difference in molecular volume for the two molecules, and since, presumably, both pack in similar polymorphic forms, it would seem likely that such a solid could be prepared. However, it is readily seen from examination of the "orthorhombic" cetyl palmitate crystal habit that "roof" structures similar to those from heated paraffin crystals are being grown from hot solution (Dorset, 1978a) (Figure 9.6). Since the impure paraffin prefers an orthorhombic structure, the comixture

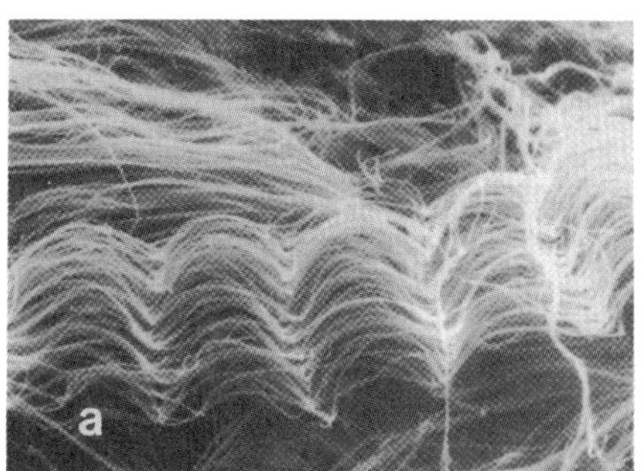

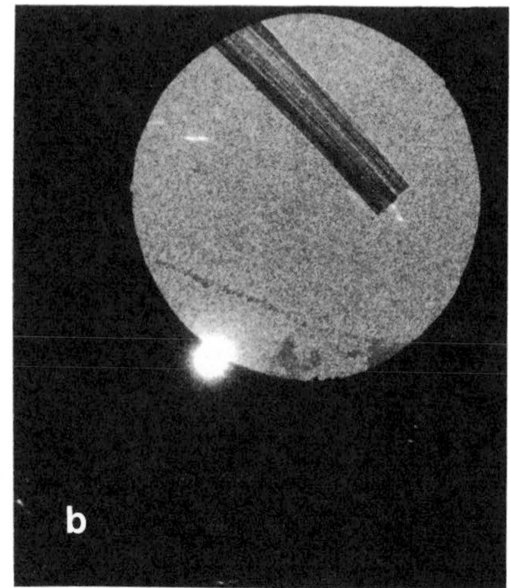

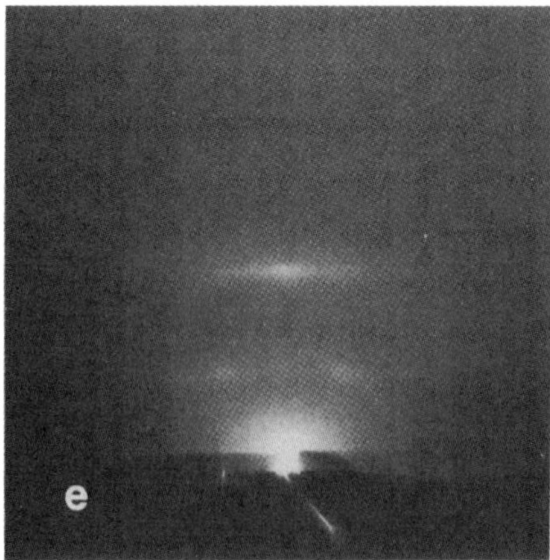

FIGURE 9.7. (a) Streamers of wax from *Prociphilus tesselatus* as extruded from the insect cuticle. (b) Section of a single streamer with its selected area electron diffraction pattern. (c) Electron diffraction pattern of a wax crystal grown from toluene solution. (d) Electron diffraction pattern from epitaxially oriented wax showing lamellar reflections. (e) RHEED pattern from wax oriented on an aluminum stub. (Reprinted from D. L. Dorset and H. Ghiradella (1983) "Insect wax secretion: the growth of tubular crystals," *Biochimica et Biophysica Acta* **760**, 136–142; with kind permission of Elsevier Science B.V.)

actually forms a eutectic solid (Dorset, 1989). Thus, the presumed orthorhombic form cannot occur and the actual crystal structure is still monoclinic. A similar interaction was found more recently for binary combination of *n*-$C_{37}H_{76}$ with dicetyl ether. Electron diffraction measurements have shown that the ether packs in the same oblique layer structure as the ester, consistent with Kohlhaas' (1940) earlier x-ray measurements. On the other hand, secondary amines, e.g., didodecyl- and dioctadecyl-amine have been shown by electron diffraction to crystallize in rectangular layers, so that their interactions with alkanes of the same chain length are more complicated at room temperature, i.e., some cosolubility is observed. Also, *n*-paraffins do form solid solutions with the asymmetric wax esters (that favor the orthorhombic layer packing), as discovered in recent electron diffraction experiments.

Studies of wax esters has been extended to natural products. For example, Hurst (1948, 1950) has used electron diffraction to study insect larval waxes. A similar study carried out in this laboratory has considered the structure of the monodisperse 15-oxotetratriacontyl 13-oxodotriacontanoate wax secreted by the woolly alder aphid *Prociphilus tesselatus* as a defense against predators (Dorset, 1975; Dorset and Ghiradella, 1983). The organism produces the wax as wavy streamers (Figure 9.7a) which are actually microcylinders of wax monolayer with 0.16 to 0.25 μm diameter cross section (Figure 9.7b). When a dilute solution was made in hot xylene and evaporated onto an electron microscope grid, the familiar orthorhombic pattern was seen (Figure 9.7c) with cell constants $a = 7.52$, $b = 5.05$ Å. Use of the orthorhombic subcell structure to calculate theoretical structure factor magnitudes gave a good fit to the observed data (Table 9.5). Epitaxial orientation on benzoic acid resulted in a

Table 9.5. Observed and Calculated Structure Factors for Woolly Alder Aphid Wax Crystals Grown from Solution

hk	$\|F_o\|$	$\|F_c\|$ subcell
200	5.10	6.40
400	1.80	2.40
110	5.80	6.80
210	1.20	1.30
310	2.30	1.80
410	1.10	0.87
510	0.74	0.76
020	3.70	3.60
120	1.00	1.10
220	1.60	1.10
320	1.40	0.84
420	0.92	0.40
520	0.97	0.83
130	1.50	1.50
230	1.20	1.10
330	1.00	0.45
430	1.40	0.77

diffraction pattern (Figure 9.7d) with a lamellar spacing d_{002} = 88.1 Å corresponding to the equivalent odd-chain paraffin. RHEED measurements (Figure 9.7e) indicated also that the chains were untilted, in agreement with the transmission measurements.

The orientation of the orthorhombic chain structure within the filament was determined by selected area diffraction measurements on single filament segments. After calibrating the rotation between image and diffraction pattern, it was found that the elongation direction corresponded to the [110] vector direction of the methylene subcell. At first glance, this result is unusual, because, in the growth of orthorhombic

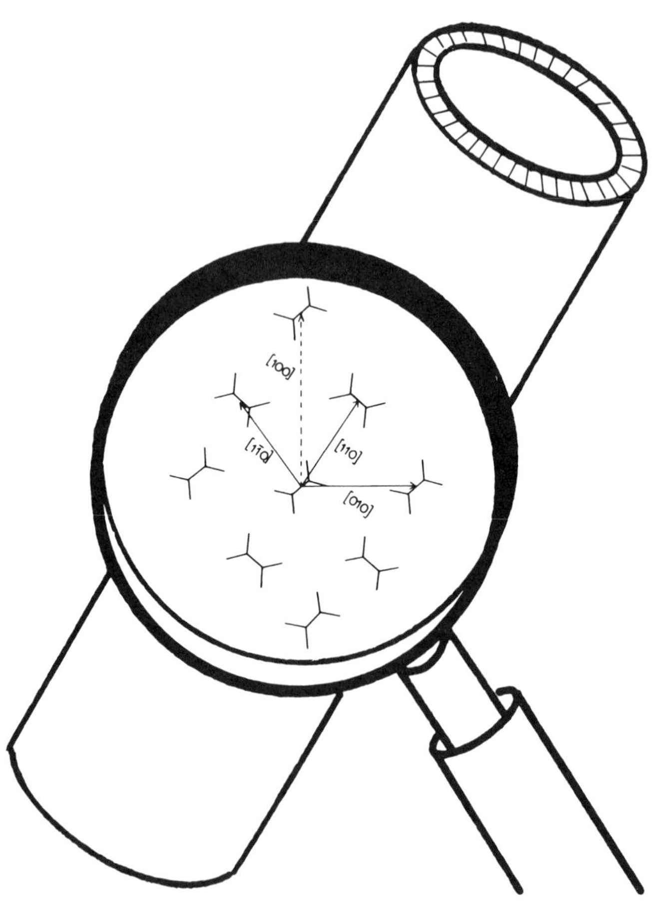

FIGURE 9.8. Growth of alder aphid wax filament with respect to the orthorhombic layer packing of the chains. (Reprinted from D. L. Dorset and H. Ghiradella (1983) "Insect wax secretion: the growth of tubular crystals," *Biochimica et Biophysica Acta* **760**, 136–142; with kind permission of Elsevier Science B.V.)

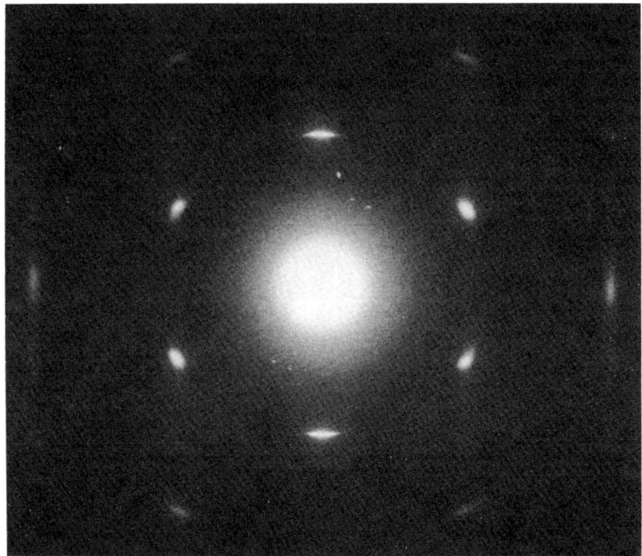

FIGURE 9.9. Electron diffraction pattern of yellow beeswax epitaxially oriented on benzoic acid.

paraffin dendrites packing in the same chain arrangement, the directions of elongation for the flat monolayers lies along [100] and [010]. The reason for this apparent discrepancy is given by an analysis of paraffin crystal growth (Boistelle and Aquilano, 1978). They found that the greatest attachment energy for this untilted chain packing occurs along the equivalent <110> directions of the unit cell, corresponding to the direction of fastest growth. The second preferred direction is [010]. Hence, as the wax is extruded from the insect cuticle surface, crystal physics dictates the optimal unit cell direction for elongation and also assures an energetically favorable helical wrapping of optimal molecular attachment sites around [1 $\bar{1}$ 0] and [010] directions to stabilize the tubular structure (Figure 9.8). The insect needs only to assemble the metabolic fragments at a surface site that is closed into a ring. More recently, electron diffraction measurements have been made on epitaxially oriented beeswax, indicating that the multicomponent mixture might have some nematiclike disordering of components across the lamellar interface (Figure 9.9).

9.2.5. Alkyl Halides

Preliminary electron diffraction studies (Dorset, 1983b) have been undertaken on various alkyl halides, viz.: 1,10-dibromodecane, 1-bromo-*n*-octadecane, and 1-iodo-*n*-octadecane. Rectangular and oblique chain packings are found but all pack in the orthorhombic methylene subcell found for *n*-paraffins. If this is used as a model for the electron scattering, there is a very good match ($R \leq 0.20$) to the observed intensity data. In other words, because of the bend distortion of the crystals, the contribution of the heavier halogen atoms to the total diffraction intensity is not detected in the orientation down the molecular chain axis.

9.2.6. Detergents

Alkyl glucosides are important nonionic detergents for the isolation and purification of integral biomembrane proteins. In the purification of the Omp F porin from the outer membrane of *E. coli*, α- and β-octyl glucosides were considered as likely candidates for this procedure (Rosenbusch et al., 1981). It is well-known that the α-anomer has a much lower solubility in water than does the β-form and the crystal structure is known only for the former compound. Electron diffraction experiments were carried out to find the differences between the two isomers in the solid state (Dorset and Rosenbusch, 1981). Using material grown from solution and by epitaxial orientation, it was possible to show that the unit cell of the α-octyl glucoside has axial lengths $a = 5.12$, $b = 7.74$, $c = 19.6$ Å, $\beta \approx 90°$, values consistent with the decyl homologue (Moews and Knox, 1976). From a projection achieved by epitaxial growth, single-crystal electron diffraction patterns of the β-anomer have measured cell spacings: 4.88×29.0 Å on a rectangular net. Although there are obvious differences in the unit cell spacings, the two materials seemed to form a metastable solid solution with one another, that eventually fractionates (Dorset, 1990f). Recently the x-ray structure of the resultant molecular compound of the two has been reported (Jeffrey and Yeon, 1992).

Another series of nonionic detergents considered for protein extraction and purification are the alkyl oligo(ethylene oxides), C_nE_m. Here electron diffraction studies have been more useful for quantitative characterization of the crystal structure than for the glucosides considered above (Dorset, 1983c). Plots of freezing points for the dodecyl series ($n = 12$) from $m = 0$ to 9, found that a straight line is established

FIGURE 9.10. Electron diffraction pattern from untilted *n*-alkyl octa(ethylene oxide) crystal. (Reprinted from D. L. Dorset (1983) "Molecular conformation and crystal packing of *n*-alkyl oligo(ethylene oxide)s," *Journal of Colloid and Interface Science* **96**, 172–181; with kind permission of Academic Press, Inc.)

Table 9.6. Observed and Calculated Structure Factors for *n*-Dodecyl Octa(ethylene Oxide)–Oligo(ethylene Oxide) Segment T_2G

hk	$\|F_o\|$	$\|F_c\|$
200	3.78	4.30
400	1.04	0.32
600	0.59	0.51
210	0.77	0.93
020	2.09	2.62
120	0.77	0.84
220	0.88	0.40

beyond E_4, corresponding to the first complete turn of the poly(ethylene oxide) helix. For untilted crystals of *n*-dodecyl octa(ethylene oxide), the electron diffraction pattern appeared to be the same as found for the native poly(ethylene oxide) (Figure 9.10), with cell constants $d_{100} = 9.04$, $d_{010} = 9.30$ Å. The T_2G geometry of the polymer (Bunn, 1942) was used to solve this projection of the polymer structure with an agreement $R = 0.24$ (Table 9.6). When the crystals are tilted by 25°, patterns of the kind shown in Figure 9.11 can be found. These are obviously from the alkyl chain packing. The

FIGURE 9.11. Electron diffraction pattern from tilted *n*-alkyl octa(ethylene oxide) crystal showing polymethylene chain diffraction. (Reprinted from D. L. Dorset (1983) "Molecular conformation and crystal packing of *n*-alkyl oligo(ethylene oxide)s," *Journal of Colloid and Interface Science* **96**, 172–181; with kind permission of Academic Press, Inc.)

Table 9.7. Observed and Calculated Structure Factors for *n*-Dodecyl Octa(ethylene Oxide)–Alkane Chain Segment (Tilted Crystals) in Previously Uncharacterized Subcell

hk	$\lvert F_o \rvert$	F_c
020	3.36	1.34
040	7.13	−8.61
120	13.02	−16.01
140	2.06	−1.19
160	3.24	3.08
200	14.08	−15.30
240	5.77	5.95
320	6.66	6.84
400	4.83	3.65
440	3.48	−1.66

orthorhombic subcell is a perfectly twinned version of $M_{\parallel}$ and the agreement of the model to the observed data is given by $R = 0.19$ (Table 9.7). These results were in accord with later x-ray studies of triblock segments based on the two components (Domszy and Booth, 1982), and later confirmed for the diblock compounds (Craven et al., 1991). It is interesting to note in this case that the diffraction incoherence caused by elastic crystal bending could be used to advantage for "dissecting out" the diffraction patterns from the two individual chain segments in this structure. Thus we know that beyond E_4 that the oligo(ethylene oxide) moiety has the same structure as the native polymer and that the chain axis is oriented perpendicular to the crystal layer surface. The alkyl chains are tilted by 25° to the layer normal to compensate for the differences in cross-sectional area and pack in a subcell similar to the $M_{\parallel}$ discussed in Chapter 10.

9.3. Conclusions

Electron diffraction measurement has been particularly important for the characterization of thin layers formed by alkane derivatives. Significant examples include the monolayers formed by amphiphiles or even more unusual arrangements of natural waxes. To date, mostly information about the chain methylene packing and orientation has been detected during studies of polymorphism of these substances. However, epitaxial orientation can be used to provide specimens suitable for intensity data collection from which features of the total unit cell contents can be visualized. Quantitative electron diffraction structure analysis of a wax ester is in progress.

10

The Lipids

10.1. The Methylene Subcell: Its Significance for Electron Diffraction from Lipids

For many amphiphilic or neutral lipid molecules, the presence of an alkyl chain residue, linked as an ester, an amide, or an ether to the "functional" moiety, often requires the polymethylene chains to be sequestered in separate domains when these substances are crystallized or even when they are transformed to the liquid crystalline state. In the crystalline state, these isolated chain regions often form a sublattice of methylene groups, oriented at some angle to the overall layer packing for the whole molecule. This sublattice is usually called the "methylene subcell" (Abrahamsson et al., 1978). The subcell concept was originally devised by Vand and Bell (1951) to solve the x-ray crystal structure of a triglyceride. The identification of subcells is useful because there are only a few of them, with three or four being most commonly expressed in the polymethylene lattices found in lipid crystal structures.

As described in Chapter 5, elastic bending of crystals can have a profound influence on electron diffraction intensities in a projection down a large unit cell axis. For lipids, the methylene subcell is often all that can be determined with observed electron diffraction intensity data from an unknown compound crystallized from solution. Further information is not available unless some epitaxial growth technique is used to provide a projection onto the longest molecular axis (Dorset, 1980). For example, if the substance is grown onto a metal surface for a RHEED experiment, the geometry of the reflection diffraction pattern can be used to determine the average chain tilt, using the relationship derived by Karle and Brockway (1947), discussed in Chapter 9. Specimens crystallized onto a carbon-covered grid for transmission diffraction experiments can be tilted to this angle and the grid can be rotated until the subcell tilt axis coincides with the goniometer axis to provide a view down the chain axes. The diffraction intensities will correspond to the Fourier transform of the methylene subcell, with no salient scattering contributions found from "functional groups" on the molecule (Dorset, 1983b). Five representative methylene subcells and their electron diffraction patterns are shown in Figure 10.1. (Another, the $M_{\parallel}$, was depicted in Figure 8.42.) Four of the common subcells had been predicted by Kitaigorodskii (1961) in his analysis of paraffin layer packing. These are listed in Table 10.1 along with the

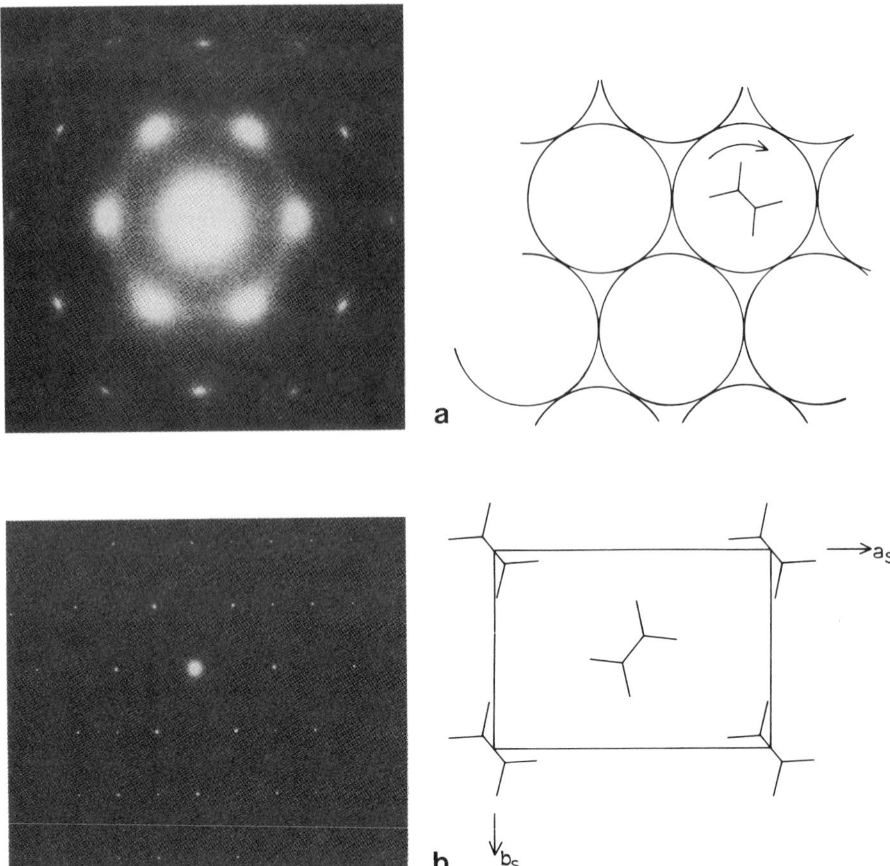

FIGURE 10.1. Electron diffraction patterns from commonly observed methylene subcells, projecting down the polymethylene chain axes: (a) *H*-subcell; (b) $O_\perp$ subcell; (c) $T_\parallel$ subcell; (d) *HS*1 subcell; (e) orthorhombic subcell with features similar to $M_\parallel$. (Reprinted from D. L. Dorset (1983) "Electron crystallography of alkyl chain lipids: identification of long chain packing," *Ultramicroscopy* **12**, 19–28; with kind permission of Elsevier Science B.V.)

permissible tilt angles for smooth methyl end-plane packing. These settings correspond well to experimental results from crystals of simple alkane derivatives and some glycerolipids. Based on electron diffraction observations, examples of compounds crystallizing in these subcells are reviewed in Table 10.2.

The prospect of using electron diffraction intensity data for the determination of lipid crystal structures was first proposed in the 1960s (Parsons and Nyburg, 1966; Parsons, 1967) as an alternative to single-crystal x-ray studies. Thin microcrystals of these lipids are prepared very easily, in contrast to the great difficulties encountered when growing samples suitable for single-crystal x-ray data collection. The enhanced

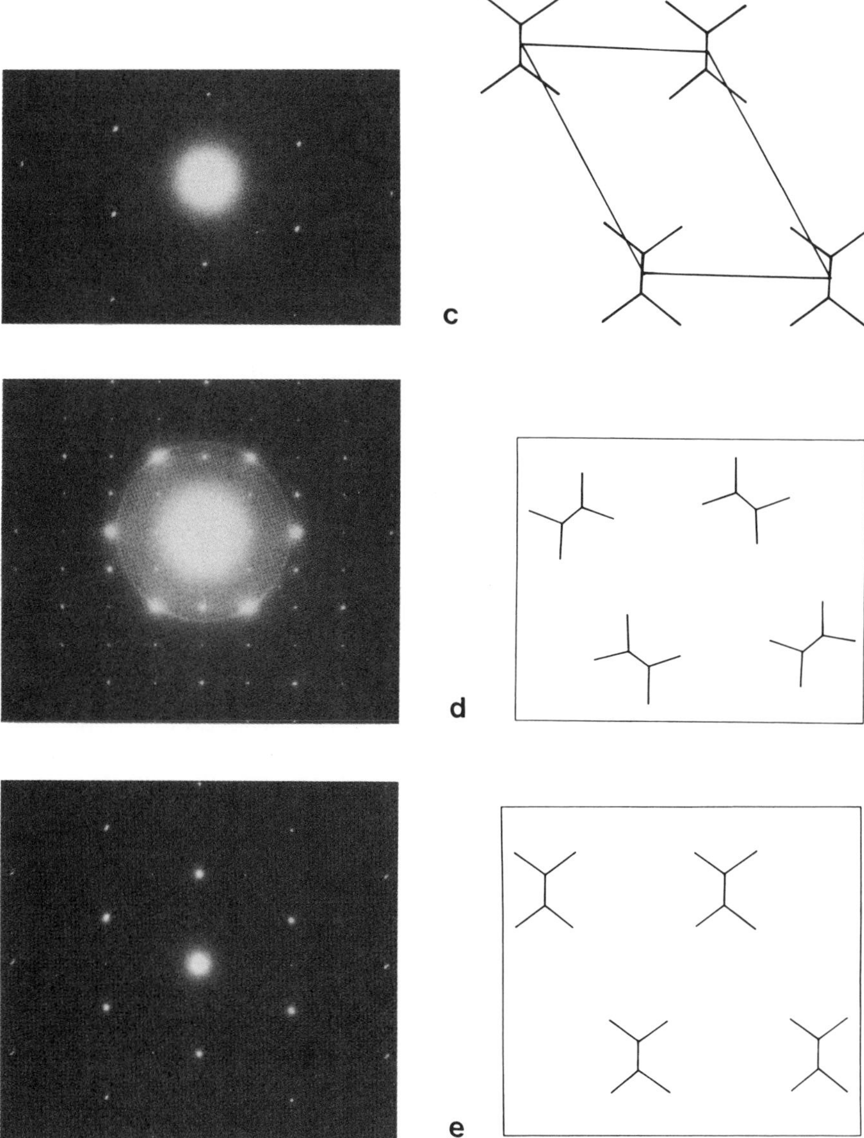

scattering cross section of matter for electrons, compared to x-rays, therefore, suggested that electron diffraction would be a perfect technique for data collection from the available thin microcrystals. However, the constraints imposed by elastic crystal bending, mentioned above and in Chapter 5, were not fully appreciated then. Since there is more to a lipid structure than just the polymethylene chain packing and its

Table 10.1. Methylene Subcells Predicted by A. I. Kitaigorodskii (1961)

Subcell*	Layer	Layer unit cell constants
H (H)	*H*[0,0]	$a = 4.8$ Å
R ($O_\perp$)	*R*[0,0]	$a = 4.96$, $b = 7.42$ Å, $\gamma = 90°$
	R[0,±2]	$a = 4.96$, $b = 9.0$ Å, $\gamma = 90°$
	R[±1,0]	$a = 5.57$, $b = 7.42$ Å, $\gamma = 90°$
	R[0,±1]	$a = 4.96$, $b = 7.85$ Å, $\gamma = 90°$
	R[±1,±1]	$a = 5.57$, $b = 7.85$ Å. $\gamma = 81.5°$ for *R*[1,1] and *R*[−1,−1] $\gamma = 98.5°$ for *R*[−1,1] and *R*[1,−1]
M ($M_\parallel$)	*M*[0,0]	$a = 4.2$, $b = 4.4$ Å, $\gamma = 111°$
	M[±1,0]	$a = 4.9$, $b = 4.4$ Å, $\gamma = 107°$
	M[0,±1]	$a = 4.2$, $b = 5.1$ Å, $\gamma = 107°$
T ($T_\parallel$)	*T*[±1/2,0]	$a = 4.3$, $b = 4.5$ Å, $\gamma = 103°$
	T[1/2,1]	$a = 4.3$, $b = 5.2$ Å, $\gamma = 109°$
	T[−1/2,1]	$a = 4.3$, $b = 5.2$ Å, $\gamma = 109°$

*Notation in parentheses is that of Abrahamsson et al. (1978).

orientation in the lamellar layer, other approaches to interpreting the data were sought, so that information about the nonpolymethylene molecular moiety could be inferred. For example, conformationally locked analogs of the glycerolipids, based on configurational isomers of cyclopentane-1,2,3-triol (Greenwald et al., 1977; Hancock et al., 1977), were used as replacements for glycerol in the synthesis of common lipids (Figure 10.2). As will be seen below, matching of diffraction from these analogs to those from natural lipids was only a partially successful solution to this problem. Eventually, epitaxial orientation techniques were found to provide some information about the projection onto the long molecular axis, although such crystals are often deformed by paracrystalline disorder.

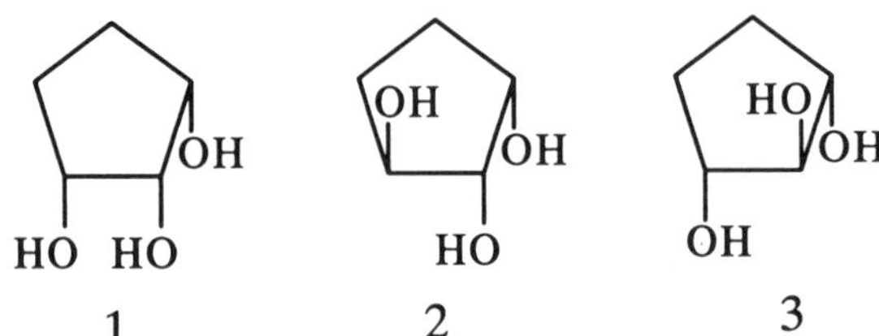

FIGURE 10.2. Conformationally fixed analogs of glycerol based upon configurational isomers of cyclopentane-1,2,3-triol. Nomenclature: [1]: 1,2,3/0; [2]: 1,2/3; [3]: 1,3/2. (Reprinted from D. L. Dorset, W. A. Pangborn, A. J. Hancock, and I. S. Lee (1978) "Influence of molecular conformation on the solid state packing of 1,2-diglycerides. Study of 1,2-dipalmitin and some structural analogs by electron diffraction, x-ray diffraction and infrared spectroscopy," *Zeitschrift für Naturforschung, Teil C* **33c**, 39–49; with kind permission of the Verlag der Zeitschrift für Naturforschung.)

Table 10.2. Compounds Packing in Common Methylene Subcells as Determined by Electron Crystallography (R-factor)

Compound (R-factor)	Reference
hexagonal subcell (*H*):	
rac-glycerol-1,2-dipalmitate (0.13)	Dorset (1974)
1,2-dimyristoyl-*sn*-glycerophosphocholine (0.03)	Dorset (1975b)
1-palmitoyl-*sn*-glycerophosphocholine (0.01)	Dorset (1975b)
N-nervonyl-4-sphingosyl-1-phosphocholine (0.01)	Dorset (1975b)
1,2-hexadecyl-*sn*-glycerophospho-ethanolamine (0.06)	Dorset et al. (1976)
1,2-dipalmitoyl-*sn*-glycerophosphocholine, hydrated bilayer (0.03)	Hui et al. (1974)
octadecanol (not given)	Frede and Precht (1974)
glycerol trilaurate (not given)	Buchheim (1970)
rotationally disordered orthorhombic subcell:	
trihexadecyl glycerol (0.13)	Dorset & Pangborn (1982)
orthorhombic perpendicular subcell ($O_\perp$):	
n-octadecane (0.22)	Vainshtein et al. (1958)
n-tritriacontane (0.28)	Vainshtein et al. (1958)
n-hexatriacontane (0.19)	Dorset (1976b)
polyethylene (0.23)	Dorset and Moss (1983)
n-octadecanol (0.26)	Dorset (1979)
cetyl palmitate (0.14)	Dorset (1976e)
stearic acid, *B*-form (0.20)	Dorset (1983b)
behenic acid, *B*-form (0.20)	Dorset (1983b)
1, 10-dibromodecane (0.19)	Dorset (1983b)
1-bromooctadecane (0.20)	Dorset (1983b)
ethylene di(11-bromoundecanoate) (0.17)	Dorset (1983b)
15-oxotetratriacontyl 13-oxotriacontanoate (0.19)	Dorset and Ghiradella (1983)
1-iodooctadecane (0.24)	Dorset (1983b)
1,2-dihexadecylglycerol (0.20)	Dorset and Pangborn (1982)
n-hentriacontane (not given)	Rigamonti (1936)
tricosanol (not given)	Li (1963)
glycerol trilaurate (not given)	Buchheim (1970)
n-dotriacontane (not given)	Cowley et al. (1951)
triclinic parallel subcell ($T_\parallel$):	
glycerol tripalmitate (0.10)	Dorset (1983a)
glycerol triheptadecanoate (0.14)	Dorset (1983a)
glycerol-1,3-dipalmitate (0.13)	Dorset and Pangborn (1979)
1,3-dihexadecylglycerol (0.13)	Dorset and Pangborn (1982)
glycerol trilaurate (not given)	Buchheim (1970)
hybrid orthorhombic subcell (*HS*1):	
1,2-dipalmitoyl-*rac*-glycerophosphoethanolamine (0.24)	Dorset (1976c)
orthorhombic, similar to monoclinic parallel:	
n-dodecyl octa(ethylene oxide) (0.19)	Dorset (1983c)
monoclinic parallel subcell ($M_\parallel$):	
cyclooctatetracontane (0.24)	Dorset and Hsu (1989)

Despite the difficulties encountered in the application of electron crystallographic methods to the study of lipid structure, there are still a number of benefits which favor use of the technique. First, selection of individual microareas is especially good for the identification of polymorphic forms for these compounds that may otherwise remain undetected in powder x-ray studies, for example. Often these can be visualized in terms of single-crystal diffraction patterns. Second, electron diffraction is a good technique for studying molecular packing in polydisperse crystals. Although micro-crystals are easily formed, the presence of homologous impurities can actually retard crystallization of samples for x-ray data collection (Albon, 1976). The investigation of various molecular species (in terms of different headgroups and chain substitutions), not characterized by any other way, is also facilitated because the constraints on crystallization are somewhat relaxed. Also, under proper conditions, it is possible to study dynamic behavior of single hydrated bilayers, as elegantly shown in pioneering work using an environmental chamber in the electron microscope (Hui et al., 1974, 1975, 1980; Hui and Parsons, 1974, 1975; Hui, 1976, 1981; Hui and He, 1983).

10.2. Structure Analyses of Glycerolipids

10.2.1. 1,3-Diglycerides

From the x-ray crystal structure of an ω-brominated analog (Hybl and Dorset, 1971), it is well known that the 1,3-diglycerides pack as an extended chain structure in a herringbonelike array, but the halogen substitution causes the chain tilt to the layer surface to be larger than found for the natural lipid. Single, solution-grown, platelike crystals of 1,3-dipalmitin, when tilted by 14° to the layer surface normal, can be used to find the projection down the chain axes (by rotation of the crystal), revealing the electron diffraction pattern from the $T_{\parallel}$ methylene subcell (Figure 10.1) for this β-polymorph. The agreement between calculated and observed data is $R = 0.13$ (Table 10.3).

Table 10.3. Observed and Calculated Structure Factors for 1,3-Dipalmitin Crystals Oriented down the Chain Axes to Find the *hk*0 Pattern from the $T_{\parallel}$ Methylene Subcell

*hk*0	$\lvert F_o \rvert$	F_c (subcell)
100	4.45	3.68
200	1.67	1.88
010	3.48	–3.38
110	1.79	–1.81
–110	3.59	–3.55
–210	2.10	–2.36
020	1.20	0.79
–120	1.28	1.50
–220	0.93	1.54

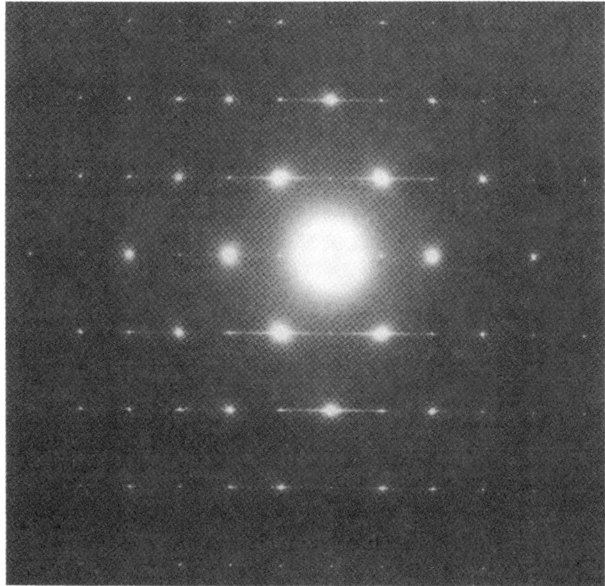

FIGURE 10.3. Electron diffraction patterns from 1,3-dihexadecyl glycerol, showing streaking due to an alternating polar lattice packing, restricted along **a**. (Reprinted from D. L. Dorset and W. A. Pangborn (1982) "Ether linkages in glycerolipids: their effect on long chain packing," *Chemistry and Physics of Lipids* **30**, 1–15; with kind permission of Elsevier Science Ireland, Ltd.)

When the chains are joined to the glycerol moiety by ether links instead of ester groups (Dorset and Pangborn, 1982), the overall structure changes. The β'-polymorph, shown in Figure 10.3, is a minor form. The intense reflections for the $O_\perp$-methylene subcell ($R = 0.22$) are streaked along the a^* direction, indicating a sequestering of the structure into elongated polarized "cigarello" substructures with limited correlation along the a-axis, perhaps due to a reversible orientation of hydrogen bonding chains formed from the unsubstituted hydroxyl group at the center of the glycerol. The major polymorph of this diether is a β-packing, as also found for the diester, with chains crystallizing in the $T_\parallel$ methylene subcell ($R = 0.16$). However, the chains are now tilted to the crystal surface normal by about 45°, similar to the case of the brominated analog structure determined by x-ray crystallography (Hybl and Dorset, 1971).

10.2.2. 1,2-Diglycerides

From electron diffraction studies of 1,2-diglyceride layers crystallized from solution, e.g., L-1,2-dipalmitin, the β'-crystal form is most often observed, with chains packing in the $O_\perp$-methylene subcell and inclined to the layer normal by about 27° (Dorset and Pangborn, 1979, 1982) (Figure 10.4). When crystallized from the melt,

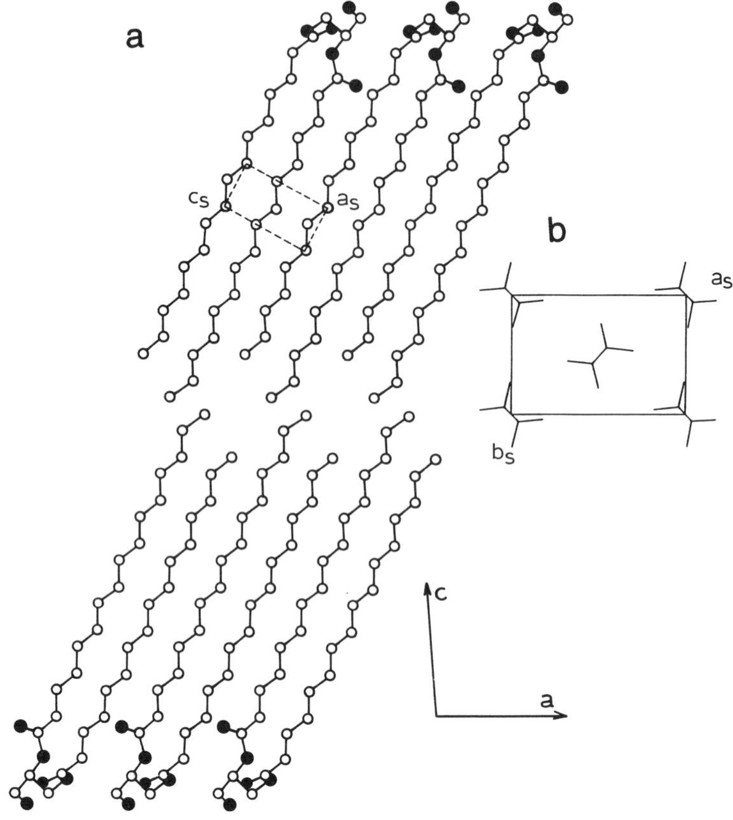

FIGURE 10.4. (a) Molecular conformation of *L*-1,2-dipalmitin determined from its x-ray crystal structure. (b) When tilted approximately 27° to project along c_S, the $O\perp$ subcell electron diffraction pattern is observed. (Reprinted from D. L. Dorset (1990) "Alkane chains in an electron beam: a crystallographic history," *Electron Microscopy Society of America Bulletin* **20**, 54–63; with kind permission of the Microscopy Society of America.)

the chains are untilted and pack in the *H* methylene subcell. In this so-called α-polymorph, the chain packing is identical to the "rotator" phase of the *n*-paraffins. (In fact, one of the first quantitative analyses of this *H* subcell packing with measured intensity data was made with the electron diffraction set from *D,L*-1,2-dipalmitin (Dorset, 1974), using a rotor model for the structure factor calculation, analogous to the one described in Chapter 8 for a perfluoroalkane.) When the chain linkages are replaced by ethers, the α-polymorph again resembles that of the ester-linked diglyceride. However, for this compound, the β'-polymorph now packs with chains inclined around the subcell $b = 4.95$ Å axis rather than the $a = 7.42$ Å axis employed for the chain tilt in the diester. (Both chain packings are common (Table 10.1) for this subcell and are also found in fatty acids.)

The orientation of the acyl chains in the crystal with respect to the glycerol moiety was not known until recently (Pascher et al., 1981). It was thought originally that either an extended structure, as seen in the crystal structure of ethylene (di-11-bromoundecanotate) (Dorset and Hybl, 1972) or some hairpin arrangement, as found for the phospholipids (see below), might be possible. Matching electron diffraction patterns from solution-crystallized layers of the lipid to those from configurational isomers based on cyclopentane-1,2,3-triol suggested that the hairpin conformation would be the most likely (Dorset et al., 1978a). However, some details of the spontaneous solid state acyl shift of this diglyceride to the 1,3-diglyceride product argued against this molecular conformation, particularly if the **intra**molecular mechanism proposed in earlier work was accepted as correct. Originally, this reaction model was assumed because mixed-chain products were never found (Mank et al., 1976) for binary combinations of the 1,2-diglycerides. Because the 1,3-diglyceride was known to have an extended chain conformation, an extended chain conformation, therefore, would also have to occur in the 1,2-diglyceride. Electron diffraction experiments on the crystalline 1,2-diglycerides could be used to monitor the progress on the acyl shift and the formation of the 1,3-diglyceride product, the reaction being driven to 100% completion because of the phase separation of reactant and product (Dorset and Pangborn, 1982).

It was only after binary phase diagrams were constructed for homologous 1,2-diglycerides that the reaction mechanism was understood, leading to a correct view of the molecular conformation (Dorset, 1987e). Binary solids of, e.g., 1,2-dimyristin and 1,2-dipalmitin grown from a solution as the β'-polymorph were found to be completely phase-separated. It was clear, therefore, that the acyl shift, which was only observed by us to occur within this polymorph, could only produce products with homogeneous chains, regardless of the reaction mechanism. With epitaxial growth (or growth from hot solution), the diglycerides crystallized in the α-polymorph, on the other hand, in which the homologues were freely cosoluble, as evidenced by the continuity of lamellar spacings with increasing concentration of the larger component (Figure 10.5). However, this form was never found by us to produce the 1,3-diglyceride end product directly. Since the x-ray crystal structures of 1,2-dilaurin and 1,2-dipalmitin had been solved by this time (Pascher et al., 1981; Dorset and Pangborn, 1988), verifying that the molecular conformation is the hairpin suggested in earlier electron diffraction work on analogs, a reaction mechanism could be proposed for the acyl shift, involving an **inter**molecular mechanism (Dorset, 1987e). The geometry of the molecular layer packing is quite favorable for a S_N2 reaction pathway of the type proposed by Bürgi et al. (1974).

Why does the α-form not produce the 1,3-diglyceride product? Lamellar electron diffraction intensity data (Figure 10.6) from epitaxially crystallized samples were used to test various 1,2-diglyceride conformational models, based on the β'-polymorph or the conformers found in various phospholipid structures (Dorset and Pangborn, 1987). The models were oriented in a one-dimensional cell, assuming centrosymmetry, with chain axes parallel to the length of this cell, in accord with the orientation of the methylene subcell found for solution-crystallized samples. The best fit to the intensity

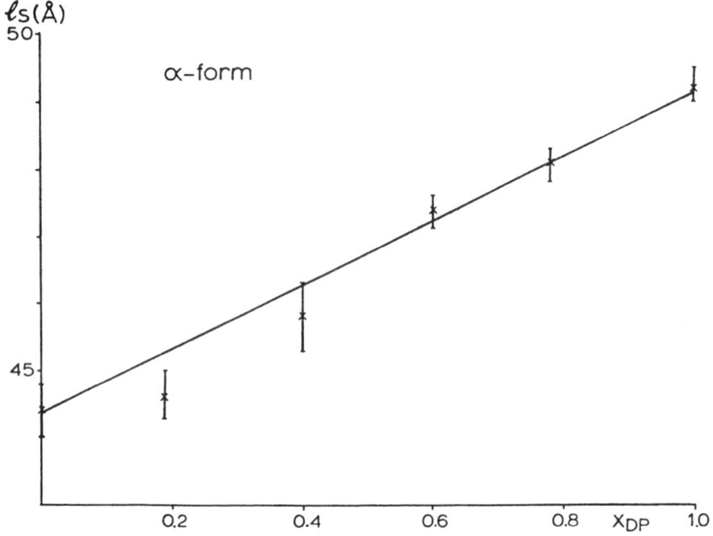

FIGURE 10.5. Continuity of lamellar spacings for *L*-1,2-dimyristin/*L*-1,2-dipalmitin solid solutions as measured from electron diffraction of epitaxially oriented crystals in the α-polymorph. (Reprinted from D. L. Dorset (1987) "Is the solid state acyl shift of 1,2-diglycerides intermolecular?," *Chemistry and Physics of Lipids* **43**, 179–191; with kind permission of Elsevier Science Ireland Ltd.)

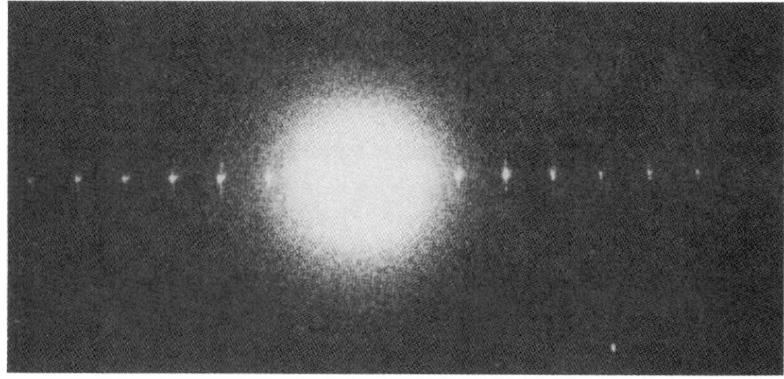

FIGURE 10.6. Electron diffraction data from epitaxially oriented *L*-1,2-dipalmitin.

Table 10.4. Observed and Calculated Structure Factors for Epitaxially Oriented 1,2-Dipalmitin Crystals in the α-Form

$00l$	$\lvert F_o \rvert$	F_c
00 1	0.91	0.30
00 2	1.03	−0.73
00 3	0.99	−0.56
00 4	0.94	−1.20
00 5	0.94	−1.03
00 6	0.78	−1.17
00 7	0.93	−1.06
00 8	0.65	−0.72
00 9	0.72	−1.05
00 10	0.44	−0.47
00 11	0.43	−0.45

data (Table 10.4) was *not* the conformer of the β'-form. Rather, the diglyceride conformation found in phospholipids such as DPPC or DPPE provided much better (but still not perfect) match to these experimental data. From this one-dimensional analysis with electron diffraction data, it is clear that the recrystallization of 1,2-diglycerides must also be accompanied by a conformational change in the molecule. Thus, if the molecular conformational and packing geometries do not correspond to an orientation appropriate for the solid state reaction, the acyl shift will not occur.

10.2.3. Triglycerides

Extensive single-crystal electron diffraction studies of homo-acid triglycerides had been carried out by a group in Kiel (Buchheim and Knoop, 1969; Buchheim, 1970; Precht, 1979; Precht and Frede, 1983), particularly in search of various polymorphic forms. Using data from solution-crystallized samples, Buchheim (1970) has published potential maps for the methylene subcell packings based on visual estimates of intensities but there was no R-factor match of observed intensity data to the theoretical model provided. In these studies of trilaurin, for example, electron diffraction patterns were published for the perpendicular chain packings in the α- and a β'-form, respectively, in the H and $O_\perp$ subcells. In addition, patterns were obtained from an oblique β'- form, a structure characterized earlier by Larsson (1965a). The layer tilt angle for the acyl chains was found to be 62–63°. The most prominent polymorphs for the triglycerides are the β-forms packing in the $T_\parallel$ methylene subcell. Other work (Knoop and Sandhammer, 1961) found the so-called β_{III}- form to be most prominent, consistent with the x-ray crystal structures of the triglycerides (Vand and Bell, 1951; Larsson, 1965b; Jensen and Mabis, 1966; Doyne and Gordon, 1968). Data have also been obtained from the β_{II}-form in this study and, in later work by Precht (1979), from a β_{I}-form. According to model studies (Precht and Frede, 1983), the chain tilt angles are respectively 61°, 75°, and 90°.

In later transmission electron diffraction measurements of tripalmitin, comparing data from conformationally fixed cyclopentane-1,2,3-triol analogs (Dorset and Han-

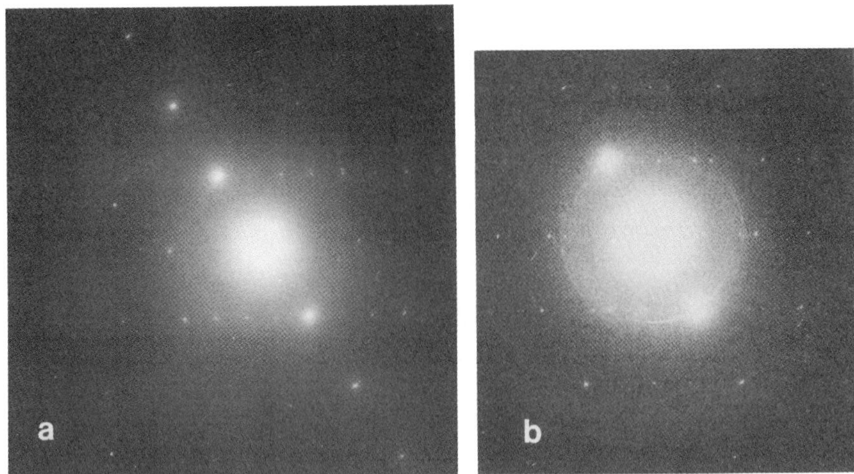

FIGURE 10.7. Electron diffraction patterns from: (a) tripalmitin; (b) tripalmitoyl cyclohexane-1,2,3-triol derivative with configuration 1,3/2 (see Figure 10.2). (Reprinted from D. L. Dorset and A. J. Hancock (1977) "Glycerol conformation and the crystal structure of lipids. I. An electron diffraction study of tripalmitin and conformationally fixed analogs," *Zeitschrift für Naturforschung, Teil C* **32c**, 573–580; with kind permission of the Verlag der Zeitschrift für Naturforschung.)

cock, 1977; Dorset et al., 1978a) (Figure 10.7), it was found that the most probable conformation of the observed β_{III}- form would be the "tuning-fork" geometry observed in the x-ray crystal structures. This was later confirmed by RHEED measurements (Dorset et al., 1978b). Other α-forms and a β'-form were also found in these studies, the latter with a chain tilt of about 69°. When the long chains are linked to glycerol through an ether oxygen (Dorset and Pangborn, 1982), the α-forms and a β'-form are predominant with no evidence for the β-structure. However, in this case, the orthorhombic subcell is rotationally distorted. Tilting 42° to project down the chain axis, a pattern with *cmm* symmetry, rather than *pgg* symmetry, is seen.

Table 10.5. Observed and Calculated Structure Factors for Solution Crystallized Tripalmitin and Triheptadecanoin, Oriented to Observe *hk*0 Intensities from the $T_{\parallel}$ subcell

*hk*0	$\lvert F_o \rvert$ (tri-C_{16})	$\lvert F_o \rvert$ (tri-C_{17})	F_c
100	3.97	3.93	3.68
200	0.89	1.89	1.88
010	3.45	3.31	−3.38
110	1.40	1.41	−1.81
−110	3.70	3.48	−3.55
−210	2.13	1.76	−2.36
020	1.28	1.55	0.79
−120	2.13	1.93	1.50
−220	1.53	1.24	1.54

Finally, a comparison was made between the even- and odd-chain saturated triglycerides (Dorset, 1983a). Electron diffraction data from solution-crystallized triheptadecanoin resemble those from, e.g., tripalmitin, i.e., the molecular packing must be similar to the β_{III}-form. Tilting the crystals by 29° orients the $T_{\parallel}$ subcell [001] projection along the beam and the *hk*0 intensity data from the odd-chain triglyceride agree well with those from the even-chain compound, as well as with the theoretical model (Table 10.5).

10.2.4. Phospholipids and Glycolipids

As mentioned above, it is very easy to obtain electron diffraction data from methylene subcells when phospholipids are crystallized from solvent as multilayers. It has been more of a challenge to orient the molecules so that diffraction intensities representing the total unit cell contents can be recorded. Epitaxial growth of these

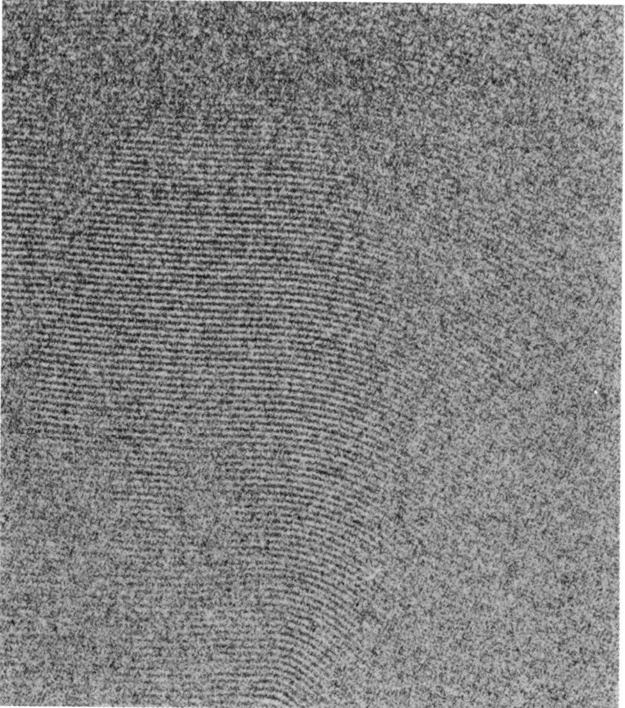

FIGURE 10.8. Experimental lattice image of 1,2-dihexadecyl-*sn*-glycerophosphoethanolamine epitaxially oriented on naphthalene. Note extreme curvature of lamellar stack due to paracrystalline disorder. (Reprinted from D. L. Dorset, A. K. Massalski, and J. R. Fryer (1987) "Interpretation of lamellar electron diffraction data from phospholipids," *Zeitschrift für Naturforschung, Teil A* **42a**, 381–391; with kind permission of the Verlag der Zeitschrift für Naturforschung.)

materials on substrates such as naphthalene or benzoic acid readily provide the correct crystal orientation but, as is shown by a typical low-dose lattice image (Figure 10.8), the lamellae are curvilinearly deformed so that the discrete electron diffraction maxima of the so-called "lamellar reflections" correspond only to the bilayer cross section (Figure 10.9) (Fryer and Dorset, 1987; Dorset et al., 1989). The nucleation of the lipid by the epitaxial substrate, therefore, is probably only one-dimensional, aligning chain axes but not the methylene repeats. Because the diffraction maxima are arced, then, as found in similar x-ray studies of multilamellar arrays (Franks, 1976), derivation of structure factor magnitudes from intensities must include a correction for a phenomenological Lorentz factor before they can be used in a structure analysis. Hence:

$$|F_{00l}|^2 = I_{00l}^{\text{obs}}\, l^n$$

where l is the order of the lamellar reflection and n takes a suitable value to compensate for the intensity not read by a densitometer scan. If only the centers of the reflection are measured then $n = 2$; if a slit is used to scan the arced intensity, $n = 1$ is often used, but sometimes this must be adjusted to a lower value.

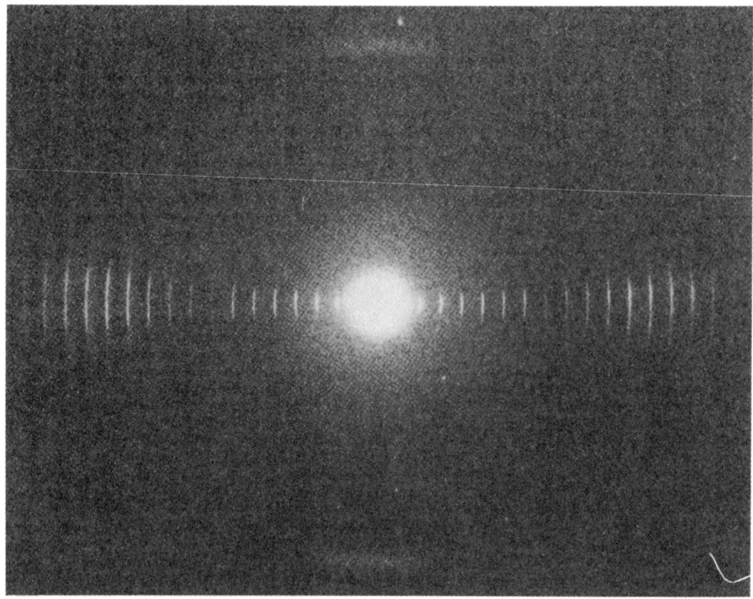

FIGURE 10.9. Electron diffraction pattern of 1,2-dihexadecyl-*sn*-glycerophosphoethanolamine epitaxially oriented on naphthalene. Note arcing of reflections. (Reprinted from D. L. Dorset, A. K. Massalski, and J. R. Fryer (1987) "Interpretation of lamellar electron diffraction data from phospholipids," *Zeitschrift für Naturforschung, Teil A* **42a**, 381–391; with kind permission of the Verlag der Zeitschrift für Naturforschung.)

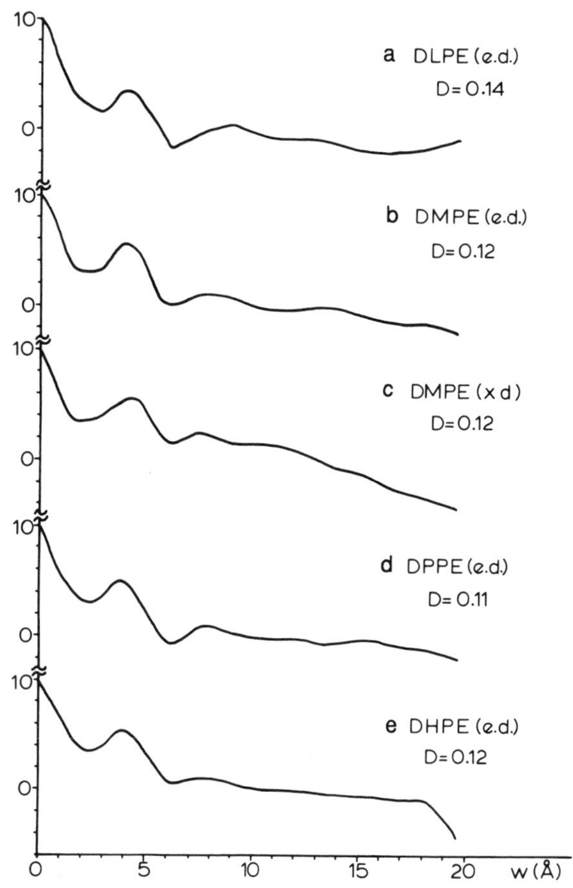

FIGURE 10.10. Experimental Patterson functions for various substituted 1,2-diradyl-*sn*-glycerophosphoethanolamines. (Reprinted from D. L. Dorset, A. K. Massalski, and J. R. Fryer (1987) "Interpretation of lamellar electron diffraction data from phospholipids," *Zeitschrift für Naturforschung, Teil A* **42a**, 381–391; with kind permission of the Verlag der Zeitschrift für Naturforschung.)

For phospholipid species that are similar to those structures determined by x-ray crystallography, it is possible, from a one-dimensional Patterson function, to ascertain whether or not the observed intensity data correspond to a known headgroup conformation (Khare and Worthington, 1971; Dorset, 1987a) (Figure 10.10). From this, and assuming that the diglyceride conformation can be approximated successfully, it is possible to construct a molecular model (e.g., based on a known phospholipid x-ray crystal structure) to be translated past an arbitrary origin in space group $P\bar{1}$ (the centrosymmetry being valid for all lamellar projections of phospholipid structures determined so far).

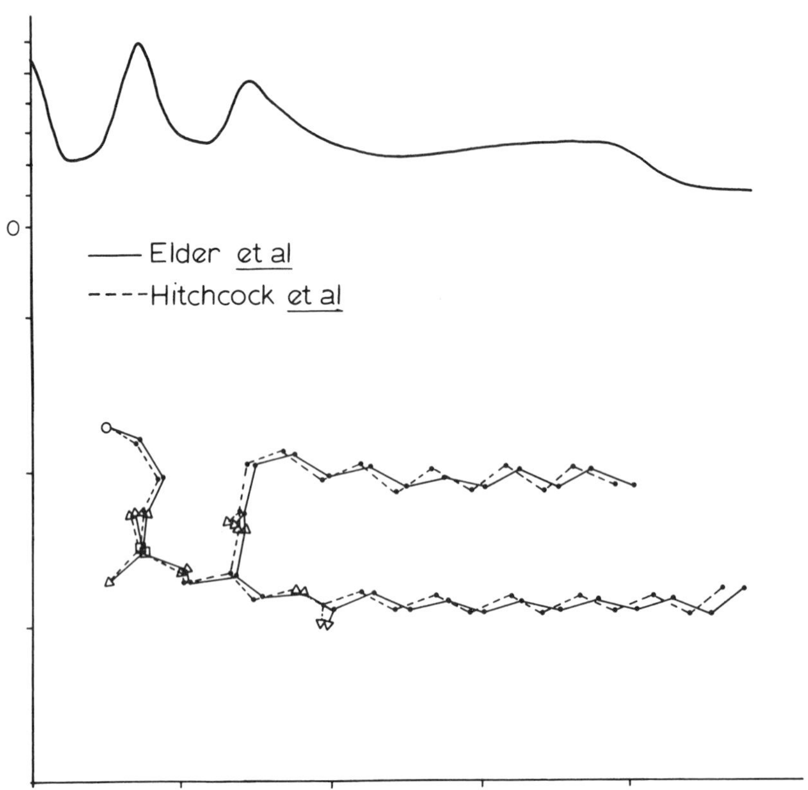

FIGURE 10.11. Molecular conformational model used for translational searches in the analyses of phosphatidylethanolamine lamellar structures with electron or x-ray diffraction data. The continuous density at the origin is from the acetic acid solvation (removed from molecular drawing). Two molecular conformations are compared resulting from the earliest (Hitchcock, et al., 1974) and refined (Elder et al., 1977) x-ray crystal structures of the racemic 1,2-dilauroyl derivative. (Reprinted from D. L. Dorset, A. K. Massalski, and J. R. Fryer (1987) "Interpretation of lamellar electron diffraction data from phospholipids," *Zeitschrift für Naturforschung, Teil A* **42a**, 381–391; with kind permission of the Verlag der Zeitschrift für Naturforschung.)

This model construction will also include appropriate adjustments for chain length and the type of chain linkage. A translational search had been carried out for 1,2-dimyristoyl-*rac*-glycerophosphoethanolamine (DL-DMPE) with measured x-ray data (Hitchcock et al., 1975). Based on the crystal structure of the racemic dilauroyl homolog (DL-DLPE) (Elder et al., 1977), the acetic acid solvate molecule was removed, and, after proper adjustments were made for chain length, a good fit to the x-ray data was found (Figure 10.11) at a geometrically reasonable position in the bilayer. Justification of one packing model instead of another depends on the statistical significance of the crystallographic *R*-factor, again, according to the criteria established by Hamilton (1964), i.e., can the lower of two *R*-factor minima be associated with a better structural model within a reasonable confidence level (see Chapter 4)?

Another approach to structure analysis has been to use direct phasing methods, which are surprisingly robust for one-dimensional data of this kind. In this case, after origin definition, generally by setting $\phi_{001} = 0$, reflections with phase values of π are found from Σ_1-triples, and the identification of reflections with the same crystallographic phase made *via* Σ_2-triples. (See Chapter 4.) This procedure was tested with simulated lamellar x-ray data from ten representative known lipid structures (Dorset, 1991a,b). Phasing difficulties occur only when solvent molecules (or counterions) occur near the unit cell origin, requiring some modifications (i.e., a π-shift) to the use of triple invariants for the determination. A partial phase determination can then be refined by density flattening (of the chain-packing region), by analogy to solvent flattening in protein crystallography (Wang, 1985), and the correct structure identified by monitoring the smoothness of the density profile. These procedures were tested successfully with published x-ray data from phospholipid bilayers (Dorset, 1991a). Obviously, as will be shown below, additional phase information from electron microscope images is beneficial for such structure determination, but may not be absolutely critical.

The structure analysis for phospholipids based on electron diffraction data, therefore, relies on two sources of information. Diffraction patterns from solution-crystallized samples are used to identify the subcell and its orientation in the bilayer. Data from epitaxially oriented samples are then used to determine the potential distribution across the bilayer, consistent with the molecular orientation predicated by the subcell orientation. Examples of such analyses follow.

10.2.4.1. Phosphatidylethanolamines. Most effort has been spent on the analysis of data from 1,2-dihexadecyl-*sn*-glycerophosphoethanolamine (L-DHPE). In a comparison of this ether-linked analog (Dorset and Pangborn, 1982) to the ester-linked 1,2-dipalmitoyl-*sn*-glycerophosphoethanolamine (L-DPPE), it was found that only one chain-packing form, in the hexagonal subcell, was observed. The more ordered *HS*1 subcell was found in the predominant polymorph of the diester phosphatides (Figure 10.1). Otherwise the molecular axes for both molecules was found to lie parallel to the long unit cell direction and the observed lamellar spacings were quite similar (Table 10.6). Patterson maps for a homologous series of the phosphatidylethanolamines closely resembled one another (Dorset et al., 1987), irrespective of whether the polymethylene chains were ester- or ether-linked (Figure 10.10).

Initially a conformational model of L-DHPE was constructed for a translational search of the crystal structure (Figure 10.11) and a final phosphorus position was found at $z{\cdot}c = 2.23$ Å, similar to the value given in the x-ray crystal structure of a lysophosphatidylethanolamine (Pascher et al., 1981). However, the crystallographic R-value was rather large (0.30) when comparing calculated and observed structure factors. An alternative phasing method was required, therefore, to prove the validity of this packing model. Low-angle phases were obtained from the Fourier transform of high-resolution electron micrographs observed at 6-Å resolution on a cryoelectron microscope (Dorset et al., 1990) (Figure 10.12). Unfortunately, there was some problem in guessing the contrast of the micrograph in the original use of the image-processing software, so that the Babinet solution was chosen inadvertently for these

Table 10.6. Comparison of Lamellar Spacings for Phosphatidylethanolamines Measured from Electron Diffraction Experiments on Epitaxially Oriented Samples or X-ray Diffraction Experiments on Multilamellar Arrays

Chain substitution	Electron diffraction*	X-ray diffraction
1,2-dilauroyl	47.7 ± 0.5 Å	45.2 Å (*rac*)
		47.8 Å (*rac*)
1,2-dimyristoyl	49.2 ± 0.4 Å	50.0 Å (*rac*)
		49.9 Å (*rac*)
		49.5 Å (*rac*)
1,2-dipalmitoyl	55.7 ± 0.5 Å	55.0 Å (*sn*)
		55.2 Å (*rac*)
		55.3 Å (*rac*)

*All measurements on chiral molecules.

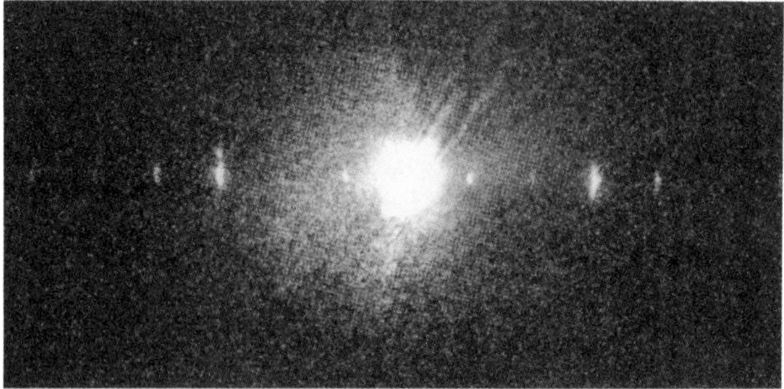

FIGURE 10.12. (a) Low-dose electron micrograph of epitaxially oriented 1,2-dihexadecyl-*sn*-glycerophosphoethanolamine; (b) its optical transform. (Reprinted from D. L. Dorset, E. Beckmann, and F. Zemlin (1990) "Direct determination of phospholipid lamellar structure at 0.34 nm resolution," *Proceedings of the National Academy of Sciences USA* **87**, 7570–7573; copyright held by the author.)

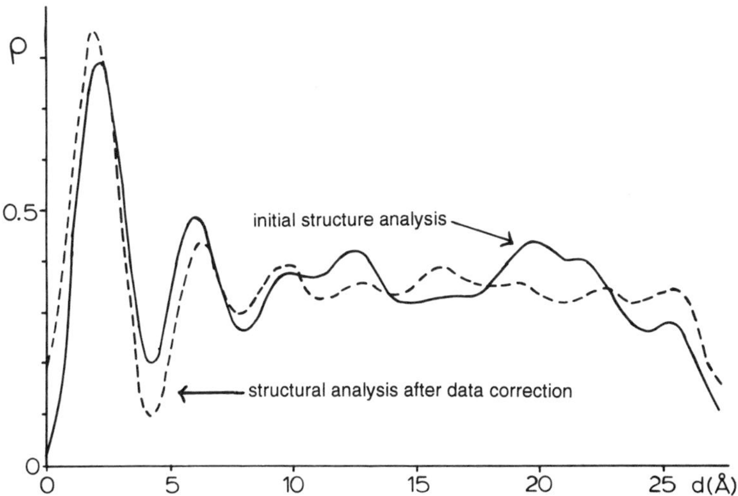

FIGURE 10.13. Potential profile for 1,2-dihexadecyl-*sn*-glycerophosphoethanolamine (1/2 bilayer). (Reprinted from D. L. Dorset, M. P. McCourt, W. F. Tivol, and J. N. Turner (1993) "Electron diffraction from phospholipids—an approximate correction for dynamical scattering and tests for a correct phase determination," *Journal of Applied Crystallography* **26**, 778–786; with kind permission of the International Union of Crystallography.)

low-angle values, giving the potential profile shown in Figure 10.13. The original analysis was also adversely affected by dynamical scattering.

An empirical correction was made for the dynamical scattering, using data collected at 1000 kV with a high-voltage electron microscope and at a lower voltage on a conventional microscope shown here in Figure 10.14 (Dorset et al., 1993). As was discussed in Chapter 5, one seeks:

$$|F_{00l}^{\text{kin}}| \approx [|F_{00l}^{HV}|^2 - \text{m}|F_{00l}^{LV}|^2]^{1/2}$$

When such an adjustment was made, the direct phasing procedure easily defined (Figure 10.15) values for all but one reflection (which is very weak) without the need for a high-resolution electron micrograph as an additional source of phases. (Details of this phase determination have already been discussed as an example in Chapter 4.) Essentially, after this new analysis, the phases of two reflections, ϕ_{002} and ϕ_{004} were changed from π to 0 to produce the revised potential map in Figure 10.13. This map was found to have a flatter density distribution across the hydrocarbon chain-packing region, a result expected for such a molecular array. Later, a new image analysis with CRISP (Hovmöller, 1992) reproduced the correct phase assignment for the low-angle reflections. After these adjustments were made, it was found that the error in the original structural model was not too significant.

FIGURE 10.14. Electron diffraction patterns from 1,2-dihexadecyl-*sn*-glycerophosphocholine: (a) obtained at 1000 kV; (b) obtained at 40 kV. (Reprinted from D. L. Dorset, M. P. McCourt, W. F. Tivol, and J. N. Turner (1993) "Electron diffraction from phospholipids—an approximate correction for dynamical scattering and tests for a correct phase determination," *Journal of Applied Crystallography* **26**, 778–786; with kind permission of the International Union of Crystallography.)

Table 10.7. Observed and Calculated Structure Factors for Epitaxially Oriented 1,2-Dimyristoyl-*sn*-glycerophosphoethanolamine

00l	$\|F_o\|$	F_c
00 1	1.38	1.92
00 2	1.59	–0.24
00 3	0.87	–0.40
00 4	1.36	–1.44
00 5	0.89	–0.79
00 6	0.78	–0.88
00 7	0.82	–0.83
00 8	0.44	–1.03
00 9	0.90	–1.51
00 10	1.34	–1.42
00 11	2.04	–2.05
00 12	1.47	–1.55
00 13	1.44	–1.96
00 14	1.40	–0.68

Translational searches were also used to determine the crystal structure of L-DMPE based on electron diffraction lamellar data (Dorset, 1988a), with the match shown in Table 10.7. Direct methods find the same solution except for an uncertain assignment for the weakest reflection. The methylene subcell packing in the HS1 form had been found earlier for the racemic form of this compound (Figure 10.1), a structure solved originally by Patterson methods (Dorset, 1976c) (see Table 10.8). Later it was redetermined by direct methods in the second application to experimental electron diffraction data (Dorset and Hauptman, 1976). The hydrated form of this lipid has also been studied (Dorset, 1988a). Epitaxially oriented crystals maintained in a high humidity have been shown to swell (Figure 10.16) and the expanded structure can be maintained in the electron microscope at low temperatures in a liquid, nitrogen-cooled specimen holder.

The structures of phospholipid binary solid solutions (Dorset and Massalski, 1987) have also been determined. Electron diffraction patterns and lattice images were used to follow the increase of lamellar spacings with concentration of the longer component for solutions of L-DLPE/L-DMPE, L-DLPE/L-DPPE, L-DMPE/L-DPPE, and L-DMPE/L-DHPE (Figure 10.17). The latter combination was observed both in crystalline and thermotropic smectic phases.

Table 10.8. Observed and Calculated Structure Factors for 1,2-dipalmitoyl-*rac*-glycerophosphoethanolamine Projected down the Chain Axes to View the (*hk*0) *HS*1 Subcell Diffraction Pattern

*hk*0	$\lvert F_o \rvert$	F_c (s)	*hk*0	$\lvert F_o \rvert$	F_c (s)
020	4.65	−5.00	320	5.90	6.24
040	4.70	−6.73	330	2.55	−2.66
060	1.65	2.94	340	2.70	−3.02
080	1.65	0.56	350	1.65	−1.93
110	2.00	−2.24	360	1.80	−0.34
120	12.00	−13.63	370	1.60	0.99
130	2.45	1.57	400	4.10	4.39
140	3.30	4.49	410	3.10	−2.84
150	2.10	1.05	420	1.95	−2.01
160	3.30	1.77	430	1.75	−2.38
170	1.60	−0.60	440	2.25	−1.00
200	11.45	−13.39	450	1.15	1.51
210	3.75	3.40	460	1.00	1.27
220	3.45	3.82	510	1.60	−2.02
230	2.25	2.68	520	2.20	−1.23
240	5.45	4.44	530	1.64	1.20
250	2.35	−1.53	540	0.85	1.60
260	1.80	−1.88	550	0.70	1.29
270	1.20	−1.06	600	0.50	−0.54
280	1.20	−0.15	610	1.20	1.23
310	3.00	3.54	620	0.70	0.16

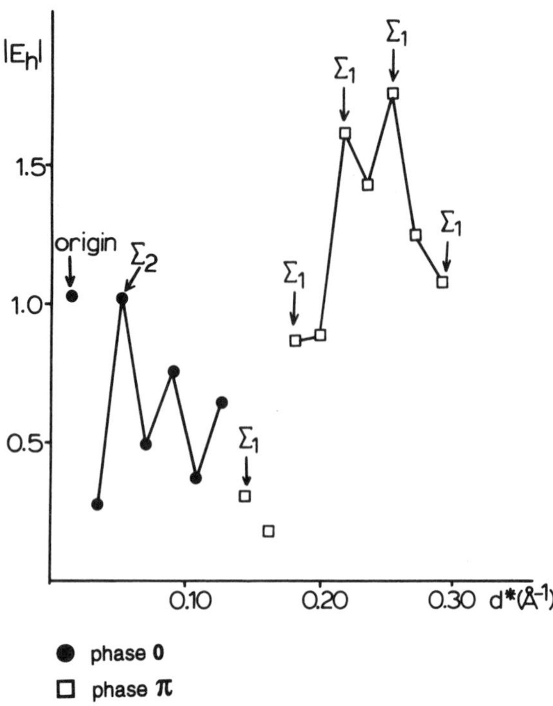

FIGURE 10.15. Direct phase determination of 1,2-dihexadecyl-*sn*-glycerophosphoethanolamine. (Reprinted from D. L. Dorset, M. P. McCourt, W. F. Tivol, and J. N. Turner (1993) "Electron diffraction from phospholipids—an approximate correction for dynamical scattering and tests for a correct phase determination," *Journal of Applied Crystallography* **26**, 778–786; with kind permission of the International Union of Crystallography.)

A 2:3 combination of L-DMPE/L-DPPE was epitaxially oriented on naphthalene to collect electron diffraction intensity data, so that its lamellar structure could be determined (Dorset, 1994b). (The methylene subcell is *H* and oriented perpendicular to the layer plane.) For 17 observed diffraction orders, direct methods determined the phases of 12 reflections. The phases of three more were accepted after the reverse Fourier transform of the initial potential map was calculated after flattening the density in the chain-packing region. Finally, after permuting the phases of the remaining two reflections, the best map was chosen on the basis of smoothness of the hydrocarbon chain packing. As shown in the final map (Figure 10.18), the polar region of the bilayer is stable, meaning that the hydrogen bonding network is not disrupted by forming the solid solution. However, lower density due to the mixing of two chain lengths can be identified at the bilayer nonpolar interface. Final calculated and observed structure factors, as well as phases determined by direct methods, are listed in Table 10.9

Finally, a comparison of subcell and lamellar spacing has been made (Dorset et al., 1983a) for phosphatidylethanolamines with their analogs based on the configurational isomers of cyclopentane-1,2,3-triol (lamellar spacings are compared in Table 10.10). Unlike the neutral lipids, it is difficult to use these data for an unequivocal prediction of most likely molecular conformations, though some similarities can be found for the native lipid and "reasonable" locked conformers.

10.2.4.2. N-*Methylphosphatidylethanolamines.* Paracrystals of 1,2-dihexdecyl-*sn*-glycerophospho-*N*-methylethanolamine (L-DHPEM) were grown by epitaxial orientation on naphthalene (Dorset, 1988b). Two polymorphic forms were found on the electron microscope grids, one of which (termed form 1) packed with untilted chains in the *H* methylene subcell and with a lamellar spacing $d_{001} = 58.4 \pm 0.3$ Å, The other, form 2, packed in an untilted $O_{\perp}$ subcell with a much shorter lamellar spacing, $d_{001} = 33.4 \pm 0.5$ Å (Figure 10.19). Comparing the lamellar spacing of the first form, which is due to an ordinary bilayer packing, to the values found for the *N,N*-dimethylphosphatidylethanolamine and phosphatidylcholine, it is clear that this first methylation accounts for most of the longitudinal increase in the bilayer cross-section for this sequence of headgroups formed from ethanolamine.

For crystal structure analysis of the form 1 structure, two headgroup conformations of the L-DMPC crystal structure (Pearson and Pascher, 1979) were used to construct trial molecular models for a translational search. The conformation where the substituted ethanolamine moiety parallels the bilayer surface arrived at the best fit

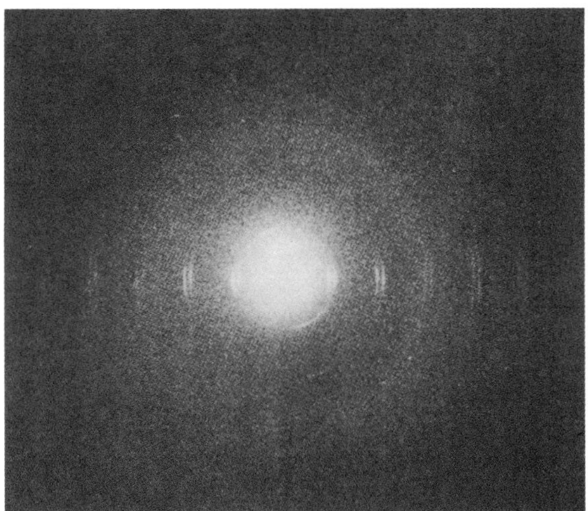

FIGURE 10.16. Electron diffraction pattern of partially swollen L-DMPE crystals. Note doubling of lamellar spacings due to two crystalline hydrations. (Reprinted from D. L. Dorset (1988) "How different are the crystal structures of chiral and racemic diacyl phosphatidylethanolamines?," *Zeitschrift für Naturforschung, Teil C* **43c**, 319–327; with kind permission of the Verlag der Zeitschrift für Naturforschung.)

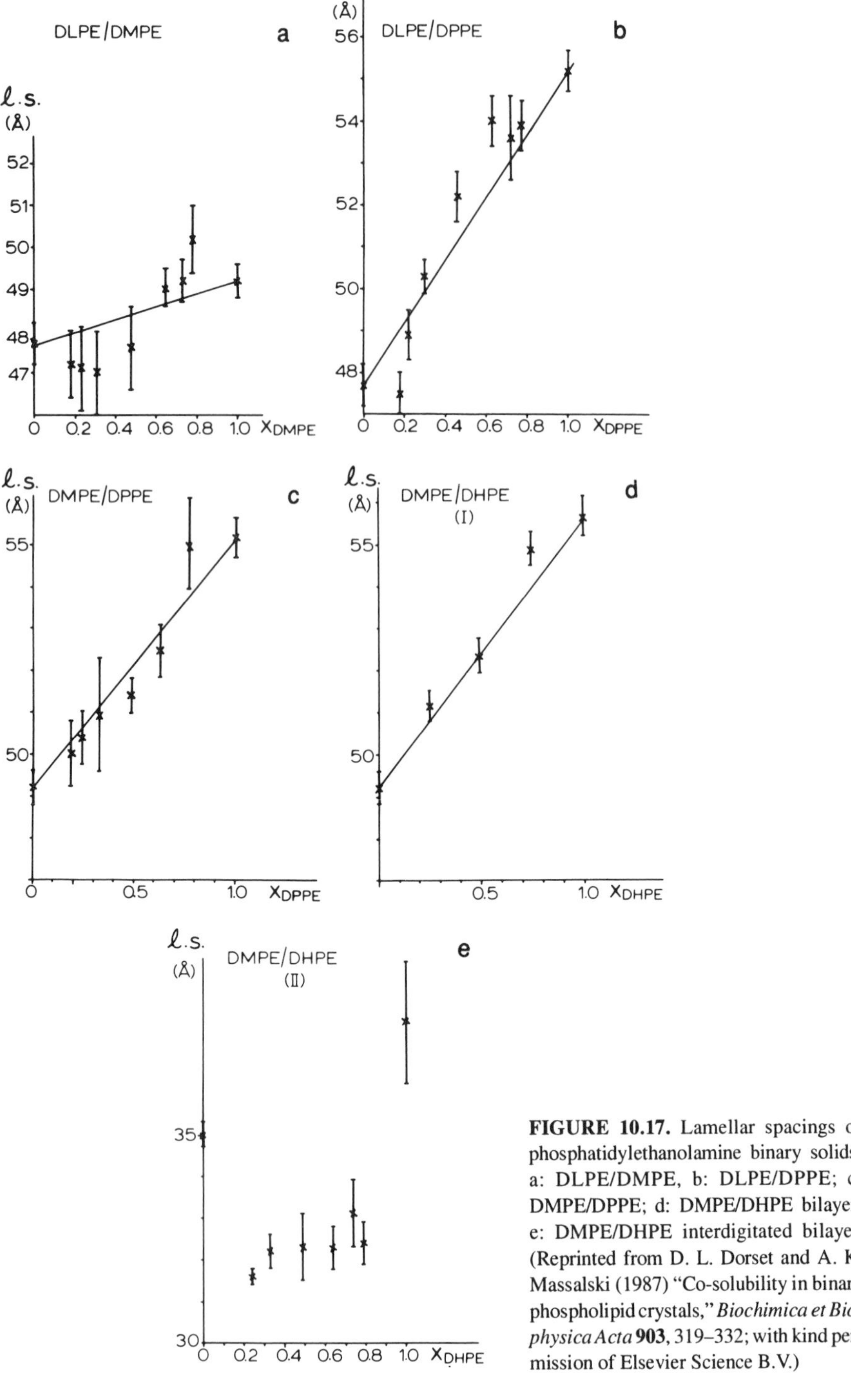

FIGURE 10.17. Lamellar spacings of phosphatidylethanolamine binary solids: a: DLPE/DMPE, b: DLPE/DPPE; c: DMPE/DPPE; d: DMPE/DHPE bilayer; e: DMPE/DHPE interdigitated bilayer. (Reprinted from D. L. Dorset and A. K. Massalski (1987) "Co-solubility in binary phospholipid crystals," *Biochimica et Biophysica Acta* **903**, 319–332; with kind permission of Elsevier Science B.V.)

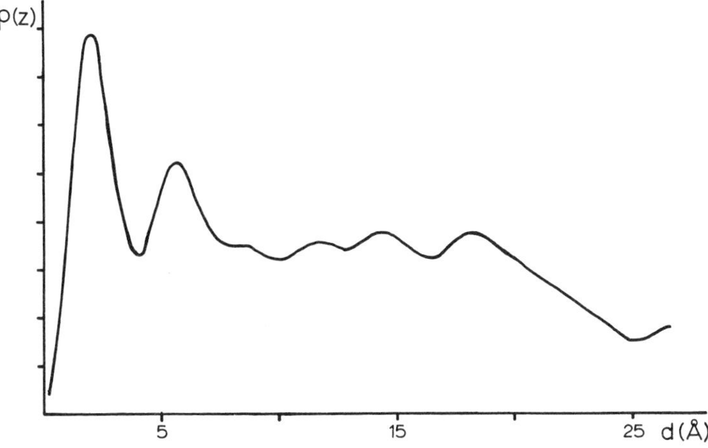

FIGURE 10.18. Layer profile for a 2:3 solid solution of L-DMPE/L-DPPE.

Table 10.9. Amplitudes and Phases for an L-DMPE/L-DPPE Solid Solution

l	$\lvert F_{obs}\rvert$	F^*_{calc}	ϕ (direct methods)
1	1.86	1.80	0
2	0.27	–0.42	—
3	0.79	0.55	0
4	0.40	–0.83	—
5	0.35	–0.20	—
6	0.39	–0.48	π
7	0.51	–0.20	—
8	0.17	–0.26	π
9	0.40	–0.44	—
10	0.64	–0.40	π
11	0.71	–0.61	π
12	0.81	–0.77	π
13	1.09	–0.93	π
14	0.81	–0.95	π
15	0.68	–1.01	π
16	0.57	–0.67	π
17	0.39	–0.32	π

Table 10.10. Lamellar Spacings for 1,2-Dipalmitoyl Phosphatidylethanolamines Measured from Electron Diffraction Experiments on Epitaxially Oriented Samples or X-ray Diffraction Experiments on Multilamellar Arrays Compared to the Conformationally Locked Derivatives Based on Configurational Isomers of Cyclohexane-1,2,3-triol

Compound	Electron diffraction	X-ray diffraction
DL-DPPE	54.3 ± 1.0 Å	53.8 Å
L-DPPE	55.0 ± 0.3 Å	55.0 Å
cyclitol 1,2,3/0-1*P*	49.0 ± 0.2 Å	48.8 Å
cyclitol 1,3/2-1*P*	54.7 ± 0.9 Å	50.8 Å
cyclitol 1,2/3-3*P*	55.6 ± 0.7 Å	53.2 Å
cyclitol 1,2/3-1*P*	55.4 ± 1.2 Å	48.5 Å

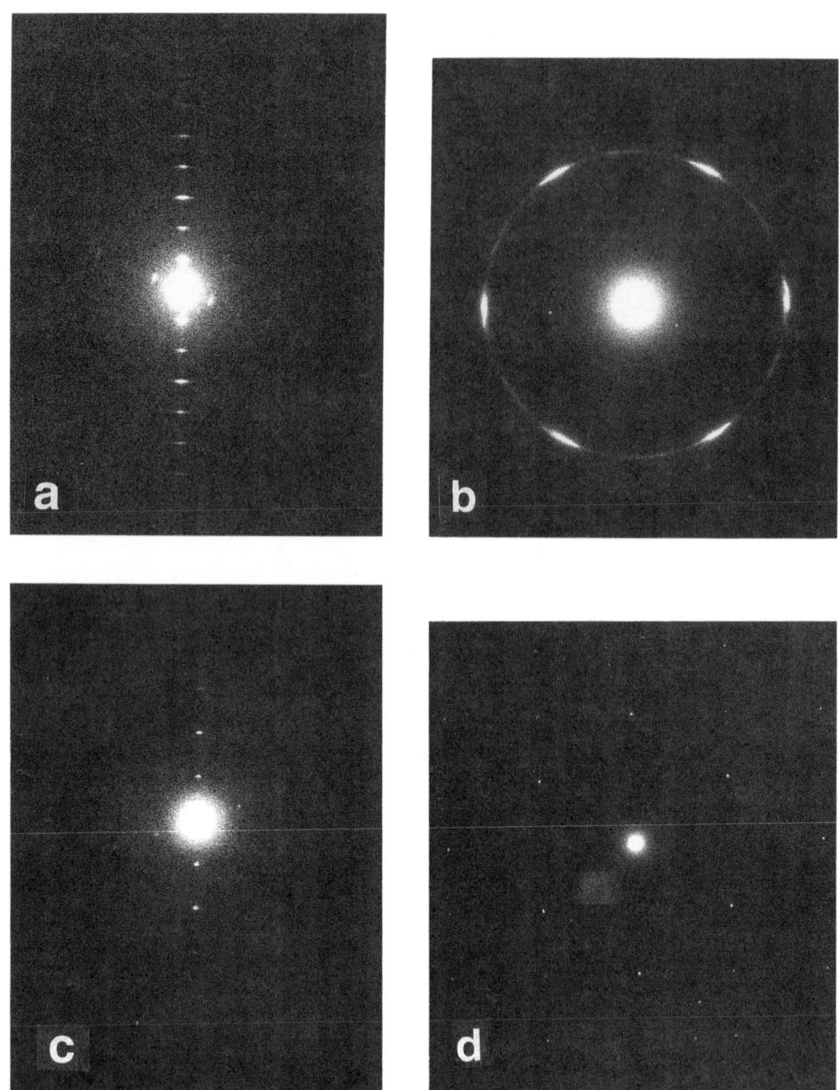

FIGURE 10.19. Electron diffraction patterns from 1,2-dihexadecyl-*sn*-glycerophospho-*N*-methylethanolamine: (a) form 1, epitaxial orientation; (b) form 1, solution growth; (c) form 2, epitaxial orientation; (d) form 2, solution growth. Form 1 is an ordinary bilayer; form 2 is interdigitated. (Reprinted from D. L. Dorset (1988) "Two untilted lamellar packings for an ether-linked phosphatidyl-*N*-methylethanolamine. An electron crystallographic study," *Biochimica et Biophysica Acta* **938**, 279–292; with kind permission of Elsevier Science B.V.)

Table 10.11. Observed and Calculated Structure Factors for Epitaxially Oriented 1,2-Dihexadecyl-*sn*-glycerophospho-N-methylethanolamine, Bilayer Form 1

00*l*	$|F_o|$	F_c
00 1	0.22	0.35
00 2	0.09	−0.11
00 3	0.24	0.16
00 4	0.18	−0.17
00 5	0.18	0.06
00 6	0.15	−0.18
00 7	0.11	−0.14
00 8	0.15	−0.10
00 9	0.14	−0.14
00 10	0.20	−0.13
00 11	0.14	−0.18
00 12	0.23	−0.23
00 13	0.14	−0.22

to the experimental data (Table 10.11). This could also be established by comparison of Patterson functions (Figure 10.20). A direct phasing analysis arrived at the same solution found with the translational search with a model (Dorset, 1990g; 1991b) (Figure 4.14c).

Analysis of the form 2 structure was based also on a headgroup conformer with the ethanolamine moiety parallel to the bilayer surface. At the minimum *R*-factor (Table 10.12), the agreement of Patterson functions for the model and experimental data was again very good (Figure 10.21). It is clear that this second form is a structure with an interdigitated chain packing. The phosphorus distance from the origin is comparable to the value found for lysolecithins (Hauser et al., 1980; Pascher et al., 1986).

10.2.4.3. N,N-*Dimethylphosphatidylethanolamines.* Epitaxial orientation of 1,2-dipalmitoyl-*sn*-glycerophospho-*N,N*-dimethylethanolamine (L-DPPEM$_2$) was

Table 10.12. Observed and Calculated Structure Factors for Epitaxially Oriented 1,2-Dihexadecyl-*sn*-glycerophospho-*N*-methylethanolamine, Interdigitated Bilayer Form 2

00*l*	$|F_o|$	F_c
00 1	0.20	−0.25
00 2	0.36	0.38
00 3	0.41	0.34
00 4	0.23	0.16
00 5	0.20	0.27
00 6	0.25	0.25
00 7	0.19	0.21
00 8	0.15	0.13

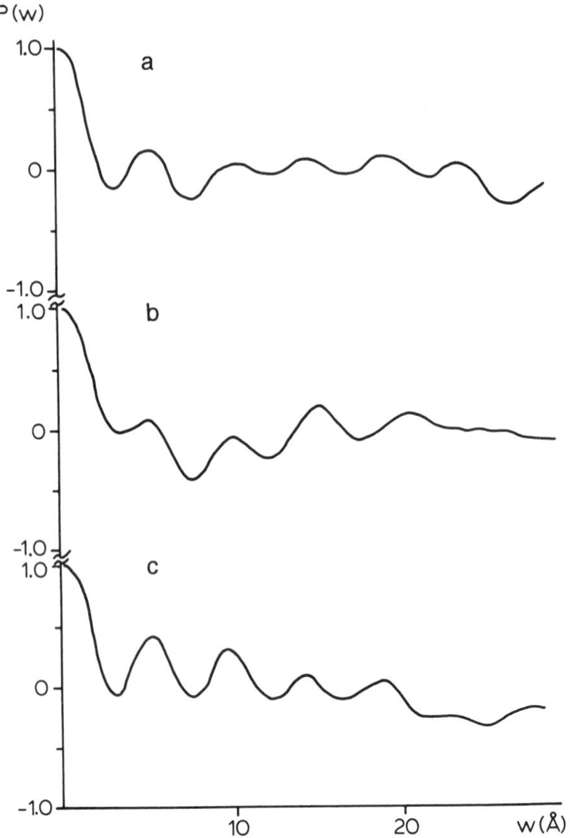

FIGURE 10.20. Comparison of Patterson functions for L-DHPEM, form 1: (a) observed data; (b) model with *N*-methylethanolamine perpendicular to bilayer surface; (c) model with *N*-methylethanolamine parallel to bilayer surface. (Reprinted from D. L. Dorset (1988) "Two untilted lamellar packings for an ether-linked phosphatidyl-N-methylethanolamine. An electron crystallographic study," *Biochimica et Biophysica Acta* **938**, 279–292; with kind permission of Elsevier Science B.V.)

achieved by growth on naphthalene when crystallizing from the co-melt. An electron diffraction lamellar spacing $d_{001} = 58.7 \pm 0.4$ Å (Dorset and Zhang, 1990) was found in experimental patterns. The acyl chains packed in an untilted *H* subcell. The one-dimensional Patterson function calculated from the lamellar intensity data compared well to the one found for a lecithin (Pearson and Pascher, 1979). Thus, a model based on the conformer of L-DMPC with a choline moiety parallel to the bilayer surface was constructed and used to seek the structure by translation. A direct phase determination was also carried out. The final agreement of calculated and observed structure factors is shown in Table 10.13. There is general agreement of the two phase determinations although two assignments are missing in the direct analysis based on

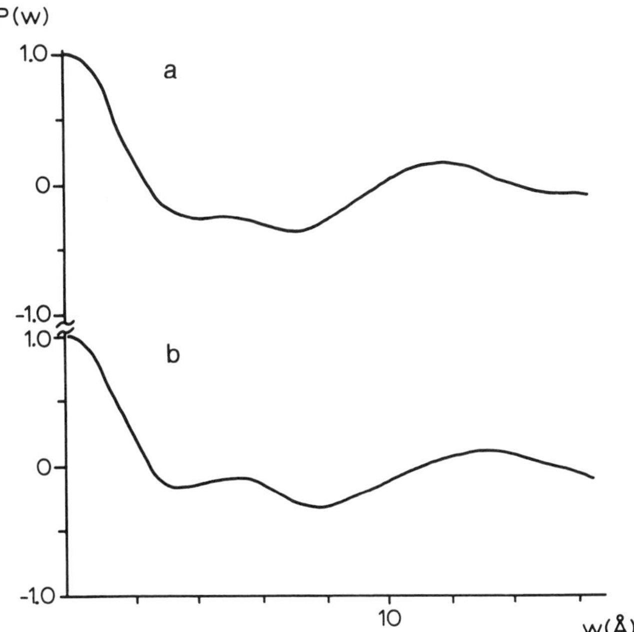

FIGURE 10.21. Comparison of Patterson functions for L-DHPEM, form 2: (a) experimental; (b) model. (Reprinted from D. L. Dorset (1988) "Two untilted lamellar packings for an ether-linked phosphatidyl-N-methylethanolamine. An electron crystallographic study," *Biochimica et Biophysica Acta* **938**, 279–292; with kind permission of Elsevier Science B.V.)

Table 10.13. Observed and Calculated Structure Factors for Epitaxially Oriented 1,2-Dipalmitoyl-*sn*-glycerophospho-*N,N*-dimethylethanolamine

00*l*	$\|F_o\|$	F_c
00 1	0.34	0.38
00 2	0.20	–0.11
00 3	0.15	0.06
00 4	0.17	–0.26
00 5	0.15	–0.11
00 6	0.14	–0.21
00 7	0.11	–0.15
00 8	0.12	–0.13
00 9	0.13	–0.18
00 10	0.17	–0.10
00 11	0.15	–0.14
00 12	0.12	–0.12
00 13	0.12	–0.13

phase invariant sums. A later refinement (Dorset, 1991b) demonstrated that one phase assignment from the model may be incorrect. Despite these small inconsistencies, it is clear that a headgroup conformation of this *N,N*-dimethylphosphatidylethanolamine similar to that of lecithin can be found in bilayers (Figure 4.14e). Such a geometry was not found in the x-ray crystal structure of a shorter chain homolog. In that form, the polar group is perpendicular, rather than parallel, to the bilayer surface (Pascher and Sundell, 1986). Hence the electron crystallographic analysis indicates that two alternative headgroup packing geometries are possible.

10.2.4.4. Phosphatidylcholines. Electron diffraction patterns from layers of 1,2-dihexadecyl-*sn*-glycerophosphocholine (L-DHPC), epitaxially crystallized on naphthalene, were observed to have a lamellar spacing $d_{001} = 59.2 \pm 0.2$ Å (Dorset, 1987b). The methylene subcell was found to be hexagonal and the chain axes were seen to lie normal to the bilayer surface. A similar lamellar distance, $d_{001} = 57.9 \pm 0.9$ Å, was observed for L-DPPC (Dorset et al., 1983). (The hexagonal subcell is also found in many ester-linked phosphatidyl cholines.) The one-dimensional Patterson function calculated from the lamellar intensity data from the ether-linked lipid (Figure 4.5) was similar to the one calculated from x-ray intensities for the lecithins.

Two headgroup conformations found in the crystal structure of L-DMPC (Pearson and Pascher, 1979) were used to construct models for translational searches. Only the search (Figure 4.3b) with a headgroup parallel to the bilayer surface was successful (Table 10.14, as shown also by comparison of Patterson functions in Figure 10.28). One crystallographic phase assignment disagreed with the one found by model translation when direct methods were used, but, as usual, this corresponded to a reflection with rather weak intensity (Dorset, 1991b). Thus the ether- and ester-linked lecithins have essentially the same layer packing (Figure 4.14d).

The above analyses assumed that all form factors were for neutral atoms. In future work, it might be worth considering the potential effect of charged groups in the headgroup zwitterion on the electron-scattering factors of the phosphorus and nitrogen

Table 10.14. Observed and Calculated Structure Factors for Epitaxially Oriented 1,2-Dihexadecyl-*sn*-glycerophosphocholine

$00l$	$\|F_o\|$	F_c
00 1	0.30	0.34
00 2	0.07	−0.08
00 3	0.22	0.12
00 4	0.15	−0.19
00 5	0.12	−0.09
00 6	0.16	−0.20
00 7	0.11	−0.14
00 8	0.14	−0.11
00 10	0.20	−0.13
00 11	0.12	−0.19
00 12	0.14	−0.14

Table 10.15. Lamellar Spacings for 1,2-Dipalmitoyl Phosphatidylcholines Measured from Electron Diffraction Experiments on Epitaxially Oriented Samples or X-ray Diffraction Experiments on Multilamellar Arrays Compared to the Conformationally Locked Derivatives Based on Configurational Isomers of Cyclohexane-1,2,3-triol

Compound	Electron diffraction	X-ray diffraction
L-DPPC	57.9 ± 0.9 Å	58.0 Å
cyclitol 1,2,3/0-1*P*	58.1 ± 0.9 Å	46.0 Å
cyclitol 1,3/2-1*P*	60.4 ± 0.9 Å, 64.2 ± 0.6 Å	51.5, 60.8 Å
cyclitol 1,2/3-3*P*	56.8 ± 0.7 Å	40.1, 46.4, 55.9 Å
cyclitol 1,2/3-1*P*	56.5 ± 1.2 Å	46.5, 56.6 Å

atoms, given the density distributions found for charged groups in a recent electron crystallographic determination of a membrane protein (Kühlbrandt et al., 1994).

A comparison of the lamellar and subcell packings has been made for the lecithins and their analogs based on the configurational isomers of cyclopentane-1,2,3-triol (lamellar spacings are given in Table 10.15) (Dorset et al., 1983). As with the phosphatidylethanolamines above, it is difficult to find direct similarities of molecular conformation for these compounds.

Preliminary studies have been carried out on the phosphatidylcholines substituted with iso-branched fatty acids of the type described by Church et al. (1986). The subcell packing is *H*. Although these compounds can be epitaxially oriented on naphthalene crystals, they have the striking feature of being unusually hygroscopic, and, unless they are kept in vacuo, the absorption of water from air will cause them to form vesicles so that they lose their specific orientation with respect to the nucleating surface.

Electron diffraction patterns have also been obtained from unsupported hydrated bilayers of L-DPPC formed across 1000-mesh electron microscope grids dipped into a Langmuir trough (Hui and Parsons, 1974). The differentially pumped environmental chamber was also equipped with a heater so that phase transitions of the lipid could be followed directly via the diffraction pattern as the chain structure changed from the hexagonal subcell to the disordered liquid crystalline chain packing. When cholesterol was added to the bilayers, domain structures could be observed at lower temperatures by diffraction contrast electron microscopy (Hui and Parsons, 1975). Finally, when multilayers of the lipid were built up, a change of chain packing from a rectangular to an oblique layer was observed (Hui, 1976), in agreement with early experiments with simpler amphiphilic molecules (see Chapter 9).

Recently there has been interest in using phospholipids composed of fatty acid chains with alkyne segments to construct thin film devices. A monolayer of the monomer lipid can be formed on a Langmuir trough and, hopefully, the reactive triple bonds can then be cross-linked in UV light to synthesize a conjugated polymer in situ with long-range conjugation parallel to the layer surface. Lando and Sudiwala (1990) have attempted a structure analysis of such a monomer (1,2-bis(10,12-tricosadiynoyl)-*sn*-glycerophosphocholine) with electron diffraction intensities measured from built-up multilayers. They reported cell constants $a = 5.18$, $b = 7.79$, $c = 78.5$ Å, $\beta = 117°$,

Table 10.16. Observed and Calculated Structure Factors for an Analysis of an Alkyne Fatty Acid–containing Lecithin Based Only on the Subcell Carbon Atom Position

hkl	$hk0$	$\lvert F_o \rvert$	$\lvert F_c \rvert$
20,–14	020	15.24	13.38
11,–7	110	20.89	23.70
21,–14	120	5.62	3.37
31,–21	130	7.16	4.10
020	200	18.71	21.03
12,–7	210	6.73	4.38
22,–14	220	9.83	8.94
32,–21	230	4.64	4.28
13,–7	310	11.27	9.25
23,–14	320	6.00	5.97
33,–21	330	2.91	1.96
14,–7	410	3.88	3.82
24,–14	420	4.20	2.36
34,–21	430	3.50	4.52
15,–7	510	3.87	0.70
25,–14	520	3.18	4.02
35,–21	530	3.18	0.17

space group $P2_1$. A set of zonal structure factor magnitudes is listed in Table 10.16 with their three-dimensional indices. After a three-dimensional structure analysis, including refinement of the headgroup conformation, a residual $R = 0.159$ was found.

The diffraction pattern of this lipid, published by these authors, strongly resembled the one already shown for the $O_\perp$ subcell and the experimental a and b cell constants were very much in the accepted range (e.g., Abrahamsson et al., 1978). If the reflections were reindexed as an $hk0$ set (according to the central section theorem, this is permitted because this is a projection), a direct phase determination could be carried out after calculating $\lvert E_h \rvert$ (here based on the reported unit cell contents). Weak reflections along one axis were reset to zero. The progress of the direct analysis is outlined in Table 10.17. The map calculated from the phased structure factors (Figure 10.22) strongly resembled those expected for the methylene subcell packing; there were no density features observed for the headgroup. If only one carbon atom position (defined by the known subcell structure) was used for a structure factor calculation, the match to the observed data was again reasonable (Table 10.16), i.e., $R = 0.22$.

Table 10.17. Direct Phase Determination with Data from Table 10.16

origin definition:
$\phi_{110} = 0, \phi_{320} = 0$
Σ_1-triples:
$\phi_{220} = 0$
Σ_2-triples:
$\phi_{310} = \phi_{420} = \phi_{530} = \phi_{430} = \phi_{330} = \phi_{200} = \phi_{020} = \phi_{130} = \phi_{210} = 0$

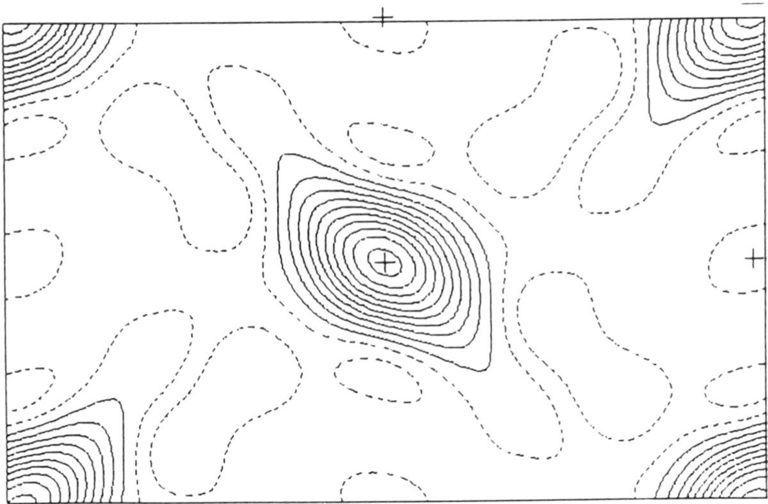

FIGURE 10.22. Direct phase determination for an alkyne fatty acid–containing lecithin. Only the methylene subcell packing is observed in the map.

While the residual found in the multiparameter model may seem to be preferable to the one found with the single-atom structure, consider the statistical significance of this figure of merit (Hamilton, 1964). For the complicated structure model, based on 19 data (including the weak reflections) and 8 fitted conformational parameters, any solution up to $R = 0.28$ cannot be distinguished from it at the 5% confidence level. The result with the simpler model falls well within this range and has the virtue of being based on only three parameters (two positional and an isotropic thermal value).

The analysis underscores the danger of using data from a projection down a long unit cell axis, particularly if the structure contains a sublattice. The layers were observed by these authors to be curved so that only the subcell Fourier transform will be expressed by the observed intensities, as explained in Chapter 5.

10.2.4.5. Phosphatidic Acids. Preliminary electron diffraction data exist for the phosphatidic acids and their dipotassium salts, as outlined in Table 10.18. Comparing these to derivatives based on the configurational isomers of cyclohexane-1,2,3-triol, it may be possible to state that the free acid lamellae pack similarly to the 1,2/3-3*P* analog. Such a comparison is not possible, however, for the salts (Dorset et al., 1983). To date, no further structure analyses have been carried out due to insufficient data resolution in the patterns.

10.2.4.6. Monogalactosyl Diglyceride. One of the most prevalent lipids in nature, the monogalactosyl diglyceride (the fatty acid residue being stearoyl) found in the cell membranes of plants has been epitaxially crystallized on naphthalene. Electron diffraction patterns and lattice images reveal that this lipid packs in bilayers with a repeat of 54 Å (Sen et al., 1987). The methylene subcell is hexagonal with untilted chains. Attempts to solve its layer structure are in progress.

Table 10.18. Lamellar Spacings for 1,2-Dipalmitoyl Phosphatidic Acids and Their Dipotassium Salts, Measured from Electron Diffraction Experiments on Epitaxially Oriented Samples or X-ray Diffraction Experiments on Multilamellar Arrays Compared to the Conformationally Locked Derivatives Based on Configurational Isomers of Cyclohexane-1,2,3-triol

Compound	Electron diffraction	X-ray diffraction
L-DPPA	55.9 ± 0.5 Å	57.4 Å
cyclitol 1,2,3/0-1*P*	52.8 ± 0.8 Å	48.4, 53.3 Å
cyclitol 1,3/2-1*P*	53.2 ± 0.8 Å	41.1 Å
cyclitol 1,2/3-3*P*	54.4 ± 0.8 Å	52.9 Å
cyclitol 1,2/3-1*P*	52.8 ± 0.4 Å	54.5 Å
K_2 (L-DPPA)	54.8 ± 0.2 Å	55.8 Å
cyclitol 1,2,3/0-1*P*	56.4 ± 0.2 Å	52.1 Å
cyclitol 1,3/2-1*P*	55.0 ± 0.7 Å	54.7 Å
cyclitol 1,2/3-3*P*	55.7 ± 0.6 Å	54.2 Å
cyclitol 1,2/3-1*P*	55.5 ± 0.5 Å	54.5 Å

10.3. Cholesteryl Esters

The x-ray crystal structures of the cholesteryl esters have been extensively studied by Craven (1986) and coworkers. For the saturated series, there are essentially three packing motifs (Sawzik and Craven, 1980). For ester chain lengths between C_6 and C_8, the so-called monolayer II packing is found, a crystal form that is also found for the physiologically significant cholesteryl oleate (Craven and Guerina, 1979). Chains from C_9 through C_{12} pack in monolayer I, a layer packing also adapted by cholesteryl palmitoleate (Sawzik and Craven, 1982). Higher saturated chain lengths pack in the so-called bilayer structure, as exemplified by cholesteryl myristate (Craven and DeTitta, 1976).

Although good transmission electron diffraction patterns can be obtained from cholesteryl ester layers crystallized from dilute solution (Dorset, 1985b) (Figure 10.23), the intensity data are not very useful for determining the crystal structure. This problem, of course, is again due to the elastic crystal bending problem for other lipid structures and also the alkane derivatives discussed in Chapter 5. For the cholesteryl esters, its effect has been verified quantitatively with a model calculation. In terms of ab initio crystal structure analysis, the situation is less favorable than for other polymethylene-containing compounds, even for the one-layer structure where an actual subcell packing exists. That is to say, the methylene packing does not dominate enough of the total unit cell volume to be represented clearly in the diffraction intensities, particularly since there are also many interatomic vectors in cholesterol with about the same length and orientation. The experimental electron diffraction patterns can be used, however, for measurement of unit cell spacings.

Epitaxial crystals of these esters can also be formed on benzoic acid (Dorset, 1985b), and have characteristic electron diffraction patterns (Figure 10.24). As was found for the glycerolipids, there seems to be some paracrystalline disorder in the layer

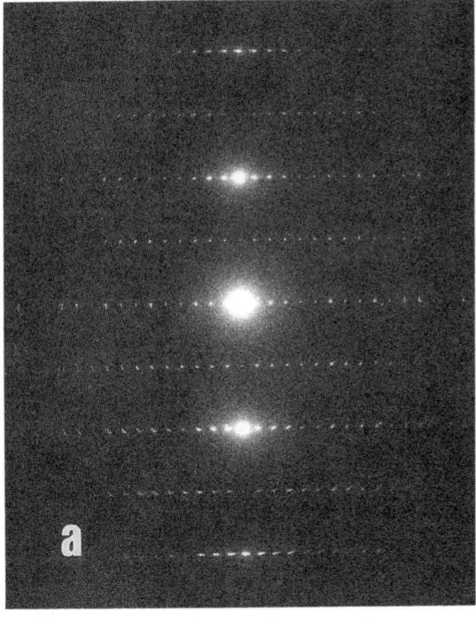

(continued)

FIGURE 10.23. Electron diffraction patterns form solution-crystallized cholesteryl esters: (a) monolayer II, cholesteryl oleate; (b) monolayer I, cholesteryl palmitoleate; (c) bilayer, cholesteryl stearate. (Reprinted from D. L. Dorset, W. A. Pangborn, and A. J. Hancock (1983) "Epitaxial crystallization of alkane chain lipids for electron diffraction analysis," *Journal of Biochemical and Biophysical Methods* **8**, 29–40; with kind permission of Elsevier Science B.V.)

packing so that intensities from a single two-dimensional net are not recorded. Again, the epitaxial nucleation is largely one-dimensional. Lattice images of the cholesteryl esters have also been recorded (Figure 10.25).

As shown by an analysis of simulated data, it is, in principle, possible to determine the layer structure from the lamellar data, but the potential distribution does not have the salient density features seen, e.g., in phospholipids, so that the one-dimensional structure is harder to interpret. On the other hand, the lamellar spacing measurements from individual microareas have confirmed the existence of previously unidentified polymorphs in the chain length favoring both monolayer I and bilayer structures. A listing of unit cell dimensions for a number of cholesteryl esters, based on the electron diffraction patterns from two orthogonal projections, is given in Table 10.19. A possible even–odd effect has also been identified from a plot of such lamellar diffraction spacings for the bilayer form (Figure 10.26).

These diffraction patterns have been most useful for the qualitative study of thermotropic mesomorphism and binary phase behavior (Dorset, 1985b, 1987f) in these compounds. For example, in solution-crystallized samples of cholesteryl myristate, it was possible to follow changes in the structure at both low and high temperature

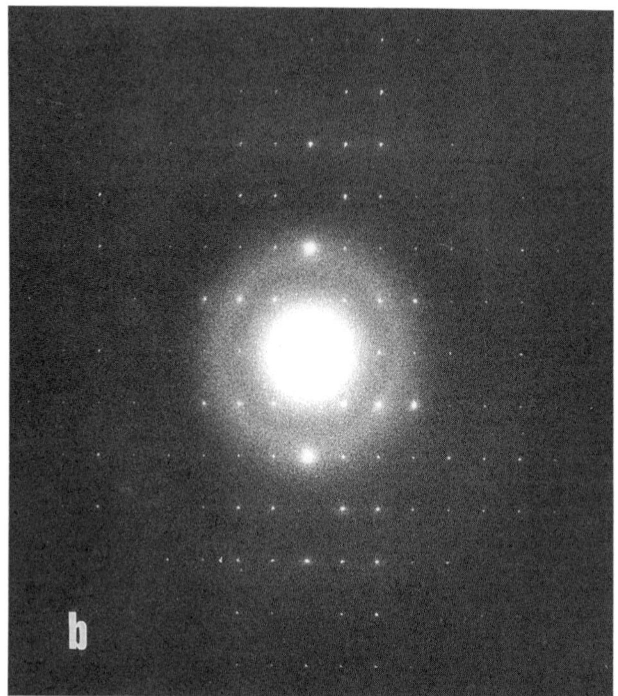

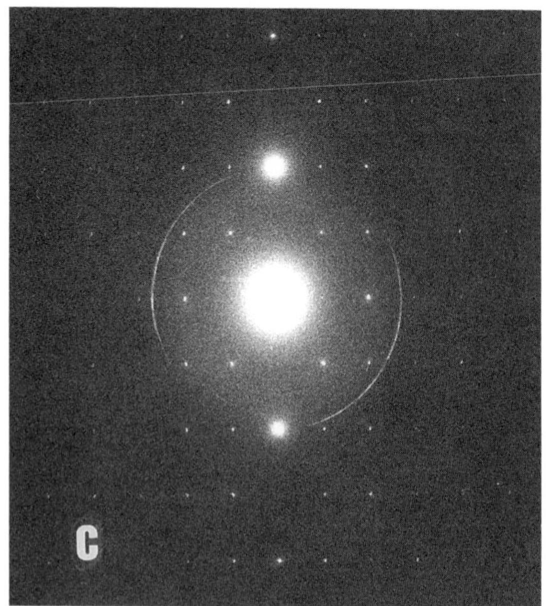

FIGURE 23. (*Continued*)

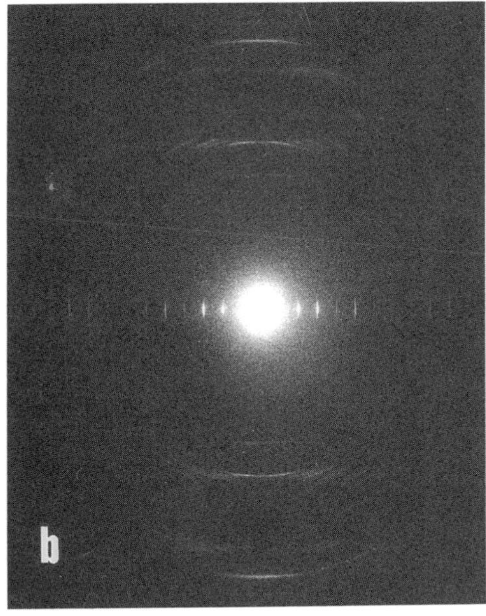

FIGURE 10.24. Electron diffraction patterns from epitaxially oriented cholesteryl esters: (a) monolayer I, cholesteryl laurate; (b) bilayer, cholesteryl myristate. (Reprinted from D. L. Dorset (1987) "Cholesteryl esters of saturated fatty acids: co-solubility and fractionation of binary mixtures," *Journal of Lipid Research* **28**, 993–1005; with kind permission of the *Journal*; copyright held by the author.)

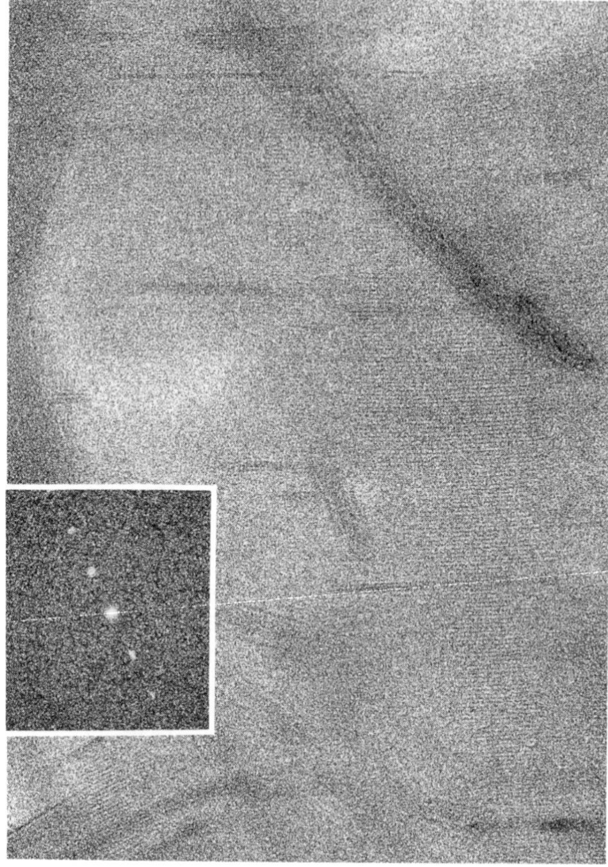

FIGURE 10.25. Low-dose electron micrograph of epitaxially oriented cholesteryl myristate. Inset is the optical transform of the lattice. (Reprinted from D. L. Dorset (1985) "Thermotropic mesomorphism of cholesteryl myristate. An electron diffraction study," *Journal of Lipid Research* **26**, 1142–1150; with kind permission of the *Journal*; copyright held by the author.)

Table 10.19. Unit Cell Parameters for Cholesteryl Esters Measured from Electron Diffraction Patterns (Solution and Epitaxially Crystallized Samples)

Compound	Cell constants (e.d.)	Cell constants (x.d.)
cholesteryl undecanoate	monolayer I	monolayer I
	$d_{100} = 13.33$	$a = 12.99$
	$d_{010} = 9.24$	$b = 9.00$
	$d_{001} = 31.43$ Å	$c = 31.03$ Å, $\beta = 90.58°$
cholesteryl laurate	monolayer I	monolayer I
	$d_{100} = 13.20$	$a = 12.99$
	$d_{010} = 9.14$	$b = 9.01$
	$d_{001} = 31.57$ Å	$c = 32.02$ Å, $\beta = 91.36°$
	bilayer	
	$d_{100} = 10.47$	
	$d_{010} = 7.83$	
	$d_{001} = 46.99$ Å	
cholesteryl myristate	bilayer	bilayer
	$d_{100} = 10.36$	$a = 10.26$
	$d_{010} = 7.68$	$b = 7.60$
	$d_{001} = 51.22$ Å	$c = 101.43$ Å, $\beta = 94.41°$
cholesteryl pentadecanoate	bilayer	
	$d_{100} = 10.47$	
	$d_{010} = 7.94$	
	$d_{001} = 54.90$ Å	
cholesteryl palmitate	bilayer	bilayer
	$d_{100} = 10.39$	$a = 10.15$
	$d_{010} = 7.71$	$b = 7.55$
	$d_{001} = 53.53$ Å	$c = 105.5$ Å, $\beta = 95.6°$
cholesteryl stearate	bilayer	bilayer
	$d_{100} = 10.54$	$a = 10.20$
	$d_{010} = 7.90$	$b = 7.55$
	$d_{001} = 56.87$ Å	$c = 57.5$ Å, $\beta = 96.0°$

(Figure 10.27), including the appearance of the diffuse ring in the smectic phase. Samples cooled through the cholesteric phase were found to have a rotational disorder which might be related to the helical twists proposed for such structures. Oriented samples heated toward the smectic phase were found to become more ordered so that the lamellar reflections became sharpened (Figure 10.28). An intermediate temperature could be reached where single-crystal and smectic diffraction patterns coexisted on the same lamellar row, and beyond this point, only the single lamellar reflection from the smectic structure was seen. This study has improved our understanding of the smectic structure, since it was not known how a bilayer structure with lamellar spacing near 50 Å could transform to a smectic layer with a spacing near 33 Å (Craven and DeTitta, 1976). The coexistence of polymorphs in a critical chain length domain (packing either in bilayer or monolayer I structures), in addition to the diffraction data in Figure 10.28, indicates that the transformation of the bilayer packing to a smectic

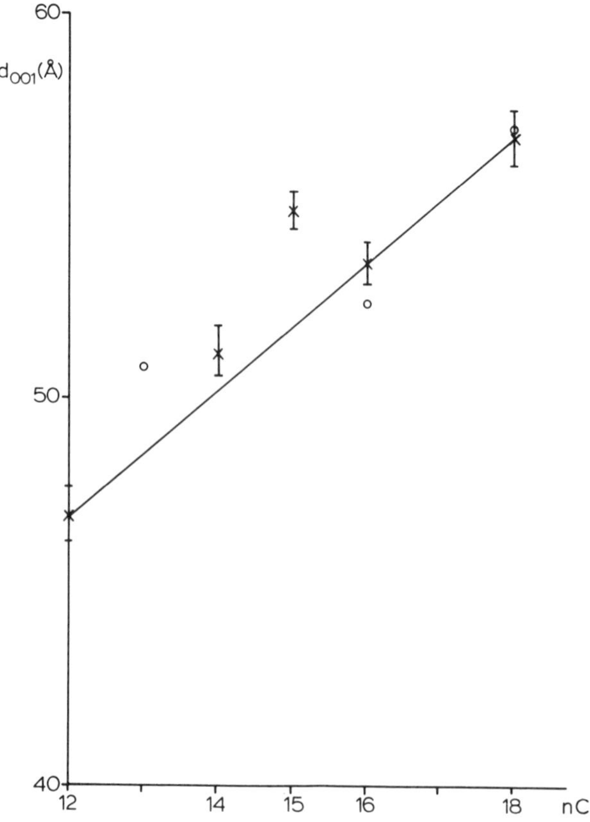

FIGURE 10.26. Lamellar spacings of cholesteryl esters packing in the bilayer form, showing a possible even–odd effect. (Reprinted from D. L. Dorset (1987) "Cholesteryl esters of saturated fatty acids: co-solubility and fractionation of binary mixtures," *Journal of Lipid Research* **28**, 993–1005; with kind permission of the *Journal*; copyright held by the author.)

array, loosely resembling the monolayer I crystal structure, might occur easily via a longitudinal molecular translations.

The study of binary solids has been especially facilitated by the electron diffraction information from epitaxially oriented material (Dorset, 1987f, 1988c). Although experimental cholesteryl ester phase diagrams had been published earlier (see review by Small, 1986), no systematic study of the effects of relative molecular volumes and space group symmetry (as discussed by Kitaigorodskii, 1961) had been carried out until recent electron diffraction and DSC studies of saturated chain series packing in the monolayer I and bilayer structures. In general, it is found that two esters which differ by even one methylene group will fractionate if each prefers a different crystal structure. (Actually the difference may be even *zero* methylene groups, as demonstrated by the fractionation of two polymorphs of the same ester.) This observation is

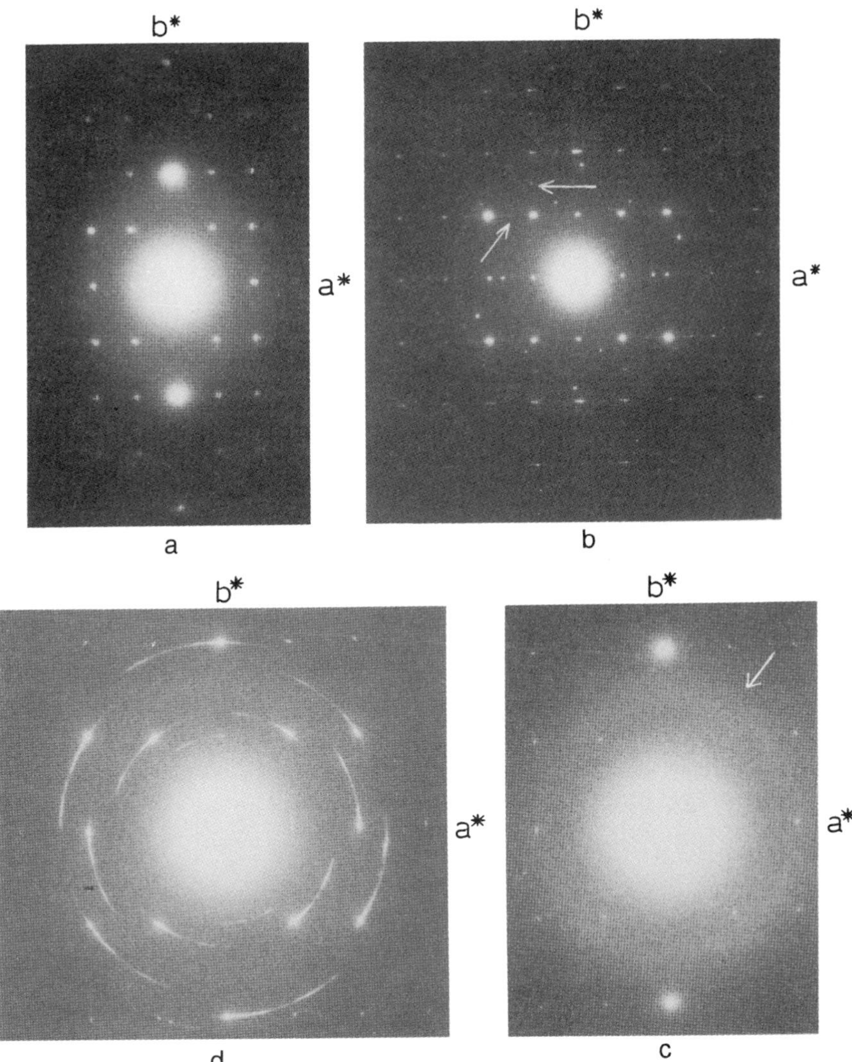

FIGURE 10.27. Thermotropic behavior of cholesteryl myristate studied with solution-crystallized samples: (a) room temperature; (b) low temperature (<–65 °C), showing presence of added lattice rows; (c) transition to smectic; (d) rotational disorder when specimen is cooled from the cholesteric (but not smectic) phase. (Reprinted from D. L. Dorset (1985) "Thermotropic mesomorphism of cholesteryl myristate. An electron diffraction study," *Journal of Lipid Research* **26**, 1142–1150; with kind permission of the *Journal*; copyright held by the author.)

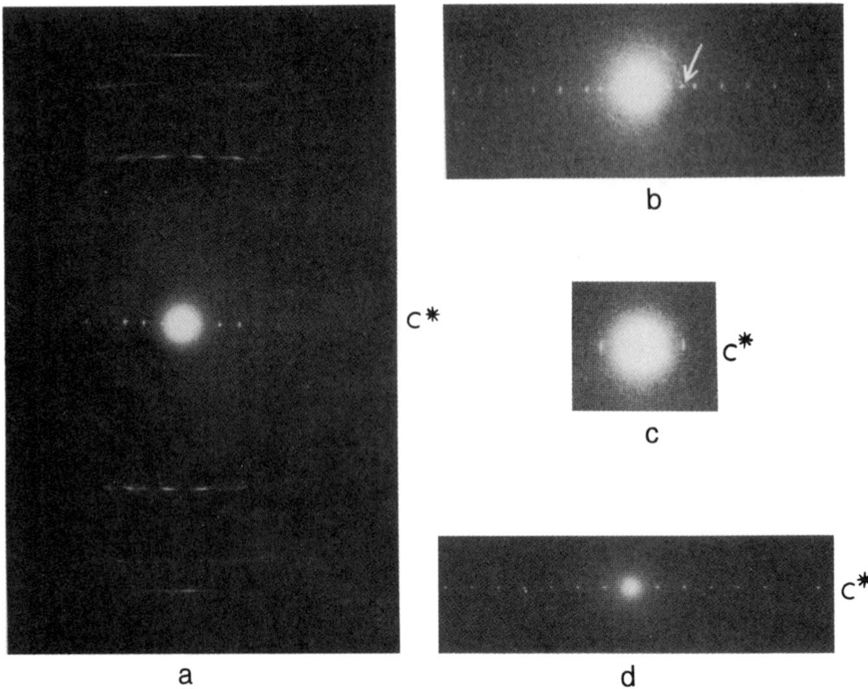

FIGURE 10.28. Thermotropic behavior of cholesteryl myristate studied with epitaxially crystallized samples: (a) room temperature; (b) heating to smectic—note superposition of sharp smectic peak with that of the original lattice row from crystalline phase; (c) smectic phase; (d) specimen cooled from smectic phase. (Reprinted from D. L. Dorset (1985) "Thermotropic mesomorphism of cholesteryl myristate. An electron diffraction study," *Journal of Lipid Research* **26**, 1142–1150; with kind permission of the *Journal*; copyright held by the author.)

consistent with one of Kitaigorodskii's rules for the formation of a stable solid solution discussed in Chapter 8. The behavior of a monolayer I/bilayer combination can be illustrated with the phase diagram for cholesteryl undecanoate/cholesteryl myristate (Figure 10.29) which was once imagined to form a solid solution. Electron diffraction data demonstrate that the two components are insoluble because only lamellar spacings from the pure components are ever observed. The cholesteryl laurate/cholesteryl myristate phase diagram is a bit more complicated (Figure 10.30). For the fraction of laurate that crystallizes in the bilayer form, a solid solution is formed with the myristate. The fraction in the monolayer I crystal is insoluble with the bilayer polymorph, however. This behavior is clearly indicated by the electron diffraction lamellar spacings.

Within a single-crystal polymorph, the rules for formation of solid solution are strictly volumetric, similar to the case of the alkanes. For example, with a difference of one methylene unit, cholesteryl undecanoate/cholesteryl laurate forms a nearly ideal

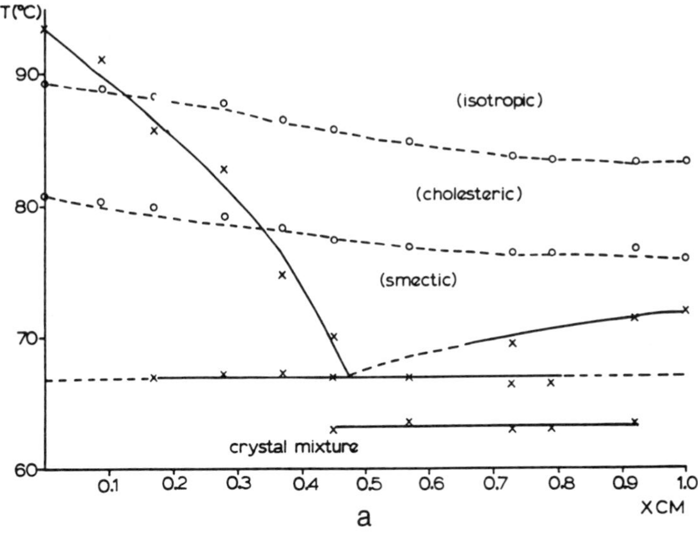

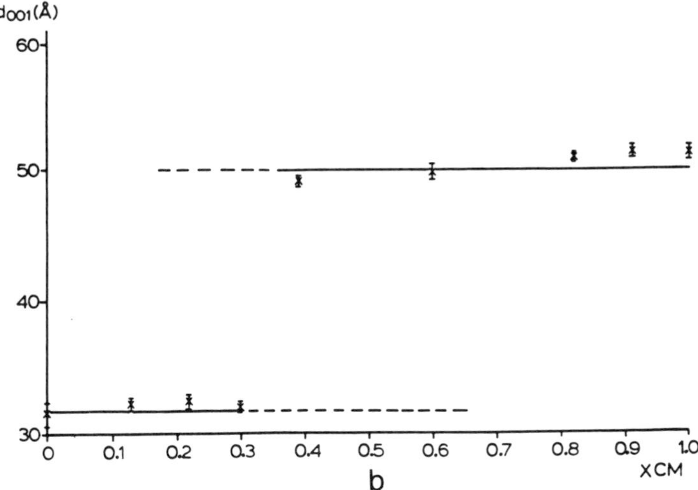

FIGURE 10.29. Binary interaction of cholesteryl undecanoate/cholesteryl myristate: (a) phase diagram; (b) lamellar spacings from epitaxially crystallized samples. The two components are fully fractionated. (Reprinted from D. L. Dorset (1987) "Cholesteryl esters of saturated fatty acids: co-solubility and fractionation of binary mixtures," *Journal of Lipid Research* **28**, 993–1005; with kind permission of the *Journal*; copyright held by the author.)

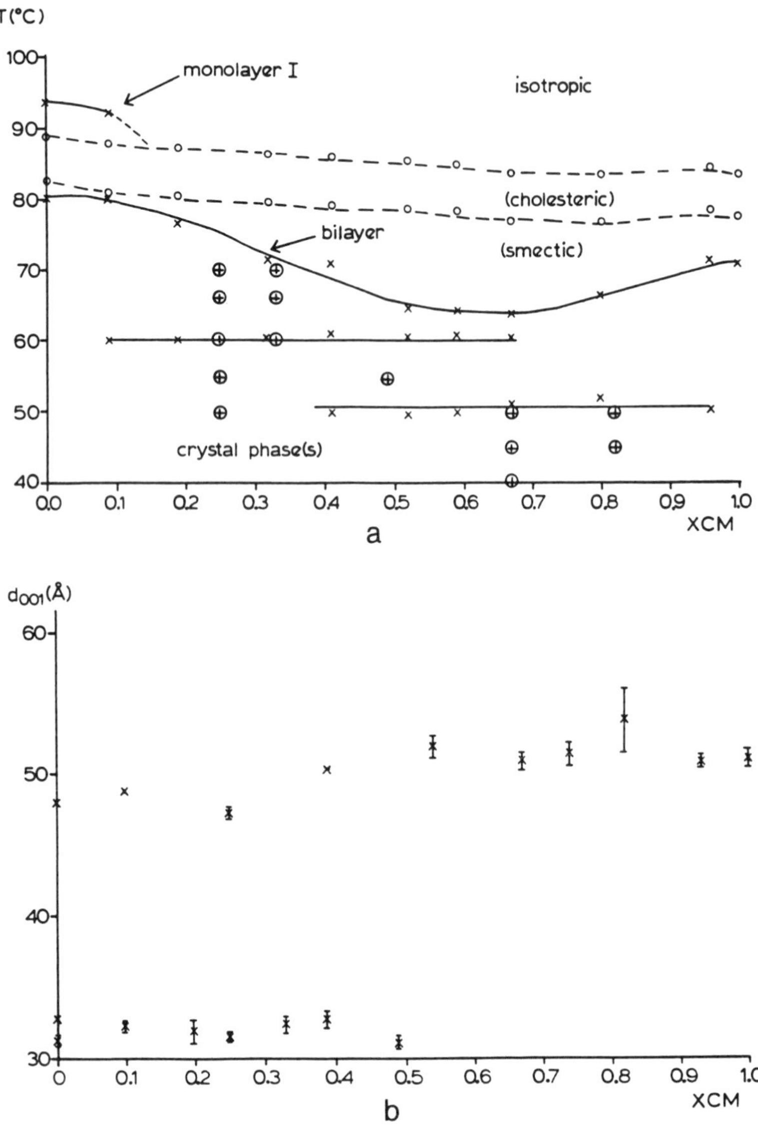

FIGURE 10.30. Binary phase behavior of cholesteryl laurate/cholesteryl myristate: (a) phase diagram; (b) experimental lamellar spacings that form epitaxially oriented samples. Cholesteryl laurate is polymorphic. The bilayer form is fully cosoluble with the myristate whereas the fraction in the monolayer I form is not. (Reprinted from D. L. Dorset (1987) "Cholesteryl esters of saturated fatty acids: co-solubility and fractionation of binary mixtures," *Journal of Lipid Research* **28**, 993–1005; with kind permission of the *Journal*; copyright held by the author.)

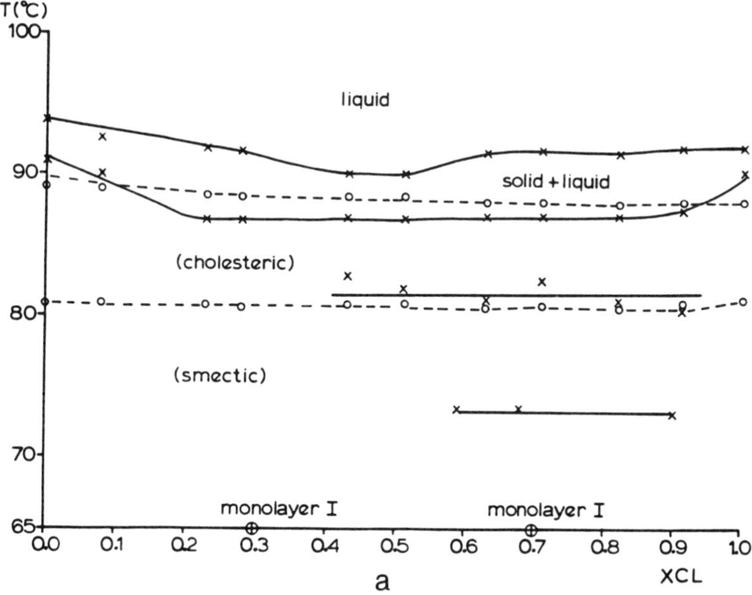

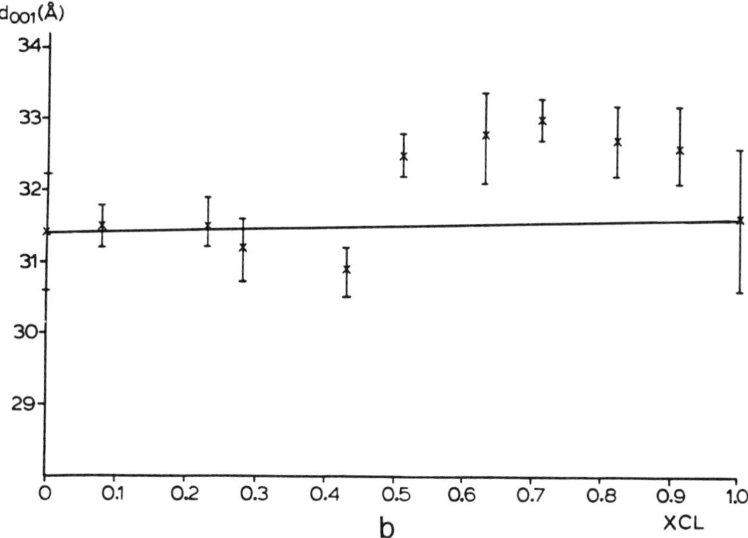

FIGURE 10.31. Binary phase behavior of cholesteryl undecanoate/cholesteryl laurate: (a) phase diagram; (b) experimental lamellar spacings. The two components, packing in monolayer I, are fully cosoluble. (Reprinted from D. L. Dorset (1987) "Cholesteryl esters of saturated fatty acids: co-solubility and fractionation of binary mixtures," *Journal of Lipid Research* **28**, 993–1005; with kind permission of the *Journal*; copyright held by the author.)

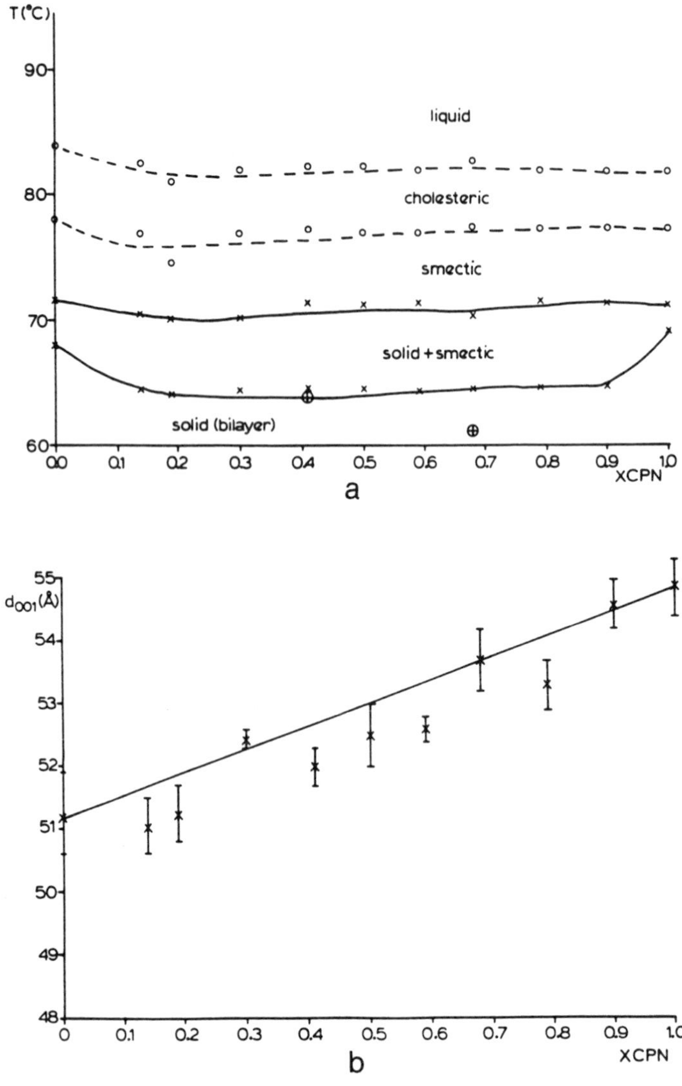

FIGURE 10.32. Binary phase behavior of cholesteryl myristate/cholesteryl pentadecanoate: (a) phase diagram; (b) lamellar spacings. The two components, packing in a bilayer, are fully cosoluble. (Reprinted from D. L. Dorset (1987) "Cholesteryl esters of saturated fatty acids: co-solubility and fractionation of binary mixtures," *Journal of Lipid Research* **28**, 993–1005; with kind permission of the *Journal*; copyright held by the author.)

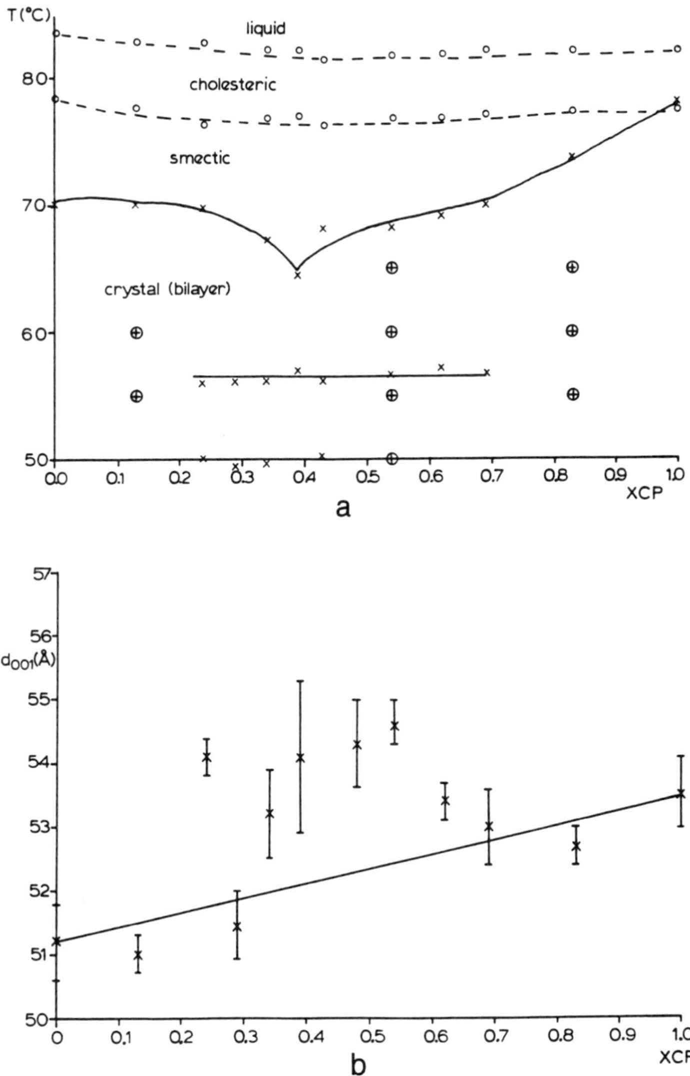

FIGURE 10.33. Binary phase behavior of cholesteryl myristate/cholesteryl palmitate: (a) phase diagram; (b) lamellar spacings. There is some evidence here for fractionation of solid solutions. (Reprinted from D. L. Dorset (1987) "Cholesteryl esters of saturated fatty acids: co-solubility and fractionation of binary mixtures," *Journal of Lipid Research* **28**, 993–1005; with kind permission of the *Journal*; copyright held by the author.)

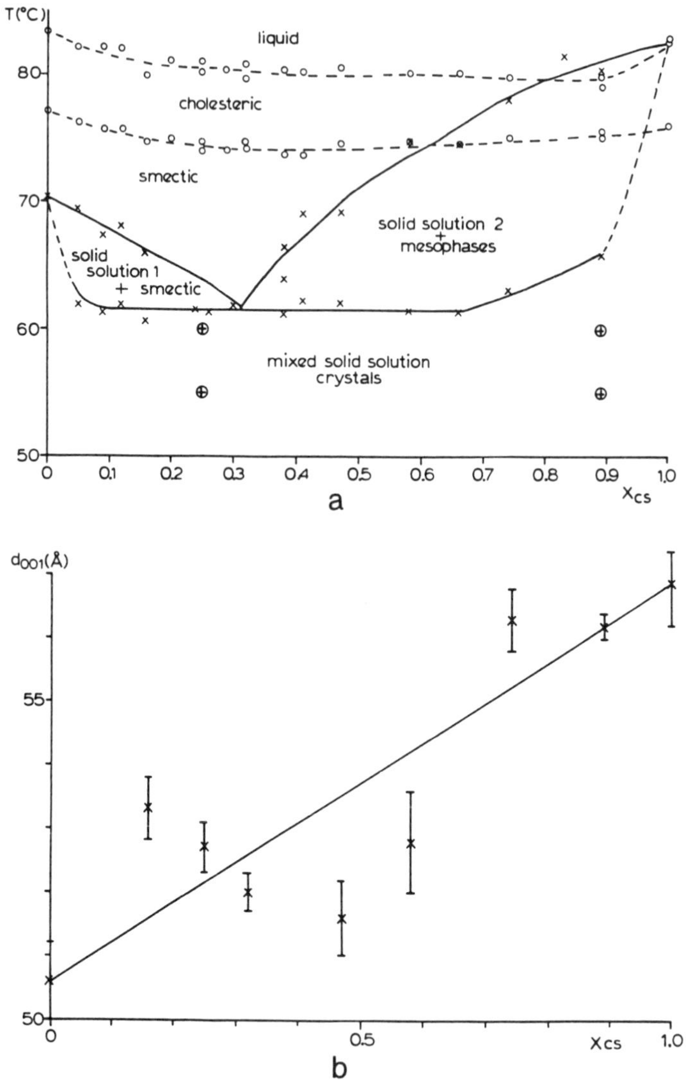

FIGURE 10.34. Binary phase behavior of cholesteryl myristate/cholesteryl stearate: (a) phase diagram; (b) lamellar spacings. Two solid solutions are fractionated. (Reprinted from D. L. Dorset (1987) "Cholesteryl esters of saturated fatty acids: co-solubility and fractionation of binary mixtures," *Journal of Lipid Research* **28**, 993–1005; with kind permission of the *Journal*; copyright held by the author.)

solid solution (Figure 10.31). In fact, encouraged by the electron diffraction observations, large single crystals of a nearly 1:1 solid solution were grown and their x-ray crystal structure determined to show that the ends of the acyl chains have fractional occupancies (Dorset and Pangborn, 1992). When the chain length differs by two methylenes, for example in the case of cholesteryl caprate/cholesteryl laurate, the solid solution is less ideal, but electron diffraction lamellar spacings still indicate the continuity of these structures over the concentration range. The x-ray crystal structure betrays a microfractionation phenomenon similar to the case of metastable *n*-paraffin solid solutions held below the binodal phase boundary (McCourt et al., 1994). Larger chain length differences begin to demonstrate eutectic behavior.

Similar observations can be made for the esters packing in the bilayer polymorph (Dorset, 1987f). For example, phase diagrams and electron diffraction measurements (Figure 10.32) demonstrate the cosolubility of the cholesteryl myristate/cholesteryl pentadecanoate binaries, which differ by one methylene unit. (Single crystals of cholesteryl myristate/cholesteryl pentadecanoate in a nearly 1:1 concentration have been grown for x-ray data collection and the crystal structure determination has recently been completed in this laboratory.) When the chains differ by two methylene units, fractionation begins (Figure 10.33), e.g., the combination cholesteryl myristate/cholesteryl palmitate, and continues for larger chain length differences, e.g., cholesteryl myristate/cholesteryl stearate (Figure 10.34).

In all of the work on binary solids, electron diffraction measurements on heated, oriented samples have been very useful for identifying the actual phase present at any given temperature. This is very important because many phase diagrams often cannot be interpreted without structural data. For example, some phase diagrams show isothermal transitions below the main disordering transitions. Without any additional knowledge, this behavior could be attributed to the appearance of a eutectic solid.

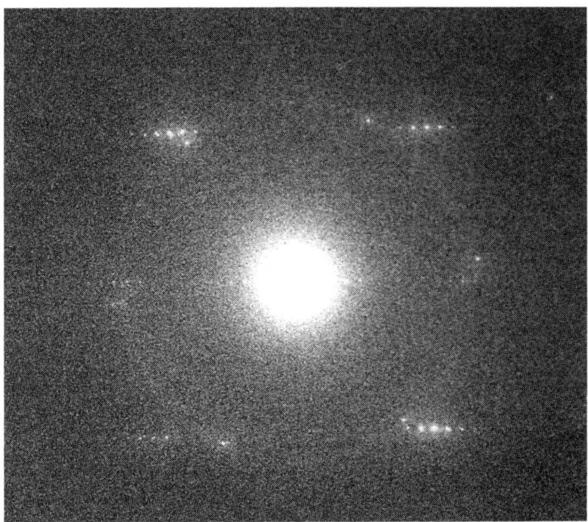

FIGURE 10.35. Electron diffraction patterns from solution-crystallized 1,2-dipalmitoyl-*sn*-glycerophosphoethanolamine. Because of the needle habit, a suitable orientation onto the long unit cell axis was observed by chance. (Reprinted from D. L. Dorset (1978) "Transmission electron diffraction intensities from real organic crystals: thin plate microcrystals of paraffinic compounds," *Zeitschrift für Naturforschung, Teil A* **33a**, 964–982; with kind permission of the Verlag der Zeitschrift für Naturforschung.)

However, in many cases, electron diffraction patterns have shown that the transitions are only due to small crystal–crystal reorientations in layers which still retain the same overall packing. On the other hand, the transition to the smectic phase, when it occurs for a binary combination, is also readily identified.

Thus, although the electron diffraction data from cholesteryl esters have not been of particular use for direct quantitative structure analysis, the qualitative characterization of microdomain structures has proven to be invaluable for the understanding of thermotropic behavior and binary phase relationships. The results of these studies encouraged successful attempts to grow larger crystals of the solid solutions for x-ray crystal structure analysis, thus opening a new research area in lipid crystallography.

10.4. Conclusions

Electron crystallography has not replaced x-ray crystallography as the method of choice for determining three-dimensional lipid structures at atomic resolution. Part of the problem with the preparation of oriented crystals is due to the high crystallization temperature used for epitaxial growth, combined with the one-dimensional nucleation by the substrate. Otherwise a useful projection of the crystal structure for 3-D data collection should be obtained. All attempts to remove the paracrystalline disorder by annealing have been frustrated so far. (However, once in the study of L-DPPE, a true crystalline form, in a projection onto the chain axis, was observed (Figure 10.34). The needle crystal habit was fortuitously tilted in an optimal direction to permit visualization of the crystal orientation that ideally should be achieved by epitaxial growth.) Nevertheless, investigation of lipid microcrystals with electron beams is a rapid way for determining chain packing and the density profile of the lamellar cross section. For quantitative studies it appears that data collected at high accelerating voltages are most useful for ab initio structure analyses since multiple scattering effects are then minimized. Various polymorphic forms are readily observed that would be overlooked in measurements on bulk samples. Oriented samples are also excellent for characterization of polydisperse lipid solids, thus facilitating the interpretation of experimental phase diagrams.

11

Linear Polymers

11.1. Background

Quantitative electron diffraction techniques have been especially important for the determination of linear polymer structures. This is because chain-folded lamellae are the only approximation to single crystals that can, in most cases, be grown. As outlined in Chapter 4, the usual method for determining crystal structures with electron diffraction data from polymers has been to collect an *hk*0 net from untilted lamellae and to use a conformational search around linkages between the rigid subunits of a chain model to find the best molecular packing. In this search, the crystallographic residual is minimized simultaneously with an atom–atom nonbonded potential energy. More recently, this technique, which is borrowed from fiber x-ray analysis (e.g., see Atkins, 1989), has been applied to three-dimensional electron diffraction data sets. Even when electron diffraction intensities are not used quantitatively for structure analysis, single-crystal patterns are of considerable benefit for identifying plane-group or space-group symmetry, measurement of cell constants, and, in general, as an aid for indexing fiber x-ray patterns. Tables of unit cell constants and symmetry based on electron diffraction measurements can be found in standard works on polymer physics (Geil, 1963; Wunderlich, 1973). A list of crystal structures determined from electron diffraction intensity data is given in Table 11.1. Specific examples based on the use of direct phase determination will be discussed in the following sections.

11.2. Crystal Structure Analyses

11.2.1. Two-Dimensional Data Sets

When structure factor values from atomic models, based on conformational searches, are compared to two-dimensional electron diffraction amplitudes, reasonable structures are found in many cases. However, it would also be useful if direct phasing methods could be used independently to provide a map with an accurate representation of the chain-packing density envelope, and also to find a reference phase set for the structural search with a model. Since considerable overlap of atomic positions occurs in projections down the chain axes, it was not certain how well this phasing technique

Table 11.1. Electron Crystallographic Analyses of Polymers

Two-dimensional determinations

Polymer	Data	R-factor
poly(tetrafluoroethylene) (oligomer) (Dorset, 1977b)	*hk*0	0.11
poly(ethylene) (Ogawa et al., 1994)	*hk*0	0.20
poly(diacetylene) (Day and Lando, 1980)	0*kl*	0.12
β-isotactic poly(propylene) (Meille et al., 1994)	*hk*0	0.10
trans-poly(acetylene) (Shimamura et al., 1981)	0*kl*	0.09
poly(ethylene sulfide) (Hasegawa et al., 1977; Moss and Dorset, 1983)	*hk*0	0.33(0.21*)
poly(3,3-bis-chloromethyloxacyclobutane) (Claffey et al., 1974)	*hk*0	0.25
cellulose triacetate (Roche et al., 1978)	*hk*0	0.26
poly(trimethylene terephthalate) (Poulin-Dandurand et al., 1974; Moss and Dorset, 1982)	*hk*0	0.35(0.25*)
poly(hexamethylene terephthalate) (Brisse et al., 1984)	*hk*0	0.17
poly(1,11-dodecadiyne) (Thakur and Lando, 1983)(macromonomer)	*h*0*l*	0.13
(cross-linked)	*h*0*l*	0.13
poly(γ-methyl-*L*-glutamate)		
α-form (Tatarinova and Vainshtein, 1962)	*hk*0	not given
β-form (Vainshtein and Tatarinova, 1967)	*h*0*l*	0.38
dextran		
low temperature (Guizard et al., 1983)	*hk*0	0.26
high temperature (Guizard et al., 1984)	*hk*0	0.20
nigeran (Perez et al., 1979)	*hk*0	0.25
chitosan (Mazeau et al., 1992)	*hk*0	< 0.20

Three-dimensional determinations

Polymer	Data	R-factor
polyethylene (Dorset, 1991; Hu and Dorset, 1989)	*hkl*	0.21
poly(ε-caprolactone) (Hu and Dorset, 1990; Dorset, 1991)	*hkl*	0.20
poly(sulfur nitride) (Boudeulle, 1973)	0*kl* + *hk*0	0.19
cis-poly(acetylene) (Chien et al., 1982)	*hkl*	0.13
Valonia cellulose (Claffey and Blackwell, 1976)	*hkl*	0.13
mannan I (Chanzy et al., 1987; Dorset and McCourt, 1993)	*hkl*	0.22
V_H-amylose (Brisson et al., 1984)	*hkl*	0.24
poly(*trans*-1,4-cyclohexanediyl dimethylene succinate) (Brisse et al., 1984)	*hkl*	0.24
γ poly(pivalolactone) (Meille et al., 1989)	*hkl*	0.14
poly(1-butene), form III (Dorset et al., 1994)	*hkl*	0.26

*Corrected data.

would succeed. As it turns out, the approach is quite useful, as will be demonstrated below with several examples.

11.2.1.1. Poly(ethylene Sulfide). Poly(ethylene sulfide) crystallizes in space group *Pbcn* with cell constants $a = 8.51$, $b = 4.94$, $c = 6.69$ Å. There are two atoms in the asymmetric unit. An initial report of its crystal structure, based on a conformational search, referring both to zonal spot and fiber electron diffraction intensities, was given by Hasegawa et al. (1977). A later correction of the model for symmetry violations (Moss and Dorset, 1983b) led to an *R*-value of 0.21.

Table 11.2. Direct Phase Determination for Poly(ethylene Sulfide)*

$hk0$	ϕ(DP)	ϕ(CS)	$hk0$	ϕ(DP)	ϕ(CS)
020	a	π	350	a	π
040	0	0	400	0	0
110	0	0	420	a	π
130	0	0	510	a	π
150	a	π	530	0	0
200	0	0	620	a	π
220	a	π	710	a	π
240	$a^{\dagger}$	0	820	a	π
330	0	0			

*$a = \pi$. DF = direct phasing; CS = crystal structure.
†Incorrect.

After calculation of $|E_h|$ from 22 observed $hk0$ reflection intensities, 26 Σ_2-triples, and 11 positive quartets were used for phase determination (Dorset, 1992e) after three phase values were accepted from Σ_1-triple estimates. The origin was defined by setting $\phi_{130} = 0$ (only one choice is allowed because of the centered projection in *cmm*). An algebraic unknown was also required to produce the set of 17 phases in Table 11.2 in which one phase assignment differed from the earlier analysis. (Comparison to the model derived from the earlier conformational search revealed also that 3/26 Σ_2 triples and 1/11 quartets were in disagreement with the resultant phase estimates.)

Since there is one algebraic phase ambiguity, two maps were generated to visualize the structure projection (Figure 11.1). Obviously, one of these corresponds to a reasonable packing model. If atomic coordinates for C and S were estimated from the map, the resulting structure factor calculation led to a phase set in total agreement with the result of the original conformational search. The results are summarized in Table 11.2.

11.2.1.2. Poly(3,3-bis(chloromethyl)oxacyclobutane (BCMO). BCMO crystallizes in the orthorhombic space group $Pna2_1$ with cell constants $a = 17.85$, $b = 8.15$, $c = 4.82$ Å. There are eight atoms in the asymmetric unit. An analysis based on 29 $hk0$ intensities was originally made via a conformational search to give a crystallographic residual $R = 0.26$ (Claffey et al., 1974). A dynamical scattering correction (Moss and Dorset, 1982) led to marginal improvement ($R = 0.24$).

After origin definition, setting $\phi_{610} = 0$, $\phi_{530} = 0$, phase values were found for 19 reflections (Table 11.3) after evaluation of 27 Σ_2-triples (containing two erroneous relationships) and 9 positive quartets (including two erroneous relationships). However, during this evaluation (Dorset, 1992e), contradictory results were found for ϕ_{130}, ϕ_{230}, and ϕ_{420}, requiring an algebraic value a to be assigned to them, thus necessitating the calculation of two potential maps (Figure 11.2). Correct atomic positions were found for the chlorines in either map, but only one began to show details of the organic residue. However, since a chain methylene and ether oxygen were nearly eclipsed in this projection, as well as two other chain methylene groups, the main chain could only

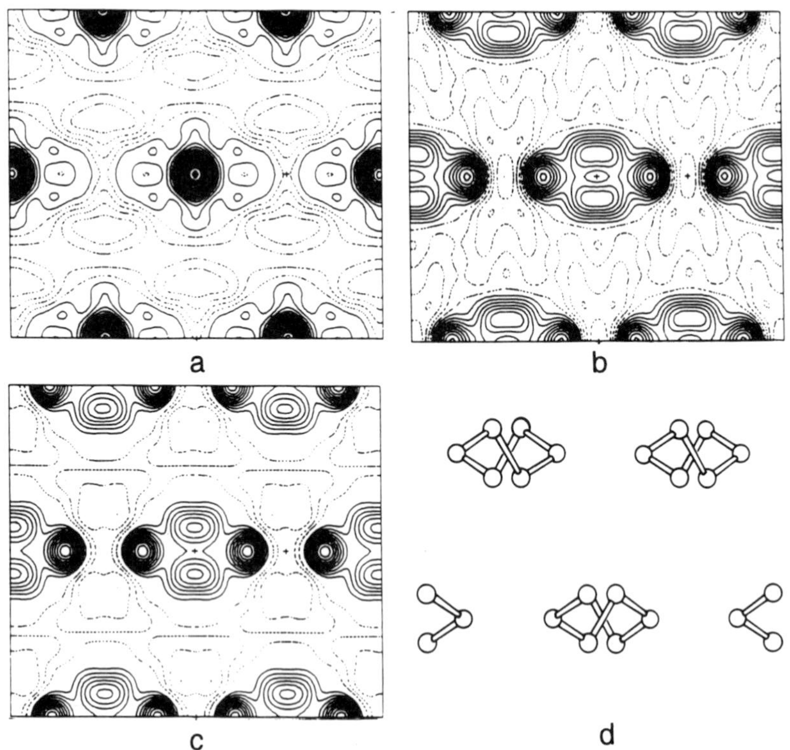

FIGURE 11.1. Direct phase determination for poly(ethylene sulfide): (a) $a = 0$; (b) $a = \pi$; (c) after Fourier refinement; (d) molecular packing. (Reprinted from D. L. Dorset (1992) "Electron crystallography of linear polymers: direct phase determination for zonal data sets," *Macromolecules* **25**, 4425–4430; with kind permission of the American Chemical Society.)

Table 11.3. Sayre Refinement for Poly(3,3-bis(chloromethyl)oxacyclobutane)*

hk	$\lvert F_o \rvert$	ϕ(DP)	ϕ(Say1)	ϕ(Say2)	*hk*	$\lvert F_o \rvert$	ϕ(DP)	ϕ(Say1)	ϕ(Say2)
20	0.24		0	0	32	1.82	π	π	π
40	2.44	π	π	π	42	2.40	a	π	0^*
60	0.24		π	π	52	1.34		0	0
80	1.42	0	0	0	62	0.24		π	π
11	0.24		0	0	72	0.95	0	0	0
21	3.82	π	π	π	82	0.71	π	π	π
31	2.17	π	π	π	13	1.42	$a^\dagger$	$\pi^\dagger$	$\pi^\dagger$
41	0.24		π	$0^\dagger$	23	1.65	a	π	$0^\dagger$
51	2.89	π	π	π	33	1.00		0	0
61	3.24	0	0	0	43	1.34	π	π	π
71	1.74	0	0	0	53	2.13	0	0	0
81	0.24		0	0	63	0.71	π	$0^\dagger$	π
02	4.69	π	π	π	73	1.00	π	π	π
12	2.66	π	π	π	83	0.24		$0^\dagger$	$0^\dagger$
22	0.45		0	0					

*DP: direct phase determination (basis set); Say 1: Sayre refinement, assuming $a = \pi$; Say 2: Sayre refinement, ignoring reflections with algebraic phase values.

†Incorrect.

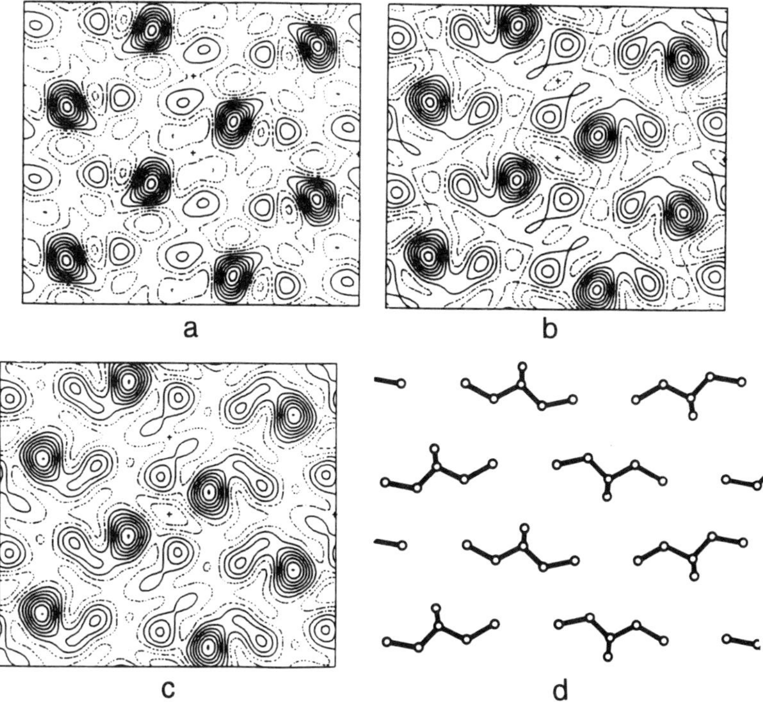

FIGURE 11.2. Direct phase determination for poly(3,3-bis(chloromethyl)oxacyclobutane): (a) $a = 0$; (b) $a = \pi$, (c) phases from earlier structure analysis; (d) molecular packing. (Reprinted from D. L. Dorset (1992) "Electron crystallography of linear polymers: direct phase determination for zonal data sets," *Macromolecules* **25**, 4425–4430; with kind permission of the American Chemical Society.)

be visualized as a large peak in this projection. In the best map ($a = \pi$), one methylene linkage to a chlorine was well resolved but the other position was not so clearly defined. This condition was not improved when the complete phase set from the previous determination was used to compute another potential map.

It was also possible to expand the partial phase set obtained from the evaluation of invariant sums with the Sayre equation. If algebraically ambiguous phases were omitted from this basis set, there were 5 errors in the complete list of 29 reflections (Dorset et al., 1995). When the correct value was given to a to expand the starting list to 19 reflections, the final set had only 3 errors, one of which was encountered in the original phase determination (Table 11.3).

11.2.1.3. γ-Poly(pivalolactone). Although there are a large number (45) of $hk0$ data for these lamellar crystals (Meille et al., 1989), the structure was difficult to solve by direct methods (Dorset, 1992e). A conformational search against three-dimensional data had, on the other hand, given a good match between observed and calculated structure factors ($R = 0.14$). The space group is $P2_12_12_1$ with unit cell constants $a = 8.23$, $b = 11.28$, $c = 6.02$ Å. There are seven atoms in the asymmetric unit.

Table 11.4. Direct Phase Determination for γ-Poly(pivalolactone)*

$hk0$	ϕ(DP)	ϕ(CS)	$hk0$	ϕ(DP)	ϕ(CS)
020	0	0	380	$-\pi/2$	$-\pi/2$
110	$-\pi/2$	$-\pi/2$	400	$b^{\dagger}$	0
120	b+$\pi/2$	$-\pi/2$	430	$\pi^{\dagger}$	0
160	$\pi/2^{\dagger}$	$-\pi/2$	440	b	π
250	0	0	470	$\pi^{\dagger}$	0
260	b	π	510	$\pi/2^{\dagger}$	$-\pi/2$
270	0	0	540	$a+\pi/2$	$\pi/2$
280	$a^{\dagger}$	π	610	$\pi-a$	π
350	$\pi/2-a$	$\pi/2$	620	0^{*}	π
360	$-\pi/2$	$-\pi/2$	630	$\pi-a$	π
370	$\pi/2-a$	$\pi/2$			

*$a = 0$, $b = \pi$. DP = direct phasing; CS = crystal structure.
†Incorrect.

Direct phase determination in plane group pgg (preserving the origin of the space group) began with origin definition, i.e., $\phi_{110} = \phi_{360} = -\pi/2$. After accepting a phase estimate from one Σ_1-triple, 19 other phases were found from an evaluation of 30 Σ_2-triples (containing five relationships that are discrepant with the previous crystal structure analysis) and three positive quartets to produce a set of phases via two algebraic unknowns (a, b) with seven values in disagreement with the original determination (Table 11.4).

It was not easy to interpret the potential map for this structure in terms of atomic positions. The best map (Figure 11.3) was used as a density profile for fitting a model. Estimated positions for six of the seven atoms were given and the remaining position was found after Fourier refinement (Figure 11.3). When the projected atom positions were used to calculate structure factor values for all 45 reflections, only 4 phase inconsistencies (with the original determination) were identified and these corresponded to rather weak reflections. Nevertheless, it cannot be claimed that the direct phase determination, in this example, was entirely satisfactory. Three possible structure solutions had been identified in the original study. A three-dimensional analysis is in progress to determine if the direct phase determination will correspond more closely to one of these alternative choices.

*11.2.1.4. Poly(*p*-xylylene).* The structure of poly(p-xylylene) in its hexagonal β-form was determined originally by averaging high-resolution electron micrographs to explain how 16 chains may pack in a hexagonal unit cell. The space group is $P3$ with unit cell constants $a = b = 20.52$, $c = 6.58$ Å, $\gamma = 120°$, and there are 7 atoms in the molecular repeat. A packing model derived from this image analysis (Tsuji et al., 1982) was used to determine its crystal structure from fiber x-ray diffraction intensities (Isoda et al., 1983a), giving a final $R = 0.21$.

An independent structure analysis was carried out based on 86 unique $hk0$ electron diffraction intensities obtained at 200 kV by Dr. W. P. Zhang. The projected plane group

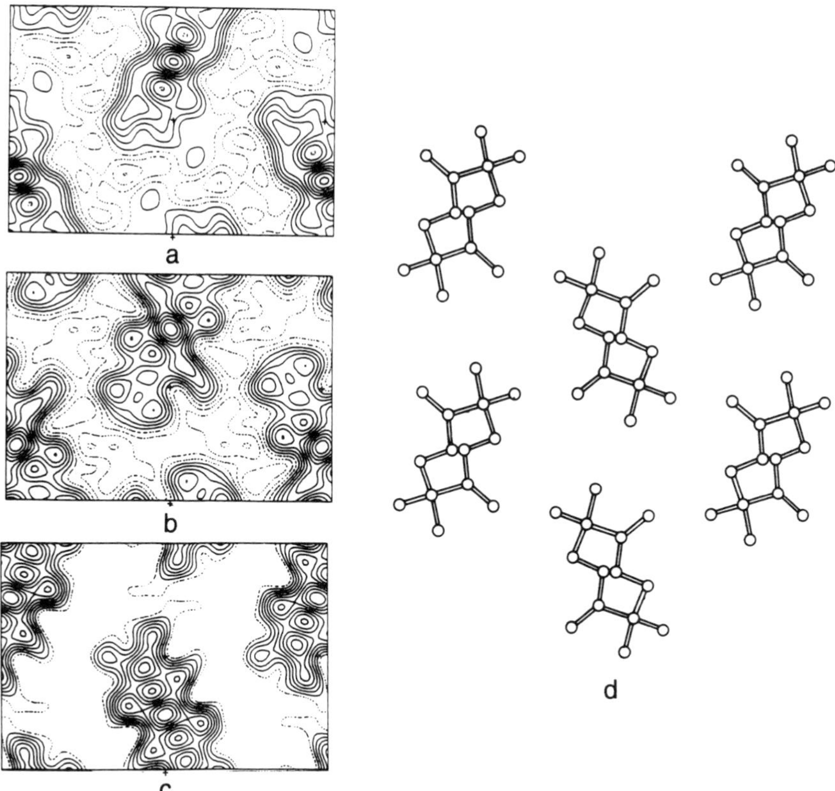

FIGURE 11.3. Direct phase determination for γ-poly(pivalolactone): (a) $a = 0$, $b = \pi$; (b) after Fourier refinement; (c) phases from previous structure analysis; (d) molecular packing. (Reprinted from D. L. Dorset (1992) "Electron crystallography of linear polymers: direct phase determination for zonal data sets," *Macromolecules* **25**, 4425–4430; with kind permission of the American Chemical Society.)

*p*6 is centrosymmetric. (Although *p*6 is not actually a projection of space group *P*3, the appearance of the electron micrographs seemed to justify this assumption of a centrosymmetric plane group.) After calculation of $|E_h|$, Σ_1- and Σ_2-triples were generated. From the former, estimates were accepted for $\phi_{040} = \phi_{440} = \phi_{080} = 0$. All *hk*0 reflections in this plane group are seminvariants (Rogers, 1980) so the first attempt to find a structure solution began with a basis set of 23 phases derived from the Fourier transform of the published electron micrographs, digitized with CRISP (Hovmöller, 1992). Phase extension with the Σ_2-triples allowed 22 more terms to be defined, resulting in a map (Figure 11.4) that was similar to the model found earlier. An independent phase determination was carried out by assigning algebraic terms to some phases, e.g., $\phi_{420} = a$ and $\phi_{150} = b$. The value $b = 0$ was soon found. The resulting phase set generated essentially the same map as before. There was no difference from the structure originally proposed.

FIGURE 11.4. Potential map for poly(p-xylylene) after direct phase determination.

High-resolution images from this polymer in two polymorphic forms provided the first direct visualization of edge dislocations at molecular resolution (Isoda et al., 1983b). The details of this defect were seen without using image-averaging techniques.

11.2.1.5. Mannan I. The crystal structure of mannan I was originally solved by a fit of a model to three-dimensional electron diffraction intensities (Chanzy et al.,

Table 11.5. Phase Determination and Refinement for Mannan I*

hk	$\|F_o\|$	ϕ(DP)	ϕ(Sayre)	hk	$\|F_o\|$	ϕ(DP)	ϕ(Sayre)
02	6.89	0	0	35	7.00	π	π
04	14.66	π	π	40	6.14	0	0
11	27.22	0	0	41	12.03	π	π
12	13.29	π	π	42	6.92	0	0
13	14.17	π	π	43	9.56	π	π
14	5.76	0	0	44	5.24	π	π
15	2.78		π	45	6.18	0	0
20	35.61	0	0	51	4.74	0	0
21	27.56	π	π	52	4.48	π	π
22	9.09	0	0	53	2.64		0
23	6.12	π	π	54	1.00		π
24	8.90	π	π	55	3.88	π	π
25	1.05		0	60	1.00		$0^{\dagger}$
31	20.13	0	0	61	1.05		$\pi^{\dagger}$
32	6.18	π	π	62	2.71		0
33	2.32		π	63	1.00		π
34	4.08	0	0	64	2.24		0

*Phase determination referred to pgg origin rather than space-group origin.
The space-group phase set can be generated by adding $\pi h/2$. DP = direct phase determination (basis set); Sayre = Sayre refinement set.
†Incorrect.

1987). Although a two-dimensional analysis based on the $hk0$ data (Dorset, 1992e) is discussed in this section, a direct phase determination with all data will also be considered below.

Mannan I crystallizes in space group $P2_12_12_1$ with cell constants $a = 8.92$, $b = 7.21$, $c = 10.27$ Å and with 12 atoms in the asymmetric unit. Values of $|E_h|$ were calculated from 34 $hk0$ intensities. After origin definition by setting $\phi_{110} = \pi/2$, $\phi_{210} = 0$, 47 Σ_2-triples (with 5 discrepancies to the earlier phase set) and 43 positive quartets (with 6 discrepancies) were evaluated. Three more phase assignments were obtained from Σ_1-triples to produce a total of 24 phases (Table 11.5). The assignment of one phase value, ϕ_{120}, was ambiguous, but statistically a value $-\pi/2$ was indicated, which

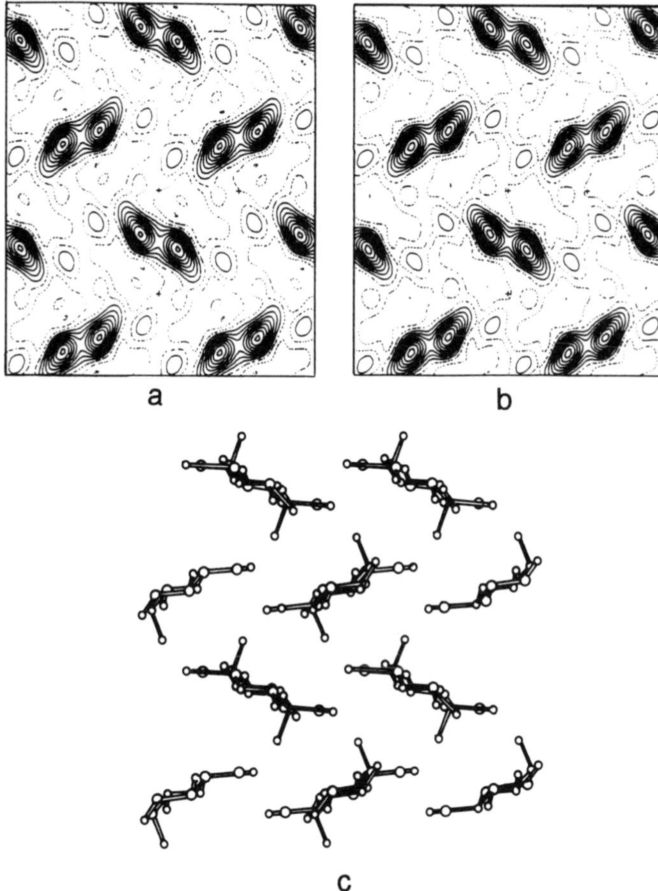

FIGURE 11.5. Direct phase determination for mannan I: (a) symbolic addition; (b) earlier conformational analysis; (c) molecular packing. (Reprinted from D. L. Dorset (1992) "Electron crystallography of linear polymers: direct phase determination for zonal data sets," *Macromolecules* **25**, 4425–4430; with kind permission of the American Chemical Society.)

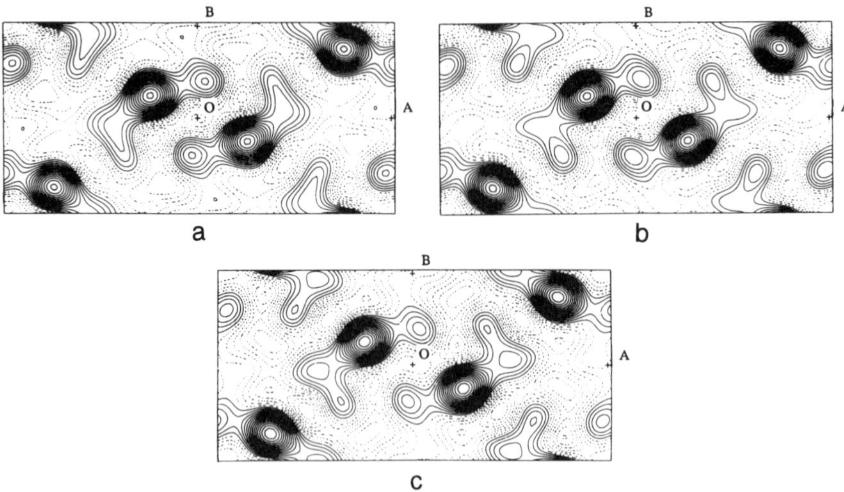

FIGURE 11.6. Direct phase determination for cellulose triacetate II: (a) symbolic addition; (b) after Sayre refinement; (c) original structure analysis.

was accepted. This resulted in a phase set with no disagreements with the earlier determination based on a model.

The potential map can only be interpreted as a density profile, even though the position of the C_6-methylene can be resolved (Figure 11.5). The experimental map is well matched by one calculated with all phased structure factors. These were obtained from atomic positions in a model resulting from the earlier conformational search (Figure 11.5). Sayre refinement of the basis phase set from the above direct analysis (Dorset et al., 1995) leads to a complete $hk0$ list with only two errors (Table 11.5).

11.2.1.6. Cellulose Triacetate II. The original structure analysis of orthorhombic cellulose triacetate was made by Roche et al. (1978) via a conformational model, based on 22 measured $hk0$ intensity data. The space group is $P2_12_12_1$ with unit cell constants $a = 24.68$, $b = 11.52$, $c = 10.54$ Å. There are 20 atoms in the asymmetric unit. In the original structure determination the final $R = 0.27$ when an isotropic temperature factor $B = 10$ Å^2 is applied to all atoms. A slight improvement of the fit is found if a correction is made for dynamical scattering (Moss and Dorset, 1982).

After defining the origin (Dorset, 1992e) via $\phi_{310} = \pi/2$, $\phi_{210} = \pi$, and accepting $\phi_{620} = 0$, $\phi_{600} = \pi$ from Σ_1-triples, 21 Σ_2-triples and 4 positive quartets were evaluated to obtain 14 phase values. (These definitions are in space group $P2_12_12_1$ and can be converted to the pgg assignments listed in Table 11.6.) These were combined with the observed $|F_o|$ to calculate a potential map, which is a good envelope for the structure determined earlier (Figure 11.6). Extension of the 14 phases to the complete $hk0$ list with the Sayre equation (Dorset et al., 1995) resulted in only 3 values that disagreed with the earlier structure factor calculation (Table 11.6).

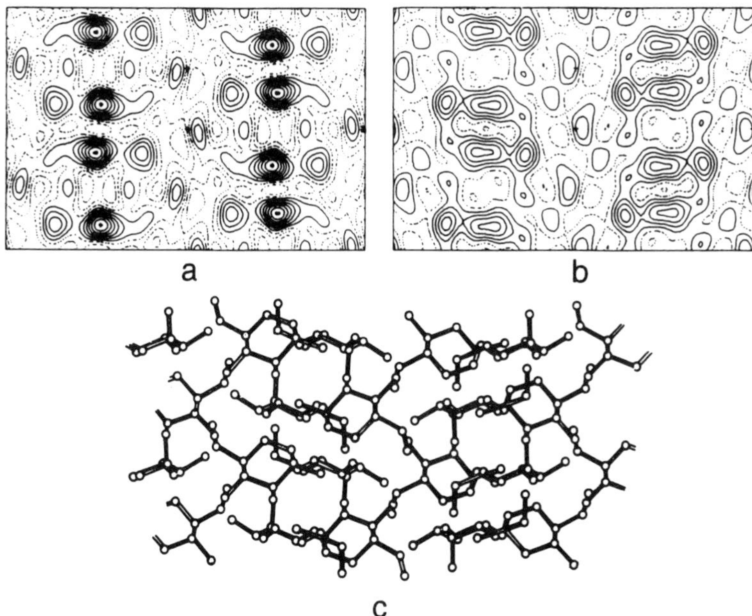

FIGURE 11.7. Direct phase determination for anhydrous nigeran: (a) phases where $a = \pi/2$ in an initial set with an ambiguity; (b) phases from previously determined structure; (c) molecular packing. (Reprinted from D. L. Dorset (1992) "Electron crystallography of linear polymers: direct phase determination for zonal data sets," *Macromolecules* **25**, 4425–4430; with kind permission of the American Chemical Society.)

11.2.1.7. Chitosan. The crystal structure of chitosan (11 atoms in the asymmetric unit) was originally solved by fitting a model to the experimental *hk*0 electron diffraction data (Mazeau et al., 1992), resulting in a residual $R < 0.20$. The space group is $P2_12_12_1$ with unit cell constants $a = 8.07$, $b = 8.44$, $c = 10.54$ Å.

Table 11.6. Phase Determination for Cellulose Triacetate II*

hk	$\lvert F_o \rvert$	ϕ(DP)	ϕ(Sayre)	*hk*	$\lvert F_o \rvert$	ϕ(DP)	ϕ(Sayre)
20	2.69		π	12	11.00	0	0
40	2.69		$\pi^{\dagger}$	22	18.20	π	π
60	9.97	0	0	32	17.30	0	0
11	18.70	0	0	42	5.50		$\pi^{\dagger}$
21	22.40	0	0	52	14.50	π	π
31	21.90	π	π	62	8.90	π	π
41	11.00	0	0	72	7.10	π	π
51	3.90		π	13	17.30	π	π
61	18.70	π	π	23	16.70	0	0
71	4.50		0	33	5.00		0
02	3.89		$\pi^{\dagger}$	43	6.30		π

*Phase determination referred to *pgg* origin rather than space-group origin. The space-group phase set can be restored by adding $\pi h/2$. DP = direct phase determination (basis set); Sayre = Sayre refinement set.
†Incorrect.

A detailed outline of the direct phase determination (Dorset, 1994), based on 7 Σ_2-triples and two positive quartets was presented in Chapter 4 to illustrate the methodology for evaluating new phase values with structure invariant sums, given the origin definition. For this example, an algebraic unknown was also required and it was, therefore, necessary to calculate two potential maps. The correct solution, identified by evaluating $\Sigma_i \rho_i^5$ to find a maximum value, is a good density envelope for the original structure (Figure 4.8). Details of the molecular model derived from a conformational search have been accepted for publication (Mazeau et al., 1994). As it turns out, all the phase assignments given by direct methods were correct for all but two weak reflections. A description of the phase extension to three dimensions will appear in a later publication.

11.2.1.8. Anhydrous Nigeran. Zonal electron diffraction data from anhydrous nigeran has been used to solve its crystal structure via a molecular model (Perez et al., 1979). This is a large structure with 22 atoms in the asymmetric unit. The orthorhombic space group is again $P2_12_12_1$ with unit cell constants $a = 17.76$, $b = 6.00$, $c = 14.62$ Å. After the original conformational search, $R = 0.25$. Attempts to correct these data for dynamical scattering did not lead to any improvement of the fit but some improvement ($R = 0.21$) was seen after a correction for crystal bending (Moss and Dorset, 1982).

Table 11.7. Phase Determination and Refinement of Anhydrous Nigeran*

hk	$\lvert F_o \rvert$	ϕ(DP)	ϕ(Sayre)	*hk*	$\lvert F_o \rvert$	ϕ(DP)	ϕ(Sayre)
20	5.80	0	0	22	4.00	π	π
40	0.50	π	$0^\dagger$	32	1.00		$0^\dagger$
60	3.50		$0^\dagger$	42	0.40		π
80	3.00	0	0	52	1.50	0	0
10 0	2.50	0	0	62	3.00	π	π
12 0	1.00	π	π	72	0.40		$0^\dagger$
14 0	2.00	$0^\dagger$	$0^\dagger$	82	1.50	$\pi^\dagger$	$\pi^\dagger$
11	2.00	π	π	92	1.50		$0^\dagger$
21	1.00		0	10 2	0.40		π
31	3.00	π	π	11 2	0.40		π
41	1.00		$\pi^\dagger$	12 2	0.60		$\pi^\dagger$
51	1.50		$\pi^\dagger$	13	3.00	0	0
61	3.00	π	π	23	2.00		π
71	2.80	π	π	33	1.50	0	0
81	2.00		π	43	1.00		$0^\dagger$
91	1.50	0	0	53	0.50		0
10 1	1.20		$\pi^\dagger$	63	0.60		0
11 1	0.40		π	73	0.20		0
12 1	0.20		$\pi^\dagger$	83	1.00	0	0
13 1	1.50	0	0	93	1.00	0	0
02	6.50	π	π	10 3	1.00		$0^\dagger$
12	0.50		$\pi^\dagger$				

*Phase determination referred to *pgg* origin rather than origin of unit cell. The space-group zonal phase set can be found by adding $\pi h/2$. DP = direct phase determination (basis set); Sayre = Sayre refinement set.

†Incorrect.

All 43 $hk0$ data were used to calculate $|E_h|$. The origin was defined (Dorset, 1992e) by setting $\phi_{130} = \pi/2$, $\phi_{610} = 0$. In all, 25 Σ_2-triples and nine positive quartets were evaluated, respectively containing 8 and 3 discrepant relationships above the A_2 and B threshold levels. The use of an algebraic unknown was required. Values for 4 phases were accepted from Σ_1-triples with suitably large A_1-values. Although 23 phases were given values in this determination, two were omitted from the final list (Table 11.7) because of inconsistent assignments that could not be resolved statistically.

A trial potential map for anhydrous nigeran is shown in Figure 11.7. This is compared to the map calculated from structure factors obtained from the atomic positions in the previous determination. Although there are some differences in the density profiles (due to the limited resolution and two phase disagreements), there are many features in common. When the Sayre equation was used for phase refinement (Dorset et al., 1995), there were 14 errors in the list (Table 11.7) so that the resultant map was not very informative (Figure 11.8).

11.2.1.9. Poly(hexamethylene Terephthalate). Zonal electron diffraction intensities had been recorded from thin lamellae of poly(hexamethylene terephthalate), form I, by Brisse et al. (1984) and its structure was solved by a conformational search with a model. Possible data perturbations were also evaluated (Moss and Brisse, 1984).

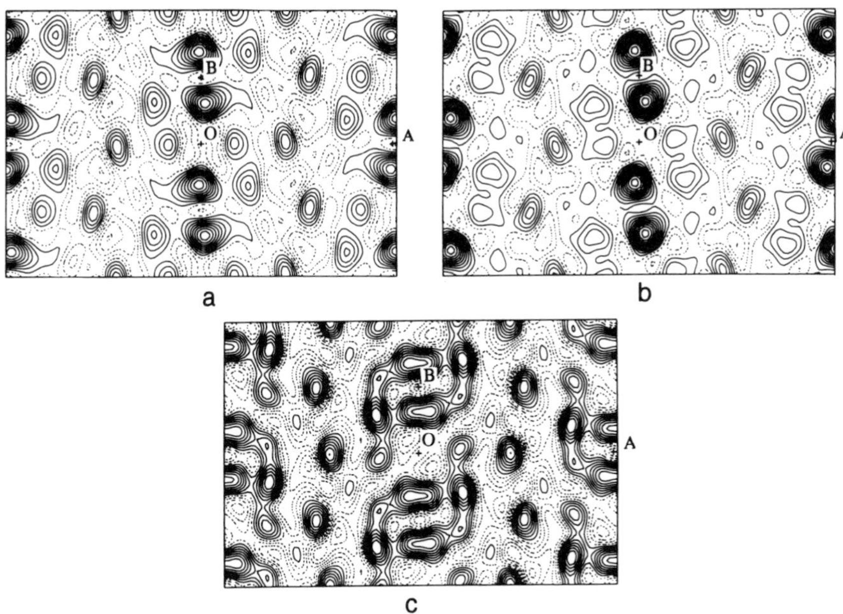

FIGURE 11.8. Direct phase determination of anhydrous nigeran without ambiguity: (a) symbolic addition set; (b) after Sayre refinement; (c) ideal phases.

Table 11.8. Phase Determination for Poly(hexamethylene Terephthalate)*

hk		$\|F_o\|$	ϕ(DP)	ϕ(Sayre)
1	0	27.64	0	0
0	1	29.00	0	0
1	–1	46.79	0	0
1	1	4.29	0	0
2	0	7.06	0	0
2	–1	12.57	0	0
1	–2	15.78	0	0
0	2	2.19	0	0
2	1	9.82	π	π
2	–2	19.00	0	0
1	2	4.76	π	π
3	–1	0.63		0
3	0	8.62	π	π
3	–2	4.00	0	0
2	2	0.32		π
0	3	1.47		π
2	–3	2.28	0	0
4	–1	1.75	π	π
0	4	0.57		π

*DP = direct phase determination (basis set); Sayre = Sayre refinement set.

The unit cell is triclinic, space group $P\bar{1}$, with unit cell constants $a = 5.217$, $b = 5.284$, $c = 15.738$ Å, $\alpha = 129.4°$, $\beta = 97.6°$, $\gamma = 95.6°$.

After normalized structure factors were calculated from the 19 observed intensity data, direct methods were used to solve the crystal structure (Dorset et al., 1995). Origin definition was achieved by setting $\phi_{-110} = \phi_{-210} = 0$ and another phase, $\phi_{-220} = 0$ was accepted from a Σ_1-triple with a large A_1-value. There were 23 Σ_2-triples where $A_2 \geq 0.15$ and 15 quartets with $|B| \geq 0.21$. After phase determination with the quartets first, and then the Σ_2-triples, it was possible to find values for 16 reflections with no discrepancies with the earlier determination. The resultant maps are shown in Figure 11.9 and the phase assignments are listed in Table 11.8. Sayre refinement, starting with 16 reflections, resulted in a complete phase list with no errors.

*11.2.1.10. β-Form of Poly-γ-methyl-*L*-glutamate.* In the previous examples of two-dimensional structure determinations, mostly the molecular envelope is found in the resulting potential map after direct phase determination for data in a projection down the polymer chain axes. One case where all but one of the atomic positions could be distinguished was in the structure analysis of β-form of poly-γ-methyl-*L*-glutamate (Vainshtein and Tatarinova, 1967). Films of the polymer were cast by slow evaporation of a solution onto a water surface. These films were then pulled off and stretched to give the β-polymorphic form, characteristic of silk. Texture electron diffraction patterns contained 19 unique $h0l$ reflections (cell constants $a = 4.725$, $c = 6.83$ Å).

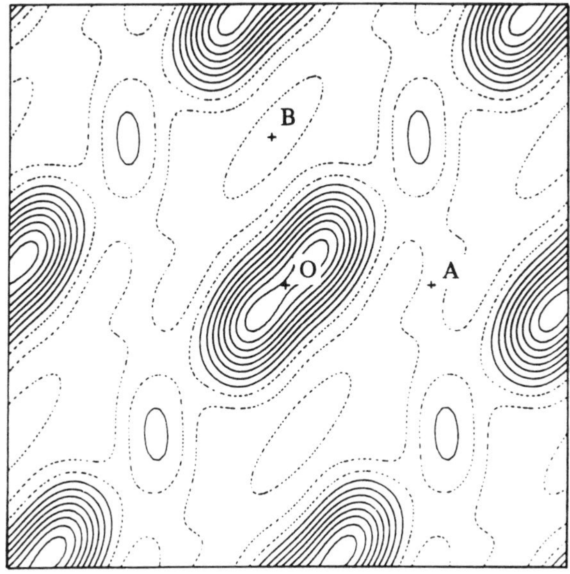

FIGURE 11.9. Direct phase determination for poly(hexamethylene terephthalate) showing molecular envelope.

Observed weak forbidden reflections along the c^*-axis were judged to be due to some antiparallel packing of chains in a disordered structure. Mostly the structure was found to be polar, packing in plane group *pg*, and the atomic positions were found with a model resulting in a final $R = 0.38$.

It was possible to determine the structure by direct methods, although the zone is noncentrosymmetric. For the data in Table 11.9, after calculation of $|E_h|$ values, Σ_1- and Σ_2-three phase invariants were generated. The $h00$ reflections still have centrosymmetric values, so $\phi_{100} = 0$ was accepted. From high probability Σ_1-triples, $\phi_{200} = \phi_{400} = 0$. Next, symbolic values were given: $\phi_{160} = a$; $\phi_{130} = b$; $\phi_{110} = c$. A convergence phasing sequence with Σ_2-triples was used to assign values in a useful pathway:

$$\phi_{060} = \phi_{160} + \phi_{-100}, \qquad \text{therefore } \phi_{060} = a$$

$$\phi_{150} = \phi_{060} + \phi_{1,-1,0}, \qquad \text{therefore } \phi_{150} = a - c + \pi$$

$$\phi_{230} = \phi_{160} + \phi_{1,-3,0}, \qquad \text{therefore } \phi_{230} = a - b + \pi$$

$$\phi_{300} = \phi_{100} + \phi_{200}, \qquad \text{therefore } \phi_{300} = 0$$

$$\phi_{020} = \phi_{130} + \phi_{-1,-1,0}, \qquad \text{therefore } \phi_{020} = b - c$$

$$\phi_{040} = \phi_{060} + \phi_{0,-2,0}, \qquad \text{therefore } \phi_{040} = a - b + c$$

Table 11.9. Structure Analysis of Poly-γ-methyl-L-glutamate in the β Form

$h0l$	$\|E_h\|$	$\|F_o\|$	$\|F_c\|$	ϕ(previous)	ϕ(this study)
002	0.48	0.72	0.57	−63°	−51°
004	0.43	0.38	0.31	49	73
006	3.01	1.47	0.88	1	−3
100	1.48	2.12	2.37	0	0
200	1.03	1.04	1.06	0	0
300	0.30	0.65	0.89	0	0
400	0.35	0.15	0.46	0	0
500	0.23	0.07	0.04	180	180
101	0.75	1.02	0.67	−169	−178
201	0.32	0.31	0.42	90	108
102	0.42	0.48	0.56	17	14
202	0.40	0.33	0.64	41	43
103	0.95	0.85	0.77	88	90
203	0.51	0.36	0.42	91	88
303	0.12	0.06	0.31	92	87
403	0.13	0.04	0.54	90	90
104	0.66	0.45	0.27	−22	−13
105	0.55	0.28	0.29	−26	−7
106	1.75	0.69	0.58	5	−5

fractional coordinates:	x*	z*	$x^{\dagger}$	$z^{\dagger}$
Cα,β	0.048	0.000	0.042	0.000
C′	0.067	0.331	0.092	0.330
O	0.281	0.335	0.300	0.330
N	0.000	0.161	−0.025	0.175

*This study.
†Vainshtein and Tatarinova (1967).

In addition, $c = \pi$ could be specified to complete origin definition for the zone. As with the tangent formula, it is possible to step a and b through successive quadrants or octants of phase and to use the test values to calculate maps. When, for example, $a = 0$, $b = \pi/2$, the map in Figure 11.10a was found. It was possible to pick off the atomic positions (assuming two carbon atoms are eclipsed near the origin) to calculate structure factors. After two cycles of Fourier refinement, the phases in Table 11.9 were found, resulting in $R = 0.32$. The corresponding potential map is shown in Figure 11.10b. The mean phase difference to the previous determination was only 6°.

11.2.2. Three-Dimensional Data Sets

Obviously, if the aim of a crystal structure analysis is to describe the molecular geometry in terms of atomic positions, the previously described use of direct methods with zonal data is satisfactory only for the simplest polymers, where the projection down the chain axis affords the least unresolved overlap of density (or, alternatively, if the crystal preparation permits projection onto the chain axis). Two options are available for the collection of three-dimensional electron diffraction data. The most

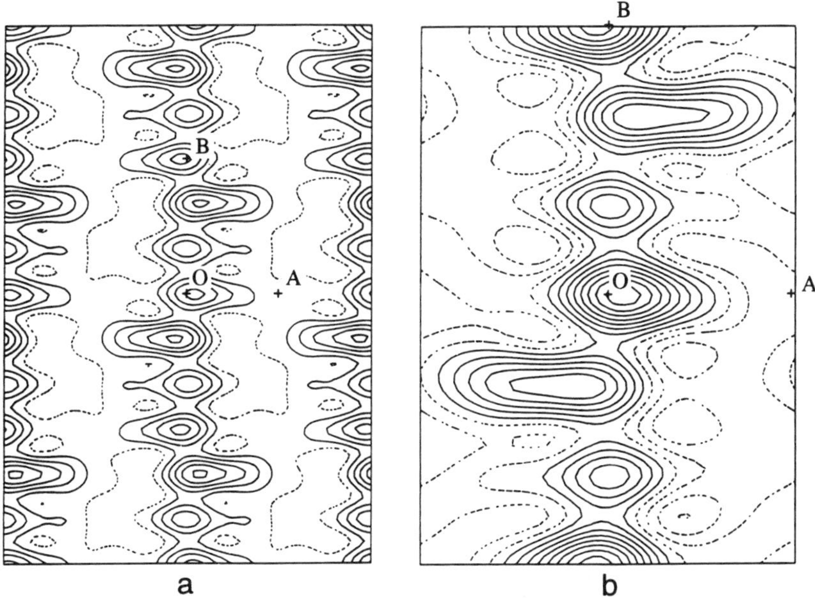

FIGURE 11.10. Potential maps for the β form of poly-γ-methyl-L-glutamate after direct phase determination: (a) based on initial set of 12 reflections; (b) after Fourier refinement (magnification doubled).

straightforward of these is to use the solution-crystallized lamellae for tilt experiments to access as much of the reciprocal lattice as possible. This data collection is restricted by the geometrical tilt limits of the electron microscope goniometer stage, and, if the data are collected at rather modest resolution, the missing information will blur the density profile along the chain direction. This is because the polymer chain axis is generally within the excluded volume element not sampled by the tilt series.

The second option is to combine tilt diffraction data from two orthogonal orientations of the polymer. The first orientation in the projection down the chain axes was already discussed, since it is obtained from solution-crystallized lamellae. The second orientation, often normal to the chain axis, is achieved by epitaxial orientation, as described above for various linear or aromatic small molecules. It is important to establish first that the two crystallizations arrive at the same polymer polymorph. Also, it may be useful to anneal the epitaxially oriented sample in the presence of the epitaxial substrate in order to improve its crystalline perfection before data collection. This will facilitate the densitometry of diffraction peaks and extraction of intensities from them.

11.2.2.1. Poly (1,4-trans-cyclohexanediyl Dimethylene Succinate) (Poly-t-CDS). Originally, the crystal structure of poly-*t*-CDS was determined by model fitting from 87 observed three-dimensional electron diffraction intensities (Table 11.10) obtained at 120 kV (Brisse et al., 1984). The monoclinic space group is $P2_1/n$ with cell constants $a = 6.49$, $b = 9.48$, $c = 13.51$ Å, $\beta = 45.9°$. After minimization of the conformational potential energy, the crystallographic R-factor was 0.20.

Table 11.10. Final Observed and Calculated Structure Factors for Poly(1,4-*trans*-cyclohexanediyl Dimethylene Succinate)

hkl	$\lvert F_o \rvert$	F_c	*hkl*	$\lvert F_o \rvert$	F_c
0 2 0	80.00	87.34	4 0 4	0.00	–9.07
0 4 0	20.00	0.40	4 1 4	7.00	–7.37
0 6 0	10.00	4.80	4 2 4	0.00	7.38
0 8 0	5.00	–3.19	4 3 4	8.00	6.02
1 1 0	140.00	142.38	4 4 4	9.00	7.16
1 2 0	32.00	–36.27	4 5 4	16.00	12.32
1 3 0	22.00	9.67	4 6 4	5.00	2.24
1 4 0	16.00	–15.06	4 7 4	9.00	3.67
1 5 0	7.00	7.04	5 0 5	12.00	10.56
1 6 0	4.00	–1.82	5 1 5	0.00	–3.68
2 0 0	65.00	69.56	5 2 5	5.00	2.24
2 1 0	19.00	–18.98	5 3 5	0.00	–1.66
2 2 0	22.00	21.47	5 4 5	0.00	0.90
2 3 0	21.00	–24.16	5 5 5	5.00	3.46
2 4 0	4.00	4.40	1 1 2	25.00	19.19
2 5 0	5.00	–3.76	1 2 2	17.00	–20.22
3 1 0	14.00	13.13	1 3 2	9.00	–6.30
3 2 0	11.00	–9.10	1 4 2	5.00	3.90
3 3 0	6.00	1.63	1 5 2	7.00	1.58
3 4 0	6.00	–5.56	1 6 2	10.00	–7.93
1 0 1	20.00	13.89	1 7 2	0.00	–0.83
1 1 1	25.00	24.96	1 8 2	7.00	4.27
1 2 1	3.00	6.48	1 9 2	0.00	–2.62
1 3 1	14.00	9.41	1 10 2	4.00	3.32
1 4 1	14.00	–16.82	2 0 4	25.00	–30.37
1 5 1	7.00	0.00	2 1 4	13.00	–15.60
1 6 1	7.00	0.56	2 2 4	14.00	–6.05
1 7 1	0.00	0.74	2 3 4	10.00	2.27
1 8 1	5.00	4.67	2 4 4	5.00	–2.24
1 9 1	0.00	–0.24	2 5 4	17.00	16.63
1 10 1	5.00	–2.29	2 6 4	10.00	10.06
2 0 2	12.00	3.14	2 7 4	9.00	7.77
2 1 2	19.00	–24.69	2 8 4	10.00	8.66
2 2 2	16.00	–9.32	3 0 5	39.00	17.93
2 3 2	7.00	–0.14	3 1 5	10.00	–14.65
2 4 2	9.00	–0.99	3 2 5	19.00	–22.58
2 5 2	3.00	3.08	3 3 5	0.00	–4.18
2 6 2	9.00	3.01	3 4 5	12.00	–17.52
2 7 2	3.00	–1.06	3 5 5	7.00	7.07
2 8 2	0.00	–1.17	3 1 6	14.00	–18.84
2 9 2	9.00	4.20	3 2 6	19.00	23.30
3 0 3	12.00	–6.38	3 3 6	27.00	–24.10
3 1 3	11.00	5.50	3 4 6	5.00	4.74
3 2 3	4.00	–5.86	3 5 6	12.00	–11.10
3 3 3	4.00	–2.78	3 6 6	0.00	3.67
3 4 3	5.00	2.72	3 7 6	7.00	3.70
3 5 3	7.00	3.68	4 1 7	4.00	4.84
3 6 3	0.00	0.57	4 2 7	7.00	–7.29
3 7 3	0.00	0.04	4 3 7	0.00	3.56
3 8 3	0.00	–1.71	4 4 7	0.00	1.26
3 9 3	9.00	–4.27	4 5 7	7.00	3.21
			4 6 7	6.00	–6.05
			4 0 8	10.00	13.24

Table 11.11. Distributions of $|E_h|$ for Poly(1,4-*trans*-cyclohexanediyl Dimethylene Succinate)

		Theory	
Experimental		Centrosymmetric	Noncentrosymmetric
$<\|E_h\|^2>$	1.000	1.000	1.000
$<E_h^2 - 1>$	1.167	0.968	0.736
$<E_h>$	0.698	0.798	0.886
% $E_h > 1.0$	24.0	32.2	36.8
% $E_h > 2.0$	6.7	5.0	1.8
% $E_h > 3.0$	1.9	0.3	0.01

It was of interest to attempt a direct phase analysis (Dorset, 1991b). Unfortunately, the goniometer tilts used to access three-dimensional reflections only were made around the unique **b**-axis, so that only one half of the unique intensity data needed for a monoclinic structure was collected. Nevertheless, a Wilson plot of the average intensity data in domains of sin θ/λ indicated that the overall $B_{iso} = 6.1$ Å^2 (see Figure 4.6). This value was used to adjust the atomic scattering factors for calculation of $|E_h|$. Distribution of these normalized structure factor magnitudes corresponded to the values expected for a centrosymmetric unit cell (Table 11.11).

For the *hkl* data, 38/42 Σ_2-triples ($A_{2,min} = 0.27$), 9/11 quartets ($B_{min} = 1.0$), and four Σ_1-triples were found to be in accord with the previous phase assignments. Restricting the analysis to the *hk*0 data, 13 Σ_2-triples and 19 quartets were found to be correct above the defined thresholds. The origin was defined by setting $\phi_{292} = 0$, $\phi_{110} = 0$, $\phi_{1,10,1} = \pi$. Use of these values in the phase invariant sums resulted in 15 unequivocally determined phases. Three algebraic unknowns were needed to build phase relationships among all the reflections. In a projection onto the chain axis, a potential map calculated from one of the phase combinations was found to contain features of an ester linkage (Figure 11.11) at $x = 0.000$. A three-dimensional map was calculated at intervals $\Delta x = 0.05$ from $0 \leq x \leq 0.2$, with the initial phase set of 51 reflections. Although the density was blurred, it was possible to find likely positions for all atoms in the structure, with knowledge of molecular architecture. The identified atomic coordinates were used to calculate structure factors for all 87 reflections and, after Fourier refinement to optimize bonding geometry, $R = 0.29$ (Table 11.10). The final structure parameters derived from the fractional coordinates (Table 11.12) are shown in Figure 10.11c.

11.2.2.2. Mannan I. As described above, two-dimensional electron diffraction intensity data from mannan I have been phased by direct methods (Dorset, 1992e) to obtain an accurate outline of the chain cross section. However, in the original study, additional data, e.g., *hkk* and *hkh* intensities were also recorded from crystals tilted, respectively, 35° and 41° to the incident beam, yielding a total set of 58 unique reflections (Table 11.13).

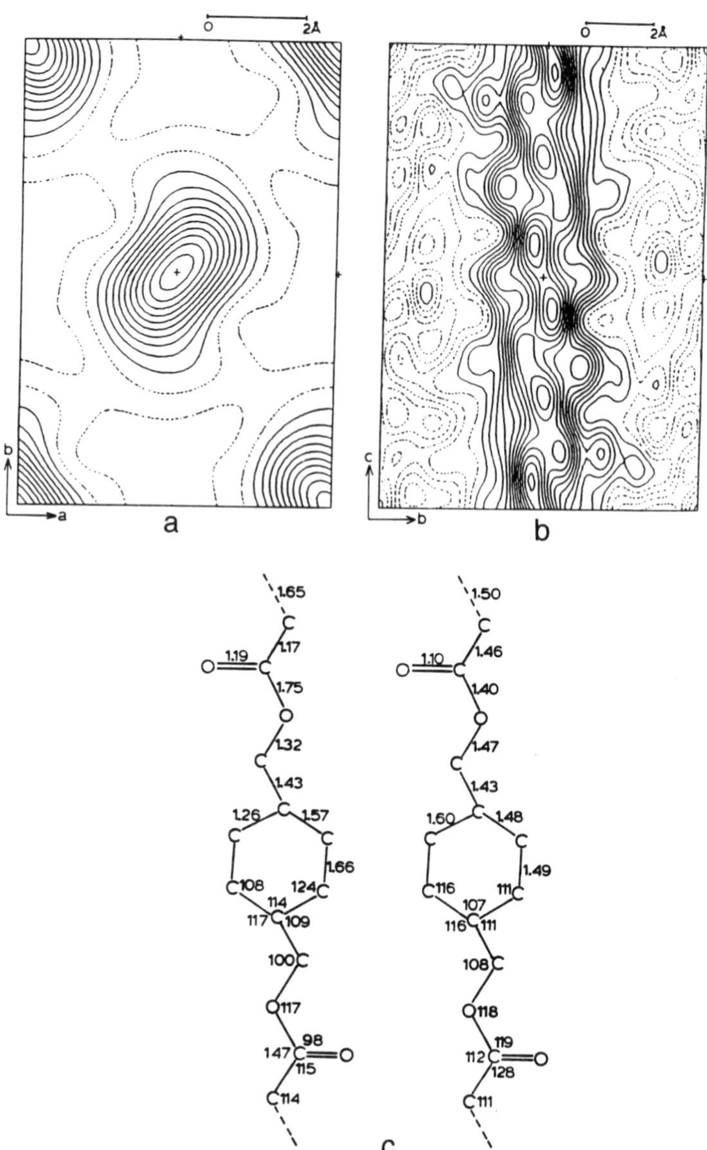

FIGURE 11.11. Potential maps for poly(1,4-*trans*-cyclohexanediyl dimethylene succinate): (a) view down chain; (b) view onto chain at $x = 0$; (c) bond distances and angles before and after Fourier refinement. (Reprinted from D. L. Dorset (1991) "Is electron crystallography possible? The direct determination of organic crystal structures." *Ultramicroscopy* **38**, 23–40; with kind permission of Elsevier Science B.V.)

Table 11.12. Atomic Coordinates for Poly(1,4-*trans*-cyclohexanediyl Dimethylene Succinate) after Fourier Refinement

Atom	x	y	z
O 1	0.025	–0.192	0.374
O 2	–0.050	–0.023	0.295
C 1	0.162	0.058	0.026
C 2	–0.075	–0.040	0.131
C 3	–0.125	–0.141	0.054
C 4	–0.038	–0.121	0.207
C 5	0.000	–0.077	0.373
C 6	–0.025	0.031	0.458

Table 11.13. Structure Factors for Mannan I

hkl	$\lvert F_o \rvert$	$\lvert F_c \rvert$	ϕ	*hkl*	$\lvert F_o \rvert$	$\lvert F_c \rvert$	ϕ
020	6.89	4.56	0	600	1.00	4.79	0
040	14.66	16.06	π	610	1.65	1.59	π
110	27.22	28.95	$\pi/2$	620	2.21	0.37	0
120	13.29	13.58	$-\pi/2$	630	1.00	1.26	0
130	14.17	15.93	$-\pi/2$	640	2.24	1.06	π
140	5.76	2.85	$\pi/2$	011	0.88	0.00	$-\pi/2$
150	2.78	2.89	$-\pi/2$	022	3.46	7.37	π
200	35.61	40.81	π	111	11.94	10.96	2.56
210	27.56	27.76	0	211	12.02	15.22	0.61
220	9.19	5.29	π	311	4.45	3.00	–0.56
230	6.12	5.68	0	322	3.02	2.88	1.82
240	8.90	7.22	0	411	6.68	9.98	3.37
250	1.08	2.17	π	511	1.77	2.53	2.85
310	20.13	14.47	$-\pi/2$	101	8.89	3.77	$-\pi/2$
320	6.18	4.24	$\pi/2$	121	0.72	2.37	3.84
330	2.32	0.81	$\pi/2$	131	6.24	4.29	0.10
340	4.08	6.25	$-\pi/2$	141	9.59	5.30	0.69
350	7.00	4.65	$\pi/2$	151	3.36	0.77	–1.35
400	6.14	3.93	0	202	9.86	0.40	π
410	12.03	11.81	π	212	9.28	5.93	0.89
420	6.92	7.28	0	222	0.72	5.47	–1.01
430	9.56	7.04	π	232	1.13	2.34	2.75
440	5.24	3.30	π	242	7.38	2.53	–1.07
450	6.18	0.98	0	252	3.11	3.37	3.85
510	4.74	4.72	$\pi/2$	303	0.72	7.58	$-\pi/2$
520	4.48	2.64	$-\pi/2$	313	8.25	3.37	0.40
530	2.64	2.58	$\pi/2$	333	1.91	2.19	3.90
540	1.00	1.26	$-\pi/2$	404	7.72	9.80	0
550	3.88	1.70	$-\pi/2$	414	1.91	3.27	–0.15

Table 11.14. Direct Phase Set for Mannan I

$0kl$	ϕ	$1kl$	ϕ	$2kl$	ϕ
020	0	110	$\pi/2$	200	π
040	π	120	$-\pi/2$	210	0
022	$a + \pi/2$	130	$-\pi/2$	220	π
		140	$\pi/2$	230	0
		150	$-a$	240	0
		101	$-\pi/2$	211	0
		111	$\pi/2$	202	$-a - \pi/2$
		131	a	212	$a + \pi/2$
		141	a	242	$a - \pi/2$
		151	$\pi - \mathrm{a}$	252	$-a + \pi/2$
$3kl$	ϕ	$4kl$	ϕ	$5kl$	ϕ
310	$-\pi/2$	400	0	510	$\pi/2$
320	$\pi/2$	410	π	520	$-\pi/2$
340	$-\pi/2$	420	0	550	$-\pi/2$
350	$\pi/2$	430	π		
311	$\pi - a$	440	π		
313	$\pi/2$	450	0		
322	$-a$	404	0		
		411	π		

After calculating $|E_h|$ for the three-dimensional set, an origin (Dorset and McCourt, 1993) was defined by setting $\phi_{110} = \pi/2$, $\phi_{210} = 0$, $\phi_{101} = -\pi/2$ and two additional reflections $\phi_{200} = \pi$ and $\phi_{420} = 0$ were assigned phases from highly probable Σ_1-triples. After these definitions, 50 zonal Σ_2-triples were evaluated, where $A_2 \geq 0.65$, along with 43 zonal quartets, where $B \geq 1.00$, yielding values for 24 reflections. Then, 47 nonzonal Σ_2-triples were generated, including three-dimensional noncentrosymmetric reflections. From this set, an additional 17 reflections were assigned phase values linked through an algebraic unknown $a = \pm\pi/2$ (including some new values that were unequivocally assigned). In principle, this ambiguity could be resolved with the choice of an enantiomorph-defining phase, but in the limited data set collected, no suitable reflection with allowable index parity could be identified. The initial phase set is listed in Table 11.14.

Resolution of the phase ambiguity required a fit of a monomer model to the three-dimensional density map. In the projection down the chain axis (Figure 4.15), only one weak reflection (150) was affected, so the choice of a made no significant change. On the other hand, a view onto the chain axis allowed a monomer structure to be fit well to the density, even though the series termination effect (from the limited three-dimensional sampling, excluding the chain axis repeat) did not permit resolution of individual atomic positions. Fractional coordinates derived from model fitting (Table 11.15) were similar to those found in the original analysis (Chanzy et al., 1987) or an earlier fiber x-ray analysis (Nieduszynski and Marchessault, 1972); the overall

Table 11.15. Atomic Coordinates of Mannan I after Direct Methods and Density Fitting

Atom	x	y	z
C 1	0.277	0.072	0.244
C 2	0.366	0.166	0.134
C 3	0.287	0.129	0.004
C 4	0.264	−0.078	0.016
C 5	0.184	−0.161	0.102
C 6	0.169	−0.370	0.093
O 2	0.512	0.096	0.130
O 3	0.377	0.202	−0.100
O (1,4)	0.174	−0.111	−0.129
O 5	0.267	−0.122	0.218
O 6	0.172	−0.455	0.218

agreement to the observed data was $R = 0.29$ when four reflections were removed from the data set. For the $hk0$ data alone, $R = 0.25$.

It is clear from this analysis that the missing cone of reflections, especially those corresponding to the chain axis direction, impeded interpretation of the structure, especially if atomic coordinates were sought in the potential maps. After comparing the results of this ab initio analysis (Dorset and McCourt, 1993) with those from the earlier conformational search (Chanzy et al., 1987), it was found that the greatest difference between the two structures was an average 0.5 Å shift of the chain along its axis, corresponding to the worst-sampled area of reciprocal space (Figure 11.12). Using the predictive power of the Sayre equation to estimate unobserved amplitudes and phases, the determination can be improved somewhat, as will be discussed in a future publication. However, while the restriction imposed by the goniometer tilt limitations may be unavoidable for certain materials, due to crystallization problems, other polymers can be oriented in two orthogonal directions to provide diffraction data sampling all directions in reciprocal space, as will be demonstrated by the following examples. (Some progress also has been made (Helbert and Chanzy, 1994) in obtaining suitably oriented crystals of polysaccharides.)

11.2.2.3. Polyethylene. Polyethylene has been the subject of crystallographic investigations for many years since the powder x-ray study carried out by Bunn (1939). Although the "stem" packing (identical to the $O\perp$ methylene subcell structure depicted in the previous three chapters) has been well understood since this early work, the possible interaction of surface chain folds with the "setting angle" of the chain stem plane to the unit cell axis (e.g., Kawaguchi et al., 1979) has motivated a more precise structural characterization.

Single crystals of polyethylene are grown by self-seeding from various solvents and may be observed as thin lozenges with the corrugated sectorization behavior (Figure 11.13) described in the work of various authors, or the center-pleated lozenges (Figure 11.13) described by others (Niegisch and Swan, 1960; Reneker and Geil, 1960; Bassett and Keller, 1961, 1962; Bassett et al., 1959, 1963; Boudet, 1984). Most recently, Revol and Manley (1986) have obtained 3.7-Å resolution micrographs of

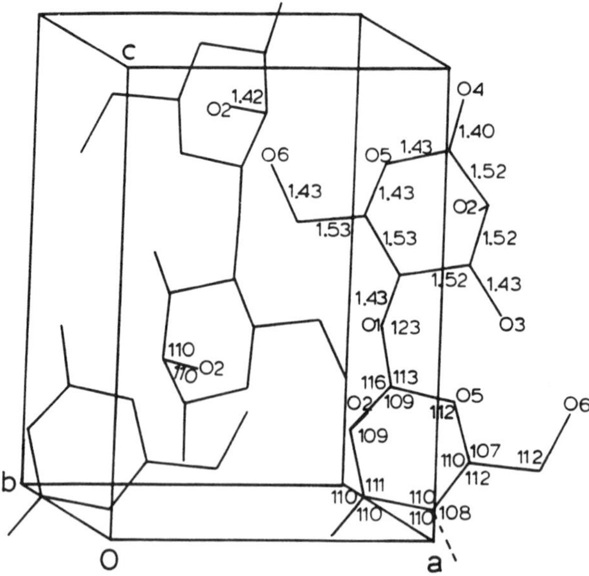

FIGURE 11.12. Molecular packing of mannan I after fitting monomer to experimental potential map generated from direct phasing set (see Chapter 4). Bond distances and angles are indicated. (Reprinted from D. L. Dorset and M. P. McCourt (1993) "Electron crystallographic analysis of a polysaccharide structure—direct phase determinations and model refinement for mannan I," *Journal of Structure Biology* **111**, 118–124; with kind permission of Academic Press, Inc.)

these lozenges, finding, at molecular scale, that the alternate light and dark bands in bright-field electron micrographs correspond, respectively, to tilted and untilted chain packing regions left after collapse of the three-dimensional pyramidal habit. (Other examples of molecular resolution images from polymers lamellar crystals are listed in Table 11.16.)

Attempts to solve the stem-packing structure from electron diffraction intensity data had been made in several laboratories. The centrosymmetric orthorhombic space group is *Pnma* with $a = 7.48$, $b = 2.55$, $c = 4.97$ Å, where **b** is the chain axis direction. (The usual definition of the **c**-axis as the chain direction for polymer crystal structures results in the equivalent space-group assignment *Pnam*.) Ingenious attempts to visualize the setting angle directly on an optical bench (Kawaguchi et al., 1979) were made when intensity-weighted masks were made to observe the intensity transform of the image—otherwise known as the Patterson function (Chapter 4). Unfortunately, the inclusion of space-group forbidden reflections in this mask, which are due to secondary scattering, distorted the transform, as it would also degrade attempts to find the structure in projection by searching for a minimum of the crystallographic residual by rigid body rotation of the chain around [010] on the twofold axis of plane group *pgg* (Dorset and Moss, 1983). It was clear that a good three-dimensional data set was required for definition of this chain packing, despite its simplicity.

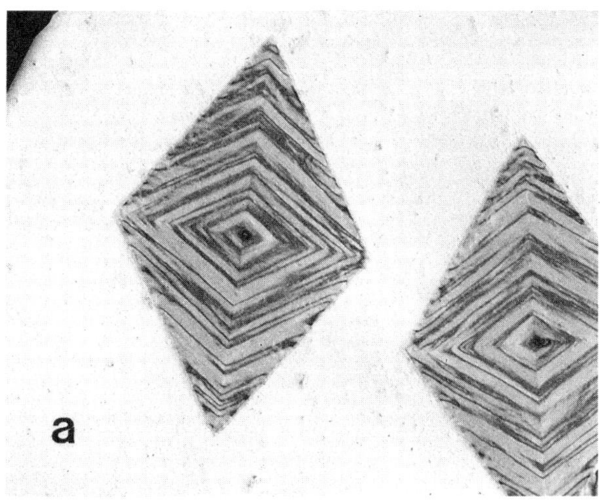

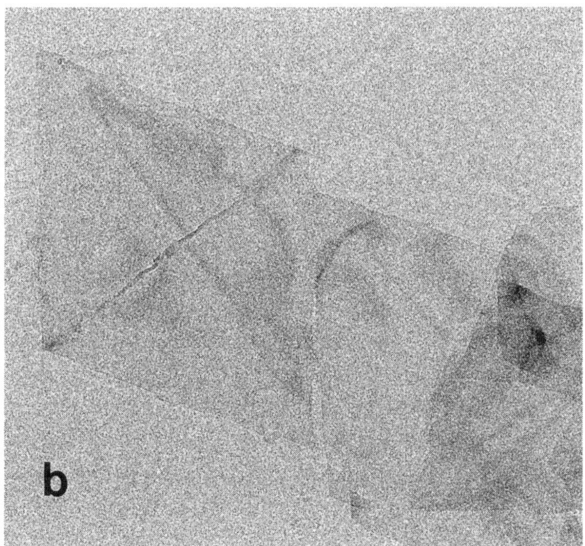

FIGURE 11.13. Morphology of polyethylene lamellae: (a) collapse of pyramidal habit showing diffraction contrast bands in sectors; (b) crystals with central pleats. (Reprinted from D. L. Dorset, J. Hanlon, C. H. McConnell, J. R. Fryer, B. Lotz, J. C. Wittmann, E. Beckmann, and F. Zemlin (1990) "Why do polyethylene crystals have sectors?," *Proceedings of the National Academy of Sciences USA* **87**, 1696–1700; copyright held by the author.)

Table 11.16. High-resolution Electron Micrographs of Polymers

Polymer	Resolution	Reference
polyethylene	3.7 Å	Revol and Manley (1986)
poly(tetrafluoroethylene)	4.9 Å	Chanzy et al. (1986)
poly(4-methyl-1-pentene)	4.3 Å	Pradere et al. (1988a,b)
poly(β-hydroxybutyrate)	3.5 Å	Revol et al. (1989)
α-chitin	5.1 Å	Revol (1989)
β-chitin	4.8 Å	Revol and Chanzy (1986)
cellulose III	4.3 Å	Sugiyama et al. (1987)
poly(p-xylylene)	2.8 Å	Tsuji et al. (1982)
		Zhang and Thomas (1991)
isotactic poly (styrene)	5.5 Å	Tsuji et al. (1984)
poly(tetramethyl-*p*-silphenylene siloxane)	3.2 Å	Tsuji et al. (1989)
poly(*p*-phenylene sulfide)	4.3 Å	Uemura et al. (1988)
poly(sulfur nitride)	2.9 Å	Kawaguchi et al. (1984)

Three-dimensional data were obtained by combining the $h0l$ set observed from the solution-grown crystals with tilt data recorded from crystals epitaxially grown on benzoic acid (Hu and Dorset, 1989) (Figure 2.17). After annealing at 80 °C in the presence of the nucleating substrate, followed by removal of the benzoic acid from the polymer sample by sublimation in vacuo, 51 unique reflections were accumulated when the diffraction from epitaxial samples (using rotation of the grid to locate $(h0l)$, (hll), (hhl), and $(h + l,h,l)$ sets) was added to intensities from the solution-grown lozenges. Initially, a structure solution was sought, using the chain rotational search, but the setting angle at 46.7° could not be distinguished from any value within the range 44.5–49.6° at an $\alpha = 0.05$ confidence level for the crystallographic R-factor, as defined by Hamilton (1964), even with the relatively large number of measured data.

Next, phases were directly determined for this data set (Dorset, 1991f). From the observed $|F_o|$ in Table 11.17, the resultant $|E_h|$ were found to have a distribution more like that expected for a noncentrosymmetric structure than the centrosymmetric unit cell identified. Nevertheless, after origin definition, setting $\phi_{101} = 0$, $\phi_{112} = \pi$, $\phi_{403} = \pi$, the values of 40/50 possible phases in the set of $|E_h|$ used for the analysis were found correctly (Table 11.17) by evaluation of Σ_2-triples to $A_{2,\min} = 1.2$ and positive quartets to $B_{\min} = 2.0$, and accepting four phase estimates from Σ_1-triples. The progress of this phase determination is reviewed in Table 11.18.

The structure was readily apparent in the potential maps based on 40 phases, for example, in the projection down the chain (Figure 11.14) or onto it (Figure 11.15). The single unique carbon position at $x = 0.033$, $y = 0.25$, $z = 0.066$ corresponds to a carbon–carbon bond distance of 1.52 Å and a C–C–C bond angle of 114.3°, consistent with x-ray structures of other typical polymethylene compounds (see also the preceding Chapters 8–10).

A least-squares refinement was attempted, starting with the coordinates from the first structure search with a chain model. If only the carbon position was used, the

Table 11.17. Normalized Structure Factors and Phase Comparison for Polyethylene

hkl	$\lvert E \rvert$	ϕ(DP)*	ϕ(M†)	*hkl*	$\lvert E \rvert$	ϕ(DP*)	ϕ(M†)
101	2.96	0	0	314	0.82	π	π
502	1.81	π	π	111	0.81	π	π
200	1.76	0	0	503	0.81	0	0
203	1.49	π	π	202	0.77	0	0
403	1.49	π	π	413	0.74	π	π
020	1.38	π	π	123	0.72	π	π
114	1.38	π	π	223	0.72	0	0
112	1.16	π	π	022	0.66	π	π
302	1.16	π	π	312	0.66	π	π
002	1.16	0	0	006	0.60		π
220	1.14	π	π	221	0.59		0
011	1.13	π	π	213	0.58	π	π
121	1.12	π	π	313	0.54		π
603	1.09	π	π	321	0.51	π	π
702	1.09	π	π	113	0.46		π
401	1.02	π	π	212	0.45		π
301	0.96	0	0	402	0.45	0	0
013	0.94	π	π	015	0.44	π	π
201	0.92	π	π	303	0.44	0	0
400	0.87	0	0	211	0.44	π	π
601	0.87	π	π	422	0.39		π
501	0.83	0	0	600	0.35		0
102	0.82	π	π	004	0.28		0
103	0.82	0	0	404	0.21		0
115	0.82	0	0	323	0.14		π

*Direct phasing.
†Model.

Table 11.18. Phasing Sequence for Polyethylene

origin definition:

$\phi_{101} = 0, \phi_{112} = \pi, \phi_{403} = \pi$

Σ_1-triples:

$\phi_{202} = \phi_{002} = \phi_{200} = \phi_{400} = 0$

Σ_2-triples:*

$\phi_{502} = \pi$	$\phi_{013} = b$	$\phi_{314} = \pi$
$\phi_{603} = \pi$	$\phi_{114} = b$	$\phi_{015} = \pi$
$\phi_{401} = \pi$	$\phi_{220} = a$	$\phi_{413} = \pi$
$\phi_{302} = \pi$	$\phi_{102} = \pi$	$\phi_{321} = a$
$\phi_{203} = \pi$	$\phi_{702} = \pi$	$\phi_{111} = \pi$
$\phi_{301} = 0$	$\phi_{201} = \pi$	$\phi_{223} = 0$
$\phi_{601} = \pi$	$\phi_{103} = 0$	$\phi_{402} = 0$
$\phi_{020} = a$	$\phi_{213} = \pi$	$\phi_{123} = \pi$
$\phi_{121} = a$	$\phi_{022} = \pi$	$\phi_{312} = \pi$
$\phi_{011} = \pi$	$\phi_{501} = 0$	$\phi_{115} = 0$

quartets:

$\phi_{121} = \pi = a$

$\phi_{211} = \pi$	$\phi_{303} = 0$	$\phi_{503} = 0$

*Can find a solution from a triple where $\phi_{013} = \pi = b$.

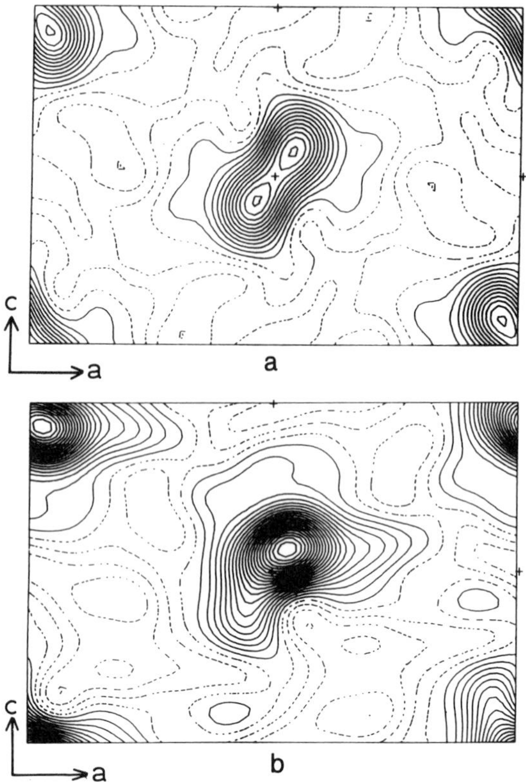

FIGURE 11.14. Potential maps for polyethylene: (a) based on $h0l$ data; (b) based on 3D data; slice at y = 0.25. (Reprinted from D. L. Dorset (1991) "Electron diffraction structure analysis of polyethylene. A direct phase determination," *Macromolecules* **24**, 1175–1178; with kind permission of the American Chemical Society.)

refinement was not stable, and produced a chemically unreasonable geometry, even if the positional shifts were dampened by 0.1 (compare to the refinement of diketopiperazine described in Chapter 6). A trial model was then constructed, including theoretical hydrogen positions. Again using a constrained shift of atomic positions in each cycle, the carbon atom movement was stabilized but the hydrogen positions had to be reset to idealized values after each shift. After completion of this refinement, the final coordinates (Table 11.19) were used to calculate structure factor values in agreement with the observed magnitudes by an R-value of 0.19 (Table 11.20). The resulting setting angle was 47°.

Of course, details of the chain stem packing say nothing about the chain folding at the surface of the monolamellar lozenges. Because of the elastic bending of each thin crystal, such details would be virtually impossible to access from the electron diffraction intensity data (see discussion of crystal bending in Chapter 5). The only

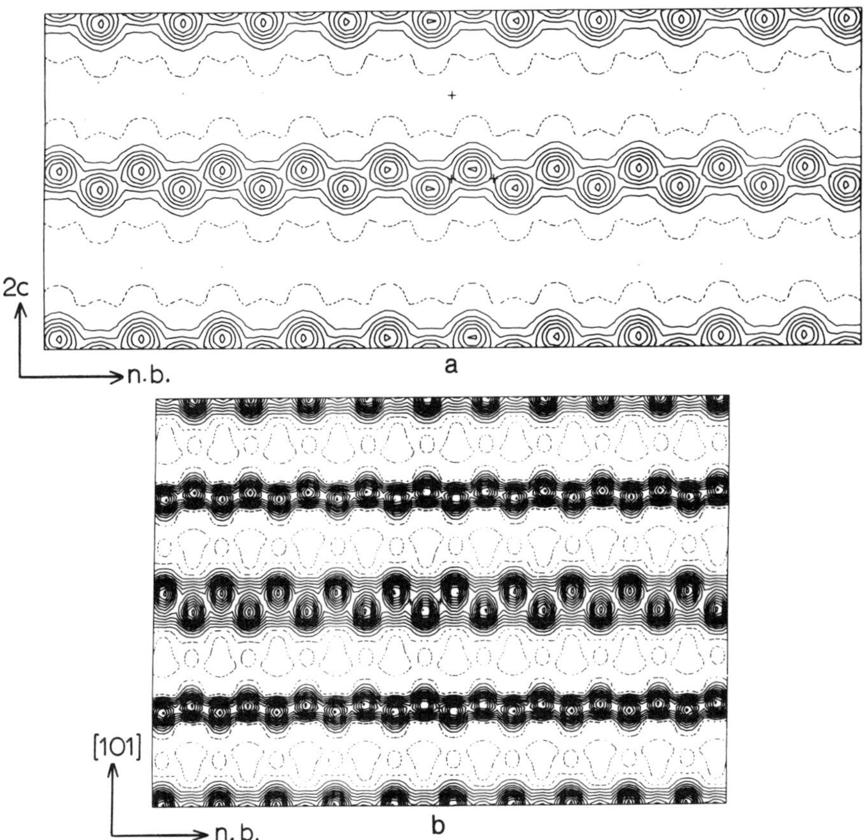

FIGURE 11.15. Two projections of the polyethylene structure onto the chain axis after direct phase determination: (a) [100]; (b) [101]. (Reprinted from D. L. Dorset (1991) "Electron diffraction structure analysis of polyethylene. A direct phase determination," *Macromolecules* **24**, 1175–1178; with kind permission of the American Chemical Society.)

information available to us is indirect. Decoration of lamellae with alkane chain fragments deposited from the vapor phase in vacuo (Wittmann et al., 1983) indicated that the surface structure approximates a common orientation and thus, presumably, this is related to the average surface fold geometry. The similarity of this result to the decoration of large-perimeter cycloalkane crystals (e.g., Ihn et al., 1990) is striking

Table 11.19. Final Atomic Coordinates and Temperature Factors for Polyethylene after Least-Squares Refinement

Atom	x/a	y/b	z/c	B_{iso}
C	0.0445	0.25	0.0617	5.2
H 1	0.1898	0.25	0.0520	7.2
H 2	0.0339	0.25	0.2607	7.2

Table 11.20. Observed and Calculated Structure Factors for Polyethylene after Least-Squares Refinement

hkl	$\|F_o\|$	$\|F_c\|$	*hkl*	$\|F_o\|$	$\|F_c\|$
002	1.43	1.56	314	0.21	0.14
004	0.27	0.10	123	0.18	0.09
011	1.28	1.29	422	0.09	0.04
020	0.74	0.72	212	0.24	0.42
022	0.29	0.25	121	0.39	0.44
111	0.62	0.80	313	0.19	0.13
220	0.54	0.40	323	0.03	0.04
101	3.55	3.44	200	2.88	3.00
202	0.56	0.57	201	0.91	0.93
303	0.19	0.15	102	0.69	0.64
006	0.15	0.07	301	0.77	0.87
013	0.57	0.64	400	0.79	0.82
015	0.13	0.18	302	0.71	0.59
112	0.68	0.86	401	0.60	0.53
113	0.20	0.14	103	0.44	0.48
221	0.19	0.12	203	0.73	0.47
114	0.41	0.35	402	0.21	0.18
223	0.18	0.10	501	0.37	0.23
115	0.17	0.02	502	0.56	0.35
211	0.29	0.70	600	0.18	0.02
312	0.31	0.28	403	0.54	0.29
213	0.23	0.32	601	0.29	0.25
413	0.22	0.11	503	0.24	0.05
321	0.15	0.16	603	0.27	0.18
404	0.05	0.02	702	0.27	0.14
222	0.21	0.12			

(see Chapter 8). Nothing definite can be said yet about whether the folds have adjacent reentry or a longer period along this average direction.

Returning to the intralamellar stem packing, analysis (Dorset et al., 1991) of continuous diffuse scattering from epitaxially oriented samples shows that the chains have slight positional shifts along their long axes, as is found also with the *n*-paraffins (see Chapter 8). This is consistent with the longitudinal motions involved in the collapse of the three-dimensional pyramids crystallized in suspension or the motions experienced when these long chain crystals are heated. Also consistent with results from longer alkanes, there is no true "rotator" phase for polyethylene at atmospheric pressure (or lower pressures). This observation was originally made by Charlesby (1945) after electron diffraction studies of the heated polymer wherein the lattice expansion was measured at different temperatures.

11.2.2.4. Poly(ε-caprolactone). The noncentrosymmetric crystal structure of this polyester was known to resemble that of polyethylene, and, indeed, the *hk*0 electron diffraction pattern from solution-crystallized samples closely matches that from the projection down the chain axis from the polyolefin (Brisse and Marchessault,

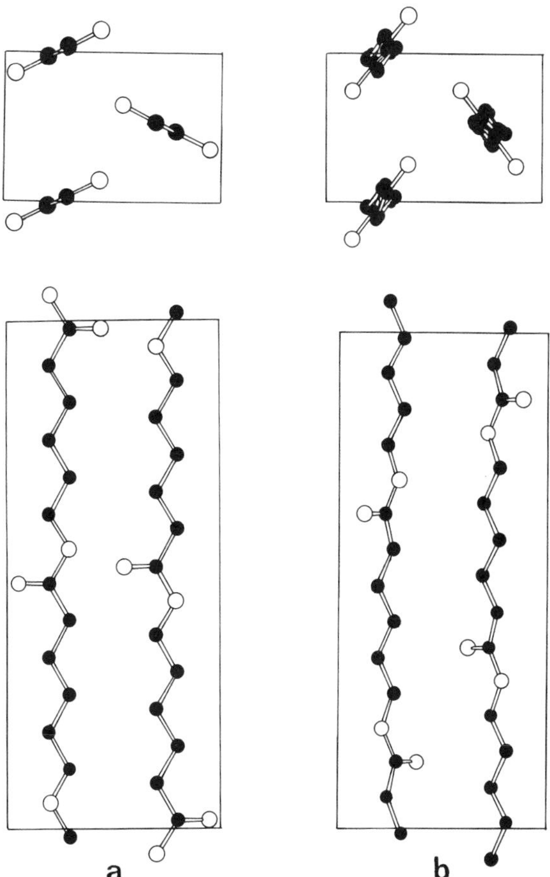

FIGURE 11.16. Two models proposed for poly(ε-caprolactone) after fiber x-ray analyses: (a) planar model; (b) conformationally twisted model. (Reprinted from H. Hu and D. L. Dorset (1990) "Crystal structure of poly(ε-caprolactone)," *Macromolecules* **23**, 4604–4607; with kind permission of the American Chemical Society.)

1980). The space group is $P2_12_12_1$ with cell constants a = 7.48(2), b = 4.98(2), c = 17.26(3) Å. Two fiber x-ray analyses have been reported, one of which claimed that the chain is flat like polyethylene (Bittiger et al., 1970). The other structural model argued in favor of a conformational twist near the ester linkage (Chatani et al., 1970) (Figure 11.16).

In order to resolve these conflicting results, crystals were epitaxially oriented on benzoic acid (Hu and Dorset, 1990) and annealed in the presence of this substrate to improve the crystallinity of the sample. After removal of the nucleating substrate, electron diffraction patterns (Figure 11.17) were collected from the [100] and [110] projections and combined with ($hk0$) data from solution-crystallized samples to give 47 unique reflections (Table 11.21). Comparing the models proposed from fiber x-ray

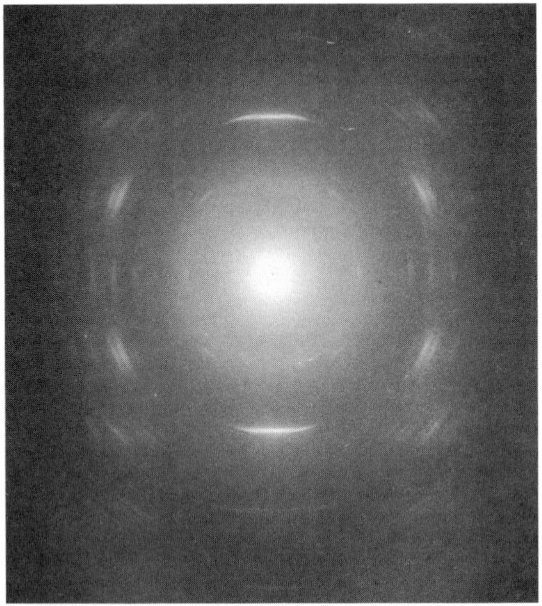

FIGURE 11.17. Electron diffraction from epitaxially oriented poly(ε-caprolactone), (0*kl*) pattern. (Reprinted from H. Hu and D. L. Dorset (1990) "Crystal structure of poly(ε-caprolactone)," *Macromolecules* **23**, 4604–4607; with kind permission of the American Chemical Society.)

analyses with these data, it was clear that the nonplanar structure is preferred ($R = 0.20$ versus 0.39). The measured intensities were then used to calculate $|E_h|$, and these values were found to have a centrosymmetric distribution, despite the noncentrosymmetric space group (Table 11.22). (This is probably due to the dominance of the observed data by zonal reflections.)

For phase determination (Dorset, 1991g), the centrosymmetric (*hk*0) and (0*kl*) reflections were examined first after defining the origin to evaluate Σ_2-triples and positive quartets, as outlined in Table 11.22. The relationships among the *hk*0 reflections are most prominent. Evaluating 45 Σ_2-triples to $A_{2,\min} = 0.20$, four invariants involving the (210) reflection were incorrect, but this reflection could be assigned a correct value if 25 quartets (all of which are correctly estimated) to $B_{\min} = 2.05$ were examined. The 0*kl* reflections are mostly involved in triples with unrestricted *hhl* reflections. Thus, a symbolic value *c* was used to link the phases in this region of reciprocal space. Two maps were required; the one at $c = 0$ (Table 11.22) was correct, and thus 30 more phases are contained in the final list.

The potential map projected down [100] clearly contains the structure (Figure 11.18) and fractional positions along *z* could be found for each atom. In the projection down [001], slices could be calculated at the indicated *z* levels. Although the positions

Table 11.21. Observed and Calculated Structure Factors for Poly(ε-caprolactone)

hkl	$\lvert F_o \rvert$	$\lvert F_c \rvert$	hkl	$\lvert F_o \rvert$	$\lvert F_c \rvert$
00 4	22.74	29.42	03 1	9.20	10.43
00 6	5.74	6.80	03 7	14.20	20.93
00 8	10.28	14.69	03 8	10.02	9.29
00 10	7.68	4.09	11 0	182.16	184.80
00 12	4.00	2.58	11 1	45.64	26.76
00 14	23.91	16.13	11 7	42.17	30.32
01 1	2.00	0.81	11 14	14.63	10.99
01 3	17.49	18.81	22 0	50.65	62.09
01 7	44.54	46.26	22 1	14.20	7.69
01 8	9.27	1.24	22 7	21.71	18.44
01 13	6.32	2.65	20 0	151.90	160.33
01 14	15.40	18.86	21 0	44.50	39.95
02 0	58.35	76.04	12 0	27.40	21.89
02 1	33.27	16.85	31 0	76.20	68.91
02 3	4.47	3.09	40 0	52.90	33.37
02 4	7.07	7.97	32 0	22.80	25.97
02 6	16.37	8.85	41 0	11.40	10.37
02 7	17.87	23.93	13 0	36.10	19.33
02 8	11.70	4.76	23 0	11.40	10.34
02 9	0.10	4.43	42 0	16.80	20.95
02 10	5.00	5.08	51 0	21.70	11.50
02 11	0.50	1.63	33 0	11.40	13.33
02 13	0.10	0.05	52 0	10.00	10.86
02 14	5.90	4.54			

Table 11.22. Direct Phase Determination for Poly(ε-caprolactone)

origin definition:

$\phi_{110} = \pi/2$	$\phi_{320} = \pi/2$	$\phi_{017} = \pi/2$	

enantiomorph:

$\phi_{038} = -\pi/2$

Σ_1-triple:

$\phi_{200} = \pi$

$hk0$ zonal Σ_2-triples and quartets:

$\phi_{020} = 0$	$\phi_{210} = 0$	$\phi_{310} = -\pi/2$	$\phi_{410} = \pi$
$\phi_{120} = -\pi/2$	$\phi_{220} = \pi$	$\phi_{330} = -\pi/2$	$\phi_{420} = 0$
$\phi_{130} = \pi/2$	$\phi_{230} = 0$	$\phi_{400} = 0$	$\phi_{510} = \pi/2$
$\phi_{520} = -\pi/2$			

hkl Σ_2-triples:

$\phi_{00,14} = 0$	$\phi_{028} = c + \pi$	$\phi_{117} = \pi/2 - c$
$\phi_{021} = \pi$	$\phi_{02,14} = 0$	$\phi_{11,14} = \pi/2$
$\phi_{026} = c - \pi$*	$\phi_{037} = \pi/2$	$\phi_{221} = 0$
$\phi_{027} = c$	$\phi_{111} = \pi/2$*	$\phi_{227} = \pi - c$

*Incorrect.

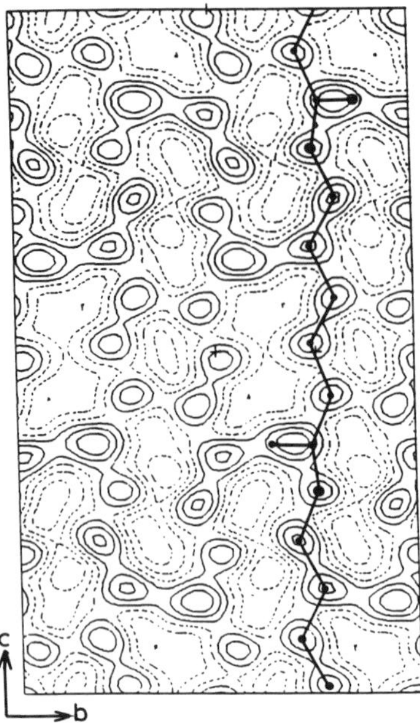

FIGURE 11.18. Potential map for poly(ε-caprolactone) after symbolic addition—projection down [100]. (Reprinted from D. L. Dorset (1991) "Electron crystallography of linear polymers: direct structure analysis of poly(ε-caprolactone)," *Proceedings of the National Academy of Sciences USA* **88**, 5499–5502; copyright held by the author.)

were not clearly resolved for the chain atoms at each interval, the direction of the projected bonds could be found from an ellipsoidal density cross section. Thus chain backbone coordinates were generated at 0.44 Å from the chain center line along the long elliptical axes. An estimate of the carbonyl position was also made on a lobe of density at $z = 0.35$. Structure factors calculated from initial atomic coordinates were found to agree reasonably well with observed magnitudes, i.e., $R = 0.30$ using all 47 reflections, and $B = 6.0$ Å^2 for all atoms.

Difference Fourier refinement based on $2\,|F_o| - |F_c|$ was used to improve these coordinates and some torsional adjustment of the C=O bond was made to optimize the match to the observed data. The final R-value calculated from the final atomic coordinates in Table 11.24 is 0.21 (Figure 11.19). The resultant bond geometries are similar to those found in the best fiber x-ray determination including a conformational chain twist. A comparison of crystallographic phases (Table 11.23) to those generated from the two x-ray models also favors the nonplanar chain conformation. The electron diffraction analysis provides a better estimate of thermal motion than does the fiber x-ray analysis (Chatani et al., 1970) since, in the latter, the carbonyl oxygen is estimated to have a rather high value, $B_{iso} = 15.0$ Å^2, compared to the values for other atoms.

Table 11.23. Comparison of Phase Sets for Poly(ε-caprolactone)*

hkl	ϕ_{DP}	ϕ_C	ϕ_B	*hkl*	ϕ_{DP}	ϕ_C	ϕ_B
00 4	π	π	π	03 1	$-\pi/2$	$-\pi/2$	$-\pi/2$
00 6	0	0	0	03 7	$\pi/2$	$\pi/2$	$-\pi/2$
00 8	0	0	0	03 8	$-\pi/2$	$-\pi/2$	$\pi/2$
00 10	π	π	π	11 0	$\pi/2$	$\pi/2$	$\pi/2$
00 12	π	0	0	11 1	–2.1	–1.9	3.1
00 14	0	0	π	11 7	2.4	2.3	–2.4
01 1	$\pi/2$	$-\pi/2$	$-\pi/2$	11 14	1.9	1.9	–0.9
01 3	$-\pi/2$	$-\pi/2$	$-\pi/2$	22 0	π	π	π
01 7	$\pi/2$	$\pi/2$	$-\pi/2$	22 1	–0.3	–0.1	2.4
01 8	$\pi/2$	$-\pi/2$	$-\pi/2$	22 7	–2.4	–2.6	–0.9
01 13	$-\pi/2$	$-\pi/2$	$-\pi/2$	20 0	π	π	π
01 14	$-\pi/2$	$-\pi/2$	$-\pi/2$	21 0	0	0	π
02 0	0	0	0	12 0	$-\pi/2$	$-\pi/2$	$\pi/2$
02 1	π	π	π	31 0	$-\pi/2$	$-\pi/2$	$-\pi/2$
02 3	0	0	π	40 0	0	0	0
02 4	0	0	π	32 0	$\pi/2$	$\pi/2$	$-\pi/2$
02 6	0	0	π	41 0	π	π	0
02 7	0	0	π	13 0	$\pi/2$	$\pi/2$	$\pi/2$
02 8	π	π	π	23 0	0	0	π
02 9	π	0	π	42 0	0	0	0
02 10	0	0	π	51 0	$\pi/2$	$\pi/2$	$\pi/2$
02 11	π	π	π	33 0	$-\pi/2$	$-\pi/2$	$-\pi/2$
02 13	π	π	π	52 0	$-\pi/2$	$-\pi/2$	$\pi/2$
02 14	0	0	π				

*DP = direct phase determination with electron diffraction data; C = fiber x-ray analysis of Chatani et al. (1970); B = fiber x-ray analysis of Bittiger et al. (1970).

Other attempts to solve this structure have been made. QTAN correctly locates the chain atom positions but cannot find the carbonyl position. The same result can be reported when the maximum entropy procedure is used (Dr. C. J. Gilmore, personal communication). More recently, a model very close to the correct structure has been retrieved by use of the minimal principle (M. McCourt, unpublished data) after a translational search along the chain axis to find the carbonyl oxygen position.

Table 11.24. Atomic Coordinates for Poly(ε-caprolactone)

Atom	*x*	*y*	*z*	B_{iso}
C 1	0.709	0.532	0.366	6.0
C 2	0.734	0.597	0.228	6.0
C 3	0.803	0.459	0.156	6.0
C 4	0.734	0.584	0.086	6.0
C 5	0.803	0.465	0.011	6.0
C 6	0.778	0.419	0.437	6.0
O 1	0.604	0.705	0.366	6.0
O 2	0.803	0.468	0.298	6.0

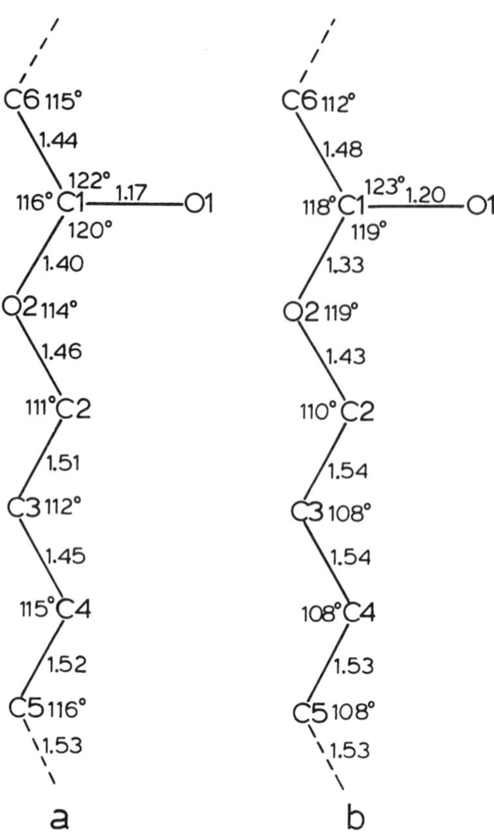

FIGURE 11.19. Final bond distances and angles for poly(ε-caprolactone): (a) electron diffraction determination; (b) best fiber x-ray determination. (Reprinted from D. L. Dorset (1991) "Electron crystallography of linear polymers: direct structure analysis of poly(ε-caprolactone)," *Proceedings of the National Academy of Sciences USA* **88**, 5499–5502; copyright held by the author.)

11.2.2.5. Poly(1-butene), Form III. Two orthogonal views of poly(1-butene) were crystallized in form III, via the addition of 2-quinoxalinol (Kopp et al., 1993). To effect a solution growth, the polymer was combined with the paraffin n-$C_{32}H_{66}$ in a 1:4 ratio and this mixture was dissolved in hot $CHCl_3$ to be evaporated onto a glass plate to form a thin film. A layer of 2-quinoxalinol was formed over this film and the sandwich heated to 140 °C, followed by cooling slowly to ambience. After washing in ethanol and chloroform, the polymer crystals remained. Epitaxial growth was achieved by a direct combination of the polymer with 2-quinoxalinol on a glass plate and heating the mixture to cause nucleation of the polymer by the diluent. Washing with chloroform left the oriented polymer behind. This crystalline form is not stable when fibers are drawn.

Electron diffraction data (Figure 11.20 and Table 11.25) were obtained at 120 kV by tilting solution-grown and epitaxially grown crystals around **a***, **b***, and **c*** axes,

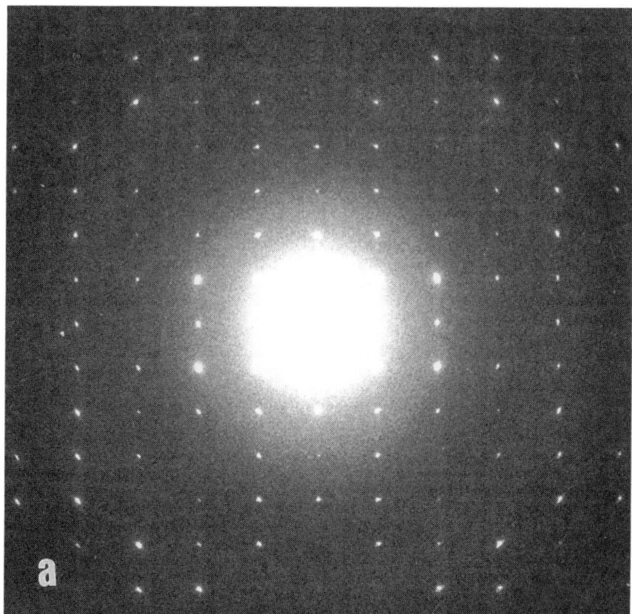

FIGURE 11.20. Electron diffraction patterns from poly(1-butene), form III: (a) *hk*0; (b) 2*h*,*h*,*l*. (Reprinted from D. L. Dorset, M. P. McCourt, S. Kopp, J. C. Wittmann, and B. Lotz (1994) "Direct determination of polymer structures by electron crystallography—isotactic poly(1-butene) form III," *Acta Crystallographica* **B50**, 201–208; with kind permission of the International Union of Crystallography.)

Table 11.25. Observed and Calculated Structure Factor Magnitudes for Poly(1-butene), Form III

hkl	$\|F_o\|$	$\|F_c\|$	*hkl*	$\|F_o\|$	$\|F_c\|$
200	1.18	1.20	611	0.49	0.38
400	0.77	0.68	711	0.29	0.24
600	0.34	0.54	021	1.26	1.54
800	0.44	0.35	121	1.35	0.99
10,00	0.33	0.12	221	0.35	0.13
110	1.42	1.75	321	0.34	0.08
210	0.82	0.81	421	1.06	0.95
310	0.65	0.41	521	0.31	0.47
410	0.79	0.77	621	0.42	0.30
510	0.76	0.52	821	0.27	0.30
610	0.29	0.32	031	0.71	0.92
710	0.22	0.09	131	0.44	0.50
810	0.26	0.16	231	0.39	0.21
10,10	0.35	0.30	331	0.87	0.97
020	0.78	0.25	431	0.76	0.49
120	1.41	1.49	531	0.23	0.26
220	0.54	0.59	041	0.32	0.38
420	0.36	0.01	141	0.53	0.50
520	0.52	0.34	241	0.52	0.27
620	0.98	0.78	341	0.32	0.41
720	0.37	0.46	541	0.91	0.73
820	0.24	0.13	641	0.18	0.13
920	0.50	0.56	741	0.19	0.31
10,20	0.32	0.18	151	0.33	0.26
130	0.62	0.36	251	0.23	0.17
230	0.34	0.20	351	0.27	0.28
330	0.48	0.26	451	0.30	0.28
430	0.33	0.04	751	0.27	0.06
530	1.04	0.63	261	0.37	0.27
630	0.80	0.97	461	0.27	0.30
730	0.21	0.22	271	0.31	0.17
830	0.17	0.15	471	0.41	0.25
930	0.28	0.26	402	0.28	0.43
040	0.58	0.87	502	0.40	0.59
140	0.58	0.35	012	0.92	1.13
240	0.80	0.44	112	1.10	1.45
340	0.64	0.75	412	0.22	0.40
440	0.90	0.73	512	0.30	0.30
540	0.71	0.41	022	0.40	0.22
740	0.28	0.35	122	0.32	0.18
840	0.29	0.32	222	0.39	0.15
940	0.26	0.14	322	0.41	0.43
250	0.23	0.24	422	0.33	0.15
350	0.52	0.50	522	0.37	0.35
450	0.59	0.59	032	0.61	0.89
060	0.45	0.37	432	0.34	0.17
160	0.32	0.15	042	0.25	0.32
360	0.42	0.41	142	0.23	0.21

(continued)

Table 11.25. *(Continued)*

hkl	$\lvert F_o \rvert$	$\lvert F_c \rvert$	*hkl*	$\lvert F_o \rvert$	$\lvert F_c \rvert$
460	0.28	0.18	242	0.38	0.41
560	0.32	0.13	342	0.27	0.17
170	0.39	0.10	442	0.29	0.23
270	0.65	0.52	642	0.23	0.11
080	0.22	0.11	742	0.22	0.37
201	2.40	2.24	113	0.85	0.82
301	1.42	1.82	223	0.35	0.20
401	0.96	0.92	033	0.35	0.11
501	0.64	0.66	133	0.32	0.30
111	1.62	1.91	233	0.33	0.26
211	0.55	0.40	333	0.28	0.45
311	0.31	0.32	004	1.60	1.98
411	0.66	0.45	114	0.49	0.40

using constantly excited reflections to scale intensities from different zones. The unit cell is $P2_12_12_1$ with cell constants $a = 12.38$, $b = 8.88$, $c = 7.56$ Å. After integrating 125 unique intensities from densitometer scans, normalized structure factor values were calculated in the usual way. For solution-crystallized samples, no Lorentz correction was needed to find the intensity values. However, because the reflections from the epitaxially oriented samples are somewhat arced (Figure 11.20) the correction:

$$|F_h| = k\,(I_h\, d_h^*)^{1/2}$$

was used. The derived $|E_h|$ values corresponded to the distribution expected for a noncentrosymmetric structure (Table 11.26).

For direct phase determination (Dorset et al., 1994b), 1883 Σ_2-three phase invariants ($A \geq 0.2$) and 72 negative quartet invariants ($B \leq -0.05$) were generated for use by the tangent formula (QTAN). However, the NQEST figure of merit was not especially for finding the best solution. Thus, 20 phase values were determined independently from zonal triples and positive quartets (Table 11.27) to compare to the

Table 11.26. Distributions of $|E_h|$ for Poly(1-butene)

	Experimental	Theory: Centrosymmetric	Theory: Noncentrosymmetric
$\langle \lvert E_h\rvert^2 \rangle$	1.000	1.000	1.000
$\langle \lvert E_h^2 - 1\rvert \rangle$	0.876	0.968	0.736
$\langle \lvert E_h\rvert \rangle$	0.889	0.798	0.886
% $\lvert E_h\rvert > 1.0$	29.4	32.2	36.8
% $\lvert E_h\rvert > 2.0$	4.0	5.0	1.8
% $\lvert E_h\rvert > 3.0$	0.0	0.3	0.01

Table 11.27. Phasing of Zonal Data for Poly(1-butene)

hkl	Phase	Source	*hkl*	Phase	Source
400	π	Σ_1	440	π	Σ_2
110	$\pi/2$	Σ_2	540	$\pi/2$	Σ_2
10,10	0	Σ_2	840	0	Σ_1
120	$\pi/2$	Σ_2	350	$\pi/2$	Σ_2
620	0	Σ_2	450	0	Σ_2
920	$-\pi/2$	Σ_2	550	$\pi/2$	Σ_2
10,20	π	Σ_2	170	$\pi/2$	Σ_2
530	$-\pi/2$	origin	270	π	origin
630	0	Σ_2	201	$-\pi/2$	origin
340	$-\pi/2$	Σ_2	301	$\pi/2$	enant.

Table 11.28. Direct Phase Refinement for Poly(1-butene)*

hk	$\lvert F_o \rvert$	ϕ(DP)	ϕ(Say1)	ϕ(Say2)	*hk*	$\lvert F_o \rvert$	ϕ(DP)	ϕ(Say1)	ϕ(Say2)
20	1.18		0	$\pi^\dagger$	43	0.33		0	0
40	0.77	π	π	π	53	1.04	π	π	π
60	0.34		$0^\dagger$	π	63	0.80	π	π	π
80	0.44		π	π	73	0.21		π	π
10 0	0.33		0	0	83	0.17		π	π
11	1.17	0	0	0	93	0.28		0	0
21	0.82		$0^\dagger$	$0^\dagger$	04	0.58		$0^\dagger$	$0^\dagger$
31	0.65		$\pi^\dagger$	$\pi^\dagger$	14	0.58		0	$\pi^\dagger$
41	0.79		π	π	24	0.80		π	$0^\dagger$
51	0.76		π	π	34	0.64	0	0	0
61	0.29		$0^\dagger$	$0^\dagger$	44	0.90	π	π	π
71	0.22		0	0	54	0.71	0	0	0
81	0.26		π	π	74	0.28		π	π
10 1	0.35	0	0	0	84	0.29	0	0	0
02	0.78		0	0	94	0.26		π	π
12	1.41	0	0	0	25	0.23		0	0
22	0.54		0	0	35	0.52	π	π	π
42	0.36		$\pi^\dagger$	$\pi^\dagger$	45	0.59	0	0	0
52	0.52		π	π	55	0.53	0	0	0
62	0.98	π	π	π	06	0.45		$0^\dagger$	$0^\dagger$
72	0.37		π	$0^\dagger$	16	0.32		0	0
82	0.24		$0^\dagger$	$0^\dagger$	36	0.42		0	0
92	0.50	π	π	π	46	0.28		0	0
10 2	0.32	π	π	π	56	0.32		0	0
13	0.62		$0^\dagger$	$0^\dagger$	17	0.39	0	0	0
23	0.34		0	0	27	0.65	0	0	0
33	0.48	(π)	π	$0^\dagger$	08	0.22		0	$\pi^\dagger$

*Phase determination referred to *pgg* origin rather than space-group origin (space-group origin found by adding $\pi h/2$). DP = direct phase determination (basis set); Say 1 = Sayre refinement including (330) reflection; Say 2 = Sayre refinement omitting (330) reflection.

†Incorrect.

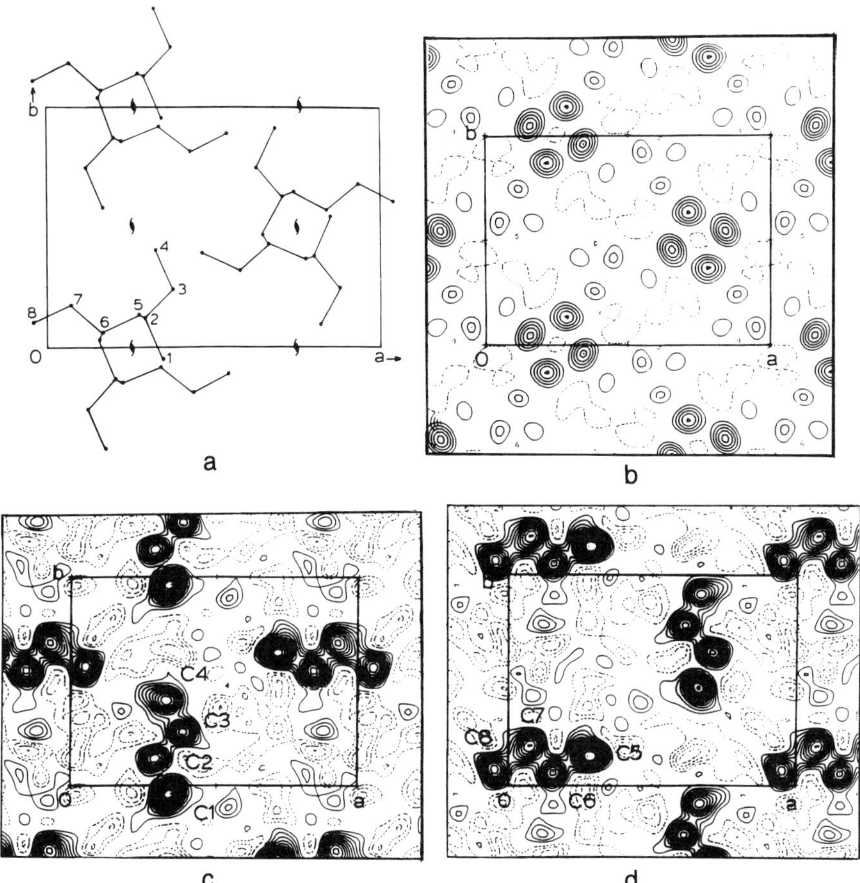

FIGURE 11.21. (a) Projected molecular packing for poly(1-butene), form III; (b) projected potential map based on *hk*0 data; (c) slice at $z = 0.12$; (d) slice at $z = 0.35$. (Reprinted from D. L. Dorset, M. P. McCourt, S. Kopp, J. C. Wittmann, and B. Lotz (1994) "Direct determination of polymer structures by electron crystallography—isotactic poly(1-butene) form III," *Acta Crystallographica* **B50**, 201–208; with kind permission of the International Union of Crystallography.)

generated three-dimensional sets from QTAN (see below). (This *hk*0 set can be expanded if the Sayre equation is used for phase refinement (Dorset et al., 1995) but there have to be enough phases in the starting set to lead to a list of 54 reflections with minimal error; see Table 11.28.) The origin was defined by setting $\phi_{201} = -\pi/2$, $\phi_{270} = \pi$, $\phi_{530} = -\pi/2$ with a possible enantiomorph sought via ϕ_{210} ($= \pi/2$). Algebraic values were given to ϕ_{004}, ϕ_{421}, ϕ_{541}, ϕ_{620}, and ϕ_{440} to generate 128 solutions. Here, centrosymmetric reflections were permuted through two allowable values and reflections with nonrestricted phases started at $\pi/4$ and cycled integrally by sequential addition

Table 11.29. Fractional Coordinates for Poly(1-butene)

	Electron crystallography			Powder x-ray study		
	x	y	z	x	y	z
C 1	0.346	–0.038	0.108	0.348	–0.034	0.107
C 2	0.298	0.125	0.163	0.301	0.121	0.159
C 3	0.381	0.248	0.103	0.379	0.247	0.102
C 4	0.325	0.402	0.120	0.324	0.401	0.123
C 5	0.279	0.134	0.358	0.274	0.136	0.357
C 6	0.160	0.067	0.413	0.163	0.071	0.409
C 7	0.072	0.183	0.353	0.073	0.180	0.352
C 8	–0.038	0.105	0.370	–0.038	0.104	0.373

of $n(\pi/2)$. Although the correct solution was found to lie within the 38 lowest NQEST values, as mentioned above, the independently determined zonal set was needed for positive identification of the three-dimensional solution containing 106 phases. This difficulty could be circumvented, however, if the phase residual for the minimal principle was used, either as a figure of merit for the tangent formula, or via the "Shake and Bake" algorithm, to find the structure (M. McCourt, unpublished data).

Initial potential maps contained the positions of all 8 carbon atoms in the asymmetric unit, revealing that a molecular 4_1-helix lay parallel to the space group

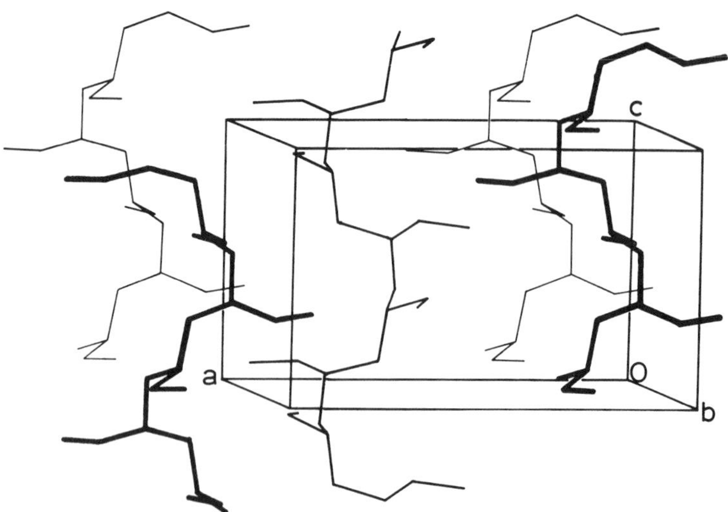

FIGURE 11.22. Molecular packing of poly(1-butene), form III based on coordinates in experimental potential maps. (Reprinted from D. L. Dorset, M. P. McCourt, S. Kopp, J. C. Wittmann, and B. Lotz (1994) "Direct determination of polymer structures by electron crystallography—isotactic poly(1-butene) form III," *Acta Crystallographica* **B50**, 201–208; with kind permission of the International Union of Crystallography.)

2_1-axes (n.b., the zero value for I_{002} in the experimental data) (Figure 11.21). Atomic peak positions, when used to calculate structure factors ($B = 4.0$ Å^2) result in a fit, $R = 0.33$, to the observed data. Fourier refinement via $2|F_o| - |F_c|$ progressed after removal of three problematic reflections to yield a chain geometry (Table 11.29) that was indistinguishable from a conformational model proposed earlier (Cojazzi et al., 1976). This direct single-crystal determination (Figure 11.22) with higher-resolution diffraction data, however, resulted in an agreement $R = 0.26$ (Table 11.25) for 122 observed data and did not require foreknowledge of the earlier structural model that was based on the analysis of only 21 powder x-ray maxima, some of which had 15 overlapping contributors.

11.3. Conclusions

Electron crystallography has become the method of choice for determining polymer structures, since it affords the only possibility for an analysis of single-crystal diffraction data. Direct methods have shown that the previously exploited analytical methods based on the conformational chain models often lead to correct results. It is even possible to extend the resultant basis zonal set with the Sayre equation. However, except for very simple polymers, the ab initio analysis of zonal data, particularly in a projection down a chain axis, is generally useless for finding atomic coordinates, due to the extensive overlap of projected positions. Atomic positions are located in three-dimensional analyses, particularly if data from both solution and epitaxially oriented crystals are combined, thus permitting a complete sampling of all of the reciprocal lattice. If only one crystal orientation is used, the cone of missing data, resulting from the physical tilt limits of the microscope goniometer stage, can lead to blurring of the potential map, particularly if the chain axis is contained in this unsampled region. In this case, fitting of the chain density in the potential map with a monomer model may be sufficient to find the correct crystal structure.

12

Globular Macromolecules

12.1. Background

Electron microscopy has often been used to visualize arrays of globular macromolecules within various cell preparations. For example, in the outer membranes of mitochondria, details were enhanced by use of a negative stain such as phosphotungstate (e.g., Parsons, 1967). A significant conceptual breakthrough occurred when crystallographic principles were adapted to the analysis of images of periodic objects, e.g., the pioneering work of DeRosier (1971) and DeRosier and Klug (1968) on T4 bacteriophage tail. (For a discussion of tubular globular protein crystals, see Vainshtein (1978).) In this work, an average image of the object was reinforced by spatially filtering the signal from the repeating motif. Such filtering could be carried out on an optical bench, using a physical mask to pass (through drilled or etched holes) the intense maxima in the Fourier transform formed at the back focal plane of the objective lens. When the continuous signal, due to the nonperiodic part of the array, was removed, a clearer image would be obtained (e.g., see Misell, 1978). Later this work was carried out digitally, using computer software that would allow calculation of forward and reverse Fourier transforms. Insertion of the mask function was applied when the reverse transform to the image was computed (e.g., Van Heel and Keegstra, 1981) or, alternatively, just the centers of the maxima were used to obtain the phase terms (e.g., Hovmöller, 1992). Correlation techniques were developed, not only for periodic objects, but also nonperiodic ones, so that small arrays could be aligned to one another to form an average by superposition (Frank, 1980; Saxton, 1980; Saxton and Baumeister, 1982). Most of the work was carried out with negatively stained objects, so that an adequate signal from the array could be obtained at "conventional" electron beam doses. It was recognized that this use of stain was itself a problem, since the stain's microcrystallinity limited most electron crystallographic analyses of macromolecules to approximately 20-Å resolution (Vainshtein, 1978). Limitations to collection of a complete data set were also apparent. If the object being examined was a thin two-dimensional crystal, then the physical tilt limits of the electron microscope goniometer stage would leave a "missing cone" of information (a restriction also recognized earlier by Vainshtein (1964a) in his description of data collection from single crystals of smaller structures—see previous chapters). It was imperative to learn whether these missing data would seriously affect the accurate reconstruction of a three-dimensional object. On the other hand, the continuous Fourier transform of a two-dimensional crystal could be sampled at very fine intervals in the direction normal to the crystal plate during the tilting experiments (Amos et al., 1982). Other details, such as the actual orientation of the sample at any given tilt (compensating for small local tilts of the grid support film), were also important for accurate three-dimensional reconstructions of the macromolecular structure (e.g., Engel and Massalski, 1984).

During the time when most studies of macromolecular arrays were made with negatively stained preparations, other workers began to consider methodology for increasing the resolution of their determinations. New preparative techniques were devised that would preserve the crystals without stain. For example, with the development of differentially pumped environmental stages for electron microscopes, electron diffraction patterns could be obtained from hydrated proteins such as ox-liver catalase to 2 Å (Matricardi et al., 1972; Dorset and Parsons, 1975a). Unfortunately, the inelastic scattering background from the water vapor surrounding the preparation frustrated attempts to obtain high-resolution images of such objects. Promising results were obtained when the same hydrated crystals were rapidly immersed in a cryogen to freeze the water into a "vitreous ice" (Taylor and Glaeser, 1974). It was soon found that, in addition to the cryostage for the microscope, an improved anticontaminator cold finger had to be added to the column to prevent the formation of crystalline ice on the sample by sublimation (Lepault et al., 1983; Echlin, 1992). Such technology is common nowadays. A simpler solution to the solvent evaporation problem was to replace the water with a less volatile hydrogen-binding substance such as a sugar (glucose, for example) (Unwin and Henderson, 1975). The dual advantages of such a solvent substitute and a negative stain could even be combined when a substance such as aurothioglucose was used. Strategies for the interpretation of low-contrast electron microscope images from such objects were also being considered at the time (Kuo and Glaeser, 1975). The importance of multiple-beam dynamical scattering has been considered both for the low-contrast phase objects (Ho et al., 1988) and negatively stained preparations (Dorset, 1984). More recently, alternative small-spot imaging techniques have been used to minimize specimen movement due to radiation damage (Downing, 1991). Perhaps of greater importance to image analysis, these small area illumination modes have been shown to minimize the difficulties experienced with phase contrast transfer function fluctuations in the image of a tilted sample, due to the effective changes in Δf in the direction normal to the tilt axis (Zemlin, 1989).

Much of the work of preparing a suitable specimen for macromolecular electron crystallography takes place in a biochemistry laboratory. Suitable detergents for the extraction and purification of membrane-bound proteins have been extensively evaluated (Garavito and Rosenbusch, 1986; Boekema, 1990; Rosenbusch, 1990). Reconstitution of the purified membrane proteins in a phospholipid bilayer to produce an ordered two-dimensional lattice is a multiple-variable problem that is beginning to be understood (Jap et al., 1992). Often, gentle digestion procedures with a phospholipase help improve the ordering of the sample (Mannella, 1984). Even when small residual in-plane paracrystalline bend distortions remain in the sample, they can be removed computationally by cross-correlation techniques to "unbend" the lattice in the image (Baldwin and Henderson, 1984; Henderson et al., 1990).

The difficulty of structure determination increases with higher resolution. Since the Fourier transforms of images are used as the primary source of crystallographic phases in such determinations, the conditions for accepting such information become more critical at resolutions where radiation damage, statistical image fluctuations, etc., have a more significant impact on the detected signal from the specimen (Henderson et al., 1986). Recently, it has been suggested that direct phasing methods based on maximum entropy techniques be used for extension of information from a more easily obtained 15-Å resolution image to the limits observed in the electron diffraction pattern (Gilmore et al., 1992, 1993b). As will be shown below, even use of the Sayre equation, coupled with solvent flattening, has been found to be effective for phase extension in the low-angle region, following the recommendation of Fan et al. (1991) that such data should be the best suited as a basis for direct analysis under most favorable circumstances. Earnest et al. (1992) have discussed the use of noncrystallographic symmetry for phase refinement in electron crystallography of proteins. Molecular replacement methods had been tried earlier for bacteriorhodopsin (Rossmann and Henderson, 1982). Although these are attractive prospects, promising to remove much of the drudgery of this high-resolution work, it should be remembered that the diffraction incoherence effects seen for other organics, when slightly bent crystals are used in the electron diffraction experiments, can be a limiting factor for extension to high resolution. This was demonstrated by model calculations on rubredoxin crystals to show that the high-resolution diffraction information may not have a simple relationship to the underlying crystal structure if significant bending occurs (Dorset, 1986). Stringent controls of specimen flatness must be imposed to guarantee that the diffraction intensities at high resolution are useful for such phase extension techniques. This point is discussed in detail in a recent review by Glaeser and Downing (1993) but, although the original work on elastic bend distortions by Cowley (1961) is also cited there, it is not apparent that its possible consequence to measured higher-resolution diffraction intensities from proteins is fully appreciated by many workers in this area.

It is not intended to give a detailed overview of the electron crystallography of macromolecules in this chapter. This has been already done quite well in a number of reviews (Glaeser, 1985; Chiu, 1986; Hovmöller et al., 1988; Jap et al., 1992). A list of membrane protein structures investigated by these techniques is given in Table 12.1, based on a recent review by Jap et al. (1992). Most of the work reported so far has been done at rather low resolution, mainly because of difficulties with specimen preparation. What will be considered in the sections below are the three highest-resolution structure analyses reported to date. In this review, this question will be kept in mind: Do these results actually mean anything or are they high-resolution fantasies?

Table 12.1. Two-dimensional Crystallization of Membrane Proteins. Representative Examples Including in situ Crystals*

bacteriorhodopsin	Unwin and Henderson (1975) Cherry et al. (1978) Michel et al. (1980) Baldwin and Henderson (1984) Popot et al. (1987)
halorhodopsin	Havelka et al. (1993)
cytochrome C oxidase	Vanderkooi et al. (1972) Deatherage et al. (1982) Frey et al. (1982)
cytochrome C reductase	Wingfield et al. (1979) Hovmöller et al. (1983)
NADH: ubiquinone reductase	Leonard et al. (1987)
(Na^+,K^+) ATPase	Skriver et al. (1981, 1988, 1989) Mohraz and Smith (1984) Mohraz et al. (1985, 1987)
(H^+,K^+) ATPase	Mohraz et al. (1990)
Ca^{++} ATPase	Buhle et al. (1983) Dux et al. (1986) Taylor et al. (1986, 1988) Stokes and Green (1990)
connexin	Zampighi and Unwin (1979) Kistler et al. (1990) Lampe et al. (1991) Yeager and Gilula (1992)
lens MIP	Kistler and Bullivant (1980)
mitochondrial porin	Mannella (1984) Mannella et al. (1986)
bacterial porins	Dorset et al. (1983) Jap (1988) Lepault et al. (1988) Stauffer et al. (1992)
bacterial space protein	Paul et al. (1992)
acetylcholine receptor	Toyoshima and Unwin (1988)
photosystem I reaction center	Ford et al. (1990)
photosystem II reaction center	Dekker et al. (1990)
bacterial reaction center	Miller and Jacob (1983) Barnakov et al. (1990)
light-harvesting complex II	Kühlbrandt (1984) Li (1985) Wang and Kühlbrandt (1991) Kühlbrandt et al. (1994)

*For a general discussion of techniques, see Jap et al. (1992).

12.2. Structure Analyses of Membrane Proteins at High Resolution

12.2.1. Bacteriorhodopsin

The three-dimensional structure analysis of the bacteriorhodopsin (MW 26,000 daltons) from the *Halobacterium halobium* purple membrane by Henderson and Unwin (1975) represents a landmark experiment in the electron crystallography of macromolecules, establishing standard procedures for the study of unstained preparations. It was these researchers who originated the idea of using dilute glucose solutions to replace the hydration of the protein so that it could be studied in the vacuum of the electron microscope. Also, low-dose imaging procedures were also developed by them as well as the use of the microscope condenser lens system to isolate the selected area from an approximate 1.0-μm crystalline patch for visualization of electron diffraction patterns (Unwin and Henderson, 1975). It is also convenient that the preparation was found in situ to be a highly ordered array of protein so that this study could be undertaken.

Using procedures established for negatively stained material by DeRosier and Klug (1968), Henderson and Unwin (1975) obtained images from tilted membrane sheets (only some 45 Å thick) and, based on the central section theorem, a common phase origin was found for the various projections to allow a three-dimensional image to be constructed at 7-Å resolution (by combining the image phases and the electron diffraction structure factor amplitudes). The structure was shown to be made up of seven α-helices spanning the lipid bilayer in the membrane, but at this diffraction limit nothing could be said about the retinal molecule responsible for the protein activity. Early attempts to obtain this protein as three-dimensional crystal sizes suitable for x-ray data collection were frustrated by a data resolution limited to 8 Å (Michel and Oesterhelt, 1980). Thus there was every reason to proceed with the electron crystallographic determination.

As mentioned in Section 12.1, striving for higher resolution involves working out instrumental and analytical details, in addition to improvement of crystalline order in the specimen. Bacteriorhodopsin crystal sheets were examined on a liquid helium–cooled cryomicroscope (the same instrument used for collection of high-resolution paraffin images described in Chapter 8). From optical micrographs, the best directly measured image resolution from untilted samples was found to be around 6 Å (Henderson et al., 1982). Correlation techniques were used for unbending the lattice deformations to achieve higher resolution phases. Using this technique, as well as a more accurate accounting for slight crystal tilts, electron beam tilt, and also the phase contrast transfer function, average images could be found from many crystals with 3.5-Å detail (Figure 12.1), which were in good agreement with one another (Henderson et al., 1990). More recently, the resolution has been extended to 2.8 Å (Baldwin et al., 1988). Finally, these procedures and controls were applied to the analysis of images from tilted crystals, i.e., half of the complete transform of three-dimensional image at this resolution. At this level, the details of the polypeptide primary structure could be fitted to the density map (Henderson et al., 1990; Downing, 1992). The retinal was

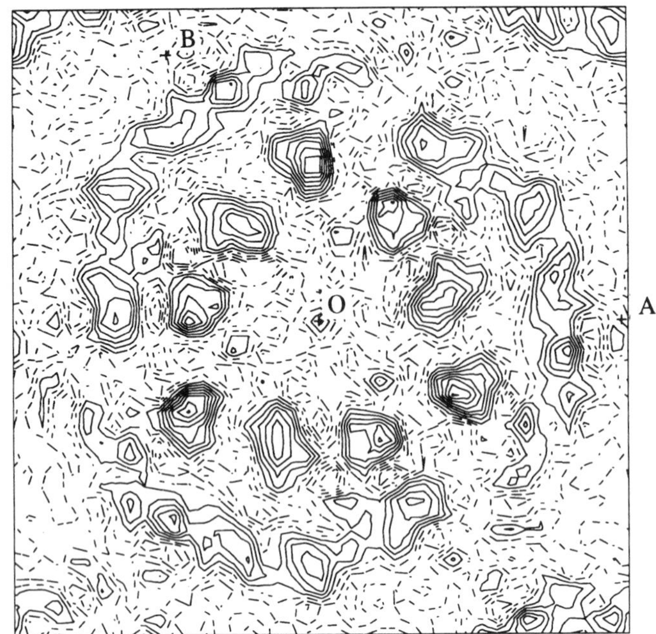

FIGURE 12.1. Projected structure of bacteriorhodopsin to 3.5-Å (diffraction) resolution, based on amplitudes and phases published by Henderson et al. (1986).

also found and speculations were made about its interactions with various amino acid residues. More recently, electron diffraction information from intermediates from the photocycle have been collected (Glaeser et al., 1986) to look for local conformational changes associated with the protein function (Han, 1993). Eventually it will be possible to compare these electron crystallographic results with the x-ray crystal structure since the earlier difficulties experienced with crystal growth have been overcome (Schertler et al., 1993). There have been some assertions made in this work that an atomic resolution model for the structure is being found (Glaeser and Downing, 1993). Fitting a chain to a density profile is, of course, not the same thing as visualizing a structure directly at atomic resolution. There will be some error in placement of individual positions, as there was in the simpler case of the mannan I structure discussed in Chapter 11. The structure of the genetically-similar halorhodopsin has been determined to 6-Å resolution (Havelka et al., 1993), after expressing it in a mutant form of *Halobacterium*. This protein was also found to have a 7-helix monomer.

12.2.2. Outer Membrane Porins from Gram-Negative Bacteria

Porins are transmembrane proteins in the outer membranes of gram-negative bacteria that function as molecular sieves for water-soluble substances, passing those with a molecular weight of, e.g., < 700 daltons. The matrix porin of *E. coli*, which was later termed the Omp F porin, was isolated by Rosenbusch (1974), who also found that it was denatured by ionic detergents such a sodium dodecyl sulfate but the structure was preserved by a variety of nonionic detergents such as the octyl glucosides and *n*-alkyl oligo(ethylene oxides). The protein was always extracted as a stable trimer with a monomer molecular weight of 36,500 daltons. Unlike the bacteriorhodopsin discussed above, spectroscopic measurement revealed that the porins had almost no α-helical content but consisted of mainly β-sheet (Vogel and Jähnig, 1986).

Initial electron microscopic observations were made by Steven and coworkers (1977) on intact outer membranes which were either dissociated from the peptidoglycan cytoskeleton or kept in contact with it. A few micrographs of negatively stained material revealed a hexagonal array of molecules with a threefold array of stain density. It was suggested that these dense regions might be the openings of transmembrane pores. Reconstitution experiments were in progress with the purified proteins and, soon, similar arrays were observed in negatively stained preparations in L-DMPC bilayers that often diffracted to 20-Å resolution on an optical bench. Depending on the amount of lipid that was used for the reconstitution, either of two hexagonal forms were found in the images (Figure 12.2) with different lattice constants (Dorset et al., 1983). Later it was seen that the larger lattice would transform spontaneously to the smaller one by loss of lipid. A rare rectangular form was also seen in this early work with lattice constants related to the smaller hexagonal form. Three-dimensional image data were obtained from the smaller hexagonal preparation and used to construct the first three-dimensional view of a transmembrane pore (Engel et al., 1985). It was thought that the pores from one side of the membrane would fuse into a single channel (Figure 12.3). Later, when electron diffraction amplitudes (obtained at lower beam doses) were combined with the phase information (Dorset and Massalski, 1989), the proposed confluence into a single pore was no longer found (Figure 12.4).

Experiments were also underway on two other porins from *E. coli*. The Omp C porin was found to have a structure similar to that of Omp F (Chang et al., 1985). Using an innovative fusion technique for crystallization, Jap (1989) was able to obtain highly ordered arrays of the Pho E porin expressed under phosphorus starvation. Three-dimensional reconstructions of the negatively stained preparation agreed well with the findings from the Omp F porin, in that three large channel openings could be found at one end of the trimer, but disagreed with the earlier statement that these would form a single channel at the other side of the trimer. This work soon was extended to 3.5-Å resolution using low-dose techniques on glucose-embedded preparation (Jap et al., 1990). The high-resolution image from the untilted preparation revealed the details of the β-sheet in projection.

For a time, high-resolution studies of the Omp F porin were frustrated by paracrystalline disorder of the samples. For example, electron diffraction experiments of glucose-embedded samples could, at best, produce a 10-Å pattern. Eventually, the preparation was improved by using Mannella's (1984) phospholipase digestion to remove excess phospholipid from the surrounding bilayer, thus reducing the hexagonal lattice to 72 Å. These preparations often diffracted to 3.2-Å resolution and, using the liquid helium-cooled cryomicroscope employed for the bacteriorhodopsin studies,

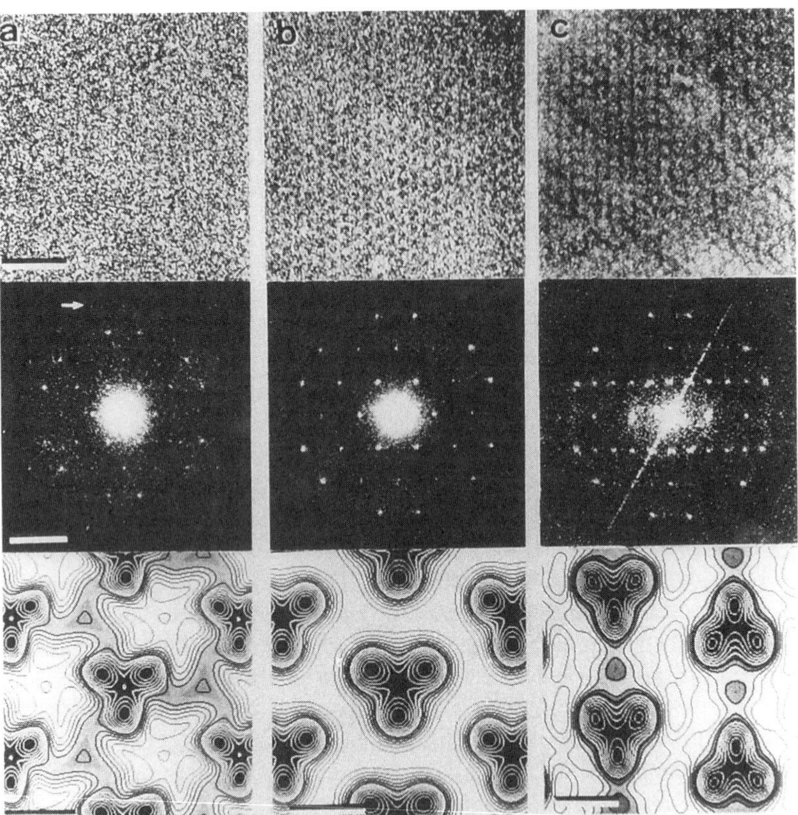

FIGURE 12.2. Three two-dimensional polymorphs of Omp F porin (negatively stained in uranyl acetate), and their average images after Fourier filtration: (a) hexagonal, excess lipid, $a = 93$ Å; (b) hexagonal, less lipid, $a = 79$ Å; (c) rectangular, $a = 79$ Å; $b = 137$ Å. (Reprinted from D. L. Dorset, A. Engel, M. Häner, A. Massalski, and J. P. Rosenbusch (1983) "Two-dimensional crystal packing of matrix porin. A channel-forming protein in *Escherichia coli* outer membrane," *Journal of Molecular Biology* **165**, 701–710; with kind permission of Academic Press, Ltd.)

3.5-Å images were obtained (Sass et al., 1989) with results similar to those from the Pho E porin which appeared somewhat later (Figure 12.5). Jap and his coworkers continued the study of Pho E porin by collecting images and electron diffraction data from tilted membrane sheets to high resolution and a three-dimensional map was published (Walian and Jap, 1990; Jap et al., 1991). More recently the density of the map has been fitted with an amino acid sequence.

Plausible as these results seem, do they, in fact, have any structural meaning? For the porins, it is fortunate that detergent-stabilized crystals suitable for x-ray crystallographic studies had been obtained as early as 1980, so that a direct comparison might be possible (Garavito et al., 1983). For the longest time, this study was hindered by

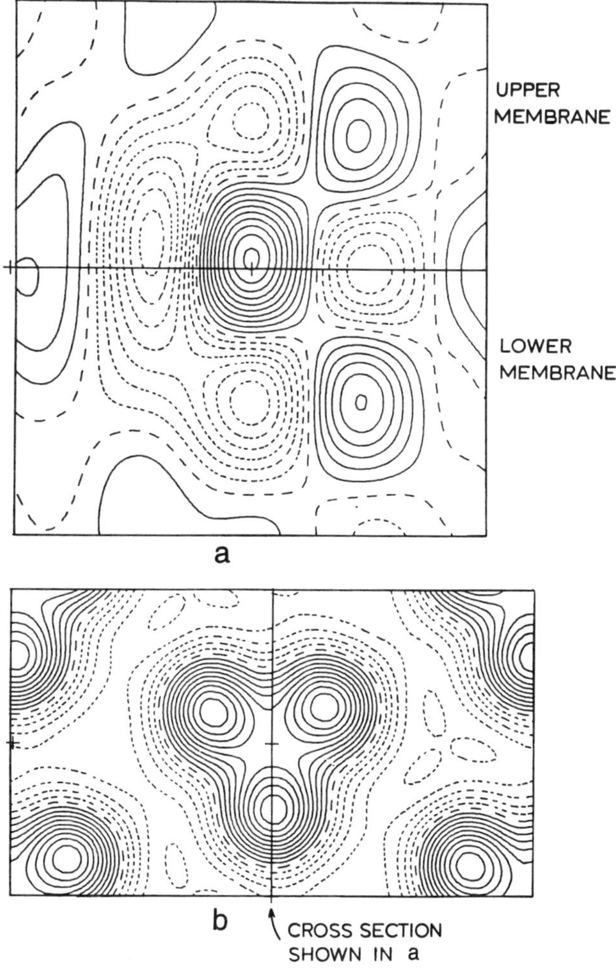

FIGURE 12.3. Three-dimensional model of negatively stained Omp F porin from image amplitudes and phases: (a) negative stain density across membrane stack for one pair of channels; (b) projection down the channel trimer axis. (Reprinted from D. L. Dorset, A. Engel, A. Massalski, and J. P. Rosenbusch (1984) "Three-dimensional structure of a membrane pore. Electron microscopical analysis of *Escherichia coli* outer membrane matrix porin," *Biophysical Journal* **45**, 128–129; with kind permission of the *Biophysical Journal*.)

the lack of good heavy-atom derivatives. The first high-resolution x-ray structure of an outer membrane porin was reported by Weiss et al. (1990) for a protein from *Rhodobacter capsulatus*. Although there is very little sequence identity between this porin with those from *E. coli*, the crystal structure has features that significantly match the results of the electron crystallographic determinations. When the *Rhodobacter* structure was used to solve the x-ray structures of the *E. coli* Omp F and Pho E porins

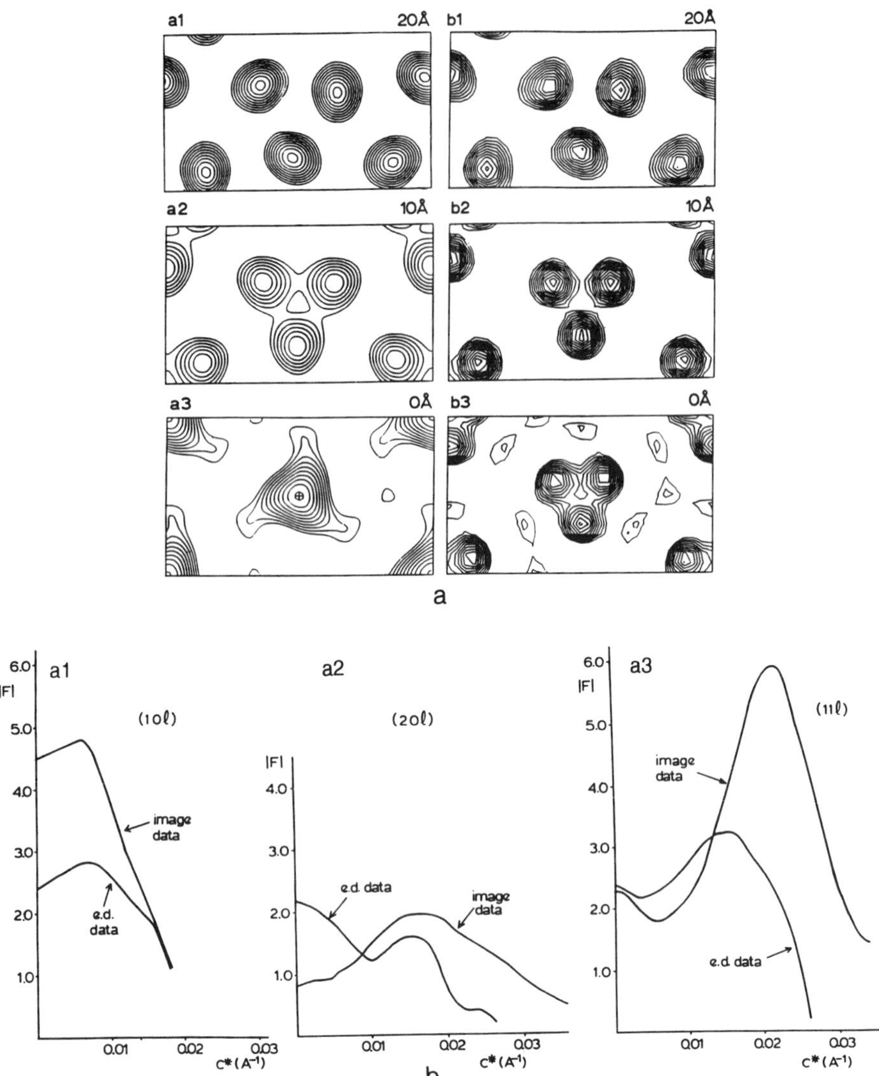

FIGURE 12.4. Comparison of image reconstructions for negatively stained Omp F porin. (a) When these reconstructions are based, respectively, on (a1)–(a3) image phases and amplitudes and (b1)–(b3) (low-dose) electron diffraction amplitudes with image phases, the channels appear to merge in the first case, whereas their integrity is maintained in the second case. (b) The difference is reflected in comparative plots of continuous electron diffraction and image transform amplitudes along the reciprocal lattice rods for the same crystalline polymorph. (Reprinted from D. L. Dorset and A. K. Massalski (1989) "Electron beam damage and the 3-D structure of negatively-stained Omp F porin," *Proceedings of the 47th Annual Meeting of the Electron Microscopy Society of America*, San Francisco Press, San Francisco, pp. 120–121; with kind permission of San Francisco Press, Inc.)

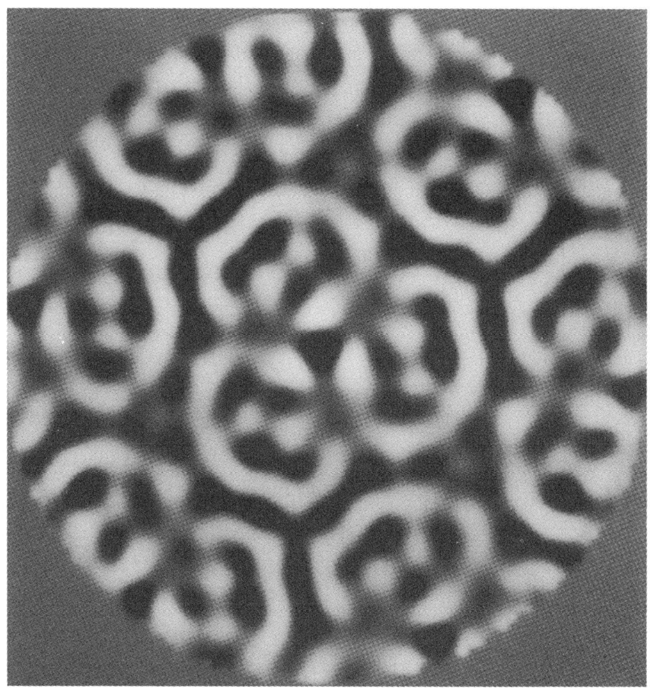

FIGURE 12.5. Projected structure of glucose-embedded Omp F porin. (Original image obtained and processed by Dr. A. Massalski.)

(Pauptit et al., 1991) by molecular replacement, the match was also apparent. Eventually, suitable heavy-atom derivatives were found and the structures were determined independently, again confirming the earlier results (Cowan et al., 1992). The β-barrel structure was described in terms of amino acid sequence and regions thought to be associated with the gating function of the channel were located.

The essential agreement between x-ray and electron crystallographic studies of the same group of proteins is significant since it serves as the first benchmark to prove that the high-resolution electron crystallographic studies can arrive at the correct structure solution. (Similar inferences have been made for the earlier study of negatively stained preparations.) However, and nota bene(!), the electron crystallographic results, in this case, were obtained first before any "prejudicial" judgements could be made from an x-ray crystal structure.

There was another aspect of the study on porins that might serve as a caution to those tempted to use a procedure for image averaging in a rather routine fashion. When carrying out a Fourier peak filtration analysis of an electron micrograph of a Omp F porin sheet in the small (a = 79 Å) hexagonal structure one day (Dorset et al., 1989b), a curious diffuse signal was observed in the computed transform (Figure 12.6a). From work on molecular crystals, it was thought that this might be due to a lattice disorder

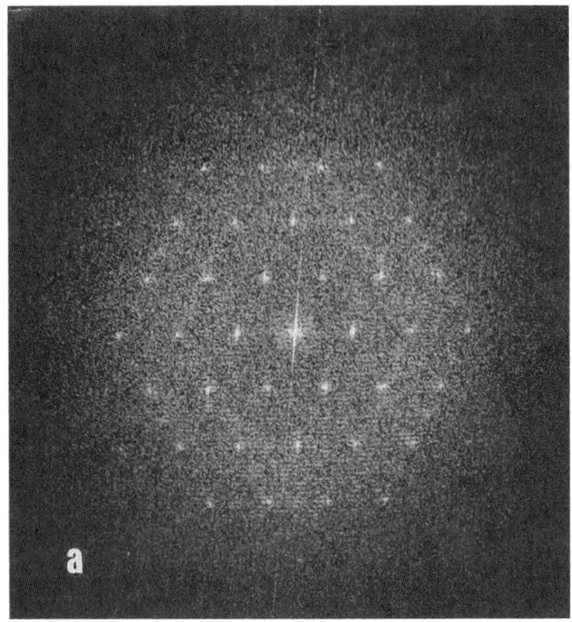

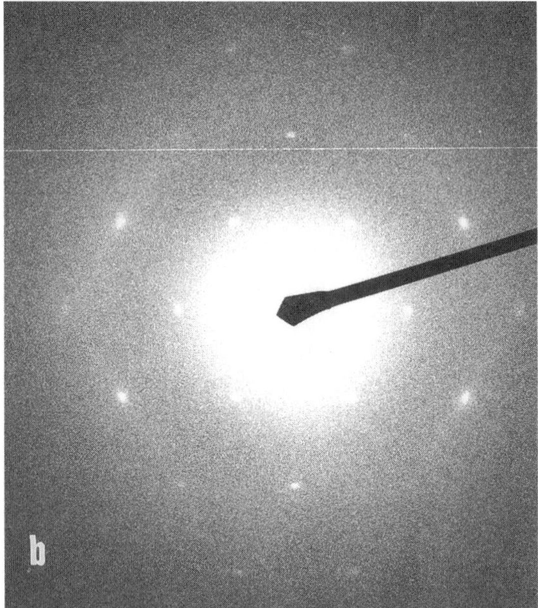

FIGURE 12.6. (a) Computed transform of negatively stained Omp F porin image with diffuse streaks; (b) electron diffraction pattern from the same preparation with diffuse streaks;

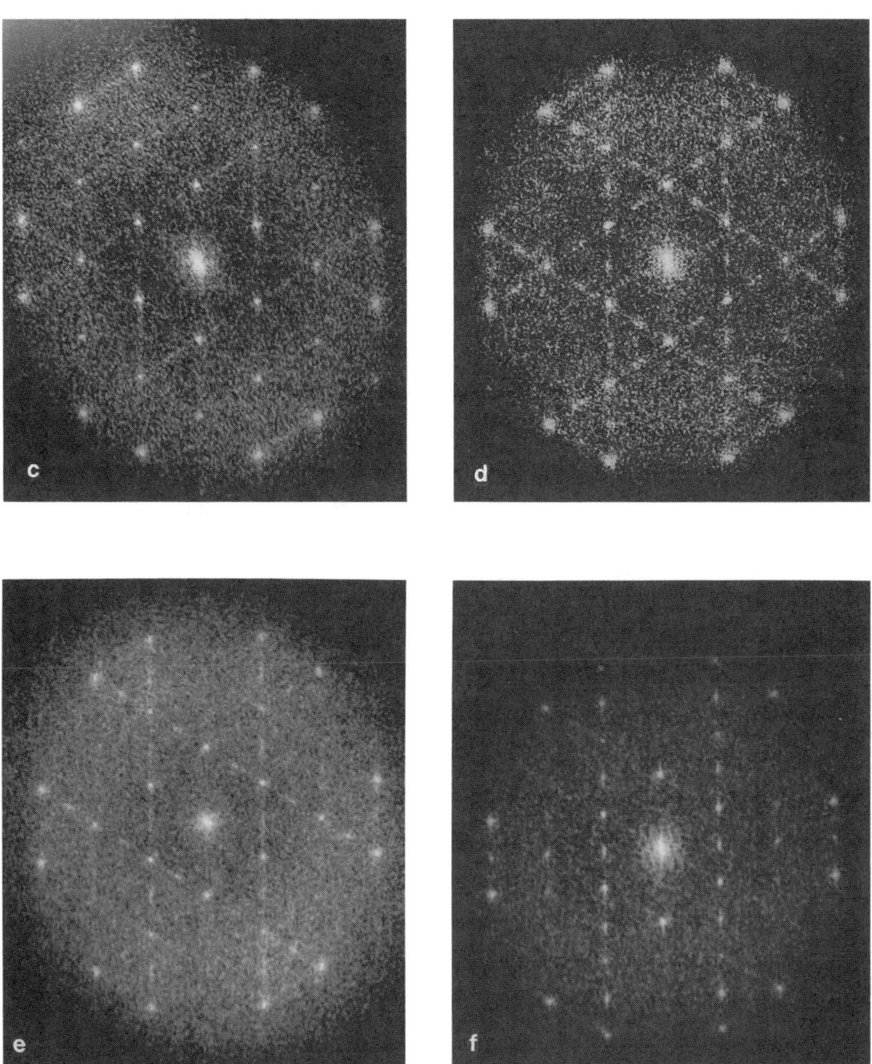

FIGURE 12.6. (*Continued*) (c) computed transform of negatively stained image with new pattern of continuous streaks; (d) streaks breaking up into spots; (e) twofold distribution of spots; (f) single distribution of spots corresponding to rectangular polymorph. (Reprinted from D. L. Dorset, A. K. Massalski, and J. P. Rosenbusch (1989) "In-plane phase transition of an integral membrane protein: nucleation of the Omp F matrix porin rectangular polymorph," *Proceedings of the National Academy of Sciences USA* **86**, 6143–6147; copyright held by the author.)

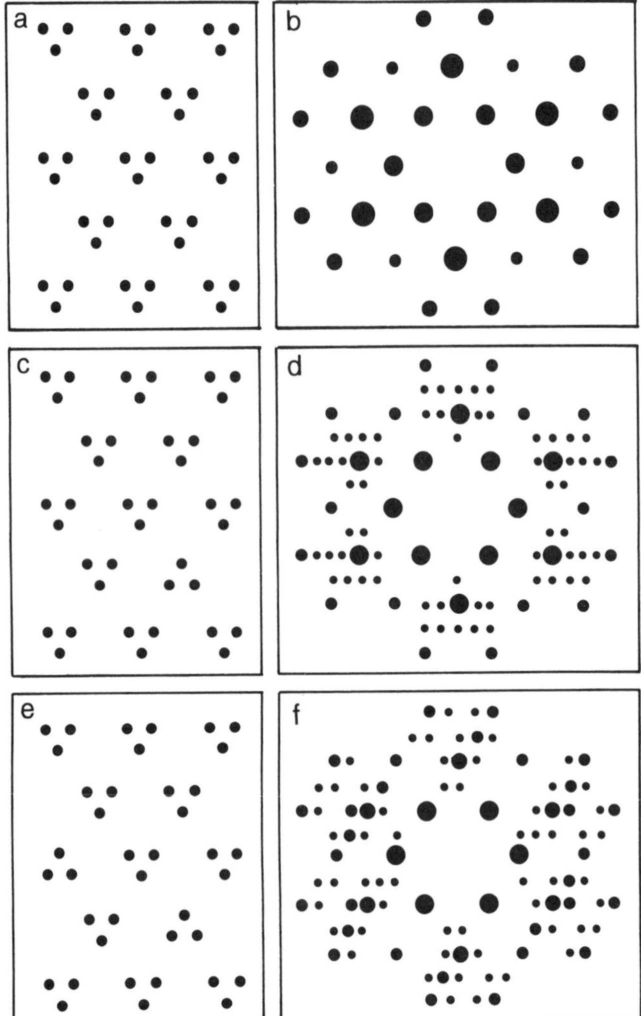

FIGURE 12.7. Model for the phase transition of Omp F porin: (a, b) perfect hexagonal packing and its diffraction pattern; (c, d) one rotationally disordered trimer site leading to diffuse scattering; (e, f) beginnings of a rectangular patch, giving streaks due to the shape transform of a narrow domain; (g, h) continued growth of the rectangular area and its diffraction pattern; (i, j) perfect rectangular lattice and its diffraction pattern. (Reprinted from D. L. Dorset, A. K. Massalski, and J. P. Rosenbusch (1989) "In-plane phase transition of an integral membrane protein: nucleation of the Omp F matrix porin rectangular polymorph," *Proceedings of the National Academy of Sciences USA* **86**, 6143–6147; copyright held by the author.)

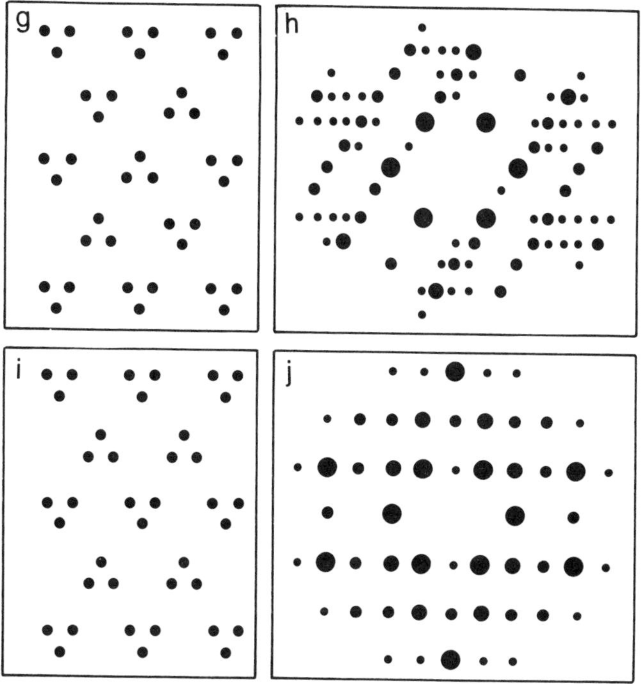

FIGURE 12.7. (*Continued*)

but attempts to model this with a Gaussian displacement of trimers or a random rotational disorder were not very successful, although the latter results were most promising. This diffuse signal was soon found in electron diffraction patterns (Figure 12.6b).

The preparation stored at 4 °C was removed periodically from the refrigerator over the span of a year, and a sample was negatively stained for observation in the electron microscope. The transform of the image soon obtained a new continuous streak (Figure 12.6c) that eventually became a row of diffraction spots oriented in three, two, or one directions (Figure 12.6d,e,f). The uniquely oriented pattern was seen to match the one from the rare rectangular form. It was apparent, therefore, that the original diffuse signal was the initial sign of a spontaneous phase transition within the two-dimensional crystal. This was successfully modeled with a superlattice (Figure 12.7). If one porin trimer was rotated by 60°, a diffuse signal would appear at the place found in the original diffraction pattern. If the start of a rectangular packing was observed as a thin strip of trimers embedded in the hexagonal array, a sharp streak was found, corresponding to the shape transform of a finite crystal. Eventually, when enough unit cells grew in this form, the pattern became an array of spots. The rectangular lattice, being more polar than the original hexagonal lattice, could be nucleated in any of three equivalent directions, explaining the threefold orientation.

This could be proved if a subarea was used for image averaging by spatial filtration. The average unit cell from one rectangular array could be used as a reference for correlational searches to find where the same orientation occurred in the membrane patch. If the reference was rotated by 120° and then moved across the experimental image, a new area was located. Taking the boundary of this subarea, a spatial filtration resulted in the same structure, but appropriately rotated. To our knowledge, this is the first direct observation of spontaneous rotational diffusion of a protein in a bilayer membrane.

12.2.3. Light-Harvesting Chlorophyll *a/b*-Protein Complex

A third major application of high-resolution electron crystallography to membrane proteins has been in the study of the light-harvesting chlorophyll *a/b*-protein complex from pea chloroplasts by Kühlbrandt and coworkers. Initially the protein was grown as two-dimensional sheets by dialysis in the presence of detergent. Electron micrographs of negatively stained preparations diffracted on an optical bench to 16-Å resolution and tilt series up to 82° were used to construct the first view of the protein trimer outline (space group P321, a = 125 Å) (Kühlbrandt, 1984). The investigation was extended to 6 Å with unstained preparations (Kühlbrandt and Wang, 1991) and then to 3.5 Å on a cryoelectron microscope (Kühlbrandt et al., 1994) using specimen tilts up to 60 Å (representing a resolution reduction of 1.3 in the direction perpendicular to the crystal plane). The resultant potential map was sufficient to trace the polypeptide chains of the structure, revealing details of the three major transmembrane α-helices. It was possible to visualize the positions of the major chlorophyll and carotenoid chromophores, so that a model for the complex function in the photosynthetic process could be proposed. The final $R = 0.33$ in this highly interesting study, which represents a significant accomplishment.

It is particularly interesting to note that in the fitting of the chain to the map, the investigators noted that deficiencies or abundances of density occurred on side chain portions that could correspond to the presence, respectively, of negative and positive charged groups. It has been long recognized that charged atoms will have different scattering factors than those of the neutral species (e.g., see Vainshtein, 1964a), so these discrepancies, in fact, have a sound theoretical basis. Also, since the reciprocal lattice of protein crystals is very finely sampled in the low-angle region where the form factors of charged and neutral atoms differ greatly, it is imagined that such local effects could be studied quite effectively by electron crystallography.

12.3. Prospects for the Use of Direct Methods in Protein Electron Crystallography

Until recently, high-resolution electron micrographs have been the sole source of crystallographic phases in the structure analysis of membrane proteins. After extensive control studies, Henderson et al. (1986) have obtained a realistic view of the difficulties inherent in such an approach to phase determination and have stated clearly, at the outset, that perhaps only 10% of experimental electron micrographs are suitable for extraction of phase terms. Nevertheless, with appropriate constraints imposed on the experiment, it is clear (Henderson et al., 1986; Baldwin et al., 1988) that the errors in these phase measurements can be kept to a minimum. Also, problems with earlier high-resolution phase measurements (e.g., Hayward and Stroud, 1981; Rossmann and Henderson, 1982) have been recognized after these exacting studies.

What are the prospects of reducing the tedium of high-resolution phase determination, e.g., by the use of direct methods? Gilmore et al. (1993) have already discussed promising results of phase extension via maximum entropy and likelihood methods with bacteriorhodopsin, but apparently disagree with Henderson et al. (1986) on the error estimate for high-resolution phases, stating that the image-derived phases should have a mean error near 89°, i.e., equivalent to a random phase estimate. For this reason, we have carried out an independent feasibility study for phase extension based on the Sayre (1952) equation.

With amplitude and phase data from bacteriorhodopsin published by Henderson et al. (1986) to 3.5-Å resolution, a basis set was considered at 10 Å (Figure 12.8), as if an electron micrograph had been taken to this resolution, and electron diffraction intensities were recorded to the limit of the published data set. Expansion of this basis set of 18 reflections to 6 Å (49 unique reflections) by the Sayre convolution (Dorset et al., 1995) was not especially difficult (Table 12.2), despite the noncentrosymmetry of the hexagonal (001) zone (plane group $p3$, $a = 62$ Å). After resetting the values of the 10-Å set, the final overall mean phase error was only 32.5°, a figure somewhat better than that found when low-angle MIR x-ray phases were expanded by matricial direct methods for another structure (Podjarny et al., 1981). For the most part, the 6-Å potential map agrees with the one calculated from image phases (Figure 12.9), although there are some deviations in the outer ring of helices. On the other hand, the

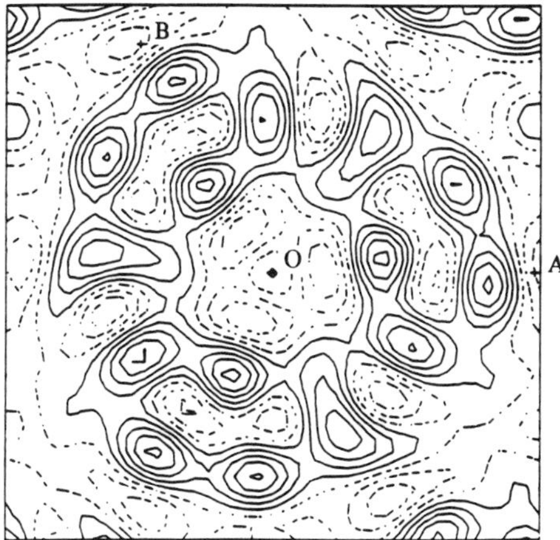

FIGURE 12.8. Bacteriorhodopsin map calculated from amplitudes and phases at 10-Å resolution.

Table 12.2. Bacteriorhodopsin. Phase Extension from 10 to 6 Å (Phases in Degrees)*

$hk0$	ϕ^1	ϕ^2	ϕ^H	$hk0$	ϕ^1	ϕ^2	ϕ^H	$hk0$	ϕ^1	ϕ^2	ϕ^H
100†	−11	−11	−11	300†	108	108	108	520	131	87	129
110†	167	167	167	310†	176	176	176	530	−135	−165	157
120†	−32	−32	−32	320†	15	15	15	540	−174	−127	116
130†	110	110	110	330†	−4	−4	−4	550	69	25	87
140†	−45	−45	−45	340	64	−166	144	600	−20	−21	−13
150	−120	−97	−96	350	32	1	−87	610	−108	−76	−24
160	146	154	27	360	−78	−10	−75	620	−36	−26	−40
170	76	122	150	370	−148	−75	88	630	36	87	62
180	66	66	38	400†	−67	−67	−67	640	85	119	−11
200†	−175	−175	−175	410†	−46	−46	−46	700	−104	−86	−102
210†	−132	−132	−132	420†	87	87	87	710	−138	−153	−108
220†	112	112	112	430	112	129	127	720	−68	−38	−81
230†	−37	−37	−37	440	−136	−128	−76	730	−70	−69	10
240†	−118	−118	−118	450	40	6	−27	800	−40	−170	−137
250	161	−146	−118	460	−13	−53	−35	810	122	103	118
260	−139	−146	−158	500†	−14	−14	−14				
270	−146	−131	117	510	68	−24	36				

*ϕ^1: Expansion of basis set (3 cycles) by Sayre convolution;
ϕ^2: renewed basis set after Fourier refinement and reexpansion;
ϕ^H:phases listed by Henderson et al. (1986).
†10 Å basis set.

map calculated with random phases for reflections with resolution better than 10 Å is severely degraded (Figure 12.9b).

That such medium-resolution phase enhancement is not just an interesting anomaly was tested further by attempting a similar resolution expansion for halorhodopsin (Havelka et al., 1993) from 15 to 6 Å. In this case the zonal projection for the tetragonal structure is centrosymmetric (space group $P42_12$, a = 102 Å). The basis set was comprised of 20 reflections and the 6-Å set contained 101 unique values. Although acceptable results (Dorset et al., 1995) were obtained from the Sayre equation (overall mean phase error of 60.6°, but only 32.0° if only the 45 most intense reflections in the projection are considered), the best agreement was found when the F_h terms in the convolution were replaced by E_h (Table 12.3). This corresponded to a mean phase error of 55.2° for all reflections but only 24.0° for the 45 most intense reflections. The resulting potential map agrees reasonably well with the one based on image-derived phases (Figure 12.10,11), but not with the map calculated with random phases beyond a resolution of 15 Å (Figure 12.11b). Prospects for ab initio structure analyses of this projection have also been found to be quite favorable.

More difficulty is experienced when an expansion to higher resolution is attempted. With the 3.5-Å data from bacteriorhodopsin, for example, the barrier to a straightforward resolution enhancement seems to be caused by an intensity minimum in the plot of average intensity versus resolution, a characteristic of many protein data sets (Figure 12.12). If this minimum can be considered to be a "phase node," so that

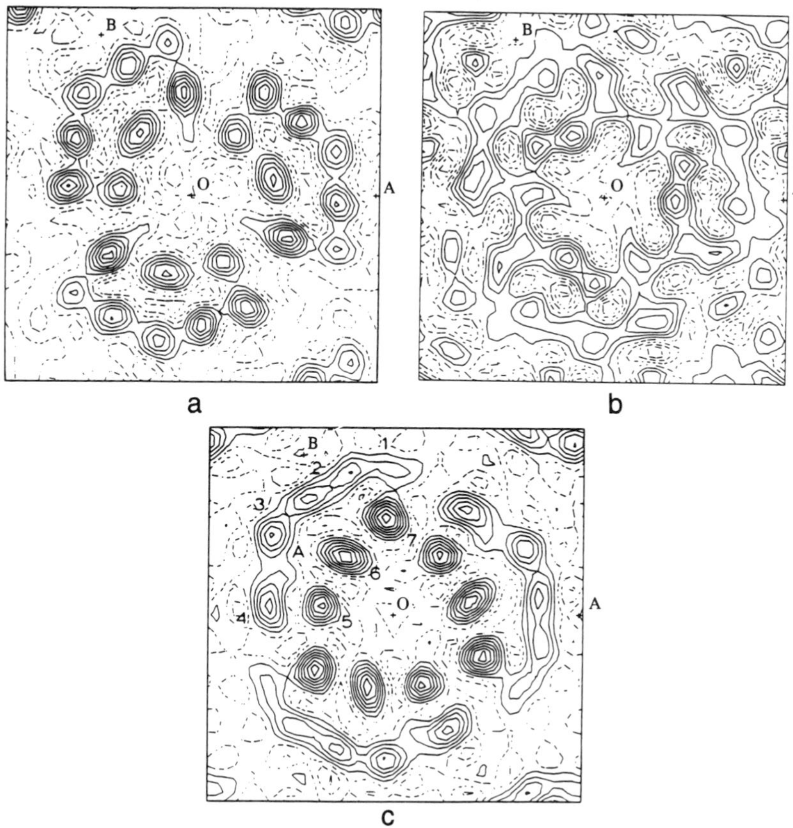

FIGURE 12.9. Bacteriorhodopsin phase extension from 10 to 6 Å: (a) result of the phase extension; (b) random phases beyond 10-Å resolution; (c) image amplitudes and phases.

a phase shift must be applied to all values generated by the Sayre equation beyond the 5.2-Å minimum in the intensity plot, the agreement of phase values is much better. (So far we have only evaluated the consequence of an overall π-shift.) As in the analysis of lipid lamellar structures (see Chapter 10), this phase shift must be justified, however, by an independent figure of merit. The measure of density flatness in the two possible maps, suggested by Luzzati et al. (1988), turns out to be satisfactory for choosing one solution over the other.

When ideal (i.e., image-derived) 6-Å phases were expanded to 3.5 Å, the overall mean phase error was only 35.4° for the 74 most intense reflections, after application of the π-shift. If the 10-Å data are expanded to the resolution limit (hence a smaller basis set for which some phases can be reset to ideal values), this value is 50°. Nevertheless, the match of high-resolution maps is acceptable (Figure

Table 12.3. Final Phases for Halorhodopsin after Sayre-Hughes Expansion (15 to 6Å)*

hk	ϕ	ϕ_H	*hk*	ϕ	ϕ_H	*hk*	ϕ	ϕ_H	*hk*	ϕ	ϕ_H
02†	$\underline{\pi}$	π	26†	$\underline{0}$	0	49	0	0	79	0	π
04†	π	π	27	π	π	410	0	π	710	$\underline{0}$	0
06†	$\underline{0}$	0	28	$\underline{\pi}$	π	412	π	0	711	0	π
08	$\underline{\pi}$	π	29	0	0	413	0	0	712	π	π
010	$\underline{0}$	0	210	$\underline{0}$	0	414	0	0	714	0	π
012	$\underline{\pi}$	π	211	π	π	415	π	0	715	0	π
014	0	π	214	π	π	416	0	π	88	$\underline{0}$	0
016	0	π	215	π	π	55	$\underline{0}$	π	89	0	π
11†	$\underline{\pi}$	π	216	0	π	56	$\underline{\pi}$	π	810	0	0
12†	$\underline{0}$	0	33†	$\underline{\pi}$	π	57	$\underline{0}$	0	811	π	π
13†	$\underline{0}$	0	34†	$\underline{0}$	0	58	π	π	812	0	0
14†	$\underline{\pi}$	π	35†	$\underline{0}$	0	59	$\underline{\pi}$	0	813	0	π
15†	$\underline{0}$	0	36†	$\underline{0}$	0	510	0	π	814	0	π
16†	$\underline{0}$	0	38	$\underline{\pi}$	π	511	$\underline{0}$	0	815	π	π
17	$\underline{\pi}$	π	39	$\underline{\pi}$	π	513	π	0	99	π	π
18	$\underline{\pi}$	π	310	$\underline{0}$	0	514	0	0	910	0	0
19	$\underline{0}$	0	311	0	π	66	$\underline{0}$	0	911	0	0
111	0	0	313	0	π	67	$\underline{0}$	0	912	π	0
112	π	π	315	0	0	68	$\underline{0}$	π	913	0	π
113	0	0	316	π	π	69	π	0	1010	π	π
114	π	0	44†	$\underline{\pi}$	π	610	π	π	1012	0	0
115	0	π	45†	$\underline{\pi}$	π	611	π	π	1013	0	0
22†	$\underline{\pi}$	π	46	$\underline{0}$	0	612	π	0	1111	0	0
23†	$\underline{\pi}$	π	47	$\underline{0}$	0	77	$\underline{0}$	π	1112	0	π
24†	$\underline{0}$	0	48	$\underline{0}$	π	78	π	π	1212	0	π
25†	$\underline{\pi}$	π									

*Underlined phases correspond to $|F_h| \geq 1.0$, scaling amplitudes of Havelka et al. (1993) by 0.01.
†Reflections in 15 Å basis set.

12.13) and the generated phase errors are still better than those found in some direct x-ray determinations of macromolecular structures in the same resolution range, or when random phases are used for resolution better than 10 Å (Figure 12.13d).

From these preliminary studies, it is clear that direct methods can play a significant role in the future for phase refinement and extension of macromolecular structures. The Sayre equation is not the best method for this task and, indeed, the earlier trials with the maximum entropy procedure seemed to produce better high-resolution maps, judged by comparison with those obtained by Henderson et al. (1986). It is unfortunate that Gilmore et al. (1993) did not compute the reverse Fourier transform of their final maps to make a direct comparison of the final phases to the values published earlier, because it is clear that the control experiments carried out by Henderson et al. (1986) gave a clear experimental view of the phase errors in the high-resolution range.

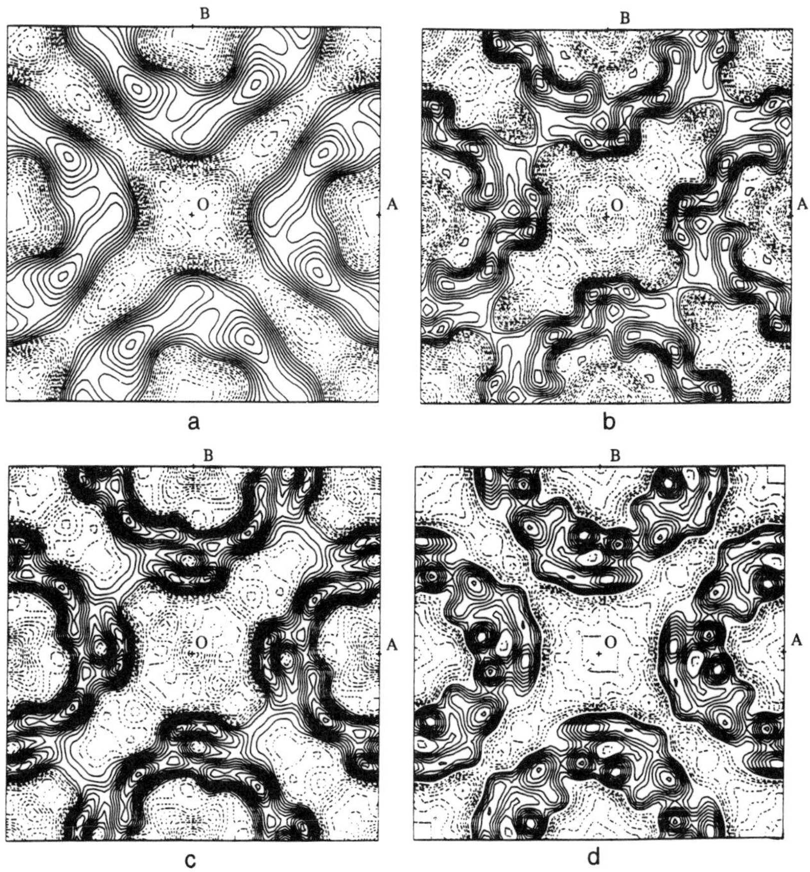

FIGURE 12.10. Phase extension for halorhodopsin from 15 to 6 Å: (a) map based on 15-Å phases; (b) 6-Å resolution form Sayre equation, in two steps; (c) 6-Å resolution from Sayre equation, three steps; (d) image amplitudes and phases.

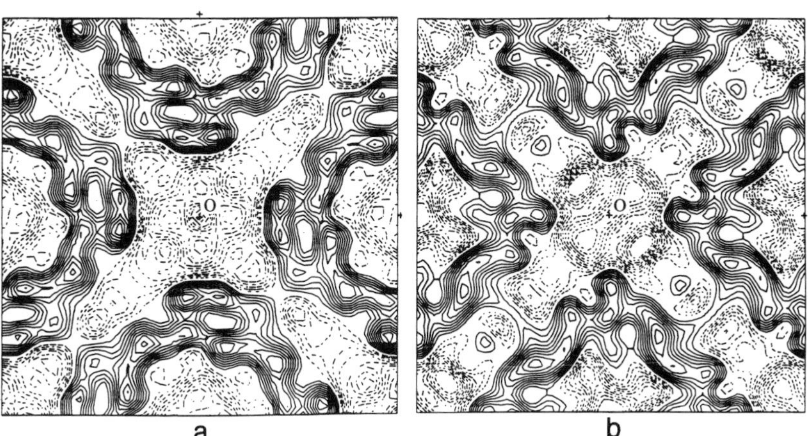

FIGURE 12.11. (a) Halorhodopsin phase extension based on Sayre–Hughes equation; (b) random phases used beyond 15 Å.

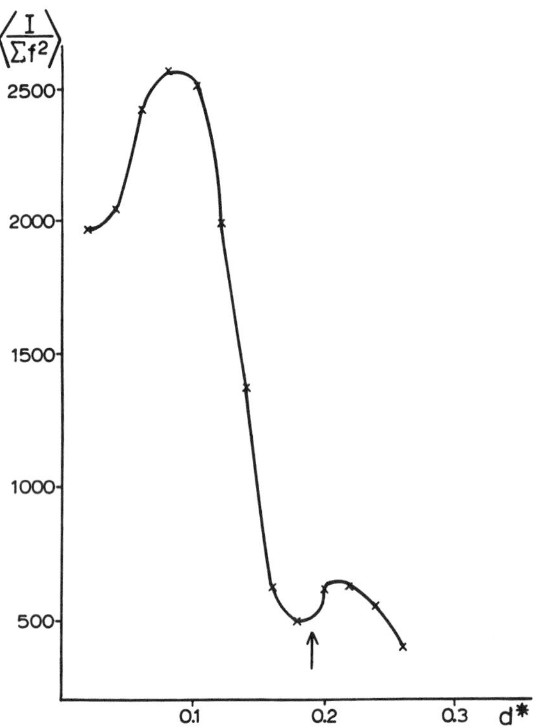

FIGURE 12.12. Average intensity versus resolution for bacteriorhodopsin.

12.4. Conclusions

The use of electron crystallographic techniques to study the structures of proteins is an impressive chapter in modern crystallography. Accurate structural results have been obtained from a membrane protein before the x-ray structure has been determined to verify them, thus quieting any criticism that the technique is useless for determining previously unsolved structures. Although crystallization techniques have been improved in recent years, opening up the study of integral membrane proteins by x-ray crystallography, the globular macromolecules in these larger crystals are stabilized in a large amount of detergent and do not pack in the two-dimensional arrays found in lipid bilayers that best mimic their aggregation in cell membranes. Thus, while x-ray analyses may be useful for highly accurate depictions of the polypeptide chain folding, this information must be related to the environment at the cell surface, a study open only to electron crystallographic techniques. It is also apparent that electron crystallography is a much more convenient way to study changes in the membrane packing induced by changes in lipid environment, etc., than is x-ray crystallography. There are still important developments being made in the use of electron crystallographic

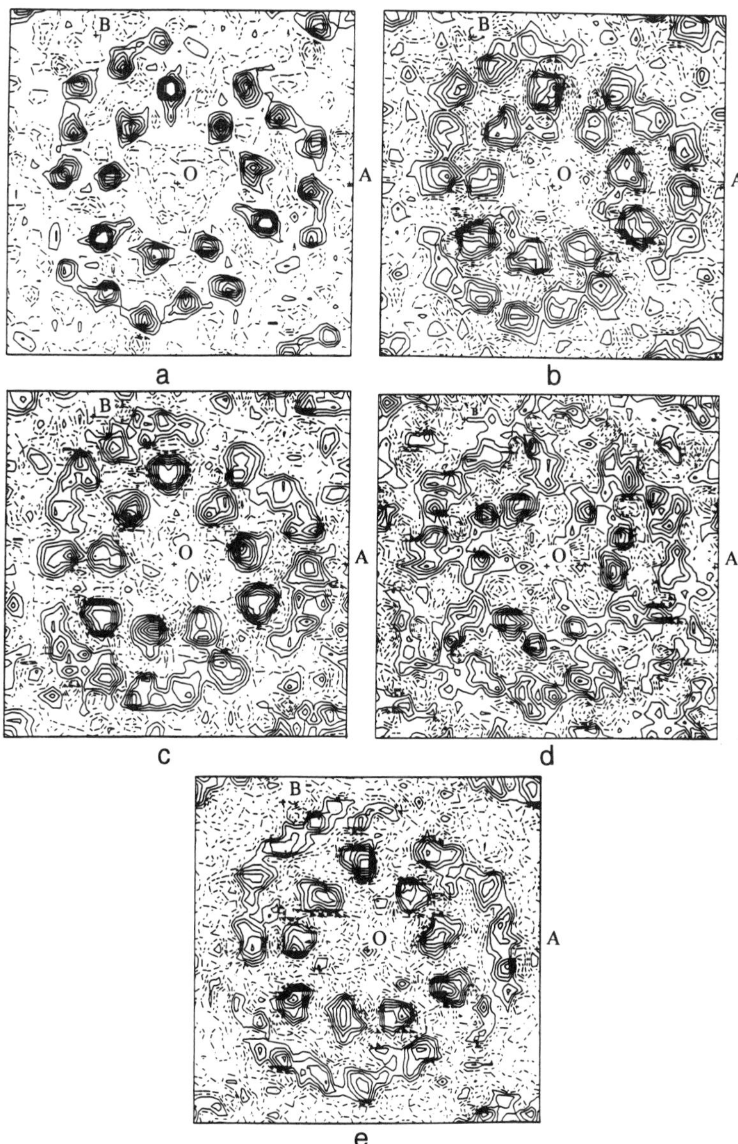

FIGURE 12.13. Phase extension for bacteriorhodopsin from 10 to 3.5 Å: (a) no π-shift; (b) with π-shift; (c) after one cycle of Fourier refinement; (d) random phases beyond 10 Å; (e) image amplitudes and phases.

techniques to study thin macromolecular crystals. One of these is the introduction of direct techniques for phase refinement that, hopefully, will eliminate much of the tedium experienced when high-resolution images are used to phase electron diffraction amplitudes.

References

Abrahamsson, S., Larsson, G., and Sydow, E. v. (1960) *Acta Cryst.* **13**, 770–778.

Abrahamsson, S., Dahlen, B., Löfgren, H., and Pascher, I. (1978) *Progr. Chem. Fats Other Lipids* **16**, 125–143.

Aebi, F. (1948) *Helv. Chim. Acta* **31**, 369–378.

Aebi, F. (1950) *Acta Cryst.* **3**, 320–322.

Albon, N. (1976) *J. Cryst. Growth* **35**, 105–109.

Aleby, S., Fischmeister, I., and Iyengar, B. T. R. (1970) *Lipids* **6**, 421–425.

Amelinckx, S., and Van Dyck, D. (1993a) in *Electron Diffraction Techniques*, Vol. 2, J. M. Cowley, ed., Oxford, Oxford Univ., Chapter 1.

Amelinckx, S., and Van Dyck, D. (1993b) in *Electron Diffraction Techniques*, Vol. 2, J. M. Cowley, ed., Oxford, Oxford Univ., pp. 309–372.

Amos, L. A., Henderson, R., and Unwin, P. N. T. (1982) *Progr. Biophys. Molec. Biol.* **39**, 183–231.

André, D., Dworkin, A., Swarc, H., Céolin, R., Agafonov, V., Fabre, C., Rassat, A., Straver, L., Bernier, P., and Zahab, A. (1992) *Mol. Phys.* **76**, 1311–1317.

Andrew, L. T. (1936) *Trans. Faraday Soc.* **32**, 607–616.

Asbach, G. I., and Kilian, H. G. (1970) *Ber. Bunsenges. Phys. Chem.* **74**, 814–823.

Asbach, G. I., Geiger, K., and Wilke, W. (1979) *Colloid Polym. Sci.* **257**, 1049–1059.

Ashida, M. (1991) in *Electron Crystallography of Organic Molecules*, J. R. Fryer and D. L. Dorset, eds., Dordrecht, Kluwer, pp. 227–240.

Atkins, E. D. T. (1989) in *Comprehensive Polymer Science, Volume 1, Polymer Characterization*, C. Booth and C. Price, eds., Oxford, Pergamon, pp. 613–650.

Baldwin, J., and Henderson, R. (1984) *Ultramicroscopy* **14**, 319–336.

Baldwin, J. M., Henderson, R., Beckmann, E., and Zemlin, F. (1988) *J. Mol. Biol.* **202**, 585–591.

Barnakov, A. N., Demin, V. V., Kuzin, A. P., Zergarov, A. A., Zolotarev, A. S., and Abdulaev, N. G. (1990) *FEBS Lett.* **265**, 126–128.

Bassett, D. C., and Keller, A. (1961) *Phil. Mag.* **6**, 345–358.

Bassett, D. C., and Keller, A. (1962) *Phil. Mag.* **7**, 1553–1584.

Bassett, D. C., Frank, F. C., and Keller, A. (1959) *Nature* **184**, 810–811.

Bassett, D. C., Frank, F. C., and Keller, A. (1963) *Phil. Mag.* **8**, 1753–1787.

Beurskens, G., and Beurskens, P. T. (1993) in *Modern Crystallography. Proceedings of the Seventh Chinese International Summer School of Physics, Beijing International Workshop*, Z. H. Mai, G. H. Rao, and B. R. Zhu, eds., Beijing, Institute of Physics, Chinese Academy of Sciences, pp. 21–32.

Beurskens, P. T., and Smykalla, C. (1991) in *Direct Methods of Solving Crystal Structures*, H. Schenk, ed., New York, Plenum, pp. 281–290.

Beurskens, P. T., Beurskens, G., Lam, E. J. W., Smaalen, S. Van, and Fan, H. F. (1993) in *Modern Crystallography. Proceedings of the Seventh Chinese International Summer School of Physics, Beijing International Workshop*, Z. H. Mai, G. H. Rao, and B. R. Zhu, eds., Beijing, Institute of Physics, Chinese Academy of Sciences, pp. 33–42.

Bittiger, H., Marchessault, R. H., and Niegisch, W. (1970) *Acta Cryst.* **B26**, 1923–1927.

Blackman, M. (1939) *Proc. Roy. Soc. (London)* **A173**, 68–82.

Blackman, R. B., and Tukey, J. W. (1959) *The Measurement of Power Spectra from the Point of View of Communications Engineering*, New York, Dover.

Blundell, D. J., Keller, A., and Kovacs, A. J. (1966) *J. Polym. Sci.—Polym. Lett.* **4**, 481–486.

Blundell, T. L., and Johnson, L. N. (1976) *Protein Crystallography*, New York, Academic Press, Chapter 3.

Boekema, E. J. (1990) *Electron Microsc. Rev.* **3**, 87–96.

Boistelle, R., and Aquilano, D. (1978) *Acta Cryst.* **A34**, 406–413.

Boudet, A. (1984) *J. Mater. Sci.* **19**, 2989–2996.

Boudeulle, M. (1973) *Cryst. Struct. Commun.* **4**, 9–12.

Bricogne, G., and Gilmore, C. J. (1990) *Acta Cryst.* **A46**, 284–297.

Brink, J., and Chiu, W. (1991) *J. Microsc.* (Oxford) **161**, 279–295.

Brisse, F. (1989) *J. Electr. Microsc. Techn.* **11**, 272–279.

Brisse, F., and Marchessault, R. H. (1980) in *Fiber Diffraction Methods*, A. D. French and K. H. Gardner, eds., ACS Symposium Ser. Vol. 141, Washington, D.C., pp. 267–277.

Brisse, F., Remillard, B., and Chanzy, H. (1984) *Macromolecules* **17**, 1980–1987.

Brisse, F., Palmer, A., Moss, B., Dorset, D., Roughead, W. A., and Miller, D. P. (1984) *Eur. Polym. J.* **20**, 791–797.

Brisson, J., Chanzy, H., and Winter, W. T. (1991) *Int. J. Biol. Macromol.* **13**, 31–39.

Brock, C. P., and Dunitz, J. D. (1994) *Chem. Mater.* **6**, 1118–1127.

Brown, C. J. (1968) *J. Chem. Soc.* **A**, 2488–2993.

Buchheim, W. (1970) *Kiel. Milchwirtsch. Forschungsber.* **23**, 3–61.

Buchheim, W., and Knoop, E. (1969) *Naturwiss.* **56**, 560–561.

Buerger, M. J. (1959) *Vector Space and Its Applications in Crystal-structure Investigation*, New York, Wiley.

Buhle, E. L., Knox, B., Serpersu, E., and Aebi, U. (1983) *J. Ultrastruct. Res.* **85**, 186–203.

Bunn, C. W. (1939) *Trans. Faraday Soc.* **35**, 482–491.

Bunn, C. W. (1942) *Proc. Roy. Soc. (London)* **A180**, 67–81.

Bunn, C. W., and Howells, E. R. (1954) *Nature* **174**, 549–551.

Bürgi, H. B., Dunitz, J. D., and Shefter, E. (1974) *Acta Cryst.* **B30**, 1517–1527.

Buxton, B. F., Eades, J. A., Steeds, J. W., and Rackham, G. M. (1976) *Phil. Trans. Roy. Soc. (London)* **281**, 171–194.

Cascarano, G., Giacovazzo, C., and Viterbo, D. (1987) *Acta Cryst.* **A43**, 22–29.

Champeney, D. C. (1973) *Fourier Transforms and their Physical Applications*, London, Academic Press.

Chang, C. F., Mizushima, S., and Glaeser, R. M. (1985) *Biophys. J.* **47**, 629–639.

Chanzy, H., Guizard, C., and Vuong, R. (1977) *J. Microsc. (Oxford)* **111**, 143–150.

Chanzy, H., Folda, T., Smith, P., Gardner, K. H., and Revol, J. F. (1986) *J. Mater. Sci. Lett.* **5**, 1045–1047.

Chanzy, H., Perez, S., Miller, D. P., Paradossi, G., and Winter, W. T. (1987) *Macromolecules* **20**, 2407–2413.

Charlesby, A. (1945) *Proc. Phys. Soc. (London)* **57**, 510–518.

Chatani, Y., Okita, Y., Tadokoro, H., and Yamashita, Y. (1970) *Polym. J.* **1**, 555–562.

Cherry, R. J., Müller, U., Henderson, R., and Heyn, M. P. (1978) *J. Mol. Biol.* **121**, 283–298.

Chichakli, M., and Jessen, F. W. (1967) *Ind. Eng. Chem.* **59**, 86–98.

Chien, J. C. W., Karasz, F. E., and Shimamura, K. (1982) *Macromolecules* **15**, 1012–1017.

Chiu, W. (1986) *Ann. Rev. Biophys. Biophys. Chem.* **15**, 237–257.

Church, S. E., Griffiths, D. J., Lewis, R. N. A. H., McElhaney, R. N., and Wickman, H. H. (1986) *Biophys. J.* **49**, 597–605.

Claffey, W., and Blackwell, J. (1976) *Biopolymers* **15**, 1903–1913.

Claffey, W., Gardner, K., Blackwell, J., Lando, J., and Geil, P. H. (1974) *Phil. Mag.* **30**, 1223–1232.

Clark, E. S., and Muus, L. T. (1962a) *Z. Krist.* **117**, 108–118.

Clark, E. S., and Muus, L. T. (1962b) *Z. Krist.* **117**, 119–127.

Cochran, W. (1955) *Acta Cryst.* **8**, 473–478.

Cojazzi, G., Malta, V., Celotti, G., and Zannetti, R. (1976) *Makromol. Chem.* **177**, 915–926.

Corey, R. B. (1938) *J. Amer. Chem. Soc.* **60**, 1598–1604.

Cotton, F. A. (1971) *Chemical Applications of Group Theory* (2d ed.), New York, Wiley-Interscience.

Coumoulos, G. D., and Rideal, E. (1941) *Proc. Roy. Soc. (London)* **A178**, 421–428.

Cowan, S. W., Schirmer, T., Rummel, G., Steiert, M., Ghosh, R., Pauptit, R. A., Jansonius, J. N., and Rosenbusch, J. P. (1992) *Nature* **358**, 727–733.

Cowley, J. M. (1948) *Trans. Faraday Soc.* **40**, 60–68.

Cowley, J. M. (1953a) *Acta Cryst.* **6**, 522–529.

Cowley, J. M. (1953b) *Acta Cryst.* **6**, 846–853.

Cowley, J. M. (1956) *Acta Cryst.* **9**, 391–395.

Cowley, J. M. (1961) *Acta Cryst.* **14**, 920–927.

Cowley, J. M. (1967) *Progr. Mater. Sci.* **13**, 267–321.

Cowley, J. M. (1981) *Diffraction Physics* (2d rev. ed.), Amsterdam, North-Holland.

Cowley, J. M. (1988) in *High Resolution Transmission Electron Microscopy and Associated Techniques*. P. Buseck, J. Cowley, and L. Eyring, eds., New York, Oxford Univ. Press, p. 20.

Cowley, J. M., and Bridges, R. E. (1979) *Ultramicroscopy* **4**, 419–427.

Cowley, J. M., and Goswami, A. (1961) *Acta Cryst.* **14**, 1071–1079.

Cowley, J. M., and Kuwabara, S. (1962) *Acta Cryst.* **15**, 260–269.

Cowley, J. M., and Moodie, A. F. (1957) *Acta Cryst.* **10**, 609–619.

Cowley, J. M., and Moodie, A. F. (1959) *Acta Cryst.* **12**, 360–367.

Cowley, J. M., Rees, A. L. G., and Spink, J. A. (1951) *Proc. Phys. Soc. (London)* **A64**, 609–619.

Craievich, A. F., Denicolo, I., and Doucet, J. (1984a) *Phys. Rev.* **B30**, 4782–4787.

Craievich, A. F., Doucet, J., and Denicolo, I. (1984b) *J. Phys. (Paris)* **45**, 1473–1477.

Craven, B. M. (1986) in *Handbook of Lipid Research, Volume 4. The Physical Chemistry of Lipids from Alkanes to Phospholipids*, D. M. Small, ed., New York, Plenum Press, Chapter 6.

Craven, B. M., and DeTitta, G. T. (1976) *J. Chem. Soc. Perkin* **II**, 814–822.

Craven, B. M., and Guerina, N. G. (1979) *Chem. Phys. Lipids* **24**, 91–98.

Craven, B. M., and Sabine, T. M. (1966) *Acta Cryst.* **20**, 214–219.

Craven, J. R., Hao, Z., and Booth, C. (1991) *J. Chem. Soc. (Faraday Trans.)* **87**, 1183–1186.

Crowther, R. A., DeRosier, D. J., and Klug, A. (1970) *Proc. Roy. Soc. (London)* **A317**, 319–340.

Dawson, I. M. (1952) *Proc. Roy. Soc. (London)* **A214**, 72–79.

Dawson, I. M., and Vand. V. (1951) *Proc. Roy. Soc. (London)* **A206**, 555–562.

Day, D., and Lando, J. B. (1980) *Macromolecules* **13**, 1483–1487.

Deatheridge, J. F., Henderson, R., and Capaldi, R. A. (1982) *J. Mol. Biol.* **158**, 487–499.

Degeilh, R., and Marsh, R. E. (1959) *Acta Cryst.* **12**, 1007–1014.

Dekker, J. P., Betts, S. D., Yocum, C. F., and Boekema, E. (1990) *Biochemistry* **29**, 3220–3225.

Denicolo, I., Craievich, A. F., and Doucet, J. (1984) *J. Chem. Phys.* **80**, 6200–6203.

DeRosier, D. J. (1971) *Contemp. Phys.* **12**, 437–452.

DeRosier, D. J., and Klug, A. (1968) *Nature* **217**, 130–134.

DeTitta, G. T., Edmonds, J. W., Langs, D. A., and Hauptman, H. (1975) *Acta Cryst.* **A31**, 472–479.

DeWael, J., and Havinga, E. (1940) *Rec. Trav. Chim.* **49**, 770–778.

D'Ilario, L., and Giglio, E. (1974) *Acta Cryst.* **B30**, 372–378.

Domszy, R. C., and Booth, C. (1982) *Makromol. Chem.* **183**, 1051–1070.

Dong, W., Baird, T., Fryer, J. R., Gilmore, C. J., MacNicol, D. D., Bricogne, G., Smith, D. J., O'Keefe, M. A., and Hovmöller, S. (1992) *Nature* **355**, 605–609.

Dorset, D. L. (1974a) *Chem. Phys. Lipids* **13**, 791–806.

Dorset, D. L. (1974b) *Biochim. Biophys. Acta* **380**, 257–263.

Dorset, D. L. (1975) *Chem. Phys. Lipids* **14**, 291–296.

Dorset, D. L. (1976a) *J. Appl. Cryst.* **9**, 142–144.

Dorset, D. L. (1976b) *Acta Cryst.* **A32**, 207–215.

Dorset, D. L. (1976c) *Biochim. Biophys. Acta* **424**, 396–403.

Dorset, D. L. (1976d) *J. Appl. Phys.* **47**, 780–782.

Dorset, D. L. (1976e) *Bioorg. Khim.* **2**, 781–788.
Dorset, D. L. (1977a) *Z. Naturforsch.* **32a**, 1166–1172.
Dorset, D. L. (1977b) *Chem. Phys. Lipids* **20**, 13–19.
Dorset, D. L. (1978a) *Z. Naturforsch.* **33a**, 964–982.
Dorset, D. L. (1978b) *Z. Naturforsch.* **33a**, 1090–1092.
Dorset, D. L. (1979) *Chem. Phys. Lipids* **23**, 337–347.
Dorset, D. L. (1980) *Acta Cryst.* **A36**, 592–600.
Dorset, D. L. (1983a) *Z. Naturforsch.* **38c**, 511–514.
Dorset, D. L. (1983b) *Ultramicroscopy* **12**, 19–28.
Dorset, D. L. (1983c) *J. Colloid Interface Sci.* **96**, 172–181.
Dorset, D. L. (1984) *Ultramicroscopy* **13**, 311–324.
Dorset, D. L. (1985a) *Macromolecules* **18**, 2158–2162.
Dorset, D. L. (1985b) *J. Lipid Res.* **26**, 1142–1150.
Dorset, D. L. (1986a) *Ultramicroscopy* **19**, 311–316.
Dorset, D. L. (1986b) *Macromolecules* **19**, 2965–2973.
Dorset, D. L. (1987a) *J. Electr. Microsc. Techn.* **7**, 35–46.
Dorset, D. L. (1987b) *Biochim. Biophys. Acta* **898**, 121–128.
Dorset, D. L. (1987c) *Proc. Electron Microsc. Soc. America, 45th Annual Meeting*, San Francisco, San Francisco Press, pp. 434–437.
Dorset, D. L. (1987d) *Macromolecules* **20**, 2782–2788.
Dorset, D. L. (1987e) *Chem. Phys. Lipids* **43**, 179–191.
Dorset, D. L. (1987f) *J. Lipid Res.* **28**, 993–1005.
Dorset, D. L. (1988a) *Z. Naturforsch.* **43c**, 319–327.
Dorset, D. L. (1988b) *Biochim. Biophys. Acta* **938**, 279–292.
Dorset, D. L. (1988c) *Biochim. Biophys. Acta* **963**, 88–97.
Dorset, D. L. (1989) *J. Polym. Sci. B. Polym. Phys.* **27**, 1161–1171.
Dorset, D. L. (1990a) *J. Phys. Chem.* **94**, 6854–6858.
Dorset, D. L. (1990b) *Proc. Natl. Acad. Sci. USA* **87**, 8541–8544.
Dorset, D. L. (1990c) *Macromolecules* **23**, 623–633.
Dorset, D. L. (1990d) *Chemtracts—Macromol. Chem.* **1**, 311–319.
Dorset, D. L. (1990e) *Macromolecules* **23**, 894–901.
Dorset, D. L. (1990f) *Carbohydrate Res.* **206**, 193–205.
Dorset, D. L. (1990g) *Biophys. J.* **58**, 1077–1087.
Dorset, D. L. (1991a) *Biophys. J.* **60**, 1356–1365.
Dorset, D. L. (1991b) *Ultramicroscopy* **38**, 23–40.
Dorset, D. L. (1991c) *Biophys. J.* **60**, 1366–1373.
Dorset, D. L. (1991d) *Acta Cryst.* **A47**, 510–515.
Dorset, D. L. (1991e) *Macromolecules* **24**, 6521–6526.
Dorset, D. L. (1991f) *Macromolecules* **24**, 1175–1178.
Dorset, D. L. (1991g) *Proc. Natl. Acad. Sci. USA* **88**, 5499–5502.
Dorset, D. L. (1992a) *Ultramicroscopy* **41**, 349–357.
Dorset, D. L. (1992b) *Ultramicroscopy* **45**, 357–364.
Dorset, D. L. (1992c) *Acta Cryst.* **A48**, 568–574.
Dorset, D. L. (1992d) *Ultramicroscopy* **45**, 5–14.
Dorset, D. L. (1992e) *Macromolecules* **25**, 4425–4430.
Dorset, D. L. (1994a) *J. Chem. Cryst.* **24**, 219–224.
Dorset, D. L. (1994b) *Proc. Natl. Acad. Sci. USA*, **91**, 4920–4924.
Dorset, D. L. (1994c) *Adv. Electronics Electron Phys.* **88**, 111–197.
Dorset, D. L. (1995) *Acta Cryst.*, in press.
Dorset, D. L., and Ghiradella, H. (1983) *Biochim. Biophys. Acta* **760**, 136–142.
Dorset, D. L., and Hancock, A. J. (1977) *Z. Naturforsch.* **32c**, 573–580.
Dorset, D. L., and Hauptman, H. A. (1976) *Ultramicroscopy* **1**, 195–201.

Dorset, D. L., and Hsu, S. L. (1989) *Polymer* **30**, 1596–1602.
Dorset, D. L., and Hybl, A. (1972) *Science* **176**, 806–808.
Dorset, D. L., and Massalski, A. K. (1987) *Biochim. Biophys. Acta* **903**, 319–332.
Dorset, D. L., and Massalski, A. K. (1989) *Proc. Electron Microsc. Soc. America, 47th Annual Meeting*, San Francisco, San Francisco Press, pp. 120–121.
Dorset, D. L., and McCourt, M. P. (1992) *Trans. Amer. Cryst. Assoc.* **28**, 105–113.
Dorset, D. L., and McCourt, M. P. (1993) *J. Struct. Biol.* **111**, 118–124.
Dorset, D. L., and McCourt, M. P. (1994a) *Acta Cryst.* **A50**, 287–292.
Dorset, D. L., and McCourt, M. P. (1994b) *Acta Cryst.* **A50**, 344–351.
Dorset, D. L., and Moss, B. (1983) *Polymer* **24**, 291–294.
Dorset, D. L., and Pangborn, W. A. (1979) *Chem. Phys. Lipids* **25**, 179–189.
Dorset, D. L., and Pangborn, W. A. (1982) *Chem. Phys. Lipids* **30**, 1–15.
Dorset, D. L., and Pangborn, W. A. (1988) *Chem. Phys. Lipids* **48**, 19–28.
Dorset, D. L., and Pangborn, W. A. (1992) *Proc. Natl. Acad. Sci. USA* **89**, 1822–1826.
Dorset, D. L., and Parsons, D. F. (1975a) *Acta Cryst.* **A31**, 210–215.
Dorset, D. L., and Parsons, D. F. (1975b) *J. Appl. Phys.* **46**, 938–940.
Dorset, D. L., and Parsons, D. F. (1975c) *J. Appl. Cryst.* **8**, 12–14.
Dorset, D. L., and Rosenbusch, J. P. (1981) *Chem. Phys. Lipids* **29**, 299–307.
Dorset, D. L., and Turner, J. N. (1976) *Naturwiss.* **63**, 145.
Dorset, D. L., and Zemlin, F. (1985) *Ultramicroscopy* **17**, 229–236.
Dorset, D. L., and Zemlin, F. (1987) *Ultramicroscopy* **21**, 263–270.
Dorset, D. L., and Zhang, W. P. (1990) *Biochim. Biophys. Acta* **1028**, 299–303.
Dorset, D. L., and Zhang. W. P. (1991) *J. Electr. Microsc. Techn.* **18**, 142–147.
Dorset, D. L., Hui, S. W., and Strozewski, C. M. (1976) *J. Supramol. Struct.* **5**, 1–16.
Dorset, D. L., Pangborn, W. A., Hancock, A. J., and Lee, I. S. (1978a) *Z. Naturforsch.* **33c**, 39–49.
Dorset, D. L., Pangborn, W. A., Hancock, A. J., Van Soest, T. C., and Greenwald, S. M. (1978b) *Z. Naturforsch.* **33c**, 50–55.
Dorset, D. L., Jap, B. K., Ho, M. S., and Glaeser, R. M. (1979) *Acta Cryst.* **A35**, 1001–1009.
Dorset, D. L., Pangborn, W. A., and Hancock, A. J. (1983a) *J. Biochem. Biophys. Meth.* **8**, 29–40.
Dorset, D. L., Engel, A., Häner, M., Massalski, A., and Rosenbusch, J. P. (1983b) *J. Mol. Biol.* **165**, 701–710.
Dorset, D. L., Holland, F. M., and Fryer, J. R. (1984a) *Ultramicroscopy* **13**, 311–324.
Dorset, D. L., Moss, B., Wittmann, J. C., and Lotz, B. (1984b) *Proc. Natl. Acad. Sci. USA* **81**, 1913–1917.
Dorset, D. L., Moss, B. and Zemlin, F. (1985) *J. Macromol. Sci.—Phys.* **B24**, 87–97.
Dorset, D. L., Zemlin, F., Reuber, E., Beckmann, E., and Zeitler, E. (1986) *Proc. Electr. Microsc. Soc. America, 46th Annual Meeting*, San Francisco, San Francisco Press, pp. 182–183.
Dorset, D. L., Massalski, A. K., and Fryer, J. R. (1987) *Z. Naturforsch.* **42a**, 381–391.
Dorset, D. L., Hanlon, J., and Karet, G. (1989a) *Macromolecules* **22**, 2169–2176.
Dorset, D. L., Massalski, A. K., and Rosenbusch, J. P. (1989b) *Proc. Natl. Acad. Sci. USA* **86**, 6143–6147.
Dorset, D. L., Beckmann, E., and Zemlin, F. (1990) *Proc. Natl. Acad. Sci. USA* **87**, 7570–7573.
Dorset, D. L., Hu, H., and Jäger, J. (1991a) *Acta Cryst.* **A47**, 543–549.
Dorset, D. L., Strauss, H. L., and Snyder, R. G. (1991b) *J. Phys. Chem.* **95**, 938–940.
Dorset, D. L., Tivol, W. F., and Turner, J. N. (1991c) *Ultramicroscopy* **38**, 41–45.
Dorset, D. L., Alamo, R. G., and Mandelkern, L. (1992a) *Macromolecules* **25**, 6284–6288.
Dorset, D. L., Tivol, W. F., and Turner, J. N. (1992b) *Acta Cryst.* **A48**, 562–568.
Dorset, D. L., McCourt, M. P., Tivol, W. F., and Turner, J. N. (1993) *J. Appl. Cryst.* **26**, 778–786.
Dorset, D. L., McCourt, M. P., Fryer, J. R., Tivol, W. F., and Turner, J. N. (1994a) *Microsc. Soc. America Bull.* **24**, 398–404.
Dorset, D. L., McCourt, M. P., Kopp, S., Wittmann, J. C., and Lotz, B. (1994b) *Acta Cryst.* **B50**, 201–208.
Dorset, D. L., Kopp, S., Fryer, J. R., and Tivol, W. F. (1995) *Ultramicroscopy* **57**, 59–89.
Downing, K. H. (1983) *Ultramicroscopy* **11**, 229–238.
Downing, K. H. (1991) *Science* **251**, 53–59.

Downing, K. H. (1992) *Trans. Amer. Cryst. Assoc.* **28**, 115–128.
Downing, K. H., and Glaeser, R. M. (1986) *Ultramicroscopy* **20**, 269–278.
Downing, K. H., Hu, M. S., Wenk, H. R., and O'Keefe, M. A. (1990) *Nature* **348**, 525–528.
Doyle, P. A., and Turner, P. S. (1968) *Acta Cryst.* **A24**, 390–397.
Doyne, T. H., and Gordon, J. T. (1968) *J. Amer. Oil Chemists Soc.* **45**, 333–334.
Dubochet, J., Lepault, J., Freeman, R., Berriman, J. A., and Homo, J. C. (1982) *J. Microsc. (Oxford)* **128**, 219–237.
Dux, L., Pikula, S., Mullner, N., and Martonosi, A. (1986) *J. Biol. Chem.* **262**, 6439–6442.
Dvoryankin, V. F., and Vainshtein, B. K. (1960) *Sov. Phys. Crystallogr.* **5**, 564–574.
Dvoryankin, V. F., and Vainshtein, B. K. (1962) *Sov. Phys. Crystallogr.* **6**, 765–772.
D'yakon, I. A., Kairyak, L. N., Ablov, A. V., and Chapurina, L. F. (1977) *Dokl. Akad. Nauk SSSR* **236**, 103–105.
Eades, J. A. (1992) in *Electron Diffraction Techniques* (Vol. 1), J. M. Cowley, ed., Oxford, Oxford Univ. Press, pp. 313–359.
Eades, J. A. (1994) *Acta Cryst.* **A50**, 292–295.
Earnest, T. N., Walian, P. J., Gehring, K., and Jap, B. K. (1992) *Trans. Amer. Cryst. Assoc.* **28**, 159–164.
Echlin, P. (1992) *Low-temperature Microscopy and Analysis*, New York, Plenum Press.
Edwards, A. M., Darst, S. A., Hemming, S. A., Li, Y., and Kornberg, R. D. (1994) *Nature Struct. Biol.* **1**, 195–197.
Elder, M., Hitchcock, P., Mason, R., and Shipley, G. G. (1977) *Proc. Roy. Soc. (London)* **A354**, 157–170.
Engel, A., and Massalski, A. (1984) *Ultramicroscopy* **13**, 71–84.
Engel, A., Massalski, A., Schindler, H., Dorset, D. L., and Rosenbusch, J. P. (1985) *Nature* **317**, 643–645.
Epstein, H. T. (1951) *J. Phys. Colloid Chem.* **54**, 1053–1069.
Erickson, H. P. (1973) *Adv. Optical Electron Microsc.* **5**, 163–199.
Fan, H. F. (1991) in *Direct Methods of Solving Crystal Structures*, H. Schenk, ed., New York, Plenum Press, pp. 265–272.
Fan, H. F. (1993) in *Modern Crystallography. Proceedings of the Seventh Chinese International Summer School of Physics. Beijing International Workshop*. Z. H. Mai, G. H. Rao, and B. R. Zhu, eds., Beijing, Institute of Physics, Chinese Academy of Sciences, pp. 1–10.
Fan, H. F., Zhong, Z. Y., Zheng, C. D., and Li, F. H. (1985) *Acta Cryst.* **A41**, 163–165.
Fan, H. F., Xiang, S. B., Li, F. H., Pan, Q., Uyeda, N., and Fujiyoshi, Y. (1991a) *Ultramicroscopy* **36**, 361–365.
Fan, H. F., Quan, H., and Woolfson, M. M. (1991b) *Z. Krist.* **197**, 197–208.
Farnall, G. C., and Flint, R. B. (1969) *J. Microsc. (Oxford)* **89**, 37–41.
Federov, E. S. (1949) *Symmetry of Crystals* (trans. David and Katherine Harker, 1971), American Crystallographic Monograph No. 7.
Ferrier, R. P. (1969) *Adv. Optical Electron Microsc.* **3**, 155–216.
Fischer, E. W. (1971) *Pure Appl. Chem.* **26**, 385–421.
Fleming, R. M., Siegrist, T., Marsh, P. M., Hessen, B., Kortan, A. R., Murphy, D. W., Haddon, R. C., Tycko, R., Dabbagh, G., Mujsce, A. M., Kaplan, M. L., and Zahurak, S. M. (1991) *Mat. Res. Soc. Symp. Proc.* **206**, 691–695.
Ford, R. C., Hefti, A., and Engel, A. (1990) *EMBO J.* **9**, 3067–3075.
Frank, J. (1977) *Optik* **49**, 81–92.
Frank, J. (1980) in *Computer Processing of Electron Microscope Images*, P. W. Hawkes, ed., Berlin, Springer, pp. 187–222.
Franks, N. P. (1976) *J. Mol. Biol.* **100**, 345–348.
Frede, E., and Precht, D. (1974) *Kiel. Milchwirtsch. Forschungsber.* **26**, 325–332.
Frey, T. G., Costello, M. J., Karlson, B., Haselgrove, J. C., and Leigh, J. S. (1982) *J. Mol. Biol.* **162**, 113–130.
Fryer, J. R. (1978) *Acta Cryst.* **A34**, 603–607.
Fryer, J. R. (1979a) *The Chemical Applications of Transmission Electron Microscopy*, London, Academic Press.

Fryer, J. R. (1979b) *Acta Cryst.* **A35**, 327–332.
Fryer, J. R. (1981) *Inst. Phys. Conf. Ser. No. 61*, Chapter 1, 19–22.
Fryer, J. R. (1984) *Ultramicroscopy* **14**, 227–236.
Fryer, J. R. (1987) *Ultramicroscopy* **23**, 321–328.
Fryer, J. R. (1989) *J. Electr. Microsc. Techn.* **11**, 310–325.
Fryer, J. R. (1993) *Microsc. Soc. America Bull.* **23(1)**, 44–56.
Fryer, J. R., and Dorset, D. L. (1987) *J. Microsc. (Oxford)* **145**, 61–68.
Fryer, J. R., and Ewins, C. (1992) *Phil. Mag.* **A66**, 889–898.
Fryer, J. R., and Gilmore, C. J. (1992) *Trans. Amer. Cryst. Assoc.* **28**, 57–75.
Fryer, J. R., and Holland, F. (1983) *Ultramicroscopy* **11**, 67–70.
Fryer, J. R., McConnell, C. H., Zemlin, F., and Dorset, D. L. (1992) *Ultramicroscopy* **40**, 163–169.
Fujiwara, K. (1959) *J. Phys. Soc. Japan* **14**, 1513–1524.
Fujiyoshi, Y., Kobayashi, T., Ishizuka, K., Uyeda, N., Ishida, Y., and Harada, Y. (1980) *Ultramicroscopy* **5**, 459–468.
Garavito, R. M., and Rosenbusch, J. P. (1986) *Meth. Enzymol.* **125**, 309–328.
Garavito, R. M., Jenkins, J., Jansonius, J. N., Karlsson, R., and Rosenbusch, J. P. (1983) *J. Mol. Biol.* **164**, 313–327.
Garrido, J., and Hengstenberg, J. (1932) *Z. Krist.* **82**, 477–480.
Gaskill, J. D. (1978) *Linear Systems, Fourier Transforms, and Optics*, New York, Wiley.
Gassmann, J. (1976) in *Crystallographic Computing Techniques*, F. H. Ahmed, ed., Copenhagen, Munksgaard, pp. 144–154.
Gassmann, J., and Zechmeister, K. (1972) *Acta Cryst.* **A28**, 270–280.
Gavezzotti, A. (1991) in *Electron Crystallography of Organic Molecules*,. J. R. Fryer and D. L. Dorset, eds., Amsterdam, Kluwer, pp. 77–83.
Geil, P. H. (1963) *Polymer Single Crystals*, New York, Wiley.
Germain, G., Main, P., and Woolfson, M. (1971) *Acta Cryst.* **A27**, 368–376.
Germer, L. H., and Storks, K. H. (1937) *Proc. Natl. Acad. Sci. USA* **23**, 390–397.
Germer, L. H., and Storks, K. H. (1938) *J. Chem. Phys.* **4**, 280–293.
Gerson, A., and Nyburg, S. C. (1994) *Acta Cryst.* **B50**, 252–256.
Gilmore, C. J., Bricogne, G., and Bannister, C. (1990) *Acta Cryst.* **A46**, 297–308.
Gilmore, C. J., Shankland, K., and Fryer, J. R. (1992) *Trans. Amer. Cryst. Assoc.*, in press.
Gilmore, C. J., Shankland, K., and Bricogne, G. (1993a) *Proc. Roy. Soc. (London)* **A442**, 97–111.
Gilmore, C. J., Shankland, K., and Fryer, J. R. (1993b) *Ultramicroscopy* **49**, 132–146.
Gjønnes, J., and Moodie, A. F. (1965) *Acta Cryst.* **19**, 65–67.
Gjønnes, J., Olsen, A., and Matuhata, H. (1989) *J. Electr. Microsc. Techn.* **13**, 98–110.
Glaeser, R. M. (1975a) *J. Ultrastruct. Res.* **36**, 466–482.
Glaeser, R. M. (1975b) in *Physical Aspects of Electron Microscopy and Microbeam Analysis*, B. M. Siegel and D. R. Beaman, eds., New York, Wiley, pp. 205–229.
Glaeser, R. M. (1985) *Ann. Rev. Phys. Chem.* **36**, 243–275.
Glaeser, R. M., and Ceska, T. A. (1989) *Acta Cryst.* **A45**, 620–628.
Glaeser, R. M., and Downing, K. H. (1992) *Ultramicroscopy* **47**, 256–265.
Glaeser, R. M., and Downing, K. H. (1993) *Ultramicroscopy* **52**, 478–486.
Glaeser, R. M., and Thomas, G. (1969) *Biophys. J.* **9**, 1073–1099.
Glaeser, R. M., Baldwin, J., Ceska, T. A., and Henderson, R. (1986) *Biophys. J.* **50**, 913–920.
Glusker, J. P. and Trueblood, K. N. (1985) *Crystal Structure Analysis. A Primer* (2d ed.), New York, Oxford Univ. Press.
Goldsmith, G. J., and White, J. G. (1959) *J. Chem. Phys.* **31**, 1175–1187.
Goodman, P. (1976) *Acta Cryst.* **A32**, 793–798.
Goodman, P., and Moodie, A. F. (1974) *Acta Cryst.* **A30**, 280–290.
Green, E. A., and Hauptman, H. (1976) *Acta Cryst.* **A32**, 43–49.
Greenwald, S. M., Hancock, A. J., Sable, H. Z., D'Esposito, L., and Koenig, J. L. (1977) *Chem. Phys. Lipids* **18**, 154–169.

Grubb, D. T. (1974) *J. Mater. Sci.* **9**, 1715–1736.

Guizard, C., Chanzy, H., and Sarko, A. (1984) *Macromolecules* **17**, 100–107.

Guizard, C., Chanzy, H., and Sarko, A. (1985) *J. Mol. Biol.* **183**, 397–408.

Hagemann, H., Strauss, H. L., and Snyder, R. G. (1987) *Macromolecules* **20**, 2810–2819.

Hahn, T., ed. (1992) *International Tables for Crystallography*, (Vol. A), Dordrecht, Kluwer.

Hall, C. E. (1966) *Introduction to Electron Microscopy* (2d ed.), New York, McGraw-Hill.

Hamilton, W. C. (1964) *Statistics in Physical Sciences. Estimation, Hypothesis Testing, and Least Squares*, New York, Ronald, pp. 157–162.

Han, B. G. (1993) *Proc. Microsc. Soc. America, 51st Annual Meeting*, San Francisco, San Francisco Press, pp. 680–681.

Han, F. S., Fan, H. F., and Li, F. H. (1986) *Acta Cryst.* **A42**, 353–356.

Hancock, A. J., Stokes, M. H., and Sable, H. Z. (1977) *J. Lipid Res.* **18**, 81–92.

Harburn, G., Taylor, C. A., and Welberry, T. R. (1975) *Atlas of Optical Transforms*, Ithaca, N.Y., Cornell Univ. Press.

Harker, D., and Kasper, J. (1947) *J. Chem. Phys.* **12**, 882–884.

Harker, D., and Kasper, J. (1948) *Acta Cryst.* **1**, 70–75.

Hasegawa, H., Claffey, W., and Geil, P. H. (1977) *J. Macromol. Sci.—Phys.* **B13**, 89–100.

Hauptman, H. A. (1972) *Crystal Structure Determination. The Role of the Cosine Seminvariants*, New York, Plenum Press.

Hauptman, H. A. (1993) *Proc. Roy. Soc. (London)* **A442**, 3–12.

Hauptman, H. A., and Karle, J. (1953) *Solution of the Phase Problem. I. The Centrosymmetric Crystal*, American Crystallographic Association Monograph No. 3.

Hauser, H., Pascher, I., and Sundell, S. (1980) *J. Mol. Biol.* **137**, 249–264.

Havelka, W. A., Henderson, R., Heymann, J. A. W., and Oesterhelt, D. (1993) *J. Mol. Biol.* **234**, 837–846.

Havinga, E., and DeWael, J. (1937a) *Rec. Trav. Chim.* **56**, 375–381.

Havinga, E., and DeWael, J. (1937b) *Chemisch. Weekblad* **34**, 694–701.

Haywood, S. B., and Stroud, R. M. (1981) *J. Mol. Biol.* **151**, 491–517.

Heavens, J. W., Keller, A., Pope, J. M., and Rowell, D. M. (1970) *J. Mater. Sci.* **5**, 53–62.

Heidenreich, R. D. (1964) *Fundamentals of Transmission Electron Microscopy*, New York, Wiley-Interscience.

Helbert, W., and Chanzy, H. (1994) *Carbohyd. Polym.* **24**, 119–122.

Henderson, R., and Glaeser, R. M. (1985) *Ultramicroscopy* **16**, 139–150.

Henderson, R., and Unwin, P. N. T. (1975) *Nature* **257**, 28–32.

Henderson, R., Baldwin, J. M., Downing, K. H., Lepault, J. and Zemlin, F. (1986) *Ultramicroscopy* **19**, 147–178.

Henderson, R., Baldwin, J. M., Ceska, T., Zemlin, F., Beckmann, E., and Downing, K. H. (1990) *J. Mol. Biol.* **213**, 809–929.

Hengstenberg, J., and Garrido, J. (1932) *An. Soc. Esp. Fis. Quim.* **30**, 175–181.

Henry, N. F. M., and Lonsdale, K., eds. (1969) *International Tables for X-ray Crystallography. Volume I. Symmetry Groups*. Birmingham, Kynoch.

Hirsch, P. B., Howie, A., Nicholson, R. B., Pashley, D. W., and Whelan, M. J. (1971) *Electron Microscopy of Thin Crystals.* London, Butterworths, pp. 18–23.

Hitchcock, P. B., Mason, R., Thomas, K. M., and Shipley, G. G. (1974) *Proc. Natl. Acad. Sci. USA* **71**, 3036–3040.

Hitchcock, P. B., Mason, R., and Shipley, G. G. (1975) *J. Mol. Biol.* **94**, 297–299.

Ho, M. S., Jap, B. K., and Glaeser, R. M. (1988) *Acta Cryst.* **A44**, 878–884.

Honjo, G., and Kitamura, N. (1957) *Acta Cryst.* **10**, 533–534.

Hoppe, W. (1956) *Z. Krist.* **107**, 406–432.

Hoppe, W., and Baumgärtner, F. (1957) *Z. Krist.* **108**, 328–334.

Hoppe, W., and Gassmann, J. (1968) *Acta Cryst.* **B24**, 97–107.

Hoppe, W., and Rauch, R. (1960) *Z. Krist.* **115**, 141–155.

Hoppe, W., Lenne, H. U., and Moranti, G. (1957) *Z. Krist.* **108**, 321–327.

Hosemann, R., and Bagchi, S. N. (1962) *Direct Analysis of Diffraction by Matter*, Amsterdam, North-Holland.
Hoshino, A., Isoda, S., and Kobayashi, T. (1991) *J. Cryst. Growth* **115**, 826–830.
Hovmöller, S. (1981a) *Rotation Matrices and Translation Vectors in Crystallography. (IUCr Pamphlet No. 9)*, C. A. Taylor, ed., Cardiff, Wales, University College Cardiff.
Hovmöller, S. (1981b) *Acta Cryst.* **A37**, 133–135.
Hovmöller, S. (1992) *Ultramicroscopy* **41**, 121–135.
Hovmöller, S., Slaughter, M., Berriman, J., Karlson, B., Weiss, H., and Leonard, K. (1983) *J. Mol. Biol.* **165**, 401–406.
Hovmöller, S., Sjögren, A., and Wang, D. N. (1988) *Progr. Biophys. Molec. Biol.* **51**, 131–163.
Hovmöller, S., Sjögren, A., Farrants, G., Sundberg, M., and Marinder, B. O. (1992) *Nature* **311**, 238–241.
Howie, A., and Whelan, M. J. (1961) *Proc. Roy. Soc. (London)* **A263**, 217–237.
Hu, H., and Dorset, D. L. (1989) *Acta Cryst.* **B45**, 283–290.
Hu, H., and Dorset, D. L. (1990) *Macromolecules* **23**, 4604–4607.
Hu, H., Dorset, D. L., and Moss, B. (1989) *Ultramicroscopy* **27**, 161–170.
Hu, J. J., Li, F. H., and Fan, H. F. (1992) *Ultramicroscopy* **41**, 387–397.
Hui, S. W. (1976) *Chem. Phys. Lipids* **16**, 9–18.
Hui, S. W. (1981) *Biophys. J.* **34**, 383–395.
Hui, S. W. (1989) *J. Electron Microsc. Techn.* **11**, 286–297.
Hui, S. W. (1993) *Microsc. Soc. America Bull.* **23(1)**, 20–27.
Hui, S. W., and He, N. B. (1983) *Biochemistry* **22**, 1159–1164.
Hui, S. W., and Parsons, D. F. (1974) *Science* **184**, 77–78.
Hui, S. W., and Parsons, D. F. (1975) *Science* **190**, 383–384.
Hui, S. W., and Parsons, D. F. (1978) in *Advanced Techniques in Biological Electron Microscopy* (Vol. 2), J. K. Koehler, ed., Berlin, Springer, pp. 215–232.
Hui, S. W., Parsons, D. F., and Cowden, M. (1974) *Proc. Natl. Acad. Sci. USA* **71**, 5068–5072.
Hui, S. W., Cowden, M., Papahadjopoulos, D., and Parsons, D. F. (1975) *Biochim. Biophys. Acta* **382**, 265–275.
Hui, S. W., Hausner, G. G., and Parsons, D. F. (1976) *J. Phys. E. Scientific Instruments.* **9**, 69–72.
Hui, S. W., Stewart, C. M., Carpenter, M. P., and Stewart, T. P. (1981) *J. Cell Biol.* **85**, 283–291.
Humphreys, C. J., and Bithell, E. G. (1992), in *Electron Diffraction Techniques* (Vol. 1), J. M. Cowley, ed., Oxford, Oxford Univ. Press, Chapter 2.
Hurst, H. (1948) *Discuss. Faraday Soc.* **3**, 193–210.
Hurst, H. (1950) *J. Exp. Biol.* **27**, 238–252.
Hybl, A., and Dorset, D. (1971) *Acta Cryst.* **B27**, 977–986.
Ihn, K. J., Tsuji, M., Isoda, S., Kawaguchi, A., Katayama, K., Tanaka, Y., and Sato, H. (1990) *Macromolecules* **23**, 1781–1787.
Imamov, R. M., and Pinsker, Z. G. (1965) *Sov. Phys. Crystallogr.* **10**, 148–152.
Ishizuka, K., and Uyeda, N. (1977) *Acta Cryst.* **A33**, 740–749.
Ishizuka, K., Miyazaki, M., and Uyeda, N. (1982) *Acta Cryst.* **A38**, 408–413.
Isoda, S., Tsuji, M., Ohara, M., Kawaguchi, A., and Katayama, K. (1983a) *Polymer* **24**, 1155–1161.
Isoda, S., Tsuji, M., Ohara, M., Kawaguchi, A., and Katayama, K. (1983b) *Makromol. Chem. Rapid Commun.* **4**, 141–144.
Isoda, S., Saito, K., Moriguchi, S., and Kobayashi, T. (1991) *Ultramicroscopy* **35**, 329–338.
James, R. W. (1950) *The Optical Principles of the Diffraction of X-rays*, London, Bell, pp. 400 ff.
Jap, B. K. (1988) *J. Mol. Biol.* **199**, 229–231.
Jap, B. K. (1989) *J. Mol. Biol.* **205**, 407–419.
Jap, B. K., and Glaeser, R. M. (1980) *Acta Cryst.* **A36**, 57–67.
Jap, B. K., Downing, K. H., and Walian, P. J. (1990) *J. Struct. Biol.* **103**, 57–63.
Jap, B. K., Walian, P. J., and Gehring, K. (1991) *Nature* **350**, 167–170.
Jap, B. K., Zulauf, M., Scheybani, T., Hefti, A., Baumeister, W., Aebi, U., and Engel, A. (1992) *Ultramicroscopy* **46**, 45–84.

Jeffrey, G., and Yeon, Y. (1992) *Carbohydrate Res.* **237**, 45–55.
Jensen, L. H. (1970) *J. Polym. Sci.* **C29**, 47–63.
Jensen, L. H., and Mabis, A. J. (1966) *Acta Cryst.* **21**, 770–781.
Karle, I. L., Dragonette, K. S., and Brenner, S. A. (1965) *Acta Cryst.* **19**, 713–716.
Karle, J., and Brockway, L. O. (1947) *J. Chem. Phys.* **19**, 213–225.
Karle, J., and Hauptman, H. (1956) *Acta Cryst.* **9**, 635–651.
Karle, J., and Karle, I. L. (1966) *Acta Cryst.* **21**, 849–859.
Karpov, V. L. (1941) *Zh. Fiz. Khim.* **15**, 577–591.
Kawaguchi, A., Ohara, M., and Kobayashi, K. (1979) *J. Macromol. Sci.—Phys.* **B16**, 193–212.
Kawaguchi, A., Isoda, S., Petermann, J., and Katayama, K. (1984) *Colloid Polym. Sci.* **262**, 429–434.
Kawaguchi, A., Tsuji, M., Moriguchi, S., Uemura, A., Isoda, S., Ohara, M., Petermann, J., and Katayama, K. (1986) *Bull. Inst. Chem. Res. Kyoto Univ.* **64**, 54–65.
Kay, H. F., and Newman, B. A. (1968) *Acta Cryst.* **B24**, 615–624.
Keller, A. (1961) *Phil. Mag.* **6**, 329–343.
Kerr, H. W., and Lewis, M. H. (1971) in *Advances in Epitaxy and Endotaxy*, H. G. Schneider and V. Ruth, eds., Leipzig, VEB Deutscher Verlag für Grundstoffindustrie, pp. 147–164.
Khare, R. S., and Worthington, C. R. (1971) *Biochim. Biophys. Acta* **514**, 239–254.
Kihlborg, L. (1990) *Progr. Solid State Chem.* **20**, 101–134.
Kim, Y., Strauss, H. L., and Snyder, R. G. (1989) *J. Phys. Chem.* **93**, 7520–7526.
Kirkland, E. J., Siegel, B. M., Uyeda, N., and Fujiyoshi, Y. (1985) *Ultramicroscopy* **17**, 87–104.
Kistler, J., and Bullivant, S. (1980) *FEBS Lett.* **111**, 73–78.
Kistler, J., Berriman, J., Evans, C. W., Gruitjers, W. T., Christie, D., Corin, A., and Bullivant, S. (1990) *J. Struct. Biol.* **103**, 204–211.
Kitaigorodskii, A. I. (1961) *Organic Chemical Crystallography*, New York, Consultants Bureau.
Kitaigorodskii, A. I. (1973) *Molecular Crystals and Molecules*, New York, Academic Press.
Kitaigorodskii, A. I., Mnyukh, Yu. V., and Nechitailo, N. A. (1958) *Sov. Phys. Crystallogr.* **3**, 303–307.
Klug, A. (1978–1979) *Chem. Scripta* **14**, 245–256.
Klug, H., and Alexander, L. E. (1971) *X-ray Diffraction Procedures for Polycrystalline and Amorphous Molecules* (2d ed.), New York, Wiley-Interscience.
Knapek, E. (1982) *Ultramicroscopy* **10**, 71–86.
Knoop, E., and Sandhammer, E. (1961) *Milchwissensch.* **16**, 201–209.
Kobayashi, K., and Sakaoku, K. (1965) *Lab. Invest.* **14**, 1097–1114.
Kobayashi, T., Fujiyoshi, Y., Iwatsu, F., and Uyeda, N. (1981) *Acta Cryst.* **A37**, 692–697.
Kohlhaas, R. (1938) *Z. Krist.* **98**, 418–438.
Kohlhaas, R. (1940) *Chem. Ber.* **73B**, 189–200.
Kopp, S., Wittmann, J. C., and Lotz, B. (1993) *Polymer*, **35**, 908–915.
Kornberg, R. D., and Darst, S. A. (1991) *Curr. Opin. Struct. Biol.* **1**, 642–646.
Ku, A. C., Darst, S. A., Kornberg, R. D., Robertson, C. R., and Gast, A. P. (1992) *Langmuir* **8**, 2357–2360.
Kühlbrandt, W. (1984) *Nature* **307**, 478–480.
Kühlbrandt, W., and Wang, D. N. (1991) *Nature* **350**, 130–134.
Kühlbrandt, W., Wang, D. N., and Fujiyoshi, Y. (1994) *Nature* **367**, 614–621.
Kuo, I. A. M., and Glaeser, R. M. (1975) *Ultramicroscopy* **1**, 53–66.
Kuwabara, S. (1959) *J. Phys. Soc. Japan* **14**, 1205–1216.
Ladd, M. F. C., and Palmer, R. A. (1980) in *Theory and Practice of Direct Methods in Crystallography*, M. F. C. Ladd and R. A. Palmer, eds., New York, Plenum Press, pp. 93–150.
Lampe, P. D., Kistler, J., Hefti, A., Bond, J., Müller, S., Johnson, R. G., and Engel, A. (1991) *J. Struct. Biol.* **107**, 281–290.
Lando, J., and Sudiwala, R. V. (1990) *Chem. Mater.* **2**, 594–599.
Langs, D. A., and DeTitta, G. T. (1975) *Acta Cryst.* **A31**, S16.
Larsson, K. (1965a) *Ark. Kemi* **23**, 35–56.
Larsson, K. (1965b) *Ark. Kemi* **23**, 1–15.
Lebedeff, A. (1931) *Nature* **128**, 491.

Lefranc, G., Knapek, E., and Dietrich, I. (1982) *Ultramicroscopy* **10**, 111–124.

Leonard, K., Haiker, H., and Weiss, H. (1987) *J. Mol. Biol.* **194**, 277–286.

Leonowicz, M. E., Lawton, J. A., Lawton, S. L., and Rubin, M. K. (1994) *Science* **264**, 1910–1913.

Lepault, J., Booy, F. P., and Dubochet, J. (1983) *J. Microsc. (Oxford)* **129**, 89–102.

Lepault, J., Dargent, B., Tichelaar, W., Rosenbusch, J. P., Leonard, K., and Pattus, F. (1988) *EMBO J.* **7**, 261–268.

Li, D. X., and Hovmöller, S. (1988) *J. Solid State Chem.* **73**, 5–10.

Li, F. H. (1963) *Acta Phys. Sinica* **19**, 735–740.

Li, F. H. (1991) in *Electron Crystallography of Organic Molecules*, J. R. Fryer and D. L. Dorset, eds., Dordrecht, Kluwer, pp. 153–167.

Li, F. H. (1993) in *Modern Crystallography. Proceeding of the Seventh Chinese International Summer School of Physics. Beijing International Workshop*, Z. H. Mai, G. H. Rao, and B. R. Zhu, eds., Beijing, Institute of Physics, Chinese Academy of Sciences, pp. 82–94.

Li, J. (1985) *Proc. Natl. Acad Sci. USA* **82**, 386–390.

Lieser, G., Lee, K. S., and Wegner, G. (1988) *Colloid Polym. Sci.* **266**, 419–428.

Lipson, H., and Cochran, W. (1966) *The Determination of Crystal Structures* (rev. and enlarged ed.), Ithaca, Cornell Univ. Press, pp. 377–381.

Lipson, S. G., and Lipson, H. (1969) *Optical Physics*, Cambridge, Cambridge Univ. Press.

Liu, Y. W., Fan, H. F., and Zheng, C. D. (1988) *Acta Cryst.* **A44**, 61–63.

Liu, Y. W., Xiang, S. B., Fan, H. F., Tang, D., Li, F. H., Pan, Q., Uyeda, N., and Fujiyoshi, Y. (1990) *Acta Cryst.* **A46**, 459–463.

Lobachev, A. N. (1954) *Trud. Inst. Krist. SSSR* **10**, 71–75.

Lobachev, A. N., and Vainshtein, B. K. (1961) *Sov. Phys. Crystallogr.* **6**, 313–317.

Lösche, M., Rabe, J., Fischer, A., Rucha, U., Knoll, W., and Möhwald, H. (1984) *Thin Solid Films* **117**, 269–280.

Lüth, H., Nyburg, S. C., Robinson, P. M., and Scott, H. G. (1974) *Mol. Cryst. Liq. Cryst.* **27**, 337–357.

Luzzati, V. (1972) in *International Tables for X-ray Crystallography*, Volume II, J. S. Kasper and K. Lonsdale, eds., Kynoch Press, Birmingham, England, pp. 355–356.

Luzzati, V., Mariani, P., and Delacroix, H. (1988) *Makromol. Chem. Macromol. Symp.* **15**, 1–17.

McCourt, M. P., Strong, P. S., Pangborn, W. A., and Dorset, D. L. (1994) *J. Lipid Res.* **35**, 584–591.

McCoy, N. H. (1960) *Introduction to Modern Algebra*, Boston, Allyn and Bacon, Chapter 9.

McKie, D., and McKie, C. (1986) *Essentials of Crystallography*, Oxford, Blackwell.

McMillan, M. (1994) *Amer. Cryst. Assoc. Abstr., Sect. 2*, **22**, 171.

Malta, V., Cojazzi, G., Zannetti, R., and Amati, L. (1974) *Gazz. Chim. Ital.* **104**, 921–928.

Mank, A. P. J., Ward, J. P., and Van Dorp, D. A. (1976) *Chem. Phys. Lipids* **16**, 107–114.

Mannella, C. A. (1984) *Science* **224**, 165–166.

Mannella, C. A., and Frank, J. (1982) *Biophys. J.* **37**, 3–4.

Mannella, C. A., Ribierto, A., and Frank, J. (1986) *Biophys. J.* **49**, 307–318.

Maroncelli, M., Qi, S. P., Strauss, H. L., and Snyder, R. G. (1982) *J. Amer. Chem. Soc.* **104**, 6237–6247.

Maroncelli, M., Strauss, H. L., and Snyder, R. G. (1985) *J. Chem. Phys.* **82**, 2811–2824.

Mason, S. J., and Zimmerman, H. J. (1960) *Electronic Circuits, Signals and Systems*, New York, Wiley, Chapter 6.

Massalski, A., Sass, H. J., Zemlin, F., Beckmann, E., Van Heel, M., Büldt, G., Dorset, D. L., Zeitler, E., and Rosenbusch, J. P. (1987) *Proc. Electron Microsc. Soc. America, 45th Annual Meeting*, San Francisco, San Francisco Press, pp. 788–789.

Mathiesen, R. R., Jr., and Smith, P. (1985) *Polymer* **26**, 288–292.

Matricardi, V., Moretz, R. C., and Parsons, D. F. (1972) *Science* **177**, 268–270.

Mazeau, K., Chanzy, H., and Winter, W. T. (1992) *Amer. Chem. Soc. Div. Polym. Chem., Polymer Preprints* **33(1)**, 244.

Mazeau, K., Winter, W. T., and Chanzy, H. (1994) *Macromolecules* **27**, 7606–7612.

Mazee, W. M. (1958) *Amer. Chem. Soc. Div. Petro. Chem., Preprints* **3(4)**, 35–47.

Meille, S. V., Brückner, S., and Lando, J. B. (1989) *Polymer* **30**, 786–792.

Meille, S. V., Ferro, D. R., Brückner, S., Lovinger, A. J., and Padden, F. J. (1994) *Macromolecules* **27**, 2615–2622.

Menter, J. W. (1950) *Research* **3**, 381–382.

Menter, J. W., and Tabor, D. (1951) *Proc. Roy. Soc. (London)* **A214**, 512–524.

Michel, H., and Oesterhelt, D. (1980) *Proc. Natl. Acad. Sci. USA* **77**, 1283–1285.

Michel, H., Oesterhelt, D., and Henderson, R. (1980) *Proc. Natl. Acad. Sci. USA* **77**, 338–342.

Miller, K. R., and Jacob, J. S. (1983) *J. Cell Biol.* **97**, 1266–1270.

Miller, R., DeTitta, G. T., Jones, R., Langs, D. A., Weeks, C. M., and Hauptman, H. A. (1993) *Science* **259**, 1430–1433.

Misell, D. L. (1978) *Image Analysis, Enhancement, and Interpretation*, Amsterdam, North-Holland.

Mnyukh, Yu. V. (1960) *Zh. Strukt. Khim.* **1**, 370–388.

Mo, Y. D., Cheng, T. Z., Fan, H. F., Li, J. Q., Sha, B. D., Zheng, C. D., Li, F. H., and Zhao, Z. X. (1992) *Supercond. Sci. Technol.* **5**, 69–72.

Moews, P. C., and Knox, J. R. (1976) *J. Amer. Chem. Soc.* **98**, 6628–6683.

Mohraz, M., and Smith, P. R. (1984) *J. Cell. Biol.* **98**, 1836–1841.

Mohraz, M., Yee, M., and Smith, P. R. (1985) *J. Ultrastruct. Res.* **93**, 17–26.

Mohraz, M., Yee, M., and Smith, P. R. (1987) *J. Cell Biol.* **105**, 1–8.

Mohraz, M., Saithe, S., and Smith, P. R. (1990) *Proc. 12th Int. Congr. Electron Microsc.*, San Francisco, San Francisco Press, p. 94.

Moss, B., and Brisse, F. (1984) *Macromolecules* **17**, 2202–2204.

Moss, B., and Dorset, D. L. (1982) *J. Polym. Sci. Polym. Phys. Ed.* **20**, 1789–1804.

Moss, B., and Dorset, D. L. (1983a) *Acta Cryst.* **A39**, 609–615.

Moss, B., and Dorset, D. L. (1983b) *J. Macromol. Sci.—Phys.* **B22**, 69–77.

Müller, A. (1930) *Proc. Roy. Soc. (London)* **A127**, 417–430.

Müller, A. (1932) *Proc. Roy. Soc. (London)* **A138**, 514–530.

Murata, Y., Fryer, J. R., and Baird, T. (1976) *J. Microsc. (Oxford)* **108**, 261–275.

Natta, E., and Rigamonti, R. (1938) *Atti della Reale Acc. Naz. Lincei* **22**, 342–348.

Natta, E., Baccaredda, M., and Rigamonti, R. (1935) *Gazz. Chim. Ital.* **65**, 182–198.

Newman, B. A., and Kay, H. F. (1967) *J. Appl. Phys.* **38**, 4105–4109.

Nieduszynski, I., and Marchessault, R. H. (1972) *Canad. J. Chem.* **50**, 2130–2138.

Niegisch, W. D., and Swan, P. R. (1960) *J. Appl. Phys.* **31**, 1906–1910.

Nowacki, W. (1967) *Crystal Data. Systematic Tables* (2d ed.), Washington, D.C., Williams and Heintz Map Corp.

Nyburg, S. C., and Potworowski, J. A. (1973) *Acta Cryst.* **B29**, 347–352.

Ogawa, T., Moriguchi, S., Isoda, S., and Kobayashi, T. (1994) *Polymer* **35**, 1132–1136.

O'Keefe, M. A., Fryer, J. R., and Smith, D. J. (1983) *Acta Cryst.* **A39**, 838–847.

Organ, S. J., and Keller, A. (1987) *J. Polym. Sci. B. Polym. Phys.* **25**, 2409–2430.

Pan, M., and Crozier, P. A. (1993) *Ultramicroscopy* **52**, 487–498.

Parsons, D. F. (1967) *Canad. Cancer Conf.* **7**, 193–246.

Parsons, D. F., and Nyburg, S. C. (1966) *J. Appl. Cryst.* **37**, 3920.

Parsons, D. F., Matricardi, V., Moretz, R. C., and Turner, J. N. (1974) *Adv. Biol. Med. Phys.* **15**, 161–270.

Pascher, I., and Sundell, S. (1986) *Biochim. Biophys. Acta* **855**, 68–78.

Pascher, I., Sundell, S., and Hauser, H. (1982a) *J. Mol. Biol.* **153**, 791–806.

Pascher, I., Sundell, S., and Hauser, H. (1982b) *J. Mol. Biol.* **153**, 807–824.

Pascher, I., Sundell, S., Eibl, H., and Harlos, K. (1986) *Chem. Phys. Lipids* **39**, 53–64.

Patel, G. N. (1975) *J. Polym. Sci.—Polym. Phys. Ed.* **13**, 351–359.

Paul, A., Engelhardt, H, Jakubowski, U., and Baumeister, W. (1992) *Biophys. J.* **61**, 172–188.

Pauptit, R. A., Schirmer, T., Jansonius, J. N., Rosenbusch, J. P., Parker, N. W., Tucker, A. D., Tsernoglou, D., Weiss, M. S., and Schulz, G. E. (1991) *J. Struct. Biol.* **107**, 136–145.

Pearson, R. H., and Pascher, I. (1979) *Nature* **281**, 499–501.

Perez, S., and Chanzy, H. (1989) *J. Electron Microsc. Techn.* **11**, 280–295.

Perez, S., Roux, M., Revol, J. F., and Marchessault, R. H. (1979) *J. Mol. Biol.* **129**, 113–133.

Pertsin, A. J., and Kitaigorodsky, A. I. (1987) *The Atom–Atom Potential Method. Applications to Organic Molecular Solids*, Berlin, Springer.
Peterson, I. R. (1987) *J. Molec. Electronics* **3**, 103–111.
Peterson, I. R., and Russell, G. J. (1984) *Phil. Mag.* **A49**, 463–473.
Peterson, I. R., Russell, G. J., Earls, J. D., and Girling, I. R. (1988) *Thin Solid Films* **161**, 325–331.
Piesczek, W., Strobl, G. R., and Malzahn, K. (1974) *Acta Cryst.* **B30**, 1278–1288.
Pinsker, Z. G. (1953) *Electron Diffraction*, London, Butterworths.
Podjarny, A. D., Schevitz, R. W., and Sigler, P. B. (1981) *Acta Cryst.* **A37**, 662–668.
Popot, J. L., Gerchman, S. E., and Engelman, D. M. (1987) *J. Mol. Biol.* **198**, 655–676.
Poulin-Dandurand, S., Perez, S., Revol, J. F., and Brisse, F. (1974) *Polymer* **20**, 419–426.
Pradere, P., Revol, J. F., and Manley, R. St. J. (1988a) *Macromolecules* **21**, 2747–2751.
Pradere, P., Revol, J. F., Nguyen, L., and Manley, R. St. J. (1988b) *Ultramicroscopy* **25**, 69–80.
Precht, D. (1976a) *Fett. Seif. Anstrichm.* **78**, 145–149.
Precht, D. (1976b) *Fett. Seif. Anstrichm.* **78**, 189–192.
Precht, D. (1979) *Fett. Seif. Anstrichm.* **81**, 227–233.
Precht, D., and Frede, E. (1983) *Acta Cryst.* **B39**, 381–388.
Prince, E. (1982) *Mathematical Techniques in Crystallography and Materials Science*, New York, Springer, pp. 17–19.
Rainville, E. D. (1963) *The Laplace Transform. An Introduction*. New York, Macmillan.
Reimer, L. (1965) *Lab. Invest.* **14**, 227–236.
Reimer, L. (1984) *Transmission Electron Microscopy. Physics of Image Formation and Microanalysis*, Berlin, Springer.
Remillard, B., and Brisse, F. (1982a) *Acta Cryst.* **B38**, 1220–1224.
Remillard, B., and Brisse, F. (1982b) *Polymer* **23**, 1960–1964.
Reneker, D., and Geil, P. (1960) *J. Appl. Phys.* **31**, 1916–1925.
Revol, J. F. (1989) *Int. J. Biol. Macromol.* **11**, 233–235.
Revol, J. F. (1991) in *Electron Crystallography of Organic Molecules*, J. R. Fryer and D. L. Dorset, eds., Amsterdam, Kluwer, pp. 169–187.
Revol, J. F., and Chanzy, H. (1986) *Biopolymers* **25**, 1599–1601.
Revol, J. F., and Manley, R. St. J. (1986) *J. Mater. Sci. Lett.* **5**, 249–251.
Revol, J. F., Chanzy, H. D., Deslandes, Y., and Marchessault, R. H. (1989) *Polymer* **30**, 1973–1976.
Rigamonti, R. (1936) *Gazz. Chim. Ital.* **66**, 174–182.
Roche, E., Chanzy, H., Boudeulle, M., Marchessault, R. H., and Sundararajan, P. (1978) *Macromolecules* **11**, 86–94.
Rogers, D. (1950) *Acta Cryst.* **3**, 455–464.
Rogers, D. (1980) in *Theory and Practice of Direct Methods in Crystallography*, M. F. C. Ladd and R. A. Palmer, eds., New York, Plenum Press, pp. 23–92.
Rosenbusch, J. P. (1974) *J. Biol. Chem.* **249**, 8019–8029.
Rosenbusch, J. P. (1990) *J. Struct. Biol.* **104**, 134–138.
Rosenbusch, J. P., Garavito, R. M., Dorset, D. L., and Engel, A. (1981) in *Protides of the Biological Fluids, Proceedings of the Twenty-ninth Colloquium*, H. Peeters, ed., Oxford, Pergamon, pp. 171–174.
Rossmann, M. G., and Blow, D. M. (1962) *Acta Cryst.* **15**, 24–31.
Rossmann, M. G., and Henderson, R. (1982) *Acta Cryst.* **A38**, 13–20.
Rupp, E. (1934) *Kolloid Z.* **69**, 369–378.
Sands, D. E. (1969) *Introduction to Crystallography*, New York, Benjamin, pp. 13–43.
Sass, H. J., Büldt, G., Beckmann, E., Zemlin, F., Van Heel, M., Zeitler, E., Rosenbusch, J. P., Dorset, D. L., and Massalski, A. K. (1989) *J. Mol. Biol.* **209**, 171–175.
Saunders, J. V., and Tabor, D. (1951) *Proc. Roy. Soc. (London)* **A204**, 525–533.
Sawzik, P., and Craven, B. M. (1980) in *Liquid Crystals*, S. Chandrasekhar, ed., London, Heyden, pp. 171–180.
Sawzik, P., and Craven, B. M. (1982) *Acta Cryst.* **B38**, 1777–1781.

Saxton, W. O. (1980) in *Electron Microscopy at Molecular Dimensions*, W. Baumeister and W. Vogell, eds., Berlin, Springer, pp. 187–222.
Saxton, W. O., and Baumeister, W. (1982) *J. Microsc. (Oxford)* **127**, 127–138.
Sayre, D. (1952) *Acta Cryst.* **5**, 60–65.
Scaringe, R. P. (1991) in *Electron Crystallography of Organic Molecules*, J. R. Fryer and D. L. Dorset, eds., Amsterdam, Kluwer, pp. 85–113.
Scaringe, R. P. (1992) *Trans. Amer. Cryst. Assoc.* **28**, 11–23.
Schertler, G. F. X., Bartunik, H. D., Michel, H., and Oesterhelt, D. (1993) *J. Mol. Biol.* **234**, 156–164.
Schevitz, R. W., Podjarny, A. D., Zwick, M., Hugher, J. J., and Sigler, P. B. (1981) *Acta Cryst.* **A37**, 669–677.
Schoon, Th. (1938) *Z. Phys. Chem.* **B39**, 385–410.
Schröder, R. R., and Burmester, C. (1993) *Proc. Microsc. Soc. America, 51st Annual Meeting*, San Francisco, San Francisco Press, pp. 666–667.
Schwickert, H., Strobl, G., and Kimming, M. (1991) *J. Chem. Phys.* **95**, 2800–2806.
Self, P. G., and O'Keefe, M. A. (1988) in *High Resolution Transmission Electron Microscopy and Associated Techniques*, P. Busek, J. Cowley, and L. Eyring, eds., New York, Oxford Univ. Press, pp. 247–307.
Sen, A., Hurley, E. L., Hui, S. W., Dorset, D. L., and McConnell, C. M. (1987) *Proc. Electron Microsc. Soc. America, 45th Annual Meeting*, San Francisco, San Francisco Press, pp. 494–495.
Sha, B. D., Fan, H. F., and Li, F. H. (1993) *Acta Cryst.* **A49**, 877–880.
Shearer, H. M. M., and Vand, V. (1956) *Acta Cryst.* **9**, 379–384.
Sherman, M. B., Orlova, E. V., Terzyan, S. S., Kleine, R., and Kiselev, N. A. (1981) *Ultramicroscopy* **7**, 131–138.
Shimamura, K., Karasz, F. E., Hirsch, J. A., and Chien, J. C. W. (1981) *Makromol. Chem. Rapid Commun.* **2**, 473–480.
Siegel, G. (1972) *Z. Naturforsch.* **27a**, 325–332.
Sim, G. A. (1959) *Acta Cryst.* **12**, 813–815.
Skriver, E., Maunsbach, A. B., and Jorgensen, P. L. (1981) *FEBS Lett.* **131**, 219–222.
Skriver, E., Maunsbach, A. B., Hebert, H., and Jorgensen, P. L. (1988) *Meth. Enzymol.* **156**, 80–87.
Skriver, E., Maunsbach, A. B., Hebert, H., Scheiner-Bobis, G., and Schoner, W. (1989) *J. Ultrastruct. Mol. Struct. Res.* **102**, 189–195.
Small, D. M. (1986) in *Handbook of Lipid Research, Volume 4, The Physical Chemistry of Lipids from Alkanes to Phospholipids*, D. M. Small, ed., New York, Plenum Press.
Smith, A. E. (1953) *J. Chem. Phys.* **21**, 2229–2231.
Smith, D., and Fryer, J. R. (1981) *Nature* **291**, 481–482.
Smith, J. V. (1968) *Amer. Miner.* **53**, 1139–1155.
Snyder, R. G., Goh, M. C., Srivatsavoy, V. J. P., Strauss, H. L., and Dorset, D. L. (1992) *J. Phys. Chem.* **96**, 10008–10019.
Snyder, R. G., Conti, G., Strauss, H. L., and Dorset, D. L. (1993) *J. Phys. Chem.* **97**, 7342–7350.
Spence, J. C. H. (1980) *Experimental High-resolution Electron Microscopy*, Oxford, Clarendon.
Spence, J. C. H., and Zuo, J. M. (1992) *Electron Microdiffraction*, New York, Plenum Press.
Spence, J. C. H., and Zuo, J. M. (1993) *Microsc. Soc. America Bull.* **23(1)**, 80–90.
Spink, J. A. (1950) *Nature* **165**, 612–613.
Stanley, E. (1986) *Acta Cryst.* **A42**, 297–299.
Starkweather, H. W., Jr. (1986) *Macromolecules* **19**, 1131–1134.
Stauffer, K. A., Hoenger, A., and Engel, A. (1992) *J. Mol. Biol.* **223**, 1155–1165.
Stehling, F. C., Ergos, E., and Mandelkern, L. (1971) *Macromolecules* **4**, 672–677.
Stephens, J. F., and Tuck-Lee, C. (1969) *J. Appl. Cryst.* **2**, 1–10.
Steven, A., Ten Heggler, B., Müller, R., Kistler, J., and Rosenbusch, J. P. (1977) *J. Cell. Biol.* **72**, 292–301.
Stokes, D. L., and Green, N. M. (1990) *J. Mol. Biol.* **213**, 529–538.
Storks, K. H., and Germer, L. H. (1936) *Phys. Rev. (Ser. 2)* **50**, 676.
Storks, K. H., and Germer, L. H. (1937) *J. Chem. Phys.* **5**, 131–134.

Stout, G. H., and Jensen, L. H. (1968) *X-ray Structure Determination. A Practical Guide*, New York, Macmillan.
Sturkey, L. (1962) *Proc. Phys. Soc. (London)* **80**, 321–354.
Sugiyama, J., Harada, H., and Saiki, H. (1987) *Int. J. Biol. Macromol.* **9**, 122–130.
Sumner, J. B., and Dounce, A. L. (1955) *Meth. Enzymol.* **2**, 775–781.
Sutton, L. E., ed. (1958) *Tables of Interatomic Distances and Configuration in Molecules and Ions*, Special Publication No. 11, London, The Chemical Society, p. 512.
Sutula, C. L., and Bartell, L. S. (1962) *J. Phys. Chem.* **66**, 1010–1014.
Takamizawa, K., Ogawa, Y., and Oyama, T. (1982) *Polym. J.* **14**, 441–456.
Tanaka, K. (1938) *Mem. Kyoto Univ. Ser. A* **21**, 85–88.
Tanaka, K. (1939) *Mem. Kyoto Univ. Ser. A.* **22**, 377–380.
Tanaka, K. (1940) *Mem. Kyoto Univ. Ser. A.* **23**, 195–205.
Tang, D., and Li, F. H. (1988) *Ultramicroscopy* **25**, 61–68.
Tatarinova, L. I., and Vainshtein, B. K. (1962) *Visokomolek. Soed.* **4**, 261–269.
Taylor, C. A., and Lipson, H. (1964) *Optical Transforms. Their Preparation and Application to X-ray Diffraction Problems*. Ithaca, N.Y., Cornell Univ. Press.
Taylor, K. A., and Glaeser, R. M. (1974) *Science* **186**, 1036–1037.
Taylor, K. A., Dux, L., and Martinosi, A. (1986) *J. Mol. Biol.* **187**, 417–427.
Taylor, K. A., Dux, L., Varga, S., Ting-Beall, H. P., and Martinosi, A. (1988) *Meth. Enzymol.* **157**, 271–289.
Teare, P. W. (1959) *Acta Cryst.* **12**, 294–300.
Thakur, M., and Lando, J. B. (1983) *Macromolecules* **16**, 143–146.
Thiessen, P. A., and Schoon, Th. (1937) *Z. Phys. Chem.* **B36**, 216–231.
Thomas, E. L., and Ast, D. G. (1974) *Polymer* **15**, 37–41.
Thomas, E. L., and Sass, S. L. (1973) *Makromol. Chem.* **164**, 333–341.
Thomson, G. P., and Murison, C. A. (1933) *Nature* **131**, 237.
Tivol, W. F., Dorset, D. L., McCourt, M. P., and Turner, J. N. (1993) *Microsc. Soc. America Bull.* **23(1)**, 91–98.
Toyoshima, C., and Unwin, P. N. T. (1988) *Nature* **336**, 247–250.
Trillat, J. J., and Hirsch, Th. v. (1933) *J. Phys.* **7**, 38–43.
Trillat, J. J., and Motz, H. (1935) *Trans. Faraday Soc.* **31**, 1127–1136.
Truter, M. R. (1967) *Acta Cryst.* **22**, 556–559.
Trzebiatowski, T., Dräger, M., and Strobl, G. R. (1982) *Makromol. Chem.* **183**, 731–744.
Tsuji, M., Isoda, S., Ohara, M., Kawaguchi, A., and Katayama, K. (1982) *Polymer* **23**, 1568–1574.
Tsuji, M., Roy, S. K., and Manley, R. St. J. (1984) *Polymer* **25**, 1573–1576.
Tsuji, M., Ohara, M., Isoda, S., Kawaguchi, A., and Katayama, K. (1989) *Phil. Mag.* **B54**, 393–403.
Turner, J. N., Valdre, U., and Fukami, A. (1989) *J. Electron Microsc. Techn.* **11**, 258–271.
Turner, J. N., Barnard, D. P., McCauley, P., and Tivol, W. F. (1991) in *Electron Crystallography of Organic Molecules*, J. R. Fryer and D. L. Dorset, eds., Dordrecht, Kluwer, pp. 55–62.
Turner, P. S., and Cowley, J. M. (1969) *Acta Cryst.* **A25**, 475–481.
Udalova, V. V., and Pinsker, Z. G. (1964) *Sov. Phys. Crystallogr.* **8**, 433–440.
Ueda, Y., and Ashida, M. (1980) *J. Electron Microsc.* **29**, 38–44.
Uemura, A., Tsuji, M., Kawaguchi, A., and Katayama, K. (1988) *J. Mater. Sci.* **23**, 1506–1509.
Ungar, G., Stejny, J., Keller, A., Bidd, I., and Whiting, M. C. (1985) *Science* **229**, 386–389.
Unwin, P. N. T., and Henderson, R. (1975) *J. Mol. Biol.* **94**, 425–440.
Uyeda, N., Kobayashi, T., Suito, E., Harada, Y., and Watanabe, M. (1972) *J. Appl. Phys.* **43**, 5181–5189.
Uyeda, N., Kobayashi, T., Ishizuka, K., and Fujiyoshi, Y. (1978–1979) *Chem. Scripta* **14**, 47–61.
Uyeda, N., Kobayashi, T., Ishizuka, K., and Fujiyoshi, Y. (1980) *Nature* **285**, 95–97.
Vainshtein, B. K. (1955) *Zh. Fiz. Khim.* **29**, 327–344.
Vainshtein, B. K. (1956a) *Sov. Phys.—Crystallogr.* **1**, 15–21.
Vainshtein, B. K. (1956b) *Sov. Phys.—Crystallogr.* **1**, 117–122.
Vainshtein, B. K. (1963) *Sov. Phys.—Crystallogr.* **8**, 127–130.

Vainshtein, B. K. (1964a) *Structure Analysis by Electron Diffraction*, Oxford, Pergamon.

Vainshtein, B. K. (1964b) in *Advances in Structure Research by Diffraction Methods* (Vol. 1), R. Brill, ed., New York, Wiley-Interscience, pp. 24–54.

Vainshtein, B. K. (1978) *Adv. Optical Electron Microsc.* **7**, 281–377.

Vainshtein, B. K. (1981) *Modern Crystallography. Vol. 1. Symmetry of Crystals, Methods of Structural Crystallography*. Berlin, Springer, pp. 120–121.

Vainshtein, B. K., and Klechkovskaya, V. V. (1993) *Proc. Roy. Soc. (London)* **A442**, 73–84.

Vainshtein, B. K., and Lobachev, A. N. (1956) *Sov. Phys. Crystallogr.* **1**, 370–371.

Vainshtein, B. K., and Pinsker, Z. G. (1950) *Dokl. Akad. Nauk SSSR* **72**, 53–56.

Vainshtein, B. K., and Tatarinova, L. I. (1967) *Sov. Phys. Crystallogr.* **11**, 494–498.

Vainshtein, B. K., and Zvyagin, B. B. (1993) in *International Tables for Crystallography, Volume B, Reciprocal Space*, U. Shmueli, ed., Kluwer, Dordrecht, pp. 310–314.

Vainshtein, B. K., Lobachev, A. N., and Stasova, M. M. (1958) *Sov. Phys. Crystallogr.* **3**, 452–459.

Vainshtein, B. K., D'yakon, I. A., and Ablov, A. V. (1971) *Sov. Phys. Doklady* **15**, 645–647.

Vainshtein, B. K., Zvyagin, B. B., and Avilov, A. S. (1992) in *Electron Diffraction Techniques* (Vol. 1), J. M. Cowley, ed., Oxford, Oxford Univ. Press, Chapter 6.

Vand, V., and Bell, I. P. (1951) *Acta Cryst.* **4**, 465–469.

Vanderkooi, G., Senior, A. E., Capaldi, R. A., and Hayashi, H. (1972) *Biochim. Biophys. Acta* **274**, 38–48.

Van Heel, M., and Keegstra, W. (1981) *Ultramicroscopy* **7**, 113–130.

Van Tendeloo, G., Van Heurck, C., Van Landuyt, J., Amelinckx, S., Verheijen, M. A., Van Loosdrecht, P. H. M., and Meijer, G. M. (1992) *J. Phys. Chem.* **96**, 7424–7430.

Vaughan, P., and Donohue, J. (1952) *Acta Cryst.* **5**, 530–535.

Vincent, R. (1985) *Inst. Phys. Conf. Ser. No. 78*, Chapter 11, pp. 427–428.

Vincent, R., and Exelby, D. R. (1991) *Phil. Mag. Lett.* **63**, 31–38.

Vincent, R., and Exelby, D. R. (1993) *Phil. Mag.* **B68**, 513–528.

Vincent, R., and Midgley, P. A. (1994) *Ultramicroscopy* **53**, 271–282.

Vogel, H., and Jähnig, F. (1986) *J. Mol. Biol.* **190**, 191–199.

Voronova, A. A., and Vainshtein, B. K. (1958) *Sov. Phys.—Crystallogr.* **3**, 445–451.

Walian, P. J., and Jap, B. K. (1990) *J. Mol. Biol.* **215**, 429–438.

Wang, B. C. (1985) *Adv. Enzymology* **115**, 90–113.

Wang, D. N., and Kühlbrandt, W. (1991) *J. Mol. Biol.* **217**, 691–699.

Weiss, M. S., Wacker, T., Weckesser, J., Welte, W., and Schulz, G. E. (1990) *FEBS* **267**, 268–272.

Welberry, T. R., and Butler, B. D. (1994) *J. Appl. Cryst.* **27**, 205–231.

Wenk, H. R., Downing, K. H., Hu, M. S., and O'Keefe, M. A. (1992) *Acta Cryst.* **A48**, 700–716.

White, J. W., Dorset, D. L., Epperson, J. E., and Snyder, R. (1990) *Chem. Phys. Lett.* **166**, 560–564.

Wilson, A. J. C. (1942) *Nature* **150**, 151–152.

Wilson, A. J. C. (1949) *Acta Cryst.* **2**, 318–321.

Wilson, A. J. C. (1950a) *Acta Cryst.* **3**, 258–261.

Wilson, A. J. C. (1950b) *Acta Cryst.* **3**, 397–398.

Wingfield, P., Arad, T., Leonard, K., and Weiss, H. (1979) *Nature* **280**, 696–697.

Wittmann, J. C., and Lotz, B. (1981a) *J. Polym. Sci.—Polym. Phys. Ed.* **19**, 1837–1851.

Wittmann, J. C., and Lotz, B. (1981b) *J. Polym. Sci.—Polym. Phys. Ed.* **19**, 1853–1864.

Wittmann, J. C., and Lotz, B. (1989) *Polymer* **30**, 27–34.

Wittmann, J. C., and Lotz, B. (1990) *Progr. Polym. Sci.* **15**, 909–948.

Wittmann, J. C., and Manley, R. St. J. (1977) *J. Polym. Sci.—Polym. Phys. Ed.* **15**, 1089–1100.

Wittmann, J. C., and Manley, R. St. J. (1978) *J. Polym Sci.—Polym. Phys. Ed.* **16**, 1891–1895.

Wittmann, J. C., Hodge, A. M., and Lotz, B. (1983) *J. Polym. Sci.—Polym. Phys. Ed.* **21**, 2495–2509.

Wooster, W. A. (1964) *Acta Cryst.* **17**, 878–882.

Wrigley, N. G. (1968) *J. Ultrastruct. Res.* **24**, 454–464.

Wunderlich, B. (1973) *Macromolecular Physics, Volume 1, Crystal Structure, Morphology, Defects*. New York, Academic Press.

Yagi, K. (1993) in *Electron Diffraction Techniques* (Vol. 2), J. M. Cowley, ed., Oxford, Oxford Univ. Press, pp. 260–308.
Yao, J. X. (1981) *Acta Cryst.* **A37**, 642–644.
Yeager, M., and Gilula, N. B. (1992) *J. Mol. Biol.* **223**, 929–948.
Zachariasen, W. H. (1934) *Z. Krist.* **88**, 150–161.
Zampighi, G., and Unwin, P. N. T. (1979) *J. Mol. Biol.* **135**, 451–464.
Zemlin, F. (1989) *J. Electr. Microsc. Techn.* **11**, 251–257.
Zemlin, F., Reuber, E., Beckmann, E., Zeitler, E., and Dorset, D. L. (1985) *Science* **229**, 461–462.
Zhang, W. P., and Dorset, D. L. (1989a) *J. Polym. Sci. B. Polym. Phys.* **27**, 1433–1447.
Zhang, W. P., and Dorset, D. L. (1989b) *Proc. Electron Microsc. Soc. America, 47th Annual Meeting*, San Francisco, San Francisco Press, pp. 702–703.
Zhang, W. P., and Dorset, D. L. (1990) *Macromolecules* **23**, 4322–4326.
Zhang, W. P., and Thomas, E. L. (1991) *Polym. Commun.* **32**, 482–485.
Zhang, W. P., Dorset, D. L., and Hanlon, J. (1989a) *Proc. Electron Microsc. Soc. America, 47th Annual Meeting*, San Francisco, San Francisco Press, pp. 662–663.
Zhang, W. P., Kuo, K. H., Dorset, D. L., Hou, Y. F., and Ni, J. Z. (1989b) *J. Electr. Microsc. Techn.* **11**, 326–332.
Zou, X. D., Sukharev, Yu., and Hovmöller, S. (1993a) *Ultramicroscopy* **49**, 147–158.
Zou, X. D., Sukharev, Yu., and Hovmöller, S. (1993b) *Ultramicroscopy* **52**, 436–444.
Zuo, J. M., Spence, J. C. H., and Hoier, R. (1989) *Phys. Rev. Lett.* **62**, 547–550.
Zvyagin, B. B. (1957) *Sov. Phys. Crystallogr.* **2**, 388–394.
Zvyagin, B. B. (1967) *Electron Diffraction Analysis of Clay Mineral Structures*, New York, Plenum Press.
Zvyagin, B. B. (1993) *Microsc. Soc. America Bull.* **23(1)**, 66–79.
Zvyagin, B. B., and Mischenko, K. S. (1961) *Sov. Phys. Crystallogr.* **5**, 575–579.
Zvyagin, B. B., and Mischenko, K. S. (1963) *Sov. Phys. Crystallogr.* **7**, 502–505.

Index

Acyl shift, 319
Alkyl halides, 307
Alumina, λ-phase, structure analysis, 224
Aluminum-germanium alloys, structure analysis, 231
Ammonium chloride, 209
Ammonium sulfate, 209
Autocorrelation function, 103; *see also* Patterson function

Babinet phases, 100, 327
Bacteriorhodopsin
  direct phase extension, 421
  structure analysis, 409
Bayes' theorem, 125
Bend contours, 23, 87
Bond distances and angles (calculation), 133
Boric acid, structure analysis, 211
Boundary lubricant films, 293
Bravais lattices, 33

C_{60} buckminsterfullerene
  data collection, 58
  secondary scattering from, 151
  structure analysis, 202
Camera length calibration (gold), 85
Celadonite, structure analysis, 215
Cellulose triacetate II, structure analysis, 370
Central section theorem, 43
Centrosymmetry, tests for, 63
Chitosan
  symbolic addition, 114
  structure analysis, 371
Chlorophyll (pea) a/b protein light harvesting complex, structure analysis, 420
Cholesteryl esters
  crystal forms, 344
  binary phase behavior, 350
  thermotropic phase behavior, 348
  unit cell constants, e. d. 349
Cochran formula, 111
Comb function, 15
Convergent beam electron diffraction, 30, 60, 119, 210, 231
Convolution, 11
Copper DL alaninate, 169
Copper chloride (basic), structure analysis, 225
Copper perbromopththalocyanine
  Fourier refinement, example, 127
  structure analysis, 198
Copper perchlorophthalocyanine
  high resolution images, 195
  structure analysis, 188
Correlation, 11
Cross-correlation function, 99, 409, 420
Cryo-stages (electron microscope), 80
Crystal bending (elastic)
  effect on diffraction intensities, 153
  effect on structure analysis, 157
  experimental observation, 154
Crystal growth
  epitaxial orientation, 69
  evaporation of dilute solution, 67
  inorganic samples, 77
  membrane protein crystals (table), 408
  reconstitution (proteins), 77
  self-seeding (polymers), 68
  sublimation growth, 68
  surface orientation (proteins), 77
Crystal thickness measurement, 68
Cycloalkanes, examples
  c-$(CH_2)_{34}$, 283

Cycloalkanes, examples (*cont.*)
 c-$(CH_2)_{36}$, 283
 c-$(CH_2)_{48}$, polymorphism, 284
 c-$(CH_2)_{72}$, polymorphism, 285
 c-$(CH_2)_{96}$, polymorphism, 285
 c-$(CH_2)_{120}$, 287
Cyclopentane-1,2,3-triol analogs of lipids, 314, 319, 322, 333, 341, 343

Delta function, 14
 properties, 14,15
Density modification, 120
Detergents
 alkyl oligo(ethylene oxide)s, 308
 n-dodecyl octa(ethylene oxide), layer structure, 309
 octyl glucoside structures, anomeric effect, 308
Dicetyl ether
 crystal structure, 305
 insolubility in alkanes, 305
Didodecyl amine, cosolubility with alkanes, 305
Diffraction contrast image
 bright field, 23
 dark field, 23
Diffraction pattern, 3,5
 magnification in electron microscope, 23
1,2-diglycerides
 acyl shift, 319
 binary phase behavior, 320
 conformation, 319
 cyclopentane-1,2,3-triol analogs, 319
 effect of ether linkage, 318
 layer packing, 317
 polymorphism, 317
1,3-diglycerides
 effect of ether linkage, 317
 layer structure, 316
Diketopiperazine, 120, 130
 structure analysis, 169
N,N-dimethylphosphatidylethanolamines, 1,2-dipalmitoyl-sn-glycerophospho-N,N-dimethylethanolamine, structure, 337
Dioctadecyl amine, cosolubility with alkanes, 305
Direct phase determination
 density modification, 124
 Harker–Kasper inequality, 107
 known fragment (DIRDIF), 122
 maximum entropy and likelihood, 124
 minimal principle, 123
 molecular replacement, 122
 Patterson search, 122
Direct phase determination (*cont.*)
 Sayre equation, 107
 symbolic addition, 109
 tangent formula, 119
Disorder diffuse scattering, 258
Dynamical scattering theory, 135
 n-beam model, 137
  accelerating voltage (optimal), 140, 143
  approximate correction, 147, 329
  Cowley–Moodie multislice calculation, 137
  effect on structure determination, 142
  experimental justification in electron diffraction, 139
  inaccuracy for organic crystals, 142
  phase grating approximation, 137
  space group forbidden reflections, 146
  thickness dependence, 141
  violation of Friedel symmetry, 142
 kinematical limits, 141
  small molecules, 141
  proteins, 142
 two-beam model, 136, 140
  experimental justification in electron diffraction, 144, 210

Electron diffraction, 22, 26
 convergent beam, 30, 60, 119, 210, 231
 high dispersion, 83
 intensity corrections, 86
 oblique texture, 29
 reflection (RHEED), 30
  paraffins, 239
  paraffin derivatives, 293
  glycerolipids, 311
 selected area, 23
Electron microscopy, low-dose procedures, 87
Electron scattering factors, 18
 ratios and detectability, 19
 relation to x-ray scattering factors, 18
Electron wavelength, 18
Enantiomorph-defining phase, 119
Environmental chambers (electron microscope), 78
Epitaxial substrates (crystal growth), 69
Escherichia coli outer membrane porins
 Omp F
  crystal–crystal transition, 415
  polymorphism, 411
  structure analysis, 411
 Pho E, structure analysis, 412
Ethylene di(11-bromoundecanoate), 299, 319
Eutectic solids, 281, 319, 350
Excitation error, 26, 137

Fatty acids
  behenic acid structure, 297
  polymorphism, 297
Fatty alcohols
  n-octadecanol, structure analysis, 295
  polymorphism, 294
  n-tricosanol, 294
Film measurement
  electron diffraction, 85
  electron micrographs, 88
Fourier filtration (image averaging), 96
Fourier refinement, 127
Fourier transform, 3
Fourier transform pairs
  convolution–multiplication, 14
  comb–comb, 15
  Gaussian–Gaussian, 10
  rectangle (pulse)–sinc, 8
  triangle (pulse)–sinc^2, 12
Friedel's law
  amplitudes, 4
  phases, 95, 114

Gaussian disorder, 16
Gaussian function, 11
Glide element (symmetry operator), 44
Glide plane, 50
Goniometry, 81
  calibration, 81
Graphite, structure analysis, 206
Group (definition), 36
  Abelian, 37

Halorhodopsin
  phase extension by direct methods, 422
  structure analysis, 410
Harker–Kasper inequality, 107
High Tc superconductor ($Bi_2Sr_2Ca_2O_x$) incommensurate phase, structure analysis, 231
Homometric structures, 100

Image analysis, 96
Insect waxes
  beeswax, 307
  15-oxotetratriacontyl, 13-oxodotriacontanoate (*Psylla alni* secretion), structure analysis and tubular crystal growth, 298

Ketoalkanes, 15-ketohentriacontane structure, 298

'Lattice images'
  polymers, list, 386
  small organics, list, 196
Langmuir–Blodgett films, 69
Lead carbonate (basic), 209
Lorentz correction, 85, 324, 399

Mannan I, structure analysis, 368, 379
Maximum entropy and likelihood, 124

N-methyl phosphatidylethanolamines, 1,2-dihexadecyl-sn-glycerophospho-N-methylethanolamine, structure, 333
Methylene subcells, 311
  examples of molecules packing in, 315
  identification by transmission electron diffraction, 311
Metric tensor, 35
Miller indices, 5
  range required for unique data set, 83
Minimal function, 123
Minimal principle, 123
  Shake and Bake algorithm, 123
Model fitting to potential map, 133
Molecular replacement, 122
Monogalactosyl diglycerides, layer structure, 343
Mott formula, 18
Muscovite, structure analysis, 219

Nigeran (anhydrous), structure analysis, 372

Oblique texture diffraction, 29, 56
Optical bench, 88, 96
Origin of unit cell
  definition, 112
  shift (phase term), 54, 96
Orthonormal transformation of coordinates, 36

N-paraffins
  as crystalline test object for dynamical scattering measurements, 139
  binary solids, 268
  binodal phase boundary
    isotope effect, 278, 281
    structure of solid, 278
  crystal bending and diffraction from, 154
  disorder diffuse scattering from, 256
  eutectic solids, structure, 281
  examples
    n-$C_{30}H_{62}$ 239
      perdeuterated analog, 243
    n-$C_{31}H_{64}$ 239
    n-$C_{33}H_{68}$, structure analysis, 259

N-paraffins (*cont.*)
examples (*cont.*)
$n\text{-}C_{36}H_{74}$
high-resolution images of epitaxial crystals, 252
longitudinal disorder, 256
perdeuterated analog, 243
structure analysis, 241, 243, 250
thermotropic behavior, 261
$n\text{-}C_{44}H_{90}$, lattice images of solution-grown crystals, 97, 244
$n\text{-}C_{50}H_{102}$, thermotropic behavior, 262
$n\text{-}C_{82}H_{166}$, sectorization of crystals, 247
$n\text{-}C_{94}H_{190}$, 263
$n\text{-}C_{168}H_{338}$, sectorization of crystals, 247
multicomponent waxes, 239
structure analyses, 239, 275
RHEED from, 239
rotator phase, 261
secondary scattering from, 149
solid solutions
$n\text{-}C_{32}H_{66}/n\text{-}C_{36}H_{74}$ (1:1) crystal structure, 261
nematic disorder, solids grown from vapor phase, 254
symmetry rules for stabilization, 272
thermal diffuse scattering from, 246
Patterson synthesis, 103
detectability of heavy atoms, 105
model for bend-deformed crystal, 153
phospholipids, 106, 325, 338, 339
search techniques, 122
'Peakiness' criterion for correct potential map, 126, 372
Perfluoroalkanes
$n\text{-}C_{16}F_{34}$, structure analysis, 287
$n\text{-}C_{20}F_{42}$, 287
$n\text{-}C_{24}F_{50}$
phase transition, 290
structure analysis, 290
Phase, 14, 23
effect of symmetry on, 46
Phase contrast, 22
Phase contrast transfer function, 21
determination of experimentally, 90
deconvolution, 88
Phase determination
direct methods, 105
Patterson synthesis, 103
automatic search, 122
trial and error, 99
Phase invariant sums
Σ_1, 110
Phase invariant sums (*cont.*)
Σ_2, 110
quartets, 111
Phase variance, 120
Phlogopite-biotite, structure analysis, 220
Phosphatidic acids, comparison to cyclopentane-1,2,3-triol analogs, 343
Phosphatidylcholines (lecithins)
alkyne chain lecithin–structure determination, 341
cyclopentane-1,2,3-triol analogs, 341
iso-branched chain lecithins–hygroscopic behavior, 341
electron diffraction of hydrated lecithin bilayers, 341
examples, 1,2-dihexadecyl-sn-glycerophosphocholine, structure analysis, 340
Phosphatidylethanolamines
cyclohexane-1,2,3-triol analogs, 335
examples, 1,2-dihexadecyl-sn-glycerophosphoethanolamine
lattice images, 328
Patterson function, 106
structure determination, 327
symbolic addition, 116
translational structure search with model, 101
1,2-dimyristoyl-sn-glycerophosphoethanolamine, structure analysis, 330
1,2-dipalmitoyl-sn-glycerophosphoethanolamine, undistorted crystals, 359
solid solutions (L-DMPE/L-DPPE solid solution structure), 332
swelling in water, 331
Phospholipids
direct phase determination, 116, 327, 332
lattice images, 323, 328
lamellar spacings, 328, 335, 341
Patterson synthesis, 103, 325, 338, 339
paracrystallinity of epitaxial samples, 323
phase refinement, 131
solid solutions, 332
subcell determination, 331
translational search with model, 101
Plane groups, 43
pgg example, 46
"two-sided," 55
Point groups, 32
Poly (1-butene), form III, structure analysis, 396
Poly (ε-caprolactone), structure analysis, 390
Poly (3,3-bis(chloromethyl)oxacyclobutane), structure analysis, 363

Poly (1,4-trans-cyclohexanediyl dimethylene succinate), structure analysis, 377
Polyethylene
crystal habit, 383
data collection, 61, 386
least squares refinement, 386
sectorization, 385
structure determination, 386
Poly (ethylene sulfide), structure analysis, 362
Poly (hexamethylene terephthalate), structure analysis, 373
Polymer crystal structures (table), 362
Polymethylene chains
determination of tilt by RHEED, 294
methylene subcells, 311
Poly-γ-methyl-L-glutamate, β-form, structure analysis, 374
Poly (pivalolactone), γ-form, structure analysis, 365
Poly (p-xylylene), structure analysis, 366
Potassium niobium oxide, structure analysis, 231
Potassium niobium tungsten oxide, structure analysis, 232
Potential energy (non-bonded), 100
Potential map, 22

R-factor (crystallographic residual)
expected values, 100, 126
statistical significance, 103, 343
Radiation damage, 159
effect of specimen cooling, 160
mechanism, 159
prevention, 161
Rose equation as model, 161
saturation doses for organic crystals (table), 160
Reciprocal lattice, 5, 32
Rectangle function, 8
Resolution, 21, 95
Rayleigh criterion, 21
Retene (1-methyl-7-isopropylphenanthrene), 169
Rotation matrix, 37
'Rotator' phase (paraffins), 262
chain length dependence, 261
isotope effect on, 265
structural model, 267

Sayre equation, 107, 123
Scherzer focus, 21, 96
Screw axis, 50
Secondary scattering, 149
correction for, 151
Setting angle, 239, 243, 384, 388
Sigma, 1 (Σ_1) phase invariant, 110
Sigma, 2 (Σ_2) phase invariant, 110
derivation from Sayre equation, 111
Sinc function, 8
Soaps, lead stearate structure, 298
Sodium niobium fluoroxide, structure analysis, 234
Solid solutions, 268, 319, 331, 352
metastable, 278
symmetry conditions for stability, 268, 272
volumetric conditions for stability, 268
Solvent replacement (proteins), 80
Space groups, 49
commonly occurring, 63, 65
identification of, 55
$P2_1/c$ example, 50
$P2_12_12_1$ example, 52
Space lattice, 31
Spot illumination, 88, 166, 406
Spread function, 20
Staurolite, structure analysis, 234
Structure determination
Babinet solution, 100
homometric solutions, 100
significance of crystallographic residual, 103
Structure factor, 22
normalized, 109
unitary, 107, 125
Structure identification
atomic resolution, 126
density smoothness, 126
'peakiness' criterion, 126
Structure refinement, 126
fitting model to potential map, 133
Fourier methods, 127
least squares, 130, 175, 388
Symmetry operators, 37
Symmetry groups
Hermann Mauguin notation, 37
Schoenflies notation, 37, 42

Tangent formula, 119
figures of merit, 120
Temperature factor, 110
Thermal diffuse scattering, 246
Thin lens, 3
aberrations
chromatic, 21
spherical, 21
back focal plane, 3
image plane, 3

Thiourea
  low temperature phase, structure analysis, 184
  room temperature phase, structure analysis, 179
Thon rings, 91
Trial and error structure analysis
  ab initio, 101
  conformational search, 100, 361
  translational search, 101, 326
  with atom atom potential functions, 100, 361
Triangle function, 12
Triglycerides
  cyclopentane-1,2,3-triol analogs, 321
  effect of ether linkage, 322
  examples
    triheptadecanoin, 323
    tripalmitin, 323
  polymorphism 321

Unit cell
  identification, 55
Unit cell (*cont.*)
  origin definition, 112
  types, 31
  volume, 34
Urea
  distinguishing correct structure, 115
  structure analysis, 176
Urotropine, 169

Wax (paraffin), crystal structures, 274
Wax esters, cetyl palmitate
    diffuse scattering, 299
    eutectic solid with alkanes, 305
    structure analysis, 298
  myristyl stearate, structure, 302
  stearyl myristate, structure, 302
Weak phase object approximation, 20
Wilson plot, 110

Zeolites, structure analysis, 237
Zone axis equation, 34